The Earth System

The Earth System

Lee R. Kump
*Department of Geosciences
and Earth System Science Center*

James F. Kasting
*Departments of Geosciences and Meteorology
and Earth System Science Center*

Robert G. Crane
*Department of Geography
and Earth System Science Center*

PRENTICE HALL
Upper Saddle River, New Jersey 07458

Library of Congress Cataloging-in-Publication Data
Kump, Lee R.
    The earth system/Lee R. Kump, James F. Kasting, Robert G. Crane.
      p. cm.
    Includes index.
    ISBN 0-13-177387-9
    1. Gaia hypothesis.   I. Kasting, James F.   II. Crane, Robert G.   III. Title.
QH331.K798   1999
577′.1—dc21   98-31916
                         CIP

Editor: Dan Kaveney
Production Editor: York Production Services
Production Assistant: Nancy Bauer
Marketing Manager: Leslie Cavaliere
Manufacturing Manager: Trudy Pisciotti
Development Editor: Karen Karlin
Art Manager: Gus Vibal
Art Editor: Karen Branson
Assistant Editor: Wendy Rivers
Cover Designer: Kiwi Design
Photo Editors: Lorinda Morris Nantz and Melinda Reo
Photo Researcher: Tobi Zawsner
Copy Editor: York Production Services
Art Studios: Maps: Geosystems, Illustrations: Smith & Frederick Graphics
Front Cover Photo: Reeve Photography
Cover Illustration: From "Reef and Forest" oil painting by Glynn Gorick
  (113 Hemingford Road, Cambridge UK/www.gorick.u-net.com)

Any uncredited photos were supplied by the author.

Printed in the United States of America
10 9 8 7 6 5 4 3 2 1

ISBN 0-13-177387-9

Prentice-Hall International (UK) Limited, *London*
Prentice-Hall of Australia Pty. Limited, *Sydney*
Prentice-Hall Canada Inc., *Toronto*
Prentice-Hall Hispanoamericana, S.A., *Mexico*
Prentice-Hall of India Private Limited, *New Delhi*
Prentice-Hall of Japan, Inc., *Tokyo*
Simon & Schuster Asia Pte. Ltd., *Singapore*
Editora Prentice-Hall do Brasil, Ltda., *Rio de Janeiro*

Contents

Preface   ix

1 Global Change   1

Introduction   1
Global Change on Short Time Scales   4
The Discovery of the Antarctic Ozone Hole   10
Global Change on Long Time Scales   10

2 Daisyworld: An Introduction to Systems   19

The Systems Approach   19
The Daisyworld Climate System   23
Graphs and Graph Making   26
External Forcing: The Response of Daisyworld to Increasing Solar Luminosity   28

3 Global Energy Balance: The Greenhouse Effect   34

Introduction   34
Electromagnetic Radiation   35
Temperature Scales   38
Blackbody Radiation   39
Planetary Energy Balance   41
Planetary Energy Balance   41
Atmospheric Composition and Structure   42
How the Greenhouse Effect Works: The One-Layer Atmosphere   43
Mechanisms of Heat Transfer   46
Physical Causes of the Greenhouse Effect   46
Effects of Clouds on the Atmospheric Radiation Budget   48
Introduction to Climate Modeling   50
Climate Feedbacks   51

4 The Atmospheric Circulation System   55

The Global Circulatory Subsystems   56
The Atmospheric Circulation   56

The Relationship among Temperature, Pressure, and Volume: The General Gas Law   59
Global Distributions of Temperature and Rainfall   68

5 The Circulation of the Oceans   79

The Ocean Circulation   80
Winds and Surface Currents   80
Vorticity   86
The Circulation of the Deep Ocean   88
The Salt Content of the Oceans and the Age of Earth   89

6 Circulation of the Solid Earth: Plate Tectonics   97

Introduction   98
Anatomy of Earth   98
The Principle of the Seismograph   100
The Theory of Plate Tectonics   104
Deep-Sea Life at Mid-Ocean Ridge Vents   107
Plates and Plate Boundaries   108
The Physiology of the Solid Earth: What Drives Plate Tectonics?   114
Radiometric Dating of Geological Materials   115
Recycling of the Lithosphere: The Rock Cycle   117
Plate Tectonics through Earth History   119

7 Recycling of the Elements: The Carbon Cycle   128

Systems Approach to the Carbon Cycle   128
The Concept of the Mole   132
The Short-Term Organic Carbon Cycle   133
Oxygen Minimum Zone   137
The Long-Term Organic Carbon Cycle   138
The Inorganic Carbon Cycle   141
pH   143

v

The Carbonate-Silicate Geochemical Cycle   147
Links Between the Organic and Inorganic Carbon
    Cycle   148
*Biological Enhancement of Chemical
    Weathering   149*

8 Long-Term Climate Regulation   152

Introduction   152
Introduction to Geological Time   153
Formation of the Solar System   155
Formation of the Atmosphere and Ocean   158
The Faint Young Sun Paradox Revisited   159
The Long-Term Climate Record   161
Causes of Long-Term Climate Change   164
Low-Latitude Glaciations   165
Climates of the Past 200 Million Years   167

9 The Chemical Environment: Evolution of the Atmosphere   173

Introduction   173
The Prebiotic Atmosphere   174
The Origin of Life   175
The Compounds of Life   177
Effect of Life on the Early Atmosphere   179
The Rise of Oxygen: Geologic Evidence   182
Oxidation States of Iron   183
The Rise of Ozone   185
Possible Causes of the Rise of Oxygen   187
Modern Controls on Atmospheric O_2 188

10 Biodiversity Through Earth History   194

Biodiversity   194
Diversity of Interactions   196
The Fossil Record of Biodiversity   197
Taxonomy   199
Mass Extinctions   200
The Cretaceous-Tertiary Extinction   201
The K-T Strangelove Ocean   206
Extraterrestrial Influences and Extinction   206

11 Pleistocene Glaciations   211

Geologic Evidence of Pleistocene Glaciation   212
Milankovitch Cycles   215
Kepler's Laws   216
Glacial Climate Feedbacks   219

12 Short-Term Climate Variability   229

Introduction   229
The Holocene   232
Present-Day Climate Variability   243
The 1982–1983 and 1997–1998 ENSO Events   248

13 Global Warming   253

Introduction   253
Carbon Reservoirs and Fluxes   254
CO_2 Removal Processes and Time Scales   256
The Chemistry of CO_2 Uptake   258
Projections of Future Atmospheric CO_2
    Concentrations   259
Long-Term CO_2 Projections   261
Projections of Future Climate Change   262
Sea-Level Change   269
Effects on Ecosystems   272
Policies to Slow Global Warming   273

14 Ozone Depletion   278

Introduction   278
Ultraviolet Radiation and Its Biological Effects   279
Ozone Vertical Distribution
    and Column Depth   280
The Chapman Mechanism   283
Catalytic Cycles of Nitrogen, Chlorine,
    and Bromine   284
Sources and Sinks of Ozone-Depleting
    Compounds   285
*How the Link Between Freons and Ozone Depletion
    Was Discovered   287*
The Antarctic Ozone Hole   288
Evidence of MidLatitude Ozone Depletion   290
Mechanisms for Halting Ozone Depletion   292

15 Human Threats to Biodiversity 296

Introduction   296

Ecosystems   297

The Modern Extinction   303

Other Consequences of Tropical Deforestation   306

Why We Should Care About Biodiversity   311

16 Climate Stability on Earth and Earthlike Planets 317

Introduction   317

Climate Evolution in the Distant Future   318

Main Sequence Stars and the Hertzsprung-Russell Diagram   319

Climate Evolution on Venus and Mars   320

Habitable Planets Around Other Stars   322

The Drake Equation   324

Ensuring Our Long-Term Survival   328

APPENDIX *A*   Units and Unit Conversions 331

APPENDIX *B*   Temperature Conversions 332

APPENDIX *C*   Periodic Table 333

APPENDIX *D*   Useful Facts 334

Glossary 335

Index 345

Preface

Nearly 12 years ago, we were asked to develop an introductory course for nonscience majors that embraced the concepts and subject matter of the nascent field of Earth system science. The task was daunting, especially because there was no text available for such a class and, moreover, there was no conceptual framework available to assist us in the design of such a class. We began to offer our class in the late 1980s. From the multitude of concepts and facts taught in traditional Earth science and life science courses, we spent a couple of challenging years identifying the concepts that were essential for a basic understanding of Earth as a system. It took us several years, but we now feel that we have distilled the essence of Earth system science and captured it in the material presented in this book. We use a top-down approach to the natural sciences: The details of geology, climatology, oceanography, and ecology are foregone in favor of unifying principles applied at the global scale.

Over these years we have seen enrollment in our class "Gaia—The Earth System" grow to nearly 1000 students a year. Students are attracted to the class because we cover the important problems involved in global change, including the greenhouse effect, ozone depletion, and loss of biodiversity. The systems approach tends to resonate with many students, and they find that it is used in other courses, both in the physical and social sciences.

We have been encouraged by national and international efforts to develop curricula in Earth system science. We hope that this text will serve the needs of those involved in such efforts. Most of all, we are excited to see faculty who are known internationally for their research accomplishments drawn back to the nonmajors undergraduate classroom by the challenge and appeal of teaching global change and Earth system science.

Organization

This is not a traditional Earth science textbook. Such books treat individual components of the Earth system—the solid Earth, atmosphere, and oceans—separately, with little consideration of the interplay among them or the important interactions with living organisms (the stuff of ecology texts). And, although they are the focus of this book, the modern environmental problems of global warming, ozone depletion, and loss of biodiversity are treated in a fundamentally different way here than in most texts. Here we recognize that these problems have analogues from Earth history: The geological past is the key

to the present and to the future. Chapter 1, on Global Change, is an overview of these important issues—the observational data that convince us that serious problems exist and the events in Earth's history that illuminate how the Earth system responds under stress.

The next six chapters develop the notion that processes active on Earth's surface are functioning together to regulate climate, the circulation of the ocean and atmosphere, and the recycling of the elements. Chapter 2 uses James Lovelock's Daisyworld to introduce important concepts from systems theory that help explain this regulation. In Chapter 3, the greenhouse effect, the most familiar regulatory process, is shown to be an essential part of Earth's natural climate system—a process that is intensified by, rather than the result of, human disturbance. Chapters 4 through 6 explain the connections between the atmosphere, oceans, and crust that cause heat, moisture, and materials to be distributed and recycled and that control global biogeochemical cycles. These cycles are in turn the subject of Chapter 7 (with a focus on the carbon cycle).

In Chapters 8 through 12 we explore the history of Earth for lessons that we can apply to the present and to the future. We discuss the causes of environmental change, especially climate change, that act on a variety of time scales. The evolution of the climate system over billions of years is the subject of Chapter 8. On this time scale, increasing luminosity of the Sun, changes in volcanic activity, the drift of the continents, and the making of mountains are primary factors influencing climate. Chapter 9 discusses the origin of life, early Earth environments, and perhaps the most significant environmental change of Earth history—the establishment of an oxygen-rich atmosphere. Changes in Earth's biota are the focus of Chapter 10. There we find that the diversity of life has, in general, increased over geologic time, but this increase has been punctuated by a number of catastrophic losses of biodiversity. Focusing in toward the present, Chapter 11 describes the climate of the last 2 million years, during which Earth has oscillated in and out of glaciation. On these time scales, climate change is driven by changes in Earth's orbit around the Sun. However, on these and longer time scales, changes in the amount of atmospheric carbon dioxide are also important. The more recent history of Earth's climate, discussed in Chapter 12, reveals smaller climate fluctuations, some of which seem to be driven by oscillations internal to the Earth system (for example, the phenomenon of El Niño).

The last four chapters draw on knowledge gained from previous sections to provide a unique perspective on current and future global change issues. Predictions of future climates, presented in Chapter 13, are based on the understanding of the climate system and carbon cycle developed in earlier chapters. In Chapter 14, ozone depletion is shown to be a serious problem, but one for which international cooperation appears to be heading toward a solution. In contrast, the loss of biodiversity, covered in Chapter 15, remains a poorly understood problem, far from resolution and with unclear but undoubtedly serious implications for the vitality of our planet. In Chapter 16 we use what we have learned about the Earth system in the past and in the present to speculate on what might happen to it in the distant future and to try to estimate the chances that other Earthlike planets and technological civilizations exist elsewhere in our galaxy.

Pedagogy

We have employed a number of pedagogical features to assist in the learning process. Each chapter begins with *Key Questions* (objective questions students should be able to answer after they have read the chapter) and a *Chapter Overview* (a broad preview of the chapter to come). Within each chapter are *boxed essays* that provide interesting asides, more detailed or quantitative treatments of material in the text, or recent advances in scientific understanding. *Chapter Summaries* are provided in outline form at the end of each chapter to aid in reviewing the most important concepts. *Suggested Readings* include both general readings and advanced readings for students (and instructors) interested in further information about the subject matter. These are followed by *Key Terms* lists, which consist of **boldfaced** terms that are introduced in the chapter and that appear in the *Glossary* in the back of the book. *Review Questions* focus the students' review on important concepts and require only brief answers, whereas *Critical-Thinking Questions* are thought questions or analytical exercises that require students to synthesize concepts presented in the chapter.

Many nonscience majors become uncomfortable when equations and chemical formulas appear in textbooks for introductory or general-education courses. However, cogent discussion of global change issues requires some quantification. For example, we need to understand the energy and carbon budgets before we can assess the impact of fossil fuel burning on future climates. Similarly, the issue of ozone depletion cannot be understood without at least a little knowledge of chemistry. Our experience has been that the effort that goes into mastering a challenging equation, such as the planetary energy-balance equation (Chapter 3), is more than compensated for by the satisfaction that comes from making the same sort of climate prediction that scientists such as Carl Sagan have made in the past.

Chapter Sequencing

We anticipate that this book will be used in a variety of ways. We teach a general education class at The Pennsylvania State University that covers approximately three-quarters of the book (12 or 13 chapters) during one semester. Several instructors teach this course, but not all of us choose to cover the same chapters. All instructors normally do go through the first seven chapters in order, but after that we pick and choose according to our own interests and needs. An instructor who is most interested in climate issues, for example, might use Chapters 8, 11, 12, 13, and 16. One who is most interested in biodiversity might choose Chapters 8, 9, 10, 11, and 13. The course can also be tailored to emphasize either Earth history (Chapters 8 through 12) or modern global environmental problems (Chapters 13 through 15). By providing more material than can easily be covered in a one-semester course, we provide the flexibility to emphasize topics or topic areas that are of interest to different instructors and different groups of students.

Acknowledgments

We are indebted to Prentice Hall editors Ray Henderson and Daniel Kaveney for their encouragement, assistance, and patience. We are particularly indebted to developmental editor Karen Karlin, who understands this book better than any of us do and who re-crafted a good fraction of the manuscript into something much more readable. Photo researcher Tobi Zausner worked miracles in finding the photographs we needed but couldn't provide ourselves. Production manager Gretchen Miller (York Production Services) moved the manuscript quickly and accurately through production.

Over the years, our students have provided verbal and written comments on various drafts of the manuscript, and we have valued their input. The book also benefited from careful reviews by a number of our colleagues, including the following:

Bud Alcock; *The Pennsylvania State University*
DeWayne Backhus; *Emporia State University*
Eric Barron; *The Pennsylvania State University*
Phillip D. Boger; *SUNY at Geneseo*
J. Allan Cain; *University of Rhode Island*
John A. Campbell; *Fort Lewis College*
Gene A. Carr; *Oakton Community College*
Clayton R. Chinn; *Vanderbilt University*

Lindgren Lynn Chyi; *University of Akron*
Kerry Cook; *Cornell University*
William C. Cornell; *University of Texas at El Paso*
Brent Dugolinsky; *SUNY, College at Oneonta*
William Dupré; *The University of Houston*
John J. Ernissee; *Clarion University of Pennsylvania*
W. G. Ernst; *Stanford University*
Kevin Furlong; *The Pennsylvania State University*
Fred Goldstein; *Trenton State College*
Dick Holland; *Harvard University*
Robert Jorstad; *Eastern Illinois University*
Erle Kauffman; *Indiana University*
Alan Kazin; *University of California at Los Angeles*
Kerry Kelts; *University of Minnesota*
Rudi H. Kiefer; *University of North Carolina at Wilmington*
Peter G. Larsen; *University of Virginia*
Abraham Lerman; *Northwestern University*
Judy Ann Lowman; *Chaffey College*
Michael Molitor; *Columbia University*
Archie C. Moore; *Southeastern Louisiana University*

Henry T. Mullins; *Syracuse University*
V. Rama Murthy; *University of Minnesota*
James Ogg; *Purdue University*
Mark Patzkowsky; *The Pennsylvania State University*
Frank J. Pazzaglia; *University of New Mexico*
Henry Pollack; *University of Michigan*
Walter A. Robinson; *University of Illinois at Urbana-Champaign*
John Samples; *University of Colorado at Boulder*
Anurag Sharma; *SUNY at Oswego*
Art Sussman; *West Ed*

They were a tough group to satisfy, but their input, from a wide variety of perspectives, was invaluable in defining the content and style of the book.

Lee R. Kump
James F. Kasting
Robert G. Crane

The Pennsylvania State University

Chapter 1
Global Change

Satellite photos of Amazonia in 1972 and 1992. (Top: From Earth Satellite Corporation/Science Photo Library, Photo Researchers, Inc. Bottom: From NASA/Science Photo Library, Photo Researchers, Inc.)

Key Questions

- What is meant by a "systems approach" to Earth science?
- How does global warming differ from the greenhouse effect, and is global warming actually occurring today?
- What is the Antarctic ozone hole, and what is its significance?
- Should we be concerned about tropical deforestation?
- What can understanding Earth's past tell us about Earth's future?

Chapter Overview

Earth is currently being altered at an unprecedented rate by human activity. The buildup of greenhouse gases in the atmosphere is expected to warm Earth's climate in the future—and may have done so already. The accumulation of chlorine-containing compounds in the atmosphere has damaged the ozone layer over part of the globe. Deforestation of the tropics is causing large decreases in biodiversity. How serious are these problems, and how do they compare with past changes in the Earth system? Chapter 1 lays out the evidence of these changes and explains why an integrated, systems approach is useful in analyzing them.

Introduction

Our world is changing. In fact, Earth has always been changing and will continue to do so for ages to come. Yet, there is a difference between the changes occurring now and those that occurred previously. Earth is changing faster today than it has throughout most of its 4.6-billion-year history. Indeed, it may be changing faster than it ever has, except perhaps in the aftermath of giant meteorite impacts. The cause of this accelerated pace of change is simple: human activity. Human populations have expanded in numbers and in their technological abilities to the point at which we are now exerting a significant influence on our planet. The effects of our actions are seen most clearly in the thin envelope of gases that supports our existence, the *atmosphere,* but they are observable elsewhere as well. Forests, mountains, lakes, rivers, and even the oceans exhibit the telltale signs of human activity.

To what extent are these **anthropogenic** (human-induced) changes a cause for concern? All of us can think of situations in which human influence has clearly been detrimental to the environment—for example, cities plagued with polluted air and water. But these are local problems, and they are hardly new. Humans have generated local pollution ever since they first developed agricultural societies around 10,000 years ago. Human inhabitants of Easter Island (which lies off the southwest coast of South America) may have set the stage for the demise of their culture about 700 years ago through **deforestation**—that is, by the clearing of all the trees—of the island. Advanced technology is not needed to damage one's immediate surroundings.

Today, however, because technological advances abound and because there are simply more people on Earth than ever before, human influence extends to the global environment. For example, *global climate,* the prevailing weather patterns of a planet or region over time, is being altered by the addition of greenhouse gases to the atmosphere. **Greenhouse gases** are gases that warm a planet's surface by absorbing outgoing *infrared radiation*—radiant heat—and reradiating some of it back toward the surface. This process is called the **greenhouse effect.** (The analogy is not perfect, however, because the glass walls of a greenhouse keep the air warm by inhibiting heat loss by upward air motions rather than by absorbing infrared radiation). The greenhouse effect is a natural physical process that operates in all planetary atmospheres. For example, the greenhouse effect, and not solely proximity to the Sun, is thought to account for the high surface temperature of Venus—460°C, compared with about 15°C at Earth's surface. On Earth, some greenhouse gases (such as water vapor) are natural, but others are anthropogenic. The most abundant anthropogenic greenhouse gas on Earth is carbon dioxide, CO_2, which is produced by the burning of **fossil fuels** (fuels such as coal, oil, and natural gas that are composed of the fossilized remains of organisms) and by deforestation. When trees are cut down, they decay, and the carbon in their trunks, branches, and leaves is released as CO_2. Carbon dioxide is also a component of volcanic emissions, and it is cycled rapidly back and forth by living plants and animals. Thus, its abundance is controlled by a combination of natural and human-controlled processes.

Humankind is also capable of damaging Earth's fragile ozone layer. The **ozone layer** is a chemically distinct region within the *stratosphere,* part of the atmosphere. The ozone layer protects Earth's surface from the Sun's harmful *ultraviolet radiation.* Ultraviolet radiation is what gives us suntans but also sunburns. **Ozone (O_3)** is a form of oxygen that is much less abundant than, and chemically unlike, the oxygen that we breathe (O_2). As we shall see, the **ozone hole** over Antarctica, a patch of extremely low ozone concentration in the ozone layer, is believed to be anthropogenic in origin.

We are also now deforesting parts of the planet—mainly the tropics—at a rate that was unimaginable until the 19th century. As we cut down the forests, we kill off many species of plants and animals that live there. Hence, we are now causing substantial decreases in **biodiversity,** or the number of species present in a given area.

The effects of these global environmental problems on humans are more difficult to assess than are the effects of local air and water pollution. Depletion of the ozone layer is a worrisome prospect, but serious losses of ozone have so far been confined to the region near the South Pole, where few people live. Small decreases in ozone have been observed at midlatitudes, but these are not yet thought to pose a serious hazard to health. Loss of biodiversity in the tropics has thus far only indirectly affected people who live at temperate latitudes. Tropical deforestation and fossil fuel burning could affect everyone by causing **global warming,** a warming of Earth's atmosphere due to an anthropogenic enhancement of the greenhouse effect. However, some people (those living in Alaska or Siberia, for instance) might see global warming as less of a threat than would others. How can we decide which global environmental problems are truly urgent and which may simply deserve careful, long-term study?

Three Major Themes

One major theme of ours will be global environmental issues such as these. All of us should be able to make our own decisions as to which modern environmental problems are worth worrying about and which, if any, are not. Making such decisions intelligently requires at least some knowledge of the scientific questions involved. Some of the issues, global warming in particular, are also politically contentious, because the scientific questions surrounding them are not entirely answered and because the actions needed to address them are potentially very costly. In such cases, it is important that both policymakers and citizens understand the problem at a reasonably detailed level.

To understand how humankind is disrupting the environment today, we need also to understand how the environment was changing before humans came on the scene. Otherwise, it is difficult to distinguish short-term, anthropogenic trends from longer-term, natural trends. So, a second major theme of ours is global change in the past. Climate is a good example of the overlap of short and long time scales of global change, and one to which we will return frequently. Earth's climate is predicted to warm over the next few decades to centuries as a consequence of the buildup of CO_2 and other greenhouse gases in its atmosphere. Evidence of past climates has come from cores drilled into sediments on the ocean floor. (**Sediments** are layers of unconsolidated material that is transported by water or air.) This evidence indicates that we are in the midst of a relatively short *interglacial period* (a warm interval marked by the retreat of Northern Hemisphere ice sheets) in between *glacial periods* (cold intervals marked by the buildup of these ice sheets). Hence, in the absence of anthropogenic influence, the planet would be destined over the next few thousand years to slip slowly into the next Ice Age. Which of these tendencies—global warming or the transition to a glacial period—will win out? We will argue later that warming is likely to win out in the short term, because the rate of increase of atmospheric CO_2 and other greenhouse gases is faster than the historical rate of interglacial-to-glacial climate change. Thus, the question of time scales is important. Understanding how and why climate has changed in the past can help us understand how it may change in the future.

We will be introduced to these two major themes in this chapter. A third major theme of ours is *systems*—in particular, the *Earth system*. We will examine this theme more thoroughly in Chapter 2. For now, let us say just that a **system** is a group of components that interact. The **Earth system** is composed of four parts: the atmosphere, the hydrosphere, the biota, and the solid Earth (Figure 1-1). As we have seen, the **atmosphere** is a thin envelope of gases that surrounds Earth. The **hydrosphere** is composed of the various reservoirs of water, including ice. The **biota** include all living organisms. (Some ecologists define the *biosphere* as the entire region in which life exists, but we will avoid that term here, because it overlaps our other system components). The **solid Earth** includes all **rocks,** or consolidated mixtures of crystalline materials called *minerals,* and all unconsolidated rock fragments. It is divided into three parts: the core, mantle, and crust. The **core** of any planet or of the Sun is the central part. Earth's core is a dense mixture of metallic iron and nickel and is part solid, part liquid. The **mantle** is a thick, rocky layer between the core and crust that comprises the largest fraction of Earth's mass. The **crust** is the thin, outer layer, which consists of light, rocky matter in contact with the atmosphere and hydrosphere.

One of our goals is to show how the different components of the Earth system interact in response to var-

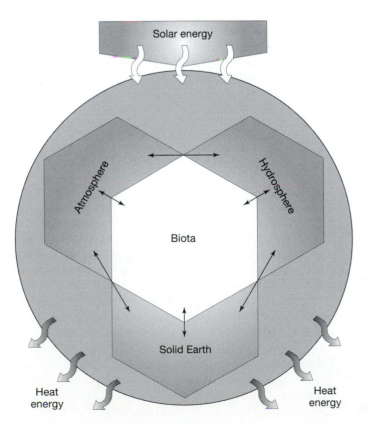

FIGURE 1-1

Schematic diagram of the Earth system, showing interactions among its four components. (From R.W. Christopherson, *Geosystems: An Introduction to Physical Geography,* 3/e, 1997. Reprinted by permission of Prentice Hall, Upper Saddle River, N.J.)

ious internal and external influences, or *forcings*. A well-known example of a forcing is the variation in the amount of sunlight received in each hemisphere during the course of a year. The response to this forcing, which is governed by the interaction between the atmosphere and the hydrosphere, is the seasonal cycle of summer and winter. But there are other, more subtle forcings at work as well that may engage all four components of the Earth system. Some examples are given later in this chapter.

Chapters 3 through 7 describe the various components of the Earth system in some detail. These chapters are not particularly distinctive; many Earth science texts do much the same thing. However, Chapter 1 and all the later chapters are devoted to problems, such as global climate history and modern global change, that cut across traditional disciplinary boundaries and that involve interactions among different parts of the Earth system. It is here that this book differs from most other introductory textbooks. The systems approach adopted here can lead to a more in-depth understanding of such problems by providing a convenient way of analyzing complex interactions and predicting their overall effect.

Global Change on Short Time Scales

We will start our discussion of the Earth system by introducing three major, global environmental changes that are occurring today: global warming, ozone depletion, and tropical deforestation. Afterward, we will backtrack to discuss how the Earth system operated in the past and how that may help us predict what will happen to it in the future.

Evidence of Global Warming

The most pervasive, and at the same time controversial, environmental change that is occurring today is global warming. This issue is extremely complex, because it involves many different parts of the Earth system. It is controversial because it is difficult to separate anthropogenic influences from natural ones and because its causes are deeply rooted in our global industrial infrastructure; hence, these causes would be difficult to eliminate. A major goal of this book, therefore, is to help the reader understand global warming and to put it in the context of past climatic change.

Although the terms "greenhouse effect" and "global warming" are sometimes used interchangeably, the two phenomena are very different. The greenhouse effect is an indisputably real, natural process that keeps the surfaces of Earth and the other terrestrial planets warmer than they would be in the absence of an atmosphere. Global warming is an increase in Earth's surface temperature brought about by a combination of industrial and agricultural activities. These activities release gases that bolster the greenhouse effect. At present, not all scientists are convinced that global warming has begun. Most researchers agree that the climate has warmed over the past century, but not all of them believe that this warming is a result of human activities. Let us now see why many researchers do believe that humans are altering Earth's climate and why even more researchers believe that we will do so in the future.

Measurements of Atmospheric CO$_2$:   The Keeling Curve. The data that have aroused much of the current concern about global warming are shown in Figure 1-2. The graph shows the atmospheric CO$_2$ concentrations measured at the top of Mauna Loa, a 4300-m-high volcano in Hawaii, over a 40-year interval. Mauna Loa was chosen as the measurement site because the air blowing over its summit—clean air from the western Pacific Ocean—is far removed from local sources of pollution. The measurements

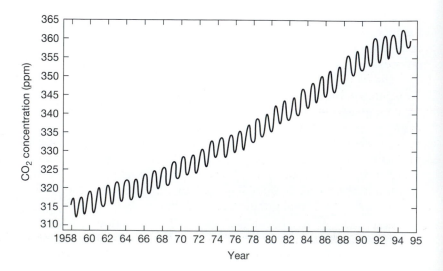

FIGURE 1-2

Measurements of atmospheric CO$_2$ concentrations at the top of Mauna Loa in Hawaii. These data are known as the "Keeling curve." (Source: C.D. Keeling and T.P. Wharf, Scripps Institute of Oceanography, La Jolla, California. (http://cdiac.esd.ornl.gov/ftp/maunaloa-c02/maunaloa.c02))

were begun in 1958 by Charles David Keeling of the Scripps Institute of Oceanography. For this reason, the data are often referred to as the "Keeling curve."

In Figure 1-2 the concentration of atmospheric gas is measured in *parts per million,* or *ppm.* A value of 1 ppm of a particular gas means that one molecule of that gas is present in every million air molecules. We shall use the abbreviation "ppm" to represent parts per million *by volume* rather than parts per million *by mass.* (In technical literature, *ppmv* is often used for parts per million by volume.) Units of mass and volume are not interchangeable, because a given gas molecule may be heavier or lighter than an average air molecule. Although one part per million may not sound like much, it represents a large number of molecules. A cubic centimeter of air at Earth's surface contains about 2.7×10^{19} molecules, so a 1-ppm concentration of a gas would have 2.7×10^{13} molecules in that same small volume. (If you are not familiar with scientific notation, refer to Appendix I for help.)

As Figure 1-2 shows, the CO_2 concentration in 1995 was about 358 ppm. We say "about" because the atmospheric CO_2 concentration varies slightly from place to place and oscillates seasonally over a range of 5 to 6 ppm. This seasonal oscillation has to do with the "breathing" of Northern Hemisphere forests. Forests take in CO_2 from the atmosphere (and give off O_2) in summer, and they release CO_2 back to the atmosphere during winter. Hawaii is in the Northern Hemisphere (latitude 19° N) and hence is influenced by this cycle. The cycle is reversed in the Southern Hemisphere, but the amount of land area is much smaller, so the magnitude of the CO_2 change is reduced.

Keeling's data show, in addition to this seasonal oscillation, that atmospheric CO_2 levels have increased significantly since 1958. The mean CO_2 concentration that year was about 315 ppm, or 43 ppm lower than today's value. The average rate of increase in CO_2 concentration since then has been 43 ppm/37 yr, or about 1.2 ppm/yr. More-detailed inspection of the curve reveals that the rate of CO_2 increase rose from 0.7 ppm/yr in the early 1960s to 1.5 ppm/yr in the late 1980s but then dropped to less than 1 ppm/yr in the early 1990s. Scientists believe that most of the increase in atmospheric CO_2 has been caused by the combustion of coal, oil, and natural gas but that tropical deforestation is also partly to blame.

The evidence that atmospheric CO_2 is increasing is indisputable. Similar measurements have been conducted at many different stations around the globe. The long-term increase in CO_2 is visible in every set of measurements and is essentially the same as that seen at Mauna Loa. (The range of the seasonal fluctuations, however, varies with the location.) For this reason, both scientists and policymakers agree that the long-term trend in atmospheric CO_2 is real rather than an artifact.

CO_2 DATA FROM ICE CORES. When did this increase in atmospheric CO_2 begin, and what was the CO_2 level before

that time? If we had to rely entirely on measurements made in the modern era, we would not be able to answer these questions. This is where analysis of the record of climate in the past can help. The composition of the atmosphere in the past can be determined by analyzing the composition of air bubbles trapped in polar ice. The bubbles are formed as snow at the top of an ice sheet is compacted, and their composition is preserved as they are buried under more snow. The age of the ice can be determined by drilling deep into the ice, removing a section of it, and counting the annual layers of snow accumulation. Figure 1-3 shows results from ice cores—cylindrical sections drilled into the ice—taken at several locations on Antarctica. This graph compares the CO_2 composition of the air bubbles in the ice with a "smoothed" version of the Keeling curve (the dashed curve, from which the seasonal oscillation has been removed). The fact that the ice core measurements match up well with the direct atmospheric measurements in 1958 is convincing evidence that the ice core technique for determining atmospheric CO_2 concentrations yields reliable results.

According to these measurements, the buildup of atmospheric CO_2 began early in the 19th century—well before the dawn of the Industrial Age, which started in earnest around 1850. The rise in CO_2 levels between 1800 and 1850 has been attributed to the deforestation of North America by westward-expanding settlers and is thus known as the *pioneer effect.* The ice core measurements show that the *preindustrial CO_2 concentration* (the value circa 1800) was about 280 ppm. Evidently, humans have been responsible for almost a 30% increase in atmospheric CO_2 concentration over the past two centuries.

Other Greenhouse Gases. Carbon dioxide is not the only greenhouse gas whose concentration is currently on the rise. Methane (CH_4), nitrous oxide (N_2O), and certain **chlorofluorocarbon** compounds (**CFCs**) have also been increasing as a result of human activities. Also called *freons,* CFCs are synthetic compounds containing chlorine, fluorine, and carbon. Collectively, such gases that are present in the atmosphere in very low concentrations, called **trace gases,** are thought to have contributed almost as much additional greenhouse effect over the past few decades as has CO_2. (Because CO_2 is much less abundant than N_2 or O_2 it is also classified as a trace gas, but it is more than 200 times as plentiful as any of the other gases mentioned here and hence deserves to be in a class by itself.) The CFCs have also been implicated in the destruction of stratospheric ozone, as we shall discuss later in this chapter. For now, we simply note that the evidence for an increase in anthropogenic greenhouse gases is unequivocal: Humans are indeed modifying the composition of Earth's atmosphere. The extent to which we should be concerned about it remains to be determined.

Observed Changes in Surface Temperature. The observed rise in greenhouse gases is quite well documented,

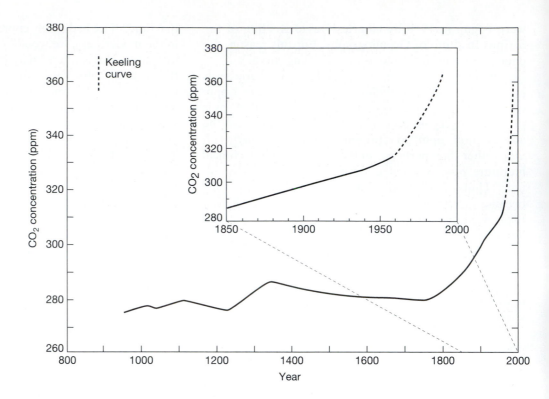

FIGURE 1-3

Atmospheric CO_2 concentrations over the past 1000 years, as determined from ice cores and from direct atmospheric measurements. (The dashed line is the Keeling curve.) (After *Climate Change, 1994,* Intergovernmental Panel on Climate Change, Cambridge University Press, Cambridge)

but what about the effects of this rise? Is there any direct evidence that climate is changing as a result?

The answer to this question is yes and no. Historical data indicate that Earth's surface temperature is on the increase. The data are not as easy to interpret as are the greenhouse gas data discussed earlier, but they are considered to be reliable. At a number of stations around the world, scientists have made accurate atmospheric temperature measurements that date back more than a century. Ocean-crossing ships have also routinely measured sea-surface temperatures during most of this time. Figure 1-4 illustrates the combined data from both types of historical measurements for the entire globe. The mean surface temperature from 1951 to 1980 has been subtracted from the data. The global mean surface tempera-

ture has increased from about 0.3°C below this mean value prior to 1900 to about 0.2°C above this mean value today. The overall temperature increase during the 20th century was thus approximately 0.5°C (or 0.9°F). This increase is broadly consistent with the warming expected from a 30% rise in atmospheric CO_2, although some climate models predict that the surface temperature should have increased about twice that much.

Problems do exist with these historical temperature data. For example, weather stations located near cities are subject to a well-documented "heat island" effect: As a city grows and as more area becomes covered with dark surfaces such as asphalt, more sunlight is absorbed and the local air temperature can increase by as much as 3°C. This systematic error has been removed from the

FIGURE 1-4

Change in global average surface temperature since 1861. The data are expressed as deviations from the 1951 to 1980 mean value. (After *Climate Change: The IPCC Scientific Assessment,* J.T. Houghton et al., eds., Cambridge Univ. Press, Cambridge, 1990)

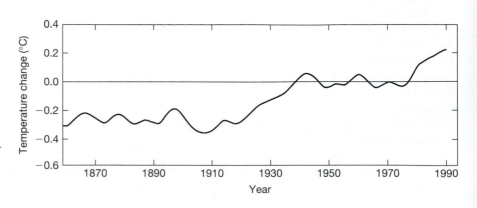

data shown in Figure 1-4, but it is still a source of uncertainty, because it is difficult to remove accurately. (*Systematic* errors exhibit a regular pattern. *Random* errors do not follow any pattern.) Sea-surface temperature measurements are also subject to systematic errors. Prior to the mid-1900s, water temperatures were determined by the "bucket method." A crewmember dropped a bucket over the side of the ship, then hauled it back up and measured its temperature with a thermometer. Since then, water temperatures have generally been measured with flow-through devices located on the ship's hull. The two methods do not yield exactly the same results, because the samples may be taken at different water depths and because buckets can warm or cool as they are being examined. Furthermore, the current procedure draws water up through the ship (normally near the engines) and can heat it up. These effects, too, can be corrected for, but not without creating additional uncertainties.

A second problem with the temperature data is that the coverage in time and space is much better in some parts of the world than in others. Populated areas of Europe and North America have been monitored most closely and for the longest time, so the coverage is best in these regions. Most land areas in the Southern Hemisphere have shorter and less-consistent temperature records. And the coverage over some regions of the ocean, particularly remote parts of the Southern Ocean where few ships travel regularly, is sparse indeed. Because sea-surface temperatures can now be monitored from satellites, the oceanic database should improve in the future. But it may well require several decades of such measurements to establish reliable trends.

Despite such difficulties, climatologists who collect and analyze these surface temperature data are confident that the observed half-degree warming trend over the past century is real. This does not mean, though, that it has been caused by human activities. Evidence shows that the climate was unusually cool between about 1500 and 1850. This period has been termed the "Little Ice Age." At least part of the warming since that time may represent a recovery from that naturally cool period rather than warm-

ing produced by anthropogenic greenhouse gases. This is another illustration of why it is necessary to understand the past if we want to predict the future.

An additional puzzle in the data shown in Figure 1-4 is that the warming trend seemed to slow, or stop entirely, between about 1940 and 1970. In the Northern Hemisphere, temperatures actually declined by a few tenths of a degree during this period (Figure 1-5). The decrease over Northern Hemisphere land areas is so pronounced that, by 1970, some climatologists were concerned that Earth might be entering a new glacial period. This worry was heightened by the historical data mentioned earlier that indicated that the present interglacial period might be nearing its end.

One possible explanation for the 1940 to 1970 cooling trend is that it was caused by increased reflection (and thus decreased absorption) of sunlight by *sulfate aerosol particles*. These tiny airborne particles are formed from sulfur dioxide (SO_2) emitted by the burning of coal. Most of the coal burning has taken place in the Northern Hemisphere, so this hypothesis could also explain why that hemisphere cooled more than did the Southern Hemisphere (see Fig. 1-5). Recent climate model simulations show that the magnitude of the aerosol effect is sufficient to account for the observed trend. But coal burning also releases CO_2 and hence should contribute to global warming—just the opposite of the observed effect during this 30-year period. This situation is a good example of why it is necessary to understand the whole Earth system in some detail if we are to interpret properly the changes that are occurring.

We cannot assume, however, that even though coal burning may have cooled Earth from 1940 to 1970, it will continue to do so in the future. In the United States, SO_2 is now being removed, or "scrubbed," from smokestack emissions in order to reduce its contribution to acid rain. **Acid rain** is produced when various acids, including sulfuric acid formed from the oxidation of SO_2, dissolve in rainwater. Acid rain can kill fish and damage plants in regions downwind from strong sources of pollution. It has been a problem in parts of the northeastern United States and in eastern Canada because there are many coal-fired

FIGURE 1-5

Change in temperature recorded over land surfaces in the Northern and Southern Hemispheres since 1861. (After *Climate Change: The IPCC Scientific Assessment*, J.T. Houghton et al., eds., Cambridge Univ. Press, Cambridge, 1990.)

power plants along and northward of the Ohio River valley. Other parts of the world, notably Europe, have problems with acid rain as well. Paradoxically, cleaning up smokestack emissions to cut down on acid rain may exacerbate the problem of global warming by reducing sulfate aerosol concentrations in the atmosphere.

Even if we were to quit scrubbing SO_2 out of smokestack gases, the ultimate effect of coal burning would be to warm Earth's atmosphere. Sulfate aerosols are removed from the lower atmosphere by precipitation in a matter of weeks, whereas CO_2 lingers in the atmosphere for decades to centuries. Thus, the CO_2 effect on climate is cumulative, whereas the aerosol effect is not. This example points out the importance of being aware of the time scale on which a global change occurs.

Possible Consequences of Global Warming. Although there is still a great deal of debate about whether humans have already altered the global climate, most climatologists agree that we will do so in the future if we continue to consume large amounts of fossil fuel. Should this be a cause for concern? In terms of the change in mean global temperature, we might expect people living in hot places such as India to be worried whereas those living in Siberia would look forward to the change. But the problem is not quite so simple: A change in temperature might cause other changes as well. A rise in sea level is one frequently mentioned concern. Sea level has already risen by at least 10 cm over the past century. The likely cause is *thermal expansion* of a gradually warming ocean; like most forms of matter, water expands when it is heated (except between 0 and 4°C when, paradoxically, it contracts). But warmer temperatures could also induce melting of mountain glaciers and ice caps. Increases in sea level on the order of several meters are possible within the next few centuries, and even larger changes are possible in the very long term. Such changes could have serious consequences for people in coastal areas and would be catastrophic for those in small island states. Other climatic changes may also have a broad-scale impact on agriculture, including decreases in soil moisture in certain areas and the spread of tropical insect pests. We will return to these possible side effects of global warming later; for now, note simply that the issues are complex and that there are very few simple answers. We also note that this is another reason to study past climate: Earth has been significantly warmer at various times in its past, and we may learn something about what it could be like in the future by examining those past time periods.

Evidence of Ozone Depletion

Global warming is not the only global environmental problem that has caught the attention of the public. Since at least 1985, the potential depletion of stratospheric ozone has also been in the news. (Stratospheric ozone should not be confused with *tropospheric* ozone—ozone near ground level—which is also often in the news because it is a component of *smog*.) The **stratosphere,** where most of Earth's ozone is located, is a layer of the atmosphere that extends from about 10 to 50 km in altitude. Stratospheric ozone is important to living organisms, because it absorbs many of the Sun's harmful ultraviolet rays. Ultraviolet radiation causes skin cancer and other health problems in humans. It adversely affects other organisms as well—notably, microscopic algae that are the base of the food chain in aquatic environments.

The year 1985 was a key one in stratospheric ozone research, because it marked the discovery of the ozone hole above Antarctica. Each year since about 1976, stratospheric ozone levels near the South Pole have fallen by large amounts during October, which is springtime in the Southern Hemisphere. Figure 1-6 shows year-to-year variations of the mean ozone *column depth* above Halley Bay in Antarctica for Octobers between 1957 and 1991. The ozone column depth is the total amount of ozone per unit area above a certain location. The decrease in ozone near the South Pole during October is striking: Ozone levels during October dropped by about half between 1975 and 1991. During the rest of the year, ozone levels have remained close to normal in this region. What has been destroying half the ozone over Antarctica during one particular month?

FIGURE 1-6

Mean total ozone over Antarctica during the month of October. The units, called *Dobson units,* measure the gas per unit area between Earth's surface and the top of the atmosphere (a measurement known as the *column depth*). One Dobson unit (DU) is equivalent to a 0.001-cm-thick layer of pure ozone at the surface. (After *Scientific Assessment of Ozone Depletion: 1994,* World Meteorological Organization, 1994.)

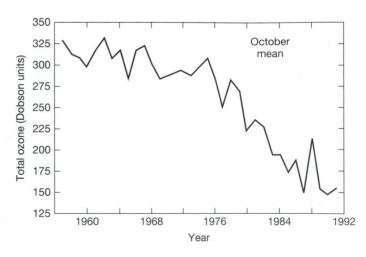

As soon as the ozone hole was discovered, atmospheric scientists guessed that chlorine compounds were to blame. By 1974, scientists had confirmed that chlorine is capable of destroying stratospheric ozone, and stratospheric chlorine levels have been increasing for the past few decades. Scientists are now fairly certain that the ozone hole is caused by chlorine compounds released from the breakdown of anthropogenic CFCs. The definitive evidence was provided in 1987, when a NASA research plane flew directly into the hole. One of the plane's instruments measured chlorine monoxide, ClO, which was thought to be a main culprit in ozone destruction; another instrument measured ozone (Figure 1-7). Outside the hole, ozone concentrations were at their normal stratospheric level, and ClO concentrations were very low. Inside the hole, ozone values were more than a factor of two lower, and ClO values were about 15 times higher, than the respective values outside the hole. Faced with such a strong inverse relationship, even scientists who had been skeptical about the connection between stratospheric chlorine and ozone depletion were driven to conclude that the chlorine was directly responsible for destroying the ozone.

The real concern about ozone depletion is not whether it is occurring over Antarctica in October but whether it might occur at hazardous levels over populated regions of the globe. (The few people living down in the far southern portions of Chile and New Zealand are already concerned, because they are so close to Antarctica.) So far, nothing as dramatic as the Antarctic ozone hole has been seen elsewhere. However, ozone does seem to be decreasing at midlatitudes in both hemispheres. So, there is no justification for complacency on this issue.

Deforestation and Loss of Biodiversity

Ever since a substantial portion of the human population switched from being hunters and gatherers to being farmers some 10,000 years ago, humans have been altering the land surface. More and more of Earth's land is being "managed" in one way or another—to the extent that it is now fairly difficult to find land areas that are pristine.

Most of these changes have tended to reduce the complexity of the landscape, such as when forested areas (or grasslands) have been cleared and replaced with a single crop species. When the natural vegetation cover is removed, it is not simply the plant species that are lost. With the plants go all the animals (mammals, birds, insects, and so on) and microorganisms that depended on that vegetation in order to live. New species may replace them, but normally the number of species decreases; that is, biodiversity is reduced. When a species is unable to move away or adapt, the change in land use can result in extinction of the species. The genetic information that is shared by—and only by—all the members of that species is thus lost permanently.

Some of the best-known examples of animal species that have gone extinct are the woolly mammoth, the

FIGURE 1-7

Simultaneous measurements of ozone (O_3) and chlorine monoxide (ClO) made from a NASA aircraft as it flew into the Antarctic ozone hole in September 1987. The hole was entered at a latitude of about 68° S. The units *ppt* and *ppb* stand for "parts per trillion" and "parts per billion," respectively. (From R.W. Christopherson, *Geosystems: An Introduction to Physical Geography*, 3/e, 1997. Reprinted by permission of Prentice Hall, Upper Saddle River, N.J.)

saber-toothed tiger, the dodo bird, and the dinosaurs. Many species that exist today, such as the mountain gorilla and the giant panda, are faced with the threat of extinction. The potential loss of these large mammals represents only the most visible of many similar threats.

The largest, and potentially the most significant, species loss occurring today is taking place in tropical rainforests. These warm, moist forests are centered around the Equator. Marked by lush vegetation, they are the

The Discovery of the Antarctic Ozone Hole

The story of the discovery of the ozone hole above Antarctica is one of the classic misadventures of modern science. Measurements of Antarctic ozone made from Earth's surface date back to 1956 and represent by far the longest continuous record of atmospheric ozone levels. But these measurements were made at only one site, Halley Bay, where a research station happened to be located. Continuous measurement of ozone levels above the entire Antarctic continent (and the rest of Earth) began in 1979 with the launching of the TOMS (Total Ozone Mapping Spectrometer) instrument on the Nimbus 7 satellite.

TOMS was a sophisticated and expensive instrument that should have been fully capable of detecting significant Antarctic ozone depletion within the first few years of going into orbit. It did not do so, however. The ozone hole was first reported at Halley Bay in 1985 by the British atmospheric scientist Joseph Farman and his colleagues, who had relied on their "old-fashioned," ground-based instruments. TOMS failed to detect the hole, it was later discovered, because the computer that processed the raw satellite data had been programmed to reject as "noise" any ozone measurements below a particular cutoff value. Values as low as those observed over Antarctica in October were considered too low to be real!

On learning about the Halley Bay measurement, the TOMS scientific investigators reanalyzed their original data, using a technique that retained the anomalously low values. The ozone hole was there, all right! Had it not been for the ground-based measurements, however, the hole might have gone undetected for years. Besides providing a wonderful illustration of the perils of having rigid preconceptions, the story of the discovery of the ozone hole shows that dedicated individuals working with relatively simple equipment can still make important contributions to modern science.

---

most biodiverse habitat on Earth. But they are rapidly disappearing due to deforestation: The trees have been cleared for grazing, farming, timber, and fuel. By 1990, the total area of tropical rainforests had been reduced to less than half the estimated prehistoric cover. The rapidity of deforestation of the Amazon rainforest is illustrated in Figure 1-8. Exactly how fast the tropical forests are disappearing is difficult to determine, but the loss rate is thought to approach 1.8% per year. If deforestation continues at such a rate, by the first quarter of the 21st century almost half the remaining rainforests will be lost, along with 5 to 10% of all the species on Earth.

Which Changes Should Concern Us the Most?

The concerns about the loss of tropical species are, in some ways, less immediate than the concerns about ozone depletion or global warming. One worry is that the tropical plants are a potential source of medicines for fighting cancer and other diseases. This concern is valid, but it does not have the urgency of the prospect of instantaneous sunburn on exposure to the Sun or of entire states or even entire nations being submerged by a rising sea level.

This does not mean, however, that species loss is not a serious problem. Indeed, it may be the most serious problem of all. One way of judging the severity of a problem is to estimate how long it would take Earth to recover. If we take this approach, ozone depletion is the least serious problem. The lifetime of chlorofluorocarbons in the atmosphere is on the order of 50 to 150 years, when they are eventually destroyed by solar ultraviolet radiation. This range is long enough to raise serious concerns, but the ozone level could be restored within a few human generations if preventive measures were taken. Indeed,

measures being taken right now may allow us to avoid any really damaging effects from this problem.

By this measure, global warming is a more serious problem, because the time scale for recovery could be much longer than 150 years. If we actually do consume an appreciable amount of the fossil fuels that are still available to us, atmospheric CO_2 levels could remain elevated for many thousands of years. Most of the excess CO_2 would be absorbed by the oceans during this time, but even then it would not be completely gone. As we will see in later chapters, it would likely take more than a million years for the excess CO_2 to be removed from the oceans and for atmospheric CO_2 to return to its preindustrial level.

Although this time scale sounds long, it is short in comparison with the time required to restore global biodiversity. Analysis of the fossil record shows that the time scale for recovery of biodiversity after a **mass extinction** (the dying out of many species within a geologically short time interval) is on the order of tens of millions of years. In fact, the system never does recover completely: Although many new species appear and flourish after a mass extinction, they are *different* from the ones that went extinct. That is why humans, instead of dinosaurs, now rule Earth! So, if we do induce a mass extinction of tropical species by deforestation, things will never again be the same.

Global Change on Long Time Scales

We have touched on three major global environmental changes that are occurring in the Earth system today: global warming, ozone depletion, and tropical deforestation. To understand fully the significance of these changes,

FIGURE 1-8
Satellite photos of Amazonia in 1972 and 1992. (Top: From Earth Satellite Corporation/Science Photo Library, Photo Researchers, Inc. Bottom: From NASA/Science Photo Library, Photo Researchers, Inc.)

however, we must understand how the Earth system operated prior to human intervention. Here, we will preview three examples of past global change—glacial–interglacial cycles, mass extinction, and changes in solar luminosity—and show how the geologic record provides evidence that allows us to study such changes.

Before we look at these examples of past global change, let us see where they occur on the *geologic time scale* (Fig. 1-9). Geologic time is divided into various intervals at several different levels. *Eons,* at the broadest level, are subdivided into *eras;* in turn, eras are broken down into *periods,* which may be further split into *epochs.*

The glacial–interglacial cycles that we will discuss, which lasted from about 2.5 million years ago until approximately 10,000 years ago, occurred during the Pliocene and Pleistocene epochs. The mass extinction that we shall talk about occurred at the boundary between the Cretaceous and Tertiary periods, approximately 65 million years ago. A *period,* typically lasting tens of millions of years, is generally a longer unit of geologic time than an *epoch.* Finally, the solar luminosity changes that we will discuss have occurred throughout the entire 4.5 billion years of Earth history.

Glacial–Interglacial Cycles

Isotopes. Earlier in this chapter, we showed how polar ice cores can be used to measure the past atmospheric concentration of CO_2. We can also use ice cores to estimate past surface temperatures by looking at isotopes. **Isotopes** are atoms of a given element that have different numbers of neutrons in their nuclei. For example, the most common isotope of hydrogen has no neutrons, whereas another hydrogen isotope, **deuterium,** has one neutron in the nucleus. Isotopes have the same **atomic number**—that is, the same number of protons in the nucleus—but a different **mass number**—the total number of protons plus neutrons in the nucleus. The hydrogen isotope with no neutrons is written as ^1H (or simply H); it has an atomic number of one and a mass number of one. The superscript preceding the element's symbol denotes the mass number. The isotope with one neutron, written as ^2H (or D for "deuterium"), has an atomic number of one and a mass number of two.

Some isotopes of a given element are **stable isotopes,** which means that they do not spontaneously change into other isotopes or into atoms of another element. **Unstable isotopes** spontaneously change into other isotopes or elements, by a process called *radioactive decay.* Both ^1H and ^2H are stable isotopes of hydrogen. **Tritium,** ^3H, which has two neutrons in its nucleus, is an unstable isotope of hydrogen. Oxygen, which has an atomic number of 8, has three stable isotopes: ^{16}O, ^{17}O, and ^{18}O. The most abundant is ^{16}O, and it is the only one we will consider for now.

Water, H_2O, may be made up of any of these isotopes of hydrogen and oxygen. By analyzing the isotopic composition of frozen water—of course, ice—in polar ice cores, a researcher can learn something about the local air temperature at the time the ice formed. Warmer water temperatures speed up the evaporation rates of both H_2O and HDO, but the effect is strongest on HDO, that is, water molecules that contain deuterium. Even more importantly, warmer air temperatures allow more of the HDO to be transported through the atmosphere to the vicinity of the polar ice caps, because the rainwater that forms along the way is not as strongly enriched in the heavier isotope. Hence, warmer temperatures correspond to a higher HDO content in the snow that falls over Antarctica and in the ice that forms from that snow.

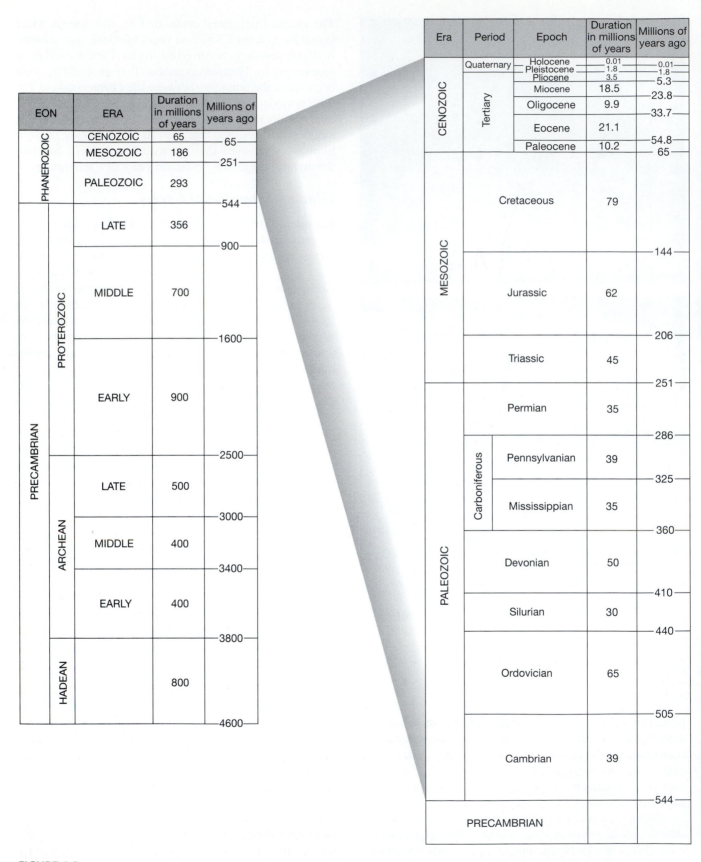

FIGURE 1-9

The geologic time scale. (From W.K. Hamblin and E.H. Christiansen: *Earth's Dynamic Systems,* 8/e, 1998. Reprinted by permission of Prentice Hall, Upper Saddle River, N.J.)

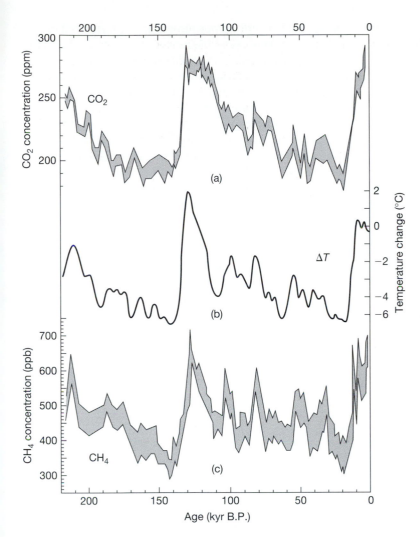

FIGURE 1-10

Measurements of (a) atmospheric CO_2, (b) temperature changes, and (c) atmospheric CH_4 determined from the Vostok ice cores. The temperature changes, ΔT, were determined from measured values of the deuterium content of the ice. The unit of age on the horizontal axis, kyr B.P., stands for "thousands of years before the present" (where k, for the prefix "kilo-," indicates "thousands"). (After *Climate Change, 1994,* Intergovernmental Panel on Climate Change, Cambridge Univ. Press, Cambridge.)

The Vostok Ice-Core Temperature Record. A set of ice cores drilled between the mid-1980s and the early 1990s at Vostok, Antarctica, near the South Pole, has provided a wealth of information. The most important results are shown in Figure 1-10. Figures 1-10a and 1-10c show the estimated range of CO_2 and CH_4 concentrations, respectively; Figure 1-10b shows the deviation in temperature, ΔT, from the present value. (The Greek capital letter delta, Δ, is often used to represent a change in a given quantity.) The value of ΔT represents the change in average air temperature over the Antarctic continent and the surrounding polar oceans. These values are derived from measured values of the deuterium content of the ice. The unit of age on the horizontal axis of Figure 1-10 is *kyr B.P.*, or thousands of years before the present.

The section of the Vostok ice core that has been fully analyzed is about 2.5 km deep and it extends back for an extraordinarily long time, almost 240,000 years. (The ice sheet is about 3.7 km thick at this location, so the record may eventually be extended back even farther.) The reason that the Vostok record extends so far back in time is that snow accumulates very slowly at that site—the equivalent of only about 2.5 cm of water per year. This value is

comparable to the mean annual precipitation over the Sahara Desert. Other parts of the polar ice sheets are approximately as thick but have faster accumulation rates. The short-term CO_2 record shown in Figure 1-3 comes from Siple Station, near the coast of Antarctica; there the snow accumulation rate is equivalent to about 50 cm of water per year. Cores from such locales cover much shorter periods of time than does the Vostok core, even if they are just as deep.

The time interval spanned by the Vostok core extends well beyond the last Ice Age. For the past 2.5 million years, Earth's climate has fluctuated between intensely cold **glacial periods,** in which ice sheets advanced across North America and Europe, and relatively warm **interglacial periods** such as the present, in which the ice sheets retreated. The present interglacial period began—and thus the last Ice Age ended—about 11,000 years ago, as an upward surge in temperature in Figure 1-10b indicates. At 21,000 years ago, Earth was in full-glacial conditions. Around 130,000 years ago, the planet was in the midst of another warm, interglacial period.

Much of the story about the advance and retreat of the glaciers was already known from other sources of data prior to the drilling of the Vostok ice core. What was new and surprising about the Vostok results was that they showed that atmospheric CO_2 and CH_4 concentrations had varied in concert with surface temperature. The Vostok data show that between 21,000 and 11,000 years ago, atmospheric CO_2 levels rose from about 200 ppm to close to its preindustrial value of 280 ppm, whereas CH_4 increased from about 350 ppb to 650 ppb. The current CH_4 concentration is about 1700 ppb, or 1.7 ppm. The same abrupt increase in CO_2 and CH_4 concentrations occurred after the previous interglacial period ended, between 140,000 and 130,000 years ago. Indeed, at a finer level, many of the smaller peaks and valleys in the temperature-change curve correspond to specific peaks and valleys in the concentration records of the two gases.

Why would atmospheric CO_2, CH_4, and temperature co-vary in this way? One part of the answer involves the greenhouse effect: As levels of the greenhouse gases CO_2 and CH_4 increased, the magnitude of the greenhouse effect also increased, and the climate became warmer. But what caused atmospheric concentrations of CO_2 and CH_4 to vary in the first place? In particular, why did those concentrations increase so abruptly just after 140,000 years ago and again just after 18,000 years ago?

These are tough questions, and we shall return to them later. Humans could not have caused these changes. Our ancestors were still making tools out of stone and tending small wood fires—and not burning fossil fuels—when these changes took place. One possible mechanism for driving changes in atmospheric CO_2 levels is a change in the circulation pattern of the deep ocean. As we will see in Chapter 5, the deep ocean circulates because cold, salty (and hence, dense) surface water sinks and is replaced by warmer, less dense water from lower latitudes. The deep ocean contains large amounts of dissolved CO_2, some of which is released to the atmosphere when deep water flows upward to the surface. So, the rate at which the deep ocean overturns can affect the concentration of atmospheric CO_2. But the circulation pattern of the deep ocean depends on climate, which is driven by changes in temperature and in evaporation rates at the sea surface. Thus, it would appear that atmospheric CO_2 levels affect climate and that climate, in turn, affects atmospheric CO_2 levels. What we have is a system in which the various components are tightly and intricately coupled. That is why a systems approach is the best way to understand global change.

Mass Extinction

Iridium and the K-T Boundary at Gubbio. Ever since dinosaur bones were first discovered, people have wondered why the dinosaurs disappeared. Dinosaurs flourished for more than 150 million years during an interval called the Mesozoic era, which ended 65 million years ago. At about the same time the dinosaurs disappeared, many other species went extinct as well. Some 60 to 80% of marine species died, as did numerous species of terrestrial plants and animals. Many possible reasons have been offered for their demise, including changes in climate, changes in vegetation, disease, destruction of the ozone layer by a nearby supernova (an exploding star), volcanic activity, and impact of an extraterrestrial body. No single hypothesis had attracted widespread support, however, until 1980.

That year, Luis and Walter Alvarez, of the University of California at Berkeley, and their colleagues published a paper about a clay layer they had studied in rocks from the mountains near Gubbio, Italy. The clay dated back 65 million years to the K-T boundary. **"K-T boundary"** stands for the transition between two time intervals: the Cretaceous period, abbreviated as "K" (to distinguish it from the Cambrian period, abbreviated as "C"), and the Tertiary period, abbreviated as "T." The Cretaceous period marked the end of the Mesozoic era and was followed by the Tertiary period, part of the Cenozoic era. The dinosaurs and other species disappeared at or just below the boundary between these two periods.

The layer of clay, only a few centimeters thick, was found between thick layers of *carbonate rock* (rock formed from the shells of certain marine organisms). The existence of this clay layer at the K-T boundary (Fig. 1-11) had puzzled geologists for decades. This clay layer had been seen at Gubbio and at numerous other spots around the world, always at the boundary between rocks of the Cretaceous and Tertiary periods. Walter Alvarez, a geologist, had journeyed to Gubbio in an effort to determine how long it had taken for the clay layer to be deposited.

Luis Alvarez, a physicist (and Walter's father), had a clever idea about how to make that determination. He reasoned that he could calculate the time required to form the clay layer by measuring the abundance of the element *iridium* (Ir). Iridium is a metal in the platinum group of elements, which are very scarce in rocks of

FIGURE 1-11

The clay layer at the K-T boundary in sediments at Gubbio, Italy. (From Prof. W. Alvarez/Science Photo Library (SPL), Photo Researchers, Inc.)

Earth's crust, because they are mostly dissolved in its molten iron core. These elements are always raining down on Earth as small particles of debris from asteroids or comets. The rate at which such debris hits Earth is known fairly accurately from measurements of its abundance in cores drilled into the ocean floor. Hence, Luis Alvarez reasoned that he could use the measured iridium abundance in the Gubbio clay layer as a kind of "cosmic clock" to determine the time needed for the clay to have been deposited.

The experiment failed, but it did so for a reason that turned out to be very informative. When the Alvarez team measured the iridium levels at Gubbio, they found the results shown in Figure 1-12. The iridium abundance in the clay layer was up to 10 ppb by mass—more than 100 times higher than what the group expected to find. The amount of iridium in the clay layer was much too large to have been supplied by debris from asteroids or comets. The time required to accumulate that much iridium would have been so long that the signal would have been swamped by the normal deposition of Earth-bound sediments. (Clay accumulates on the ocean floor at a rate of about 1 cm per thousand years as a result of wind-

blown dust that falls on the ocean surface. If the clay layer at the K-T boundary had taken more than a few thousand years to form, it should have contained a large proportion of terrestrial dust and, hence, a relatively small concentration of iridium.) The Alvarez team reasoned that the iridium must have come instead from the impact of some large, extraterrestrial object, such as an asteroid or a comet. Indeed, by calculating the amount of iridium deposited worldwide, the team estimated the mass of such an incoming body—on the order of 10^{15} kg, which corresponds to a diameter of about 10 km for a rocky asteroid. If the impacting object was a comet, it would have to have been even larger, because comets are thought to contain less iridium than do asteroids.

Asteroid Impact and the Extinction of the Dinosaurs. What would have been the effect of a 10-km-wide asteroid colliding with Earth? For such an asteroid traveling toward Earth at a speed of 20 km/s, 3×10^{23} (*joules,* or kg-m^2/s^2) of energy would be released. By comparison, 1 Mton (*megaton,* or 1 million tons) of TNT—the equivalent of a typical hydrogen bomb—releases 4.2×10^{15} J of energy. (The bomb dropped on Hiroshima during World War II was equivalent to only about 0.02 Mton of TNT.) So, the energy that would have been released by asteroid impact at the K-T boundary is equivalent to about 7×10^7 Mton of TNT, or 70 million standard-sized hydrogen bombs. In fact, at the height of the arms race in 1985, the combined nuclear arsenals of the United States and the former Soviet Union contained the explosive equivalent of about 10^4 Mton of TNT. Thus, in terms of energy release, a K-T impact would have been $(7 \times 10^7)/10^4 = 7000$ times more devastating than the simultaneous detonation of all those nuclear weapons.

We still do not know for sure if an asteroid impact is what killed off the dinosaurs and other organisms at the K-T boundary. A great deal of additional evidence corroborating a large impact 65 million years ago has been identified, including a deeply buried crater 200 km in diameter underlying the region around Chicxulub, Mexico, on the Yucatan Peninsula. Even this "smoking gun" does not prove that this impact was the cause of the mass extinction. It does demonstrate convincingly, though, that in the past the Earth system has experienced large shocks from which it has recovered, albeit slowly and in a modified form.

The changes that humans are causing in the Earth system today are less abrupt than those that occurred at the K-T boundary (assuming that the impact theory is correct), but they are still fast compared to most natural changes, and the results could still be catastrophic for certain elements of the biota. We have already noted that large land mammals such as gorillas and pandas are at risk. And with the vast majority of terrestrial species concentrated in the imperiled tropical rainforests, the potential for more-widespread mass extinctions is very high.

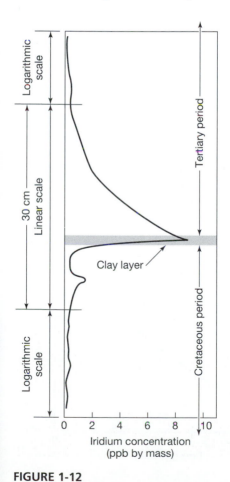

FIGURE 1-12

Iridium concentration versus depth at Gubbio. The middle portion of the depth axis is a linear scale; the upper and lower portions are logarithmic. (After L. Alvarez, *Physics Today,* July, 1987.)

A lesson learned from the K-T boundary crisis, that bio-diversity can decrease dramatically over a relatively short time interval, may therefore hold value today.

Changes in Solar Luminosity

All the examples of global change discussed thus far have been based on observational data. Observations, after all, are the cornerstone of science. Not everything of importance is observable, however. For example, we cannot see inside the Sun. Yet scientists believe that the Sun produces its energy through **nuclear fusion,** the joining of two or more light atomic nuclei to form one heavier nucleus. Specifically, four hydrogen nuclei (^{1}H) fuse to form one helium nucleus (^{4}He). This process, which is thought to occur continuously within the Sun, releases large amounts of energy. Even though we cannot observe this phenomenon directly, we are reasonably sure that the fundamental concept is correct.

The fact that the Sun produces energy in this way has important consequences for its long-term evolution. Four hydrogen nuclei take up more space, and therefore exert more pressure, than does one helium nucleus. The pressure in the Sun's core (where nuclear fusion occurs) would therefore be decreasing with time if the fusion of hydrogen into helium were the only process taking place. But what actually happens, models predict, is that the core contracts and heats up slightly as its helium content increases. The temperature rise increases the core's pressure and keeps the core from contracting further, so the Sun remains stable. As the core's temperature increases, so does the rate of nuclear fusion, just as the rates of most chemical reactions increase with increasing temperature. As a result, energy production within the Sun's core rises, and this rise is balanced by an increase in the amount of energy emitted at the surface. The more energy is emitted, the brighter the Sun appears. So, contrary to what we might intuitively expect, the Sun's **luminosity** (brightness) should gradually increase as it depletes its hydrogen fuel.

By how much has solar luminosity changed over the Sun's history? Model calculations performed by a number of different astronomers have reached essentially the same conclusion. Figure 1-13 shows a typical result, in which the unit of age on the horizontal axis is *byr B.P.,* or billions of years before the present. When the Sun first formed 4.6 billion years ago, it should have been about 30% less luminous than it is today. The Sun's luminosity increased slowly at first and then more rapidly as the buildup of helium in its core continued. At present, the Sun is thought to be brightening by about 1% every hundred million years. By the time the Sun ends its lifetime as a normal star, about 5 billion years from now, it is expected to have brightened by a factor of 2 to 3 as compared with today.

The Effects of Solar Luminosity Changes. How would reduced solar luminosity have affected the early Earth? If all other factors had remained constant, the early Earth should have been colder than it is today. Indeed, calculations (which we will do in Chapter 3) show that the entire ocean should have been ice-covered prior to 2 billion years ago. We know, however, that liquid water has existed on Earth's surface for at least the last 3.8 billion years, because sedimentary rocks (which form from sediments in liquid water) have been forming since that time. And organisms, which require liquid water to survive, have been around for at least 3.5 billion years. The early Earth could not have been a global iceball, at least not during the time for which a geologic record is available.

This apparent discrepancy is called the *"faint young Sun paradox."* We mention this paradox here because, like the Vostok CO_2 story, it is a problem that can be solved only by considering the Earth system as a whole. The most likely solution is that the level of greenhouse gases in Earth's primitive atmosphere was significantly higher than today. But why should this have been true, and why would greenhouse gas concentrations have declined as the Sun grew brighter? Does Earth's climate system have some built-in stability mechanism that has kept the mean surface temperature within survivable limits?

The Gaia Hypothesis. James Lovelock, a British biochemist, and Lynn Margulis, an American biologist, have argued that life itself has been responsible for maintaining the stability of Earth's climate. In the process of **photosynthesis,** organisms such as green plants use sunlight, CO_2, and H_2O to produce organic matter and O_2. (*Organic matter* is the carbon-rich material of which organisms are composed.) Through photosynthesis, followed by carbon burial in sediments, Earth's biota may have lowered atmospheric CO_2 levels at just the right rate to counteract the gradual increase in solar luminosity. Alternatively, the biota may have affected the rate at which atmospheric CO_2 is sequestered in carbonate rocks. Carbonate rocks form from reactions of CO_2 with elements

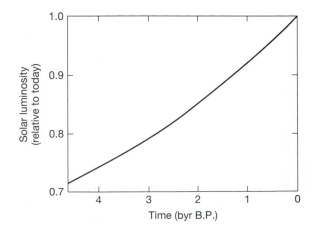

FIGURE 1-13

Estimated change in solar luminosity with time. The unit of age on the horizontal axis, byr B.P., stands for "billions of years before the present." (After D.O. Gough, Solar Physics 74, p. 21, 1981.)

(primarily calcium and magnesium) derived from other types of rocks. This process is part of the *carbonate-silicate geochemical cycle,* which we will discuss in Chapter 7. In either case, Lovelock and Margulis suggest that Earth has remained habitable precisely because it is in some sense "alive."

This theory of long-term climate stabilization is part of what Lovelock and Margulis called the Gaia hypothesis. In ancient Greek mythology, Gaia (pronounced guy-ah) was the goddess of mother Earth. In its most basic form, the **Gaia hypothesis** states that Earth is a self-regulating system in which the biota play an integral role. Some proponents of this hypothesis further suggest that the biota manipulate their environment for their own benefit or even, by optimizing the conditions for life, for the benefit of all living things. Such assertions are difficult to justify. Lovelock himself is quick to point out that the biota cannot be expected to cope with all possible disturbances. As an example, we cannot assume that we can safely emit CFCs into the atmosphere because Gaia will somehow protect the stratospheric ozone layer. But it is clear that the Gaia hypothesis is correct at some level: Organisms do play an important role in the overall functioning of the Earth system.

Some form of self-regulation must exist in order for Earth's climate to remain stable over long time scales. Higher greenhouse gas concentrations in the past are the most likely solution to the faint young Sun paradox. But whether the biota are essential to the control mechanism remains controversial. *Abiotic* (nonbiological) *feedbacks* in the carbonate–silicate cycle could have stabilized Earth's climate even if life were not present. Explaining how such a climate control mechanism might work is a recurrent topic in later chapters. Before we attempt to do so, however, we need to look more closely at how the various components of the Earth system function.

Chapter Summary

1. We will deal with three main themes: modern global environmental issues, past global change, and the behavior of Earth's systems. To understand present environmental problems, we must know something about Earth's past and something about the way different components of the Earth system interact.
2. Humans are modifying the global environment in several ways:
   a. Global warming may be the most pervasive environmental change that faces us today. The increase in concentrations of greenhouse gases, including carbon dioxide (CO_2), methane (CH_4), nitrous oxide (N_2O), and chlorofluorocarbons (CFCs), in the atmosphere is attributable to human activity. These gases are expected to warm Earth's climate over the next few decades to centuries by enhancing the natural greenhouse effect. They may have already begun to do so: Earth appears to have warmed by about 0.5°C over the past century, on the basis of surface temperature measurements made around the globe. It is still debated, however, whether this temperature rise is a consequence of increased greenhouse gas concentrations or simply a natural fluctuation in the climate system.
   b. The stratospheric ozone layer has already been severely affected by chlorine released from anthropogenic CFCs. The most dramatic impact has been confined to the Antarctic region during October. Strong regulatory steps have already been undertaken to ensure that the ozone layer will be protected in the future. Without such restrictions, the ozone layer's ability to absorb harmful ultraviolet rays from the Sun would be severely diminished.
   c. Massive deforestation is occurring in the tropics today, as it did in North America a century or more ago, when it contributed to the early rise in atmospheric CO_2. Deforestation both increases the buildup of atmospheric CO_2 and significantly decreases biodiversity. The effects of deforestation on biodiversity are permanent and irreversible.
3. Past changes in the Earth system may provide clues to how it will respond to global change in the future:
   a. Variations in surface temperature and atmospheric CO_2 concentrations recorded in ice cores illustrate the coupling between atmospheric CO_2 and climate and show how global warming today fits into the general pattern of glacial–interglacial cycles over the past 2.5 million years.
   b. Studies of the mass extinction at the end of the Cretaceous period 65 million years ago, when the dinosaurs and numerous other species forever vanished from Earth, may shed light on the loss of biodiversity that humans are causing today.
   c. Modeling studies of Earth's response to gradual increases in solar luminosity can help us understand how the climate system remains stable despite large changes in external forcing factors.

Key Terms

acid rain
anthropogenic
atmosphere
atomic number
biodiversity
biota
chlorofluorocarbons
core
crust
deforestation
deuterium
Earth system
fossil fuels
Gaia hypothesis
glacial period
global warming
greenhouse effect
greenhouse gases

hydrosphere
interglacial period
isotopes
K-T boundary
luminosity
mantle
mass extinction
mass number

nuclear fusion
ozone (O_3)
ozone hole
ozone layer
photosynthesis
rocks
sediments

solid Earth
stable isotope
stratosphere
system
trace gases
tritium
unstable isotope

Review Questions

1. a. What is meant by "anthropogenic greenhouse gases"?
   b. Name three such gases that are currently increasing in concentration in Earth's atmosphere.
2. What are the four fundamental components of the Earth system?
3. Explain the difference between global warming and the greenhouse effect.
4. a. By how much has Earth's atmospheric CO_2 concentration increased since the year 1800?
   b. How do we know this?
   c. What are thought to be the primary causes of this increase?
5. Cite two ways in which chlorofluorocarbons can affect the environment.
6. a. How far back in time do direct measurements of Earth's surface temperature extend?
   b. Why is it difficult to determine accurately the long-term temperature trend?

7. How might the burning of coal have had opposing effects on climate during the 20th century?
8. Why is stratospheric ozone important to humans?
9. To what two global environmental problems does tropical deforestation contribute?
10. How are hydrogen isotopes used to infer polar temperature records?
11. How is past surface temperature
    a. determined from the Vostok ice core?
    b. related to atmospheric CO_2 content?
12. Why is iridium a good indicator of impacts by extraterrestrial bodies?
13. a. How has solar luminosity changed during the past 4.6 billion years?
    b. What is the fundamental cause of this change?
14. What is the Gaia hypothesis, and what does it say about the importance of life on this planet?

Critical-Thinking Problems

1. Which of the three modern global change problems discussed in this chapter—global warming, ozone depletion, or loss of biodiversity—do you consider to be the most serious? Give reasons for your answer. If you wish, include information drawn from other sources.

2. How do global warming, ozone depletion, and loss of biodiversity compare with other environmental and social problems that the world faces today? You may wish to list the major problems, as you see them, in decreasing order of importance. Justify your answer with an explanation.

Further Reading

General

Lovelock, James. 1995. Gaia, *A New Look at Life on Earth.* Oxford University Press: Oxford.

Schneider, S. H. 1997. *Laboratory Earth: The Planetary Gamble We Can't Afford to Lose.* Basic Books: New York.

Chapter 2
Daisyworld: An Introduction to Systems

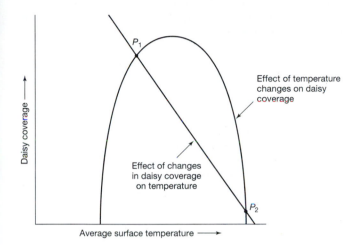

The mutual influences of average surface temperature on white-daisy coverage (the parabola) and white-daisy coverage on surface temperature (the straight line). The intersection points (P_1 and P_2) are the equilibrium states of the system.

Key Questions

- What are systems?
- What are feedback loops?
- What are equilibrium states?
- Does viewing the Earth as a system allow for deeper insight into the interrelationships among the physical and biological worlds?
- Can Earth's climate be self-regulating?

Chapter Overview

In this chapter, we develop the fundamentals of systems theory needed for the study of Earth as a system. First, examples from everyday life are used to introduce the important concepts of systems theory. Then we introduce the simplified climate system of the imaginary planet Daisyworld. Daisyworld is subjected to an increase in solar luminosity even more rapid than that which Earth has experienced over its history. Yet the hypothetical planet is able to counter the tendency for warming by increasing the reflectivity of its surface (by allowing for the spread of white daisies). We will see that this seemingly intelligent response arises without foresight or planning but rather as a natural consequence of interactions within the system. Through such feedbacks, natural systems can remain stable despite disturbances. Although the climate system of Daisyworld is oversimplified, we suspect that feedbacks like those it exhibits have played an important role in stabilizing Earth's climate over geological time.

The Systems Approach

The systems approach has been used in virtually every area of inquiry, including branches of both the natural and social sciences. Human physiology is a good example of where the systems approach is particularly illuminating. The human body is made up of a number of systems that perform the vital functions of life: a respiratory system that takes in oxygen and eliminates carbon dioxide; a cardiovascular system that circulates the blood, carrying oxygen and carbon dioxide around the body; a digestive system that processes food to fuel all body processes; a nervous system that senses changes in the internal and external environments and controls the activities of the other systems; an endocrine system that regulates ongoing processes such as growth and development; and so on. These systems are interrelated, functioning together to maintain the human body in a healthy state.

The Essentials of Systems

Each **system** is an entity composed of diverse but interrelated parts that function as a complex whole. The individual parts of a system are called **components.** A component can be a reservoir of matter (described by its mass or volume), a reservoir of energy (described by temperature, for example), an attribute of the system (such as body temperature or pressure), or a subsystem (such as the cardiovascular system, one of the interlinked subsystems of the human body). The components of the cardiovascular system itself include blood cells, blood vessels, and the heart.

The **state** of a system is the set of important attributes that characterize the system at a particular time. Body temperature, level of nutrition, and blood pressure are among the attributes that determine the state of the human body. The components of a system interact in such a way that a change in the state of the system is "sensed" throughout the system. In many systems, this linkage allows for the control of important attributes. For example, the endocrine system of the human body is capable of maintaining a nearly constant internal temperature despite large changes in the temperature of the surrounding environment. Suppose that your body temperature starts to rise as the air temperature around you rises. Your hypothalamus, a component of the endocrine system, then directs your sweat glands to increase their production of sweat, which helps cool you. If the ambient temperature then drops, your hypothalamus stops sending the signals to your sweat glands.

Couplings

It is clear from these examples from human physiology that the components of the human body "system" do not exist in isolation. They are linked, allowing for the flow of information from one component to the next. These links are called **couplings.** To understand how couplings allow for system regulation, consider an electric blanket. You set the temperature of the blanket (one system component) by adjusting a temperature controller. You adjust this controller to achieve a body temperature that is comfortable.

A *systems diagram* (Fig. 2-1) allows us to keep track of the various couplings within a system. In a systems diagram, couplings are conventionally represented by arrows. In the Earth system diagram of Chapter 1 (Figure 1-1), we linked the components of the Earth system with dashed arrows, - →. We use this convention when the sense of the coupling is unspecified but wish to indicate that one component influences another. There are two types of specified couplings. In the electric-blanket example, an increase in blanket temperature causes an increase in body temperature; such a link is called a **positive coupling** (Fig. 2-1a). In a positive coupling, a change (increase or decrease) in one component leads to a change of the same direction in the linked component. When one component increases, a positively coupled component responds by increasing. When the first component decreases, the second component responds by decreasing. A positive coupling is represented by a solid arrow with a normal arrowhead, →.

In contrast, an increase in body temperature above the comfort level would lead you to *decrease* the amount of heat by turning down the controller. This coupling, from body temperature to blanket temperature, is a **negative coupling** (Fig. 2-1b). In a negative coupling, a change in one component leads to a change of the opposite direction in the linked component. When one component increases, a negatively coupled component responds by decreasing. And when the first component decreases, the second component increases. A negative coupling is represented by an arrow with a circular arrowhead, $\perp_\circ$.

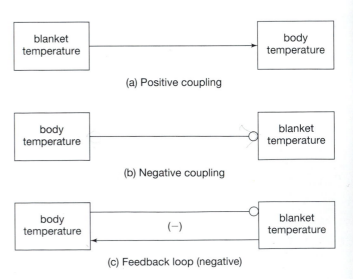

(a) Positive coupling

(b) Negative coupling

(c) Feedback loop (negative)

FIGURE 2-1

Systems diagrams; a negative feedback loop.
(a) An increase (or decrease) in blanket temperature causes an increase (decrease) in body temperature—a positive coupling.
(b) An increase (decrease) in body temperature causes you to decrease (increase) the blanket temperature—a negative coupling.
(c) A negative feedback loop is created by the two couplings.

Feedback Loops

The two couplings we described for the electric blanket create a "round trip," or **feedback loop,** between components (Figure 2-1c). *Feedback* is a self-perpetuating mechanism of change and response to that change. When you receive feedback from your friends, you are receiving their *re*action to an action of yours. Their reaction now becomes an action, and you *re*act to that action. You may modify your actions by either accentuating or suppressing them, and this modification will affect the nature of the subsequent feedback you receive. Consider Deb, who is Ed's employer. If Deb complains that Ed is dressing inappropriately at work, Ed may respond by dressing more conservatively, or he may instead dress the same or even more inappropriately. Either reaction (which is now an action) will undoubtedly cause a subsequent reaction—praise or criticism—from Deb. In terms of change and response, natural systems with feedback loops behave in a similar manner.

The feedback loop in the electric blanket example is referred to as a **negative feedback loop.** Negative feedback loops tend to diminish the effects of disturbances. An increase in body temperature, however caused, stimulates you to turn down the controller on the blanket. The blanket subsequently radiates less heat, and your body temperature then decreases.

In contrast to negative feedback loops, **positive feedback loops** amplify the effects of disturbances. To understand positive feedback loops, consider another electric-blanket example based on a real-life episode in the life of former U.S. President Jimmy Carter (Fig. 2-2). Jimmy and his wife, Rosalynn, had an electric blanket with dual controls—one for his side of the blanket and another for hers. In his autobiography, President Carter describes the irritation they were suffering:

> During each of the increasingly cold winter nights, we argued about the temperature of our electric blanket. Whenever I said it was too warm, Rosalynn said it was too cold, and vice versa.
>
> One morning I returned from an overnight trip to New York, and she met me at the front door with a warm hug and a smile. "I think our marriage is saved," she said. "I just discovered that our dual blanket controls have been on the wrong sides of the bed, and each of us was changing the temperature on the other's side." [Jimmy Carter, *Living Faith,* New York: Random House, 1996, p. 74.]

If the Carters had been thinking like systems scientists, the reason for their troubles would have immediately been obvious. A systems diagram for the proper use of the blanket is shown in Figure 2-2a. With their own controllers in hand, both persons would adjust their own blanket setting. If either person becomes chilly, he or she turns up the blanket controller and soon returns to a comfortable body temperature.

By inadvertently switching their temperature controllers, the Carters created a single, but more complicated, four-component feedback loop involving both controllers (Fig. 2-2b). The new loop is positive, consisting of two positive and two negative couplings. If Jimmy gets a bit warm, he unwittingly turns down Rosalynn's controller. She begins to feel a chill and so turns up what she thinks is her controller (but is actually Jimmy's). Jimmy gets even warmer, and then turns down Rosalynn's controller even further! This runaway response is characteristic of positive feedback loops.

A simple way to identify the "sign" of a feedback loop is to count the number of negative couplings. Negative feedback loops have an odd number of negative couplings; the rules of multiplication apply here. Recall that when two positive numbers are multiplied, the result is a positive number. When two negative numbers are multiplied, the result is also a positive number. But when a positive number and a negative number are multiplied, the result is always a negative number. Thus, if there is an odd number of negative couplings in a feedback loop, the loop is negative. If there are no negative couplings or an even number of them, the loop is positive.

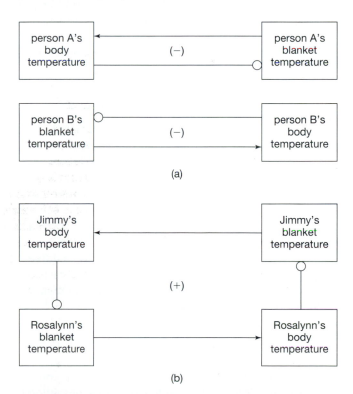

FIGURE 2-2

The consequences of combining feedback loops for a dual-control electric blanket.
(a) Proper usage: two independent, negative feedback loops.
(b) Improper usage: A single positive feedback loop formed when the Carters inadvertently exchanged temperature controllers

Equilibrium States

The normal electric-blanket feedback loop (Fig. 2-1c) acts to maintain body temperature (defining the state of the system) within a comfortable range. If your body temperature is just right, you do nothing to the blanket controller. We refer to this condition as an **equilibrium state** of the system; it will not change unless the system is disturbed. Because this state is created by a negative feedback loop, the equilibrium state is said to be **stable:** Modest disturbances from this state will be followed by system responses that tend to return the system to its equilibrium state. The equilibrium state of comfort for the Carter's feedback loop (Fig. 2-2b) was **unstable:** The slightest disturbance from a comfortable state led to system adjustments that carried the system further and further from that state.

To visualize equilibrium states better, we can represent all the possible states of a system as a hilly surface and the present state as a ball that is free to move on that surface (Fig. 2-3). The valleys represent stable equilibrium states, and the peaks represent unstable equilibrium states. After a small disturbance, a ball in a stable equilibrium state will roll back down the hill and return to its original state (Fig. 2-3a). Valleys, or regions of stability, are defined by the peaks that surround them. A large disturbance—large enough to roll the ball out of its valley and over an adjacent peak—can carry the system to a different equilibrium state (Fig. 2-3b). Thus, there are limits to the stability of stable equilibrium states.

In contrast, an unstable equilibrium has no region of stability. A ball disturbed ever so gently from its "resting" point at the top of a peak will roll down the hill and will land in a valley (Fig. 2-3c). On its own, the ball will not return to its original state. The slightest disturbance pushes the state of the system toward a new, *stable* equilibrium (if one exists). It is unlikely that a given system would remain poised for any length of time at an unstable equilibrium state.

We conclude that a system with a single feedback loop has a stable equilibrium state if the feedback loop is negative and an unstable equilibrium state if the feedback loop is positive. As we will soon see, however, more-complicated systems may involve both positive and negative feedback loops. The stability of such systems cannot be so easily determined.

Perturbations and Forcings

We can learn much by observing how a system responds to disturbances. Our understanding of human physiology, for example, has benefited from the study of patients stricken by illness or accident. The Carters learned about the problem with their electric-blanket system when their body temperatures were disturbed. Similarly, scientists

are learning about the Earth system by watching how it responds to disturbances. For example, Earth's climate system is being modified by a variety of natural and anthropogenic factors. One such **perturbation,** or temporary disturbance of a system, was the injection of sulfur dioxide (SO_2) into the atmosphere during the 1991 volcanic eruption of Mt. Pinatubo in the Philippines. This SO_2 formed sulfate aerosol particles (like those formed by the burning of fossil fuels; see Chapter 1) that prevented a small amount of sunlight from reaching Earth's surface. As a result, surface temperatures dropped about 0.5°C (1°F) globally (Fig. 2-4). The climate system had finally recovered from this perturbation 3 years later.

A more persistent disturbance of a system is called a **forcing.** In Chapter 1 we mentioned one forcing of Earth's climate: the gradual increase in the amount of sunlight Earth has been receiving over billions of years. How has the climate responded to this forcing? Many scientists

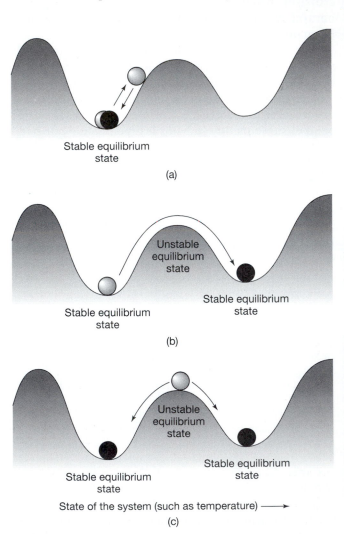

FIGURE 2-3

The equilibrium states of a system, represented as peaks (unstables) and valleys (stables). On disturbance, the system returns to stable equilibrium states but moves away from unstable equilibrium states.

FIGURE 2-4

The globally averaged temperature history from 1950 to 1996, showing the 0.5°C (1°F) cooling associated with the eruption of Mt. Pinatubo in 1991. Anomalies are defined as deviations from the 1951–1980 mean. (From R.W. Christopherson, *Geosystems: An Introduction to Physical Geography,* 3/e, 1997. Reprinted by permission of Prentice Hall, Upper Saddle River, N.J.)

argue that the tendency toward surface temperature rise that has accompanied the increased sunlight has been countered by a decrease in atmospheric CO_2 concentrations, reducing the greenhouse effect and thereby cooling the surface.

Understanding the response of the Earth system to forcings such as this is a major focus of this book. Rather than begin with the complex Earth system, however, we will consider a much simpler climate system of the hypothetical planet Daisyworld. This planet, whose only life forms are daisies, derives from the creative imaginations of James Lovelock (who, with Lynn Margulis, originated the Gaia hypothesis; see Chapter 1) and his colleague Andrew Watson, an oceanographer. When the Gaia hypothesis was first proposed, a common criticism was that the biota would need to possess the capacity for foresight or planning if the Earth system were to be self-regulating (for example, able to prevent large fluctuations in the surface environment). Lovelock and Watson used Daisyworld to demonstrate that natural systems can be self-regulating on a global scale without the need for intelligent intervention. Let us see how.

The Daisyworld Climate System

Imagine that the year is A.D. 2150. We have just determined that there is life on a nearby planet and have sent a manned mission there. On their arrival, the mission scientists observe that the planet is indeed supporting life, but only what appear to be daisies. Hence the scientists name the planet Daisyworld. These daisies are unusual, however: They are pure white in color. They appear to be getting their nutrients and water from the soil; the atmosphere has no clouds and no greenhouse gases. The daisies cover vast regions of the planet's surface; the rest of the surface is mantled in gray soil (Fig. 2-5). This means that the amount of sunlight absorbed by the planet depends on the area of darker, bare soil relative to the area of lighter daisy cover. The more sunlight absorbed, the higher the surface temperature. Experiments carried out by the mission scientists show that the growth and spread of daisies across the planet's surface depends only on the temperature around them.

growth. However, they make this calculation without considering that the daisies are part of a global climate system in which the reflectivity of the planet is affected by any change in the daisy population. Might a systems approach yield a different prediction for the survival of daisies on Daisyworld in the face of an increasingly luminous sun?

We can represent the Daisyworld climate system on the global scale as a two-component system. One component is the area of white-daisy coverage, and the other is the average surface temperature of the planet. These two components form a system because they are interdependent: The extent of daisy coverage affects the surface temperature, and the surface temperature affects the growth rate of daisies, which in turn affects the daisy coverage of the planet. Let's explore these interrelationships more fully.

Couplings in the Daisyworld Climate System

Response of Surface Temperature to Changes in Daisy Coverage. From experience, we know that on a sunny day, dark surfaces, such as asphalt roadways, tend to feel warmer than light-colored surfaces, such as concrete sidewalks. Dark surfaces absorb more (that is, they reflect less) of the incoming solar energy than do light surfaces. The reflectivity of a surface is called the surface's **albedo** (Fig. 2-6). Albedo is usually expressed as a decimal fraction of the total incoming (*incident*) energy reflected from the surface. Dark soil has a low albedo (0.05–0.15), whereas fresh snow has a high albedo (0.8–0.9). Table 2-1 lists the albedos of some common surfaces.

From the limited amount of information we have about Daisyworld, together with our intuition about

Gray soil   White daisy-covered
                 regions

FIGURE 2-5

A view of Daisyworld from outer space.

The mission scientists are alarmed because the planet's sun seems to be increasing in luminosity at a much faster rate than is our own Sun. They calculate that the planet will quickly become too hot to support daisy

FIGURE 2-6

A visual comparison of high-albedo and low-albedo surfaces. Light-colored surfaces are more reflective (that is, have a higher albedo) than dark surfaces, which absorb more sunlight.

TABLE 2-1

Albedos of Some Common Surfaces	
Type of Surface	*Albedo*
Sand	0.20–0.30
Grass	0.20–0.25
Forest	0.05–0.10
Water (overhead Sun)	0.03–0.05
Water (Sun near horizon)	0.50–0.80
Fresh snow	0.80–0.85
Thick cloud	0.70–0.80

albedo, we can graph the relationship between daisy coverage and surface temperature—in other words, the effect that changes in daisy coverage has on temperature. We know that the surface temperature on Daisyworld is determined by the amount of daisies that cover the surface: The more daisies, the more sunlight reflects off their white petals, the less sunlight absorbed, and finally the cooler the surface temperature. The graph should have a *negative slope* (that is, "runs downward" from left to right), reflecting the fact that as daisy coverage increases, temperature decreases (Fig. 2-7a). We *cannot* interpret the graph in Figure 2-7a to mean, however, that as surface temperature drops, daisy coverage rises linearly. The way changes in surface temperature

affect daisy coverage on Daisyworld is *not* the same as the way changes in daisy coverage affect surface temperature.

This graph can also be expressed as a coupling that links white-daisy coverage to temperature (Figure 2-7b). The coupling is negative: An increase in daisy coverage causes a decrease in surface temperature, and a decrease in daisy coverage causes an increase in temperature. Note that the "sign" of this coupling—negative—matches the sign of the slope in the graph.

Thus far in our discussion, the albedo of the planet has been a component of the Daisyworld climate system that we have treated only implicitly. However, we could treat albedo *explicitly* by adding it as a third component. The coupling that describes the effect of changes in white-daisy coverage on temperature is now seen as the combination of a positive and a negative coupling that links daisy coverage, albedo, and temperature (Fig. 2-8). Decreased white-daisy coverage leads to a reduction in the average albedo (a positive coupling), and reduced albedo causes an increase in temperature (a negative coupling).

When combined, the two couplings in Figure 2-8 form a negative coupling overall, like that shown in Figure 2-7. The rule for combining couplings is the same as the rule for determining the sign of feedback loops. The explicit treatment of albedo, therefore, does not change our conclusion regarding the overall sign of the coupling. For convenience and simplicity, then, we will often treat such couplings implicitly.

Response of Daisy Coverage to Changes in Temperature. In comparison with real daisies, we expect that Daisyworld daisies have an upper and lower temperature limit for survival. They must also have an *optimum,* or most favorable, temperature somewhere in between (let's specify that it is halfway between, for simplicity). A smooth curve drawn through these points is a *parabola*

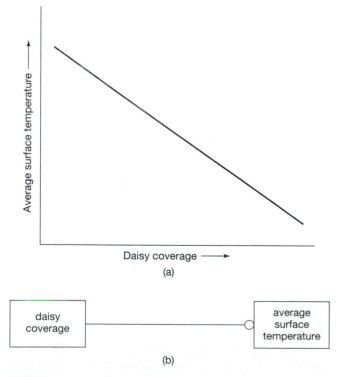

(a)

(b)

FIGURE 2-7

(a) Graph and (b) systems diagram of the effect of changes in white-daisy coverage on Daisyworld surface temperature.

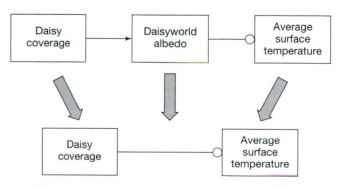

FIGURE 2-8

The same overall coupling as that in Figure 2-7b, but with albedo shown explicitly. A positive and a negative coupling combine to form a negative coupling overall.

Graphs and Graph Making

Graphs are a powerful way to convey scientific information in an economical way. If a picture is worth a thousand words, a graph can be worth a thousand data points. There are a number of ways in which graphs are used in science, and we present many of them in this textbook. All graphs convey information about the relationship between two (or more) variables. (*Variables* represent any number value.) In a conventional *x–y* graph, data are plotted with the *independent variable* on the *x*-axis and the *dependent variable* on the *y*-axis. The value of the dependent variable depends on the value of the independent variable, but the converse is not true. For example, a graph showing hourly measurements of temperature over the course of several days at a particular place would plot temperature, the dependent variable, as a function of time, the independent variable (Box Fig. 2-1a).

Graphs can be used to convey sense of relationships even when data are not available. We may, for example, convey the notion that temperature varies more or less regularly from day to night with a sketch such as Box Figure 2-1b. We make it clear that no data are being plotted by plotting only a smooth curve; in labeling the axes for such a graph, we do not even show the scale of the graph. In other cases we may wish to put some bounds on the axes, but it would be misleading to place numbers on the axes in a graph for which actual values are not available.

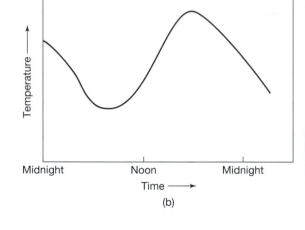

BOX FIGURE 2-1

Examples of various uses of graphs. (a) Display of data: temperature measurements made at State College, Pennsylvania, during a three-day period.

(b) Conveying a concept with no data: sinusoidal daily temperature variation.

(Fig. 2-9). The parabola is a characteristic shape for the temperature dependence of many plants on Earth. It intuitively makes sense that the abundance of an organism would be highest near the organism's optimum temperature and would drop to zero at the upper and lower limits of that organism's temperature range.

The sign of the coupling that reflects the response of daisy coverage to temperature changes depends on temperature, because the relationship is parabolic, as Figure 2-9 shows. If the temperature is below the optimum value for daisy growth, the coupling is positive. If the temperature is above the optimum value, the coupling is negative. This pattern is consistent with the slope of the parabola in Figure 2-9, which has opposite signs on either side of the optimum growth temperature for white daisies.

Equilibrium States in Daisyworld

We can determine the equilibrium states of Daisyworld by combining Figures 2-7 and 2-9. But note that tempera-

ture and daisy coverage are on opposite axes, so we cannot simply overlay the plots. Instead, we must invert the axes of Figure 2-7. This inversion does not change the nature of the coupling, it simply interchanges the positions of the two variables so that the axes will match up when we overlay the two graphs.

We can now overlay the two graphs (representing the two couplings); the resulting graph is shown in Figure 2-10. The curves intersect at two points, labeled P_1 and P_2. These points of intersection are special because they represent the only states of the system that simultaneously fall on the curves showing both the effect that white-daisy coverage has on surface temperature *and* the effect that surface temperature has on white-daisy coverage. For example, at any point other than P_1 or P_2 on the parabola, the effect of temperature on daisy coverage is properly characterized, but the effect of daisy coverage on temperature is not. Conversely, at any point other than P_1 or P_2 on the straight line, the effect of daisy coverage on tempera-

(b)

(a)

FIGURE 2-9

(a) Graph and (b) systems diagram of the effect of changes in Daisyworld surface temperature on white-daisy coverage.

ture is right, but the effect of temperature changes on daisy coverage is not.

Points P_1 and P_2 are the equilibrium states of this system, because they represent the states at which the system is said to be in *equilibrium*. If the system is already in one of these states, it will remain there unless something disturbs it. Note that neither equilibrium state corresponds to the optimum temperature.

But how will these equilibrium states respond to perturbation? We can evaluate the stability of these states

by constructing the systems diagrams for Daisyworld (Fig. 2-10). Two feedback loops characterize the Daisyworld climate system—one that applies *below* the optimum temperature, to equilibrium state P_1, and another that applies *above* the optimum temperature, to equilibrium state P_2. The diagram applicable below the optimum has a positive and a negative coupling and is thus a negative feedback loop. The feedback loop applicable above the optimum has two negative couplings and is thus a positive feedback loop. Of the two equilibrium states, then, the

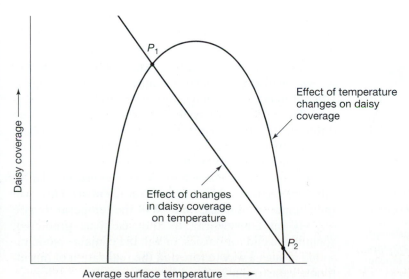

Effect of temperature changes on daisy coverage

Effect of changes in daisy coverage on temperature

FIGURE 2-10

The mutual influences of average surface temperature on white-daisy coverage (the parabola) and white-daisy coverage on surface temperature (the straight line). The intersection points (P_1 and P_2) are the equilibrium states of the system.

one that is below the optimum temperature for daisy growth is stable, and the one that is above the optimum temperature for daisy growth is unstable.

We can also use the combined graphs to evaluate the stability of the equilibrium states. Suppose, for example, that invaders uprooted enough daisies to decrease the planet's white-daisy coverage significantly from equilibrium state P_1 (perturbation a_1 in Fig. 2-11a). Based on our systems diagrams (see Figure 2-10), we surmise that the average albedo of the planet's surface would drop as daisy-covered areas are replaced by bare, gray soil. Daisyworld would then experience a temporary warming (step a_2). However, the warming would lead to an immediate increase in daisy coverage (step a_3) as long as the temperature remained below the optimum for daisy growth. The subsequent replacement of bare, gray soil with white daisies would cause an increase in average

planetary albedo. This increase would in turn tend to cool the planet (step a_4). The system would continue to adjust in this way (steps a_5 through a_9) until it eventually reached P_1 again, given enough time. As we saw before, P_1 is a stable equilibrium.

What would happen if the invaders uprooted a much larger portion of daisies (perturbation b_1 in Fig. 2-11a)? The result would be catastrophic. The reduction in albedo and subsequent warming would carry the planetary surface temperature beyond the upper limit for daisy survival (step b_2). The daisies would then go extinct. Thus, as we might have anticipated, there are limits to the stability of this stable equilibrium state.

If we apply a similar perturbation analysis to the second equilibrium state of this system, P_2, we find that even the slightest perturbation has two possible outcomes. Either the daisy population will be annihilated, or the system will approach P_1, the stable equilibrium state. This analysis confirms that P_2 is an unstable equilibrium state.

Thus, the response of Daisyworld to perturbation depends on the temperature of the planet (Fig. 2-11b). At temperatures below the optimum for daisy growth, the system is characterized by negative feedback, which will tend to maintain the temperature and daisy coverage near a stable equilibrium state. (Note that the temperature is below the optimum for daisy growth.) If temperatures are perturbed above the optimum, the system will enter a region of positive feedback without a stable equilibrium state. If the perturbation is small, the temperature will return to the cool, stable equilibrium state below the optimum. Larger perturbations will carry the system over the edge of the stable equilibrium state's "valley," and temperatures will rise above the limits for daisy growth; the daisies die.

External Forcing: The Response of Daisyworld to Increasing Solar Luminosity

We have been investigating the behavior of the Daisyworld climate system by perturbing it from its equilibrium states and analyzing its response. Many of the disturbances we will be discussing in later chapters, however, are persistent forcings that can be considered *external* to the system. The response of systems to forcings can be quite different from that to perturbations, because the system may not be able return to an original, stable equilibrium state even if negative feedback loops predominate.

The forcing on Daisyworld is the increase in solar luminosity recognized by the mission scientists. How will the climate system respond? Will the temperature rise quickly on Daisyworld, as the scientists predicted, spelling the end for daisies, or will the climate system respond in such a way to forestall the realization of this ultimate catastrophe?

(a)

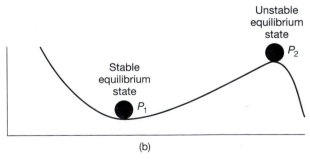

(b)

FIGURE 2-11

The response of Daisyworld to perturbation. (a) Daisyworld experiences a small (step a_1) and a large (b_1) reduction in daisy coverage. Responses (steps a_2 through a_9 and b_2) are shown as individual steps (for example, first a temperature response and then a daisy-coverage response); in reality, temperature and daisy coverage would respond simultaneously. (b) The stability of P_1 and instability of P_2.

On the basis of our experience about the ability of systems with negative feedback loops to damp perturbations, we might suspect that the system should act to extend the duration of the daisy inhabitation of the planet. Think of how the system would respond to a single, small but permanent increase in solar luminosity. The immediate response would be a warming of the planet's surface. This response, however, would be quickly followed by the spread of daisies, which, by increasing the albedo, would reduce the warming. A new equilibrium state would eventually be achieved at a temperature warmer than the original temperature. Yet it would be cooler than the temperature the planet would have achieved had the daisies not responded to the change in temperature and thereby altered the planet's albedo. Applying this line of reasoning to the problem at hand, we conclude that a persistent trend of increasing solar luminosity should lead to a gradual evolution of the equilibrium temperature of the planet to higher and higher temperatures, but at a rate that is slower than the warming that would otherwise occur in the absence of the feedback between the daisies and their environment.

Response of Daisyworld Couplings to Forcing

To predict the future climate of Daisyworld more accurately, we need to understand how the increasing intensity of Daisyworld's sun will affect the couplings in the system. Because we are assuming that the daisies respond only to temperature changes and thus not to changes in the solar luminosity itself, we would *not* expect modification of the coupling that links surface temperature changes to white-daisy coverage (the parabola in Figure 2-9). As the sun becomes more intense, surface temperature will rise, and the percentage of daisy coverage will respond according the parabolic curve, as before. However, we *would* expect a change in the coupling that relates surface temperature to the extent of daisy coverage (the straight line in Figure 2-9). For any amount of daisy coverage, the surface temperature will increase as the intensity of the sun increases. This is not to say that the temperature will rise indefinitely. Rather, for any particular value of daisy coverage, the temperature will be higher than expected from Figure 2-7. The graphical result is that the line in Figure 2-7 migrates upward as solar luminosity increases (Fig. 2-12).

Response of Equilibrium States to Forcing

Let us again consider the response to an incremental increase in solar luminosity. If we combine Figures 2-9 and 2-12 (first switching the axes of Figure 2-12), we find, as before, that two equilibrium states exist (labeled P_1' and P_2' in Figure 2-13a). We can guess that, as before, only one of them will be stable, because the diagram has not changed fundamentally. Which one is it? The tempera-

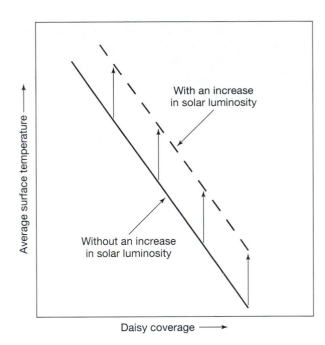

FIGURE 2-12

The effect of an increase in solar luminosity on the dependence of average surface temperature on white-daisy coverage. If the daisy coverage were fixed at a certain percentage, temperature would simply increase as shown by the arrows.

ture at point P_1' is below the optimum growth temperature for daisies, so this situation is similar to that for point P_1. The temperature at point P_2' is above the optimum temperature for daisies, so this situation is similar to that for point P_2. Thus, we determine that P_1' is stable and P_2' is unstable.

However, we see that both the temperature and the daisy coverage at the new stable equilibrium are higher. Daisyworld has apparently reacted to increased solar luminosity by increasing the daisy coverage. The accompanying increase in albedo explains why the temperature at the stable equilibrium did not rise as much as it would have without feedback. Note that the "ridge" that defines the stability limit for P_1' is lower than it was before. This new equilibrium state is apparently less resistant to perturbations; further increases in solar luminosity should eliminate the stable equilibrium state entirely.

We can determine the effectiveness of this feedback mechanism by comparing the equilibrium temperature changes with and without feedback. Without feedback—that is, without any change in the daisy coverage—the temperature change that results from the increase in solar luminosity is large. The temperature change without feedback is represented in Figure 2-13a as ΔT_0. (Recall from Chapter 1 that ΔT means the *change in temperature.*) With feedback, however, the temperature increase is smaller—but not zero. The temperature change of the new equilibrium state (with feedback) is represented as ΔT_{eq}, and the temperature change of the *feedback effect* itself is ΔT_f.

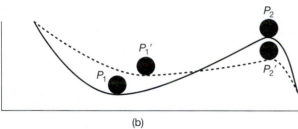

FIGURE 2-13

(a) Response of Daisyworld to an increase in solar luminosity. (b) The stability of P_1 and P_1' and the instability of P_2 and P_2'.

We can express the behavior of the Daisyworld system mathematically:

$$\Delta T_{eq} = \Delta T_0 + \Delta T_f.$$

In other words, the overall temperature change that results from increased solar luminosity is the sum of the temperature change with no feedback and the temperature change due to feedback. In our case, ΔT_{eq} is smaller than ΔT_0; the temperature effect of the feedback ΔT_f is negative. We can see this in Figure 2-13a: The arrow that represents ΔT_f points to the left—the negative direction—instead of to the right. Although we derived this equation for Daisyworld, it is a general relationship that can be applied to any stable equilibrium in a system involving feedback loops: The change in state of a system as it moves from one equilibrium to the next is the sum of the state change that would result without feedback and the effect of the feedback itself.

To quantify the strength of the feedback effect, we can define a value f, called the feedback factor. The **feed-**back factor is the ratio of the equilibrium response to forcing (the response with feedback) to the response without feedback. In our example, this ratio is as follows:

$$f = \frac{\text{temperature change with feedback}}{\text{temperature change without feedback}} = \frac{\Delta T_{eq}}{\Delta T_0}$$

Here f is less than 1, because the equilibrium response is smaller with feedback than it would have been without feedback. The value of f is between 0 and 1 whenever the feedback loop is negative but greater than 1 if the feedback loop is positive. As we mentioned previously, systems with positive feedback loops are stable only if they contain negative feedback loops as well. The feedback factor, f, can be defined only for stable systems, such as Daisyworld at point P_1. At point P_2, there is no stable equilibrium, and hence ΔT_{eq} is not defined there.

Climate History of Daisyworld

So far in our presentation of Daisyworld, we have avoided the use of actual values of temperature, daisy coverage, and albedo. However, we can assign reasonable values to the graphs to calculate the climatic response of Daisyworld to increasing amounts of sunlight. To present the details of the calculation would be premature; we shall introduce the physical laws in the next chapter. Suffice it to say that the calculations were based on typical growth curves for real daisies (see Figure 2-9), reasonable albedos for white daisies (0.9) and for gray soil (0.2), and a solar input similar to Earth's that also increases with time.

Figure 2-14a shows the history of white-daisy coverage, and Figure 2-14b shows the temperature history of Daisyworld from the time of its formation until the end of daisy inhabitation of the planet. The solid curve in Figure 2-14b represents the "actual" surface temperature change on the daisy-inhabited planet. The dashed curve in Figure 2-14b shows how that surface temperature would differ if there were no daisies (in other words, no life forms and no feedback). In the early years, temperature increases relatively rapidly. However, once the surface temperature rises above the minimum temperature for daisy survival, the white daisies begin to spread across the planet's surface. They spread slowly at first, because their growth cools the planet by increasing its albedo. As luminosity continues to increase, however, temperature begins to rise again. The daisy coverage expands rapidly in response to these increases in temperature, which are much smaller than we would predict for a lifeless (daisy-free) planet or for a planet with daisies but no feedback (and thus constant albedo).

Eventually the daisies run out of space in which to expand, and the planet's temperature begins to increase rapidly. The temperature eventually exceeds the opti-

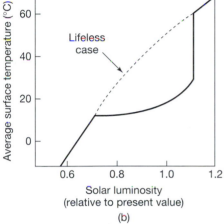

FIGURE 2–14

The response of Daisyworld to increasing solar luminosity.
(a) The change in daisy coverage of the planet in response to changes in solar luminosity (relative to the presumed present value).
(b) The change in average surface temperature of Daisyworld in response to increasing solar luminosity (solid line) and the response on a lifeless planet with fixed albedo (dashed line).

mum for daisy growth, and further increases in temperature cause reductions in daisy coverage. The feedback loop becomes positive. Once this happens, the system becomes unstable: The surface temperature rises rapidly, and the daisies go extinct. Thereafter, because it is dictated by the lower albedo of the gray soil on the lifeless planet, the temperature follows the dashed curve in Figure 2-14b.

The Lessons of Daisyworld

By studying the hypothetical planet Daisyworld from a systems perspective, we have learned some interesting things about climate systems in general. First, a planetary climate system is not passive in the face of internal or external influences. There are feedback loops that respond to perturbations and forcings (in this case, solar luminosity). Negative feedback loops in the system counter the external forcings. On Daisyworld, the consequence of this feedback is a longer lifespan for the daisies than one would predict if there were no feedback in the system. We will see in a later chapter that Earth's climate system has negative feedback loops as well that keep its climate relatively stable on both short and long time scales.

Second, the climate regulation system of Daisyworld, and, by analogy, other nonhuman systems that self-regulate, is seemingly intelligent: The response of the daisies is exactly what is needed to counter the solar warming of the planet. Yet no foresight or planning is involved. The daisies simply respond to the increase in temperature, and the planet's temperature responds to the spread of daisies. Such behavior is not restricted to contrived systems like Daisyworld. Indeed, self-regulation is a property common to many natural systems with feedback loops. Lovelock conceived of Daisyworld as a means of demonstrating to his critics that the Gaia hypothesis (which he applied to Earth) did *not* require an intelligent biota. Organisms can be components of self-regulating, natural systems simply because they influence, and are influenced by, the physical environment in which they live.

It is unlikely that the biota would be capable of optimizing their environment for their own good, as seemed to be required by the Gaia hypothesis when it was first proposed. The Daisyworld experiment indicates that it is not necessary for the biota to be capable of optimizing their environment. The Daisyworld system does not *optimize* the temperature for daisies. The stable equilibrium temperature is below the optimum for daisy growth on white-daisy Daisyworld.

Note that the self-regulation is not perfect. As the sun became more luminous, Daisyworld's climate system responded with a temperature increase, but the increase was much more gradual than would have occurred on a lifeless planet (or one with fixed daisy coverage). Systems like the Daisyworld climate system typically adjust to forcings by a slow but continual modification of their equilibrium states. This response is different from that directed by a thermostat, for instance, which is designed to maintain a constant state (temperature). In a natural self-regulating system, there is no preset state (no optimum value) that the system is programmed to "seek out."

The real Earth is not unlike Daisyworld: Its surface temperature has been maintained within the tolerance limits of living organisms for more than 3 billion years, despite substantial changes in solar luminosity. As on Daisyworld, the reason for the long-term stability of Earth's climate is the existence of strong negative feedback. The feedback loops that operate on Earth are, not surprisingly, more complicated than the one that operates on Daisyworld. Before discussing how Earth's climate system works, we will spend the next four chapters learning more about its components and their interactions.

Chapter Summary

1. Components of systems interact in ways that can either enhance or diminish the stability of the system.
   a. The components are linked by couplings, which can be either positive or negative.
   b. When couplings are arranged such that there is a round-trip flow of information, a feedback loop is formed. These feedback loops can be either positive or negative.
   c. Positive feedback loops amplify perturbations or forcings; negative feedback loops diminish them.
2. The presence of feedback loops leads to the establishment of equilibrium states.
   a. Negative feedback loops establish stable equilibrium states that are resistant to a range of perturbations; the system responds to modest perturbations by returning to the stable equilibrium state.
   b. Positive feedback loops establish unstable equilibrium states. A system that is poised in such a state will remain there indefinitely. However, the slightest disturbance carries the system to a new state.
3. The Daisyworld climate system is capable of resisting a warming trend induced by a sun that is becoming brighter with time.

   a. This capacity is the result of a negative feedback loop that involves the feedback between white-daisy coverage and temperature; it does not require foresight or planning.
   b. The key is the difference in albedo between the white daisies and the gray soil, together with the effect that temperature changes have on daisy growth and coverage.
   c. Daisyworld has two equilibrium states, but only one is stable. This equilibrium state, in general, does not coincide with the optimum temperature for daisy growth.
   d. The temperature response to an increase in solar luminosity can be thought of as a progression from one stable equilibrium state to the next. These equilibrium responses are the sum of the response that would occur without feedback plus the feedback effect itself. In the case of Daisyworld, the temperature change without feedback is larger than that from one equilibrium state to the next: The feedback effect is negative. Thus, the rate at which the planet warms is slower than it would be if there were no feedback between daisy coverage and surface temperature. The interval over which the planet is inhabited by daisies is extended because of the presence of feedback.

Key Terms

albedo
component
coupling
equilibrium state
feedback factor
feedback loop

forcing
negative coupling
negative feedback loop
perturbation
positive coupling

positive feedback loop
stable equilibrium
state
system
unstable equilibrium

Review Questions

1. A perturbation that causes a decrease in component *A* leads to a decrease in component *B*. Is the coupling between these two components positive or negative?
2. What is a feedback loop?
3. Why do negative feedback loops tend to diminish the effect of disturbances?
4. What distinguishes a forcing from a perturbation?
5. Are all equilibrium states stable? Why or why not?
6. What is albedo? How does it influence climate?
7. How are daisies on Daisyworld able to regulate the hypothetical planet's temperature?

Critical-Thinking Problems

1. a. In the Dysfunctia family, when the children get noisy, the parents get mad. When the parents get mad, the children get noisy. Draw a systems diagram for the Dysfunctia family.
   b. Is the feedback loop negative or positive?
   c. Is the family stable or unstable?
2. Earth's average temperature is determined in part by the amount of CO_2 in the atmosphere, by way of the greenhouse effect. The atmospheric CO_2 content may in turn be affected by the photosynthetic activity of plants, which convert CO_2 into plant tissue. However, the rate of photosynthesis depends on the amount of CO_2 in the atmosphere and on global air temperature. The components of this system—atmospheric CO_2 content, global temperature, and photosynthesis rate—are intimately interconnected. By increasing global photosynthesis rates, plants would tend to lower the atmospheric CO_2 level. In doing so, however, the plants would tend to cool Earth. This cooling, together with the reduced CO_2 level, might tend to reduce the photosynthetic activity of plants.
   a. On the basis of this discussion, draw a systems diagram of the photosynthetic rate–CO_2–temperature system.
   b. How many feedback loops are there?
   c. Are the feedback loops positive or negative?
   d. Describe the response of the system to the following perturbations:
      (i) an increase in atmospheric CO_2;
      (ii) a decrease in temperature.

e. Extra credit: How might the system respond to a continuous forcing—an increase in solar luminosity through time?

3. White-daisy Daisyworld has a companion planet that is similar in all ways except that the daisies are black.
   a. What is the effect of an increase in black-daisy coverage on planetary temperature? Express your answer graphically.
   b. Assuming that the effect of temperature on daisy coverage is the same on black-daisy Daisyworld as on white-daisy Daisyworld, draw a *stability diagram*—a diagram analogous to Figure 2-10—for black-daisy Daisyworld. Include two equilibrium states.
   c. Which of the two equilibrium states in part (b) is stable?
   d. Is the stable equilibrium state of part (c) cooler or warmer than that of white-daisy Daisyworld?
   e. How would this system respond to a *decrease* in solar luminosity? Express your answer graphically and in terms of the feedback factor *f*.
   f. Is *f* of part (e) greater than or less than 1?

4. The lines and curves shown in the graphs of this chapter can be converted to mathematical expressions that relate the values on the *y*-axis to those on the *x*-axis. The general equation for a straight line is $y = mx + b$, where *b* is the value of *y* when *x* is zero, and *m* is the slope of the line. To determine the slope of a line, we need only pick any two points (x_1, y_1) and (x_2, y_2) on the line; then $m = \dfrac{y_2 - y_1}{x_2 - x_1}$. A general equation for a parabola is $y = a - c(a - x)^2$, where *a* is the value of *x* when *y* is maximized, and *c* is a constant that defines the shape of the parabola (whether it is flat or peaked). The value of *c* can be found when we know the values of *x* when *y* is zero.

   a. Put reasonable numbers on the axes of the graphs for Figures 2-7 and 2-9.
   b. Give the equations for the line in Figure 2-7 and the parabola in Figure 2-9.
   c. The equilibrium states are defined as the values of *x* and *y* where the two equations in part (b) are equal. Find these equilibrium states.
   d. Find the equilibrium states for the case of higher solar luminosity on Daisyworld (Fig. 2-10).

Further Reading

General

Lovelock, James. 1991. *Healing Gaia: Practical Medicine for the Planet*. Harmony Books: New York.

Advanced

Saunders, Peter T. 1994. "Evolution without Natural Selection: Further Implications of the Daisyworld Parable." *Journal of Theoretical Biology* 166:365–373.

Levins, Richard. 1974. "The Qualitative Analysis of Partially Specified Systems." *Annals of the New York Academy of Sciences* 231:123–138.

Chapter 3
Global Energy Balance: The Greenhouse Effect

Photo of Earth from space, showing clouds. (From NASA Headquarters)

Key Questions

- What are the basic characteristics of electromagnetic radiation?
- What causes the greenhouse effect, and how is the magnitude of this effect determined?
- How do clouds affect the atmospheric radiation budget?
- What are the most fundamental feedbacks in the climate system?

Chapter Overview

Earth is heated by visible radiation from the Sun and cools by radiating infrared energy back to space. Earth's surface temperature depends on the amount of incident sunlight, the planet's reflectivity, and the greenhouse effect of its atmosphere. Certain gases in the atmosphere absorb outgoing infrared radiation and reradiate part of that energy back down to the surface. If this process did not occur, Earth's average surface temperature would be well below the freezing point of water, and life could not survive. Both the greenhouse effect and the amount of absorbed sunlight are strongly influenced by the presence of clouds. Clouds can either warm or cool the surface, depending on their altitude and thickness. Built-in feedback loops involving atmospheric water vapor and the extent of snow and ice cover are also fundamental aspects of the climate system. All these factors need to be considered in order to predict the response of Earth's climate to future increases in greenhouse gas concentrations.

Introduction

That Earth is suitable for life is largely a consequence of its temperate climate. A fundamental requirement for life as we know it is liquid water, and Earth is the only planet in our Solar System that has liquid water at its surface. Venus, our nearest neighbor toward the Sun, has an average surface temperature of 460°C (860°F), hot enough to melt lead. Mars, the closest planet away from the Sun, has an average surface temperature of −55°C (−67°F), the coldest temperatures experienced at the South Pole. Earth's average surface temperature is 15°C (59°F), the same as the mean annual temperature of San Francisco. Earth is not only habitable—it is a relatively pleasant place to live.

Why is Venus too hot, Mars too cold, and Earth just right? This question is sometimes called the "Goldilocks problem" of comparative planetology. Intuition suggests that the answer is that Earth happens to lie at the right distance from the Sun (and hence would receive exactly the right amount of sunlight), whereas Venus and Mars do not (Fig. 3-1). A closer look, however, reveals that it is not just the amount of sunlight that a planet receives that determines its surface temperature. A planet's surface is also warmed by the greenhouse effect of its atmosphere. As we saw in Chapter 1, a planet's atmosphere allows sunlight to come in but slows down the rate at which heat is lost. Without this greenhouse effect, Earth's average surface temperature would be about 33°C (59°F) colder than the observed value. Earth would be an icy, desolate world.

In this chapter, we discuss how the greenhouse effect works. We begin by considering the nature of electromagnetic radiation and why the Sun emits primarily one form of radiation (visible light), whereas Earth emits another (infrared radiation). We show that the incoming solar energy and outgoing infrared energy must be approximately in balance, and we demonstrate how this balance allows us to calculate the magnitude of the atmospheric greenhouse effect. Next, we discuss how both forms of energy are affected by atmospheric gases and by clouds, and we explain why some gases are greenhouse gases but others are not. Finally, we use the systems notation developed in Chapter 2 to introduce real climate feedback mechanisms, and we show why it is necessary to understand these feedbacks in order to estimate the climate changes that are occurring now as well as those that may occur in the future.

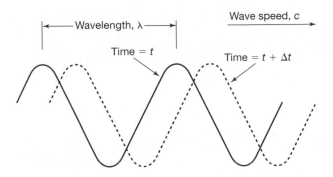

FIGURE 3-2

Simplified representation of an electromagnetic wave, illustrating the concept of wavelength. The solid curve shows the position of the wave at some time t. The dashed curve shows the wave at time $t + \Delta t$.

Electromagnetic Radiation

What exactly does it mean to say that Earth is heated by radiation from the Sun? From our sense of sight, we know that the Sun emits about 50% of its energy in the form of *visible light*. Let us start by considering what it is that makes up light and other forms of *electromagnetic radiation*.

Properties of Electromagnetic Radiation

A physicist would describe an electromagnetic wave as a propagating disturbance consisting of oscillating (regularly fluctuating) electric and magnetic fields that are perpendicular to each other. For our purposes, we can think of **electromagnetic radiation** as a self-propagating electric and magnetic wave that is similar to a wave that moves on the surface of a pond. A wave of any form of electromagnetic radiation—such as light, ultraviolet, or infrared radiation—moves at a fixed speed c (the "speed of light"). The numerical value of c for a light wave in a vacuum is $3.00 \times 10^8 \text{ m/s}$. The wave consists of a series of crests and troughs (Fig. 3-2). The distance between two adjacent crests

FIGURE 3-1

Venus, Earth, and Mars, shown roughly to scale. (Left: From NASA/Marten, Photo Researchers, Inc. Center: From NASA/Science Source, Photo Researchers, Inc. Right: From Photo Disc, Inc.)

is called the **wavelength.** It is typically denoted by the Greek letter λ (lambda). An observer standing at a fixed point in the path of the wave would be passed by a given number of crests in one second. This number is called the **frequency** of the wave. It is represented by the Greek letter ν (nu).

If we neglect complexities like polarization, an electromagnetic wave can be described by these three characteristics: speed, wavelength, and frequency. Not all of these characteristics are independent. The speed of the wave must equal the product of the number of wave crests that pass a given point each second (the frequency) and the distance between the crests (the wavelength). We can express this relationship mathematically as

$$\lambda \nu = c,$$

or, equivalently, as

$$\nu = \frac{c}{\lambda}.$$

The longer the wavelength of an electromagnetic wave, the lower must be its frequency, and vice versa. Conversely, the shorter the wavelength, the higher the frequency.

Photons and Photon Energy

Although we can think of electromagnetic radiation as a wave, at times it behaves more like a stream of particles. A single "particle," or pulse, of electromagnetic radiation is referred to as a **photon.** A photon is the smallest discrete (independent) amount of energy that can be transported by an electromagnetic wave of a given frequency. The energy E of a photon is proportional to its frequency:

$$E = h\nu = \frac{hc}{\lambda},$$

where h is a constant called *Planck's constant,* after the famous German physicist Max Planck. Its numerical value is 6.63×10^{-34} J·s (joule-seconds). Thus, high-frequency (short-wavelength) photons have high energy, and low-frequency (long-wavelength) photons have low energy. This difference in photon energy becomes important when electromagnetic radiation interacts with matter, because high-energy and low-energy photons have very different effects. High-energy photons can break molecules apart and hence cause chemical reactions to occur, whereas low-energy photons merely cause molecules to rotate faster or vibrate more strongly.

The fact that electromagnetic radiation behaves both as a particle and as a wave was one of the great discoveries of physics of the early part of the 20th century. This *wave-particle duality* is not restricted to electromagnetic waves. Rather, it is a general characteristic of matter and energy.

The Electromagnetic Spectrum

The full range of forms of electromagnetic radiation, which differ by their wavelengths (or by their frequencies), makes up the **electromagnetic spectrum** (Fig. 3-3). Wavelengths in the visible range are typically measured in nanometers (nm). One *nanometer* is one-billionth (10^{-9}) of 1 meter. **Visible radiation,** or visible light, consists of a relatively narrow range of wavelengths, from about 400 nm to 700 nm. Within this range, the color of the light depends on its wavelength. Anyone who has observed a rainbow has witnessed this phenomenon. The longest visible wavelengths appear to our eyes as the color red, whereas the shortest wavelengths register as blue to violet. The colors of the rainbow—in other words, the range of component wavelengths of visible light—are referred to as the **visible spectrum.** The term "spectrum" indicates that the light has been separated into its component wavelengths.

About 40% of the Sun's energy is emitted at wavelengths longer than the visible limit of 700 nm in a region referred to as the *infrared* region of the electromagnetic spectrum. Wavelengths of **infrared (IR) radiation** are significantly longer than those of visible light. Hence, it is

FIGURE 3-3

The electromagnetic spectrum.

convenient to keep track of them in units called *microm-eters* (μm) rather than in nanometers. One micrometer (formerly called *micron*) equals one-millionth (10^{-6}) of 1 meter. So, 1 μm equals 1000 nm. Infrared wavelengths range from 0.7 μm to 1000 μm. At even longer wavelengths within the electromagnetic spectrum, radiation is transmitted in the form of *microwaves* and *radio waves*. Radio waves can have wavelengths of many meters.

About 10% of the Sun's energy is emitted at wavelengths shorter than those of visible light as **ultraviolet (UV) radiation.** Wavelengths of the ultraviolet region extend from 400 nm down to about 10 nm. At shorter wavelengths are *X-rays* and *gamma rays*. These high-energy forms of electromagnetic radiation have little effect on our story here, but they do affect the chemistry of the uppermost atmosphere. X-rays, of course, are also used in medicine, because they can penetrate skin and muscle tissue and allow us to see the underlying bones.

The regions of the electromagnetic spectrum that are most important to climate are the visible and the infrared. The Sun emits energy in both of these spectral regions. Earth, as we shall see, emits primarily in the infrared. Solar ultraviolet radiation also affects the Earth system significantly by driving atmospheric chemistry. In addition, UV radiation would be lethal to most forms of life were it not almost totally blocked out by oxygen and ozone in Earth's atmosphere.

Flux

We will need one other basic concept from electromagnetic theory in order to proceed: the notion of flux. In general terms, *flux* (F) is the amount of energy or material that passes through a given area (perpendicular to that area) per unit time. In terms of fluid flow, for example, the flux is the volume of fluid that flows perpendicularly into or out of a unit area per unit time. Applied to electromagnetic radiation, **flux** is the amount of energy (or number of photons) in an electromagnetic wave that passes perpendicularly through a unit surface area per unit time.

To demonstrate the concept of flux, let us consider the light given off by an electric light bulb. A typical, small light bulb is labeled "60 watts." The *watt* (W) is a unit of *power*—formally, the rate at which work is done; informally, the intensity of the bulb in the SI system. Suppose that a person is standing some distance from such a light bulb and holds up a sheet of paper directly facing the light (Fig. 3-4a). The paper is illuminated by radiant energy from the bulb. The radiation crosses the paper perpendicularly from the light bulb at a certain flux, or intensity per unit area. That flux is measured in watts per square meter (W/m^2). The magnitude of the flux depends on how far from the light bulb the person is standing, but it does *not* depend on how big the paper is because flux is defined as the intensity per *unit* area.

Incoming light

Unit area

Paper is perpendicular to incoming light.

(a)

Paper is at an angle to incoming light.

(b)

FIGURE 3-4

Schematic diagram of the concept of flux. The flux of radiation into the paper is reduced when the paper is tilted at an angle to the incoming light.

The fact that flux is measured perpendicular to the direction the wave is traveling is important. Suppose that the person is holding the paper at an angle, rather than perpendicularly, to the light (Fig. 3-4b). Although the total area of the paper remains the same, the flux of radiant energy reaching its surface is less because less radiation strikes a given unit of area. This simple concept has direct, familiar consequences for Earth's climate. The polar regions are cooler than the tropics because the Sun's rays strike the ground at a higher angle at the poles. Summer temperatures are warmer than winter temperatures because the Sun is higher in the sky during summer. We will discuss these fundamental features of climate at greater length in Chapter 4. For now, we simply need to understand the concept of flux.

The Inverse-Square Law

Figure 3-4 demonstrates that the flux of radiant energy from a light bulb depends on how far away the observer (the person holding the paper) is standing. Likewise, the flux of solar energy decreases as distance from the Sun increases; that is why Venus is illuminated more strongly than Earth. The rate at which this *solar flux* decreases with increasing distance is described by a simple relationship. This relationship, called an **inverse-square law,** is expressed mathematically as

$$S = S_0 \left(\frac{r_0}{r} \right)^2,$$

where S represents the solar flux at some distance r from the source, and S_0 represents the flux at some reference distance r_0 (Fig. 3-5).

The inverse-square law has a straightforward physical interpretation: If we double the distance from the source to the observer, the intensity of the radiation decreases by a factor of $(1/2)^2$, or 1/4. Similarly, if we reduce the distance from the source to the observer by a factor of 3, the radiation intensity increases by a factor of 3^2, or 9.

As an example, consider a hypothetical planet, Planet X, located twice as far from the Sun as is Earth. What would be the solar flux hitting Planet X? Refer to Figure 3-5, and let the Sun be at the center of the two circles. Also, let the inner circle represent Earth's orbit and the outer circle represent the orbit of Planet X. Then r_0 is the average distance from the Earth to the Sun, which is 149,600,000 km, defined as one *astronomical unit* (AU), and S_0 is the solar flux at Earth's orbit, 1370 W/m². (The value of S_0 is determined by satellite measurements.) For this example $r = 2$ AU, so according to the inverse-square law, the solar flux incident at Planet X is

$$S = 1370 \text{ W/m}^2 \left(\frac{1 \text{ AU}}{2 \text{ AU}} \right)^2$$
$$= 342.5 \text{ W/m}^2$$

We would have gotten precisely the same answer if we had expressed the distances in kilometers, but the arithmetic would have been harder.

The inverse-square law is of fundamental importance to the study of planetary climates. It allows us to determine quantitatively why Earth's climate differs from that of Venus and Mars. It also plays a crucial role in our understanding of the causes of the glacial–interglacial cycles of the last 3 million years of Earth's history. As we will see later, small variations in the shape of Earth's orbit, combined with the inverse-square relationship between the distance from the Earth to the Sun and solar flux, cause large changes in the climate of the polar regions and in the size and extent of the polar ice sheets.

Temperature Scales

To understand climate, which is the prevailing weather patterns of a planet or region over time, we must first understand the concept of temperature. *Temperature* is a measure of the internal heat energy of a substance. Heat energy, in turn, is determined by the average rate of motion of individual molecules in that substance. For a solid, these motions consist of regular vibrations, whereas for a gas or liquid they are just random movements of molecules. The faster the molecules in a substance move, the higher its temperature.

Most areas of the world measure temperature (T) by the *Celsius* (formerly, *centigrade*) scale, which is measured in degrees Celsius (°C) and is part of the SI system of units. In the United States, temperature is typically measured in degrees *Fahrenheit* (°F). Scientists, particularly those studying climate, often use the **Kelvin (absolute) temperature scale,** measured in units called *kelvins* (K). (Note that temperatures in the Kelvin scale are given simply as kelvins, not as degrees kelvin.)

The Celsius temperature scale is defined in terms of the freezing and boiling points of water at sea level

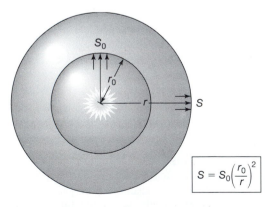

FIGURE 3-5

Diagram illustrating the inverse-square law.

TABLE 3-1

Freezing and Boiling Points of Water by Temperature Scale		
Temperature Scale	Freezing Point	Boiling Point (at sea level)
Fahrenheit	32°	212°
Celsius	0°	100°
Kelvin (absolute)	273.15	373.15

(Table 3-1). At sea-level pressure, the freezing point is 0°C, and the boiling point is 100°C. (Atmospheric pressure decreases with altitude, as we shall see later in the chapter, so it makes a difference where these temperatures are determined.) The Fahrenheit temperature scale was originally defined on the basis of the temperature of a mixture of snow and table salt (0°F) and the temperature of the human body (about 100°F). Like the Celsius scale, it is defined today in terms of the physical properties of water: The freezing point is 32°F, and the boiling point is 212°F. The following relations allow us to convert temperatures between the Celsius and Fahrenheit scales:

$$T(°C) = \frac{T(°F) - 32}{1.8}$$

$$T(°F) = [T(°C) \times 1.8] + 32$$

Note that converting a temperature *change* from one system of units to the other is easier, because the effect of the different zero points is removed. Thus, a temperature change of 1°C is equal to a change of 1.8°F. Conversely, a change of 1°F is equal to a change of 0.5556 (=1/1.8) °C.

Absolute temperature—that is, temperature on the Kelvin scale—is defined in terms of the heat energy of a substance relative to the energy it would have at a temperature of absolute zero. At *absolute zero,* the molecules of a substance are at rest (or, more precisely, are in their lowest possible energy state). A temperature change of 1 K is equal to a temperature change of 1°C. The zero point of the Kelvin scale is, however, lower than that of the Celsius scale by 273.15°. To convert temperature in degrees Celsius to kelvins, we use the following equation:

$$T(K) = T(°C) + 273.15.$$

Thus, a temperature of absolute zero corresponds to a Celsius reading of −273.15°C.

Blackbody Radiation

In order to quantify our discussion of climate and the greenhouse effect, we need one final concept from the world of physics: the concept of blackbody radiation. A *blackbody* is something that emits (or absorbs) electromagnetic radiation with 100% efficiency at all wavelengths. Consider a cast-iron ball (Fig. 3-6). At room temperature, the ball looks black because it absorbs most of the light incident on it and gives off little visible radiation of its own. If we heat the ball, however, it begins to glow a dull red. If we heat the ball further, it eventually glows white hot because it radiates at all visible wavelengths. Recall that white light is a mixture of all the colors of the spectrum.

The radiation emitted by a blackbody is called **blackbody radiation.** It has a characteristic wavelength distribution that depends on the body's absolute temperature. This distribution can be described mathematically by a relation called the Planck function. The *Planck function* relates the intensity of radiation from a blackbody to its wavelength, or frequency. When shown graphically, this relation is also known as the *blackbody radiation curve* (Fig. 3-7a). The Planck function itself is mathematically complicated and is beyond the scope of our discussion here. We can, however, use this relation to derive two simpler rules that are fundamental to an understanding of climate.

Room temperature   Hot   Hotter

FIGURE 3-6
Change in emitted radiation by a blackbody as it is warmed.

Cast-iron ball

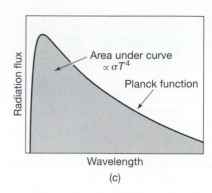

FIGURE 3-7

(a) The Planck function, or blackbody radiation curve. (b) Wien's law. (c) The Stefan–Boltzmann law.

Wien's Law

The first rule derived from the Planck function that will assist us in studying climate is called Wien's law. **Wien's law** states that the flux of radiation emitted by a blackbody reaches its peak value at a wavelength λ_{max} that depends inversely on the body's absolute temperature. According to this rule, hotter bodies emit radiation at shorter wavelengths than do colder bodies. Wien's law may be written as

$$\lambda_{max} \approx \frac{2898}{T},$$

where T is the temperature in kelvins and λ_{max} is the wavelength of maximum radiation flux in micrometers (Fig. 3-7b).

Wien's law allows us to understand why the Sun's radiation peaks in the visible part of the electromagnetic spectrum and why Earth radiates at infrared wavelengths. The Sun emits most of its energy, including visible radiation, from its surface layer, called the **photosphere.** The temperature of the photosphere is about 5780 K. Thus, according to Wien's

law, the Sun's radiation flux should maximize at 2898 μm/5780 $\approx$ 0.5 μm, or 500 nm (Fig. 3-8). This is right in the middle of the visible spectrum. (The fact that the solar radiation flux peaks in the visible is no coincidence; our sense of vision presumably evolved as it did to take advantage of the solar spectrum.) Earth, meanwhile, has a surface temperature of about 288 K, so its radiation peaks at 2898 μm/288 $\approx$ 10 μm—well into the infrared range. In reality, neither Earth nor the Sun is a perfect blackbody, so their emitted radiation flux is not exactly described by the Planck function. Nevertheless, Wien's law is still useful in predicting the wavelength at which most of their radiant energy is emitted.

The Stefan–Boltzmann Law

A second rule derived from the Planck function that will prove useful in climate studies is called the Stefan–Boltzmann law. The **Stefan–Boltzmann law** states that the energy flux emitted by a blackbody is related to the fourth power of the body's absolute temperature:

$$F = \sigma T^4,$$

where T is the temperature in kelvins and σ (the lowercase Greek letter sigma) is a constant with a numerical value of 5.67×10^{-8} W/m^2/K^4. The total energy flux per unit area is proportional to the area under the blackbody radiation curve (Fig. 3-7c).

As an example of how the Stefan–Boltzmann law can be applied, consider a hypothetical star that has a surface temperature twice that of the Sun. (We shall use stars rather than planets in this example because the radiation emitted from stars is more nearly approximated as blackbody radiation.) Our Sun has a surface temperature of about 5780 K, so the energy flux per unit area is

$$F_{Sun} = \sigma(5780 \text{ K})^4 \approx 6.3 \times 10^7 \text{ W/m}^2.$$

The other star releases energy at a rate of

$$\begin{aligned} F_{star} &= \sigma(2 \times 5780 \text{ K})^4 \\ &= 2^4 \times \sigma(5780 \text{ K})^4 \\ &= 16 \, F_{Sun}. \end{aligned}$$

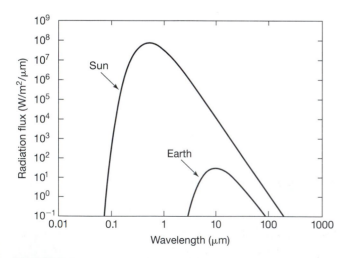

FIGURE 3-8

Blackbody emission curves for the Sun and Earth. The Sun emits more energy at all wavelengths.

Thus, the amount of energy released per unit area per unit time by the hot star is 2^4, or 16, times greater than that released by the Sun. Evidently, the amount of radiation emitted by a blackbody is a very sensitive function of its temperature.

Planetary Energy Balance

We now have all the tools necessary to analyze Earth's average climate in a quantitative manner. What we need to do next is put them together. The principle that we will apply is that of *energy balance.* To a first approximation, the amount of energy emitted by Earth must equal the amount of energy absorbed by Earth. In reality, this cannot be exactly true; if it were, Earth's average surface temperature would never change. We showed in Chapter 1 that the average surface temperature is changing—specifically, it is getting warmer. But it is getting warmer precisely because Earth's energy budget is slightly out of balance: The flux of incoming solar energy exceeds the outgoing IR flux by an almost imperceptible amount (a few hun-

dredths of a percent). The imbalance may be caused by the increase in CO_2 and other greenhouse gases in the atmosphere, or it may be caused by natural fluctuations within the climate system. When the climate system eventually reaches *steady state,* that is, when the surface temperature stops changing, the amount of energy going out will exactly equal the amount of energy coming in.

Physically, Earth's surface temperature depends on three factors: (1) the solar flux available at the distance of Earth's orbit, (2) Earth's reflectivity, and (3) the amount of warming provided by the atmosphere—that is, the greenhouse effect. The solar flux, S, as mentioned earlier, is the amount of solar energy reaching the top of Earth's atmosphere. Not all this energy is absorbed, however. About 30% of the incident energy is reflected back to space, mostly by clouds. As we saw in Chapter 2, the reflectivity of a planet is called its *albedo.* It is usually expressed as the fraction of the total incident sunlight that is reflected from the planet's surface. We shall designate albedo by the letter A.

To calculate the magnitude of the third factor, the greenhouse effect, it is convenient to treat Earth as a

---

Planetary Energy Balance

The derivation of the planetary energy balance equation is not difficult, but it does require that we consider the geometry of the Earth–Sun system. The starting point for the derivation is the relation

Energy emitted by Earth = Energy absorbed by Earth.

Let us first calculate the energy emitted by Earth. If we treat Earth as a blackbody with an effective radiating temperature T_e, the Stefan–Boltzmann law tells us that the energy emitted per unit area must be equal to σT_e^4. Earth radiates over its entire surface area, $4\pi R_{Earth}^2$, where R_{Earth} represents Earth's radius (Box Figure 3-1). Thus, the total energy emitted by Earth is

$$\text{Energy emitted} = 4\pi R_{Earth}^2 \times \sigma T_e^4.$$

Now, let us calculate the energy absorbed by Earth. From the Sun, Earth would look like a circle with radius R_{Earth} and area πR_{Earth}^2. Note that it is the area of Earth projected against the Sun's rays that enters here, not half of the surface area of Earth. (Half of Earth's surface area would be $2\pi R_{Earth}^2$, but the Sun's rays do not strike all of this area perpendicularly.) The total energy intercepted must be equal to the product of Earth's projected area and the solar flux (S), or $\pi R_{Earth}^2 S$. The reflected energy is equal to this incident energy times the albedo (A). The difference between these two quantities is the energy absorbed by Earth:

BOX FIGURE 3-1
The amount of sunlight received by and reflected by Earth.

$$\text{Energy absorbed} = \text{Energy intercepted} - \text{Energy reflected}$$
$$= \pi R_{Earth}^2 S - \pi R_{Earth}^2 SA$$
$$= \pi R_{Earth}^2 S(1 - A).$$

All that remains is for us to equate the outgoing and incoming energy. Using the expressions just calculated, we get

$$4\pi R_{Earth}^2 \times \sigma T_e^4 = \pi R_{Earth}^2 S(1 - A).$$

Cancelling out πR_{Earth}^2 on both sides of this equation and dividing both sides by 4, we obtain the desired equation,

$$\sigma T_e^4 = \frac{S}{4}(1 - A).$$

blackbody even though this is not exactly true. (As we shall discuss later, the atmosphere radiates and absorbs energy better at some wavelengths than at others because of the presence of gases such as CO_2 and H_2O.) We do this by defining a quantity T_e that represents the **effective radiating temperature** of the planet. This temperature is the temperature that a true blackbody would need to have to radiate the same amount of energy that Earth radiates. With this definition in place, we can use the Stefan–Boltzmann law to calculate the energy emitted by Earth. By balancing the energy emitted with the energy absorbed, we obtain the following formula (see the box "Planetary Energy Balance"):

$$\sigma T_e^4 = \frac{S}{4}(1 - A).$$

This formula expresses the planetary energy balance between outgoing infrared energy and incoming solar energy.

Magnitude of the Greenhouse Effect

What is the significance of the effective radiating temperature? We can think of this quantity as the temperature at the height in the atmosphere from which most of the outgoing infrared radiation derives (see Critical-Thinking Problem 4). We can also think of it as the average temperature that Earth's surface would reach if the planet had no atmosphere (assuming that the albedo remained constant). To get a better understanding, let us calculate its value for the present Earth. We can solve the planetary energy balance equation for T_e by dividing both sides of the equation by σ and then taking the fourth root of each side:

$$T_e = \sqrt[4]{\frac{S}{4\sigma}(1 - A)}.$$

If we insert the known values of S (1370 W/m^2), A (30%, or 0.3), and σ (5.67 × 10^{-8} W/m^2/K^4), we get $T_e \approx 255$ K. Thus, Earth's effective radiating temperature is a relatively chilly $-18°C$, or $0°F$.

We saw earlier, however, that the actual mean surface temperature of Earth, T_s, is 288 K, or about 15°C. The difference between the actual surface temperature and the effective radiating temperature is caused by the greenhouse effect of Earth's atmosphere. We can represent this mathematically by letting

$$\Delta T_g = T_s - T_e,$$

where ΔT_g is the magnitude of the greenhouse effect. Thus, $\Delta T_g = 15°C - (-18°C) = 33°C$.

To place this value in context, we can carry out similar calculations for Venus and Mars from known data of the albedos, surface temperatures, and orbital distances of these planets. (See Critical-Thinking Problem 2.) The results show that the solution to the Goldilocks problem posed at the beginning of this chapter is more complicated than we might have guessed. Evidently, a planet's greenhouse effect is at least as important in determining that planet's surface temperature as is its distance from the Sun.

We can also apply the planetary energy balance equation to the faint young Sun paradox mentioned in Chapter 1. Recall that solar luminosity, and thus S, is estimated to have been 30% lower early in the Solar System's history. It is easy to demonstrate that Earth's average surface temperature would have been below the freezing point of water under such circumstances, if the planetary albedo and the atmospheric greenhouse effect had remained unchanged (see Critical-Thinking Problem 5). We have already seen, though, that the early Earth had both liquid water and life on its surface. In later chapters, we shall discuss ways to resolve this apparent paradox.

Atmospheric Composition and Structure

Atmospheric Composition

To understand the greenhouse effect in more detail, along with other aspects of climate and Earth's radiation budget, we must learn a few fundamental facts about the composition and structure of Earth's atmosphere. Table 3-2 lists the main constituents of Earth's present atmosphere and their relative abundances.

As Table 3-2 indicates, the three most abundant constituents of our atmosphere are nitrogen, oxygen, and argon. Nitrogen is a relatively *inert* (chemically unreactive) gas, but when split into its constituent atoms, it plays an important role in biological cycles. Oxygen, which is highly reactive, is the essential gas that all animals must breathe; it is required by many other life forms as well. Argon is almost completely inert; it is the product of the

TABLE 3-2

Major Constituents of Earth's Atmosphere Today	
Name and Chemical Symbol	*Concentration (% by volume)*
Nitrogen, N_2	78
Oxygen, O_2	21
Argon, Ar	0.9
Water vapor, H_2O	0.00001 (South Pole)–4 (tropics)
Carbon dioxide, CO_2	0.036*

* In 1997

How the Greenhouse Effect Works: The One-layer Atmosphere

Although we calculated that Earth's greenhouse effect provides 33°C of surface warming, our method of obtaining this result (by subtracting the calculated effective radiating temperature from the observed mean surface temperature) provides little insight into the physical mechanism that causes the warming. We can remedy this by doing a simple calculation that demonstrates how the greenhouse effect actually works.

Suppose we treat the atmosphere as a single layer of gas and that this gas absorbs (and re-emits) all of the infrared radiation incident on it (Box Figure 3-2). Let us assume that it absorbs and emits infrared radiation equally well at all wavelengths, so that we can treat it as a blackbody, and that it has an albedo A in the visible spectrum, just like that of the real Earth. What are the temperatures of the gas layer and of the surface beneath it? We will call the layer temperature T_e and the surface temperature T_s, because these quantities are exactly analogous to those discussed in the text.

We can determine the values of T_e and T_s by balancing the energy absorbed and emitted by both the surface and the one-layer atmosphere. Let the amount of sunlight striking the planet be equal to $S/4$ (the globally averaged solar flux). The surface absorbs an amount of sunlight equal to $S/4 \times (1 - A)$, along with a flux of downward infrared radiation from the atmosphere equal to σT_e^4. The atmosphere absorbs an amount of upward infrared radiation from the ground equal to σT_s^4, and it emits infrared radiation in both the upward and downward directions at a rate of σT_e^4. (The real atmosphere also absorbs some of the incoming solar radiation, but we ignore that complication here.) Thus, we can write the overall energy balance in the form of two equations:

For the surface,

$$\sigma T_s^4 = \frac{S}{4}(1 - A) + \sigma T_e^4;$$

for the atmosphere,

$$\sigma T_s^4 = 2\sigma T_e^4.$$

(The factor 2 in the second equation arises because the atmosphere radiates in both the upward and downward directions.) If we now substitute the second equation into the left-hand side of the first equation and substract σT_e^4 from both sides, we obtain

$$\sigma T_e^4 = \frac{S}{4}(1 - A),$$

which is just the familiar energy-balance formula. But dividing the atmospheric energy-balance equation by σ and then taking the fourth root of both sides yields an additional result:

$$T_s = 2^{1/4} T_e$$

Thus, the surface temperature is higher than the one-layer-atmosphere temperature by a factor of the fourth root of 2, or about 1.19. For $T_e = 255$ K, as on Earth at present, we get $T_s = 303$ K, and we calculate a greenhouse effect of

$$\Delta T_g = T_s - T_e = 48 \text{ K}.$$

This is higher than the actual greenhouse effect on Earth by about 15 K.

This example is not meant to be realistic. The real atmosphere is not perfectly absorbing at all infrared wavelengths, so some of the outgoing IR radiation from the surface leaks through to space. This effect tends to make ΔT_g smaller. Conversely, a more-accurate calculation would subdivide the atmosphere into a number of different layers. This effect tends to make ΔT_g bigger and is the reason why a thick atmosphere, like that of Venus, can produce a really huge amount of surface warming. The calculation does, however, illustrate the basic nature of the greenhouse effect: By absorbing part of the infrared radiation radiated upward from the surface and re-emitting it in both the upward and downward directions, the atmosphere allows the surface to be warmer than it would be if the atmosphere were not present.

BOX FIGURE 3-2

The greenhouse effect of a one-layer atmosphere.

radioactive decay of potassium, K, in Earth's interior. These three constituents—nitrogen, oxygen, and argon—are not greenhouse gases. In other words, they do not contribute to Earth's greenhouse effect.

Although they appear at the bottom of Table 3-2, water vapor and carbon dioxide are two of the most important atmospheric constituents. Besides being directly used by organisms, they are also strong greenhouse gases. We shall soon see what makes a particular gas a greenhouse gas.

In addition to the major constituents listed in Table 3-2, Earth's atmosphere also contains a number of minor (or "trace") constituents that affect climate. The most important of these are methane, nitrous oxide, ozone, and freons. Their concentrations are generally much lower than those of the major constituents. Despite their low concen-

TABLE 3-3

Important Atmospheric Greenhouse Gases	
Name and Chemical Symbol	**Concentration (ppm by volume)**
Water vapor, H_2O	0.1 (South Pole)–40,000 (tropics)
Carbon dioxide, CO_2	360
Methane, CH_4	1.7
Nitrous oxide, N_2O	0.3
Ozone, O_3	0.01 (at the surface)
Freon-11, CCl_3F	0.00026
Freon-12, CCl_2F_2	0.00047

trations, these traces gases are important greenhouse gases. Table 3-3 lists the major greenhouse gases. (Note that water vapor and carbon dioxide are repeated here.) It is convenient to keep track of these gases in units of parts per million (ppm), which we defined in Chapter 1. Take a moment to convince yourself that the 0.00001%–4% value of water vapor and the 0.036% value of CO_2 given in Table 3-2 are equivalent to the 0.1–40,000 ppm and 360 ppm values of water vapor and of CO_2 given in Table 3-3.

Table 3-3 is by no means a complete list of greenhouse gases. Several other gases affect climate to some extent or are otherwise important in atmospheric chemistry. The gases listed in Table 3-3, however, are the ones that are most important to the modern problem of global warming, and hence they are the ones on which we will focus.

Atmospheric Structure

How Atmospheric Pressure Varies with Altitude. Other characteristics of Earth's atmosphere that influence climate and the radiation budget are its pressure and temperature structure. *Pressure* may be defined as the force per unit area exerted by a gas or liquid on some surface with which it is in contact. The pressure exerted by the atmosphere at sea level is defined as one *atmosphere* (atm). A pressure of 1 atm is equivalent to about 15 lb/in^2 in the English system and to 1.013 *bar,* or 1013 *millibars* (mbar), in the metric system. (The pressure unit in the SI system is the *Pascal,* Pa, but this unit is cumbersome in atmospheric work: 1 Pa = 1×10^{-5} bar $\approx 9.9 \times 10^{-6}$ atm.) An instrument used to measure atmospheric pressure is called a *barometer,* the name of which derives from the metric unit of measure "bar."

At higher levels in the atmosphere, the pressure decreases markedly (Fig. 3-9a). This change in pressure is what makes your ears pop in an airplane. The decrease with altitude follows the **barometric law,** which states that atmospheric pressure decreases by about a factor of 10 for every 16-km increase in altitude. Thus, the pressure is about 0.1 bar at 16 km above the surface, 0.01 bar at 32 km, and so on. In more precise terms, the barometric law says that pressure decreases *exponentially* with altitude. Note from Figure 3-9a that the exponential decrease in pressure appears almost like a straight line when pressure is plotted on a logarithmic scale. The slight deviation from linearity is caused by the variation in temperature with altitude (discussed next). Pressure decreases faster with height in regions where the air is colder.

How Atmospheric Temperature Varies with Altitude. The vertical temperature structure of the atmosphere is more complicated than the vertical pressure structure (Fig. 3-9b). This temperature profile is the basis for distinguishing four regions within Earth's atmosphere: the troposphere, the stratosphere, the mesosphere, and the thermosphere. Temperature decreases rapidly with altitude in the lowermost layer of the atmosphere, the **troposphere,** which extends from the surface up to 10 to 15 km (higher in the tropics, lower near the poles). Immediately above the troposphere is the **stratosphere,** which is located from about

(a)

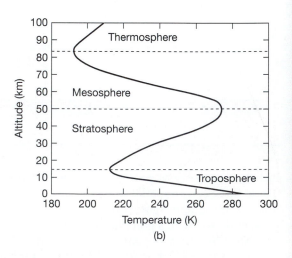

(b)

FIGURE 3-9

(a) How pressure varies with altitude in Earth's atmosphere. (b) How temperature varies with altitude in Earth's atmosphere. The different regions of the atmosphere, determined by temperature regimes, are labeled.

10 to 15 km to 50 km above the surface and in which temperature increases with altitude. Above the stratosphere, temperature decreases with altitude in the **mesosphere** (from about 50 km to 90 km) and then increases once again in the uppermost layer, the **thermosphere** (above about 90 km). These temperature-based "spheres" overlap with atmospheric layers based on other characteristics. For example, the *ionosphere* (a layer that reflects radio waves) includes parts of both the thermosphere and the mesosphere. The very outermost fringe of the atmosphere, where the gas is so tenuous that collisions between molecules become infrequent, is often termed the *exosphere*.

THE TROPOSPHERE. The atmospheric layers that are most important to climate studies are the two lowermost ones: the troposphere and the stratosphere. The troposphere is where most of the phenomena that we call weather—such as clouds, rain, snow, and storm activity—occur. It differs from the other atmospheric layers in that it is well mixed by convection. **Convection** is a process in which heat energy is transported by the motions of a *fluid* (a liquid or a gas). Such motions are generated when a fluid in a gravitational field, like that of Earth, is heated from below. (See the box "Mechanisms of Heat Transfer" for a discussion of other processes by which heat energy is transported.) A familiar example is the convective motion that occurs when a pot of water is heated on the burner of a stove (Fig. 3-10). The warm water at the bottom of the pot is less dense than the cooler water at the top. As a result of this imbalance, the fluid overturns (it circulates, or convects) and will continue to do so as long as the pot is being heated. If the water were heated uniformly or from above, convection would not occur.

The troposphere is convective because roughly half the incoming sunlight is absorbed by the ground and by the ocean surface. The energy from this light is eventual-

FIGURE 3-10

A pot of water on a stove, illustrating convection. The fluid circulates because it is heated from below. (From R.W. Christopherson, *Geosystems: An Introduction to Physical Geography*, 3/e, 1997. Reprinted by permission of Prentice Hall, Upper Saddle River, N.J.)

ly reradiated to space as IR radiation, but it cannot make its way directly from the surface in this form because IR radiation is absorbed by atmospheric greenhouse gases and by clouds. So, the energy is instead transported by fluid motions until it reaches an altitude where the atmosphere is more transparent to IR radiation. Only then can the heat energy radiate away from Earth.

As we shall see in Chapter 4, the upward convection of warm, moist air plays a major role in the global energy balance. Convection of heat in a moist atmosphere is more complicated than that in a dry atmosphere, because water can condense or evaporate. When water is evaporated from the ocean surface or from rivers and lakes, energy is taken up by the resulting vapor. This energy is referred to as the latent heat of vaporization. When the water vapor condenses to form clouds, the same amount of latent heat is released to the atmosphere. In more general terms, **latent heat** is the heat energy released or absorbed during the transition from one phase—gaseous, liquid, or solid—to another.

THE STRATOSPHERE. The stratosphere differs from the troposphere in several respects. The pressure is substantially lower in the stratosphere, in accordance with the barometric law. The two layers differ in composition as well. The stratosphere contains most of Earth's ozone. Stratospheric air is also very dry, containing less than 5 ppm of water vapor on average. Thus, condensation of water vapor does not occur, and so clouds and precipitation are absent. (An exception occurs in the polar regions during winter, where tenuous *polar stratospheric clouds* can form. These clouds play a key role in the development of the Antarctic ozone hole, as we shall see in Chapter 14.) Stratospheric air is not convective and is therefore less well mixed than tropospheric air. Indeed, the name "stratosphere" derives from the word "stratified," which means layered.

THE VERTICAL TEMPERATURE PROFILE. Why does the vertical temperature profile in Figure 3-9b exhibit all those curves? The reason has to do primarily with where the atmosphere is heated—that is, where solar energy is absorbed. The high temperatures near the ground are caused by the absorption of sunlight at Earth's surface, which then heats the atmosphere above it. The high temperatures near 50 km are caused by the absorption of solar UV radiation by ozone. The ozone concentration actually peaks some 20 km lower, in the middle stratosphere (Fig. 3-11), but the heating rate is highest in the upper stratosphere because more UV radiation is available at those altitudes. The vertical heating distribution also explains why the stratospheric is not convective: The maximum heating occurs at the top of the layer, so there is no tendency for the air to rise. Above 50 km, both the ozone concentration and the heating rate decline, so the temperature decreases with altitude in the mesophere. Finally, the temperature rise above 90 km in the thermosphere is caused by the absorption of short-wavelength UV radiation by molecular oxygen, O_2.

Mechanisms of Heat Transfer

In general, heat energy can be transferred by three different mechanisms: convection, radiation, and conduction. *Convection,* the most important mechanism for our purposes, is heat transfer by fluid movement (see the accompanying text). Radiation is obviously important to atmospheric energetics as well, as it is the subject of most of this chapter. In terms of heat transfer, **radiation** is the transfer of heat energy in the form of electromagnetic waves. **Conduction** is the transfer of heat energy by direct contact between individual molecules.

Above absolute zero, molecules are always in motion: They are constantly colliding with one another and are vibrating. For example, when we put a pot of water on a stove and turn on the burner, the bottom of the pot is heated by conduction because the molecules of the burner are in direct contact with the molecules of the pot (Box Fig. 3-3). Conduction is of relatively minor importance in the atmosphere by comparison with the two other heat-transfer mechanisms.

BOX FIGURE 3-3

A pot of water on a stove, illustrating conduction. (From R.W. Christopherson, *Geosystems: An Introduction to Physical Geography,* 3/e, 1997. Reprinted by permission of Prentice Hall, Upper Saddle River, N.J.)

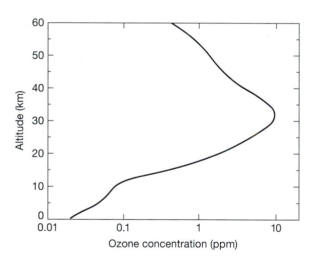

FIGURE 3-11

An approximate profile of the vertical variation of ozone concentration in Earth's atmosphere.

Physical Causes of the Greenhouse Effect

We determined earlier that Earth's greenhouse effect warms the surface by some 33°C compared with the temperature we would expect if there were no atmosphere. This warming has been attributed to the presence of greenhouse gases, especially H_2O and CO_2. Why do some gases contribute to the greenhouse effect whereas others, such as O_2 and N_2, do not?

Molecular Motions and the Greenhouse Gases H_2O and CO_2

The defining property of a greenhouse gas is its ability to absorb or emit infrared radiation. Gas molecules can absorb or emit radiation in the IR range in two different ways. One way is by changing the rate at which the molecules rotate. The theory of *quantum mechanics* describes the behavior of matter on a microscopic scale—that is, the size of molecules and smaller. According to this theory, molecules can rotate only at certain discrete frequencies, just as most house fans can operate only at certain speeds. The rotation frequency is the number of revolutions that a molecule completes per second. Consider one photon of an electromagnetic wave that is incident on an individual molecule (Fig. 3-12). If the incident wave has just the right frequency (corresponding to the difference between two allowed rotation frequencies), the

FIGURE 3-12

The rotation rate of an individual H_2O molecule increases when the molecule absorbs a photon of infrared radiation.

molecule can absorb the photon. In the process, the molecule's rotation rate increases. Conversely, the rotation rate slows down when the molecule emits a photon.

The frequency (or wavelength) of the radiation that can be absorbed or emitted depends on the molecule's structure. The H_2O molecule is constructed in such a manner that it absorbs IR radiation of wavelengths of about 12 μm and longer. This interaction gives rise to a very strong absorption feature in Earth's atmosphere called the **H_2O rotation band.** It can clearly be seen in Figure 3-13, which shows the percentage of radiation at different wavelengths that is absorbed during vertical passage through the atmosphere. Virtually 100% of infrared radiation longer than 12 μm is absorbed, although some of this absorption is caused by CO_2 (see below). The H_2O rotation band extends all the way into the microwave region of the electromagnetic spectrum (above a wavelength of 1000 μm), which is why a microwave oven is able to heat up anything that contains water.

A second way in which molecules can absorb or emit IR radiation is by changing the amplitude with which they vibrate. Molecules not only rotate, they also vibrate—their constituent atoms move toward and away from each other. Again consider an electromagnetic wave that is incident on a molecule. If the frequency at which the molecule vibrates matches the frequency of the wave, the molecule can absorb a photon and begin to vibrate more vigorously. (Similarly, a vibrating tuning fork will induce vibrations in a second tuning fork if the pitches of the two instruments are the same. The pitch is proportional to the frequency of the sound wave.)

The *triatomic* (three-atom) CO_2 molecule can vibrate in three ways. We need to concern ourselves only with the *bending mode* of vibration (Fig. 3-14). This vibration has a frequency that allows the molecule to absorb IR radiation at a wavelength of about 15 μm. It gives rise to a strong absorption feature in Earth's atmosphere called the **15-μm CO_2 band.** The 15-μm CO_2 band overlaps the H_2O rotation band and, hence, is hard to distinguish in Figure 3-13. It is, however, easily seen by satellites that look down at Earth's atmosphere from above. Because it occurs fairly near the peak of Earth's outgoing radiation,

FIGURE 3-14

The bending mode of vibration of the CO_2 molecule.

Bending mode
(15-μm band)

this absorption band is particularly important to climate. Earth's surface emits strongly in this wavelength region, but very little of this radiation is able to escape directly to space because it is absorbed by CO_2 molecules in the atmosphere. This is why CO_2 is such an important contributor to the greenhouse effect.

Other Greenhouse Gases

Water vapor and CO_2 are the most important greenhouse gases in Earth's atmosphere, but several other trace gases—notably CH_4, N_2O, O_3, and freons—also contribute to greenhouse warming (Table 3-3). These gases have more of an effect on outgoing radiation than their small concentrations would suggest because they absorb at different wavelengths than do H_2O and CO_2. Freons, for example, have absorption bands within the 8- to 12-μm *window region,* where both H_2O and CO_2 are poor absorbers (see Fig. 3-13). Thus, one molecule of Freon-11 contributes much more to the greenhouse effect than does one CO_2 molecule. Ozone also has an absorption band in this region centered at 9.6 μm. Thus, O_3 is a good greenhouse gas as well.

Now, recall that we asked the question: Why are O_2 and N_2 poor absorbers of IR radiation and, thus, do not contribute significantly to the greenhouse effect? We are now ready to answer that question. *Diatomic* (two-atom) molecules can rotate and vibrate just like the more complicated molecules, H_2O and CO_2, discussed earlier (Fig. 3-15). The O_2 and N_2 molecules, however, are perfectly symmetric: Both of their constituent atoms are identical. Hence, there is no separation of positive and negative electric charges within the molecule. As noted earlier, an electromagnetic wave actually consists of oscillating electric and magnetic fields. To a first approximation, these fields cannot interact with a totally symmetric mol-

FIGURE 3-13

Percentage of radiation absorbed during vertical passage through the atmosphere. Absorption of 100% means that no radiation penetrates the atmosphere. The nearly complete absorption of radiation longer than 13 μm is caused by absorption by CO_2 and H_2O. Both of these gases also absorb solar radiation in the near infrared (wavelengths between about 0.7 μm and 5 μm). The absorption feature at 9.6 μm is caused by ozone. (From data originally from R. M. Goody and Y. L. Yung, *Atmospheric Radiation,* 2nd ed., New York: Oxford University Press, 1989, Fig. 1.1.)

FIGURE 3-15

Rotation and vibration for a diatomic molecule, such as N_2 or O_2.

Rotation   Vibration

ecule; the electromagnetic wave passes by such a molecule without being absorbed. (Note that CO_2 is a symmetric molecule, because the three atoms are, on average, arranged in a line. However, the symmetry is broken when the molecule bends, allowing 15-μm radiation to be absorbed or emitted.)

Effect of Clouds on the Atmospheric Radiation Budget

Gases are not the only constituents of the atmosphere that affect its radiation balance; that balance is also influenced by the presence of clouds and aerosols. We will postpone the discussion of aerosols until Chapter 10, because their climatic effects are usually rather small and short term. (Sulfate aerosols, for example, cooled Earth by about 0.5°C (1°F) for a year or two after the Mt. Pinatubo eruption, as described in Chapter 2.) The effects of clouds, though, are large and cannot be ignored (Fig. 3-16). Unfortunately, these effects cannot always be calculated reliably, either, and this leads to significant problems in climate prediction.

Types of Clouds

The effect of clouds on Earth's radiation budget is difficult to calculate quantitatively, partly because there are many different types of clouds (Fig. 3-17). *Cumu-*

FIGURE 3-16

Photo of Earth from space, showing clouds. (From NASA Headquarters)

lus clouds are the familiar puffy, white clouds that look like balls of cotton. They are composed of droplets of liquid water and are formed in convective updrafts. *Cumulonimbus clouds* are big, tall cumulus clouds that give rise to thunderstorms. *Stratus clouds* are grey, low-level water clouds that are more or less continuous. They cover much of the eastern United States during winter. *Cirrus clouds* are high, wispy clouds composed of ice crystals rather than liquid water, because the temperature of the upper troposphere is well below the freezing point.

Opposing Climatic Effects of Clouds

Have you ever noticed that cloudy days are relatively cool, yet cloudy nights are relatively warm? That is because clouds affect planetary energy balance in two opposing ways. Clouds cool Earth during the daytime by reflecting incident sunlight back to space. We noted earlier that at present the planetary albedo is about 0.3. A large fraction of this is caused by clouds. In fact, without clouds, Earth's albedo would probably be closer to 0.1. According to the planetary energy balance equation, reducing the albedo from 0.3 to 0.1 would raise the effective radiating temperature of Earth by about 17°C (30°F). The increase in surface temperature on a cloud-free Earth would be much smaller than this, however, because clouds also absorb and re-emit outgoing infrared radiation and, thus, contribute significantly to the greenhouse effect. This effect dominates at night and helps keep cloudy nights warm.

To complicate matters further, the effect of any particular cloud depends on its height and thickness. Low, thick clouds, such as stratus clouds, generally cool the surface because their primary influence is to reflect incoming solar radiation. High, thin clouds, such as cirrus clouds, tend to warm the surface because they contribute more to the greenhouse effect than to the planetary albedo. The reason for the difference is twofold: First, the elongated ice crystals of which cirrus clouds are composed allow much of the incident solar radiation to pass through but absorb most of the outgoing IR radiation. In contrast, stratus clouds reflect much of the incoming visible radiation, in addition to absorbing radiation at IR wavelengths. Second, cirrus clouds occur higher in the troposphere than do stratus clouds and are therefore colder (Fig. 3-18). According to the Stefan–Boltzmann law, cirrus clouds therefore radiate less IR energy to space. Because they absorb the upward-directed IR radiation from the warm surface and reradiate it at a lower temperature, cirrus clouds make a large contribution to the atmospheric greenhouse effect. Lower-lying, stratus clouds do this as well, but their radiating temperature is higher and so their contribution to the greenhouse effect is not as large.

(a)   (b)

FIGURE 3–17

Photos of (a) stratus (From Claudia Parks, The Stock Market.) and (b) cirrus clouds. (From G.R. Roberts/Nelson Riwaka, Photo Researchers, Inc.)

Earth's Global Energy Budget

The various factors that we have just discussed can be combined to calculate a global energy budget for Earth, as in Fig. 3-19. The incident solar flux in this diagram is normalized to 100 arbitrary "units" of radiation. These 100 units of incoming energy are balanced by 30 units of reflected solar energy (25 reflected by the atmosphere and 5 reflected by the surface) and 70 units of outgoing infrared radiation. About half the incident solar radiation makes it down to the surface; the other half is either reflected or absorbed by the atmosphere. Within the atmosphere, energy is transported by a combination of radiation, convection, and the latent heat associated with the evaporation and condensation of water vapor. This latter process is a very important one: Roughly half of the

solar energy absorbed by the surface (24 out of 45 units) goes directly into evaporating water.

The greenhouse effect is shown here as an additional 88 units of downward-directed infrared radiation. Thus, the total energy flux absorbed by the surface is 133 units (=45 units of solar radiation + 88 units of IR radiation). This value is almost twice the net amount of energy absorbed by the Earth (70 units). The reason is that infrared radiation "bounces" around within the atmosphere, so the internal fluxes can actually be higher than the net input of energy. *At the top of the atmosphere, however, the net downward solar radiation flux (incoming minus reflected) must equal the outgoing infrared flux.* This statement is the principle of *planetary energy balance.*

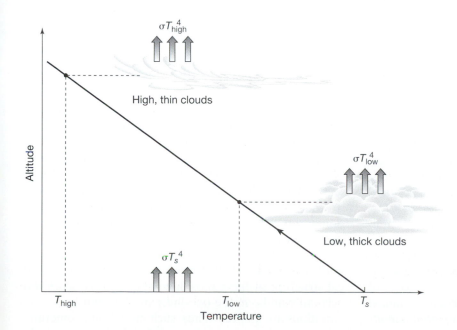

FIGURE 3-18

The different effects of high and low clouds on the atmospheric radiation budget. High, thin clouds are more transparent to incoming sunlight and radiate at a lower temperature than do low, thick clouds. The expressions σT_{high}^4, σT_{low}^4, and σT_s^4 represent the radiation flux at the temperature of high, thin clouds, at the temperature of low, thick clouds, and at the surface temperature, respectively.

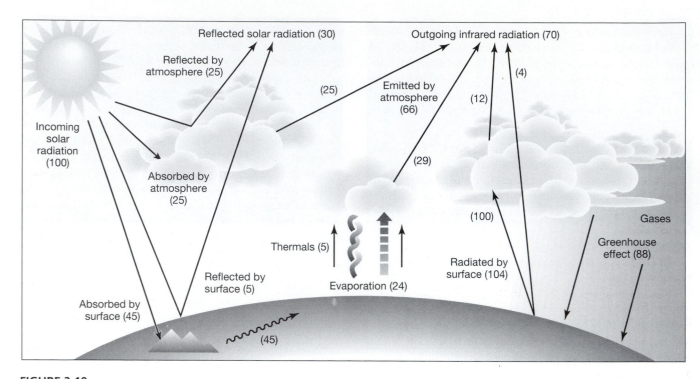

FIGURE 3-19

Earth's globally averaged atmospheric energy budget. All fluxes are normalized relative to 100 arbitrary units of incident radiation. (From S. Schneider, Climate Modeling, Scientific American 256:5, 72–80, 1987.)

Introduction to Climate Modeling

How can we utilize our knowledge of Earth's current energy budget to predict what Earth's surface temperature might have been in the past or how it might vary in the future? Because the climate system is complex, we need some sort of computer model to keep track of all the intricacies. Instead of estimating the magnitude of the greenhouse effect by subtraction, as we did earlier in this chapter, we need to be able to calculate it directly from the measured or predicted concentrations of greenhouse gases.

Such a calculation must take into account the rotational and vibrational absorption bands of all the different greenhouse gases. Doing so quantitatively requires that we combine the predictions of quantum mechanics with laboratory measurements of the strengths of different absorption bands. Fortunately for climate modelers, a great deal of effort has gone into obtaining the required parameters. As a result, a voluminous database of information on the absorption characteristics of various molecules of atmospheric interest is now available.

Armed with these data, the climate modeler must next decide how to incorporate them into a computer model of the atmosphere. The most complete type of model is called an atmospheric **general circulation model (GCM),** also sometimes referred to as a *global climate model.* These elaborate computer models include a three-dimensional representation of the atmosphere (or oceans) that simulates winds (or currents), moisture transport, and energy balance. Atmospheric GCMs are replete with clouds, winds, snow, rain, and most of the other phenomena that we call weather. Thus, they are capable of predicting how climate varies on a regional basis. But GCMs have a number of drawbacks, the most serious of which is that they require large amounts of runtime on the world's fastest computers to simulate even a few years of global climate.

One-Dimensional Climate Models—RCMs

For many purposes it is sufficient to construct simpler climate models that require less effort to program and less computer time to run. We have already seen one such model—the one-layer-atmosphere model described in the box earlier in this chapter. But that model did not produce a good, quantitative estimate of the greenhouse effect; nor did it account for the contributions of different greenhouse gases. The simplest model that is capable of doing both of these things reliably is called a **radiative-convective model (RCM).** In an RCM, the climate system is approximated by averaging the incoming solar and outgoing IR radiation over Earth's entire surface. The vertical structure of the atmosphere (Fig. 3-9) is taken into account (unlike in the one-layer model), but horizontal variations are ignored. Thus, such models are sometimes called one-dimensional climate models, in contrast with the three-dimensional GCMs. The vertical dimension (altitude) is then divided into a number of layers. The RCM calculates the temperature of each layer by taking into account the amount of energy received or emitted in the form of radiation, along with the effects of convection and latent heat release in the lowermost layers.

Radiative Effect of Doubling Atmospheric CO_2. Although RCMs are quite simple compared with the real climate system, they allow us to estimate the magnitude of the greenhouse effect as a function of the concentrations of various greenhouse gases in Earth's atmosphere. These models correctly predict that the greenhouse-induced temperature difference ΔT_g for the present atmosphere is 33°C, in agreement with the estimate derived earlier by the subtraction $\Delta T_g = T_s - T_e$. (This is *not* a trivial result. We would have to work through a lot of the relevant physics to come up with this answer.) More importantly, RCMs allow us to predict the average surface temperature increase that should result from an increase in the concentration of greenhouse gases. A commonly cited benchmark is the temperature change that would result from a doubling of the atmospheric CO_2 concentration from 300 ppm (its value near the turn of the 20th century) to 600 ppm. RCM calculations show that, all other factors being equal, such a change in CO_2 would produce an increase of about 1.2°C (2.2°F) in the global average surface temperature. In the terminology developed in Chapter 2, this value is the temperature change ΔT_0 that would result in the absence of any feedbacks in the climate system.

In reality, we would expect other factors in the climate system to change as atmospheric CO_2 increases, and so a temperature change of +1.2°C is not the best estimate we could make of the effect of CO_2 doubling. To obtain a better estimate, we must consider what those climate feedbacks might be and how strongly they affect our answer.

Climate Feedbacks

Climate feedbacks are extremely important, because they can either amplify or moderate the radiative effect of changes in greenhouse gas concentrations. That is why we devoted most of Chapter 2 to explaining how they work. There, we dealt with an imaginary feedback system involving the percentage of daisy cover on the hypothetical planet Daisyworld. Here, we discuss several feedback processes that affect climate on Earth.

The Water Vapor Feedback

One of the most important feedbacks in the climate system involves the concentration of atmospheric water vapor. As noted earlier, water vapor is an excellent ab-

sorber of IR radiation and, hence, a good greenhouse gas. Unlike CO_2, however, water vapor is typically close to its *condensation point*—the temperature at which a vapor condenses to form a liquid. If Earth's surface temperature were to decrease for some reason, water vapor would condense out in the form of rain or snow, leaving less water vapor behind in the atmosphere. This reduction in atmospheric water vapor would cause a corresponding decrease in the greenhouse effect, which, in turn, would lower the surface temperature still further. Conversely, an increase in surface temperature would cause an increase in the rate at which water vapor evaporates from the oceans. This would increase the concentration of water vapor in the atmosphere, thereby increasing the greenhouse effect and further warming Earth's surface.

The net result of this interaction between water vapor abundance and Earth's surface temperature is a positive feedback loop that tends to amplify small temperature perturbations (Fig. 3-20). This feedback loop can be incorporated in RCMs by assuming a fixed *relative humidity* profile in the troposphere. **Relative humidity** is the concentration of water vapor in an air parcel divided by the concentration that would be present if the air parcel were *saturated* with water vapor (that is, on the verge of condensation). When such a calculation is performed, the RCM predicts that the equilibrium change in surface temperature for CO_2 doubling, ΔT_{eq}, is about twice that which would have occurred otherwise. Recall from Chapter 2 that we can write

$$\Delta T_{eq} = \Delta T_0 + \Delta T_f,$$

where ΔT_0 is the temperature change with no feedbacks and ΔT_f is the change caused by the feedback. For the problem of CO_2 doubling, $\Delta T_0 = 1.2°C$ (2.2°F), $\Delta T_{eq} \approx 2.4°C$ (4.4°F), so the temperature change caused by the water vapor feedback is approximately 1.2°C. Furthermore, the feedback factor f is given by

$$f = \frac{\Delta T_{eq}}{\Delta T_0} = \frac{2.4°C}{1.2°C} = 2.$$

A feedback factor of 2 indicates that this is a strong, positive feedback on the climate system.

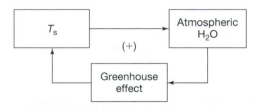

FIGURE 3-20
Systems diagram showing the positive feedback loop that includes atmospheric water vapor.

FIGURE 3-21

Systems diagram showing the positive feedback loop that includes snow and ice cover.

Snow and Ice Albedo Feedback

A second feedback loop that is expected to have some impact on modern global warming, but is especially important for glacial–interglacial variations, involves albedo changes caused by snow and ice. As Earth's climate cools, the extent of wintertime snow and ice cover increases in temperate regions. On longer time scales, the permanent ice cap in the northern polar regions expands toward the equator, resulting in the periods of glaciation known as the Ice Ages. Snow and ice have a much higher albedo than does land or water (refer to Table 2-1). Therefore, increases in snow and ice cover should cause further decreases in surface temperature. The result is a positive feedback loop that tends to amplify induced changes in Earth's surface temperature (Fig. 3-21). As snow and ice cover are restricted to middle and high latitudes, modeling this feedback loop quantitatively requires the use of two-dimensional or three-dimensional computer models.

The IR Flux/Temperature Feedback

Both of the feedbacks discussed so far are positive. But systems that contain only positive feedback loops are unstable. Does this mean that Earth's climate is unstable? No. Earth's climate system contains a very strong negative feedback that is so basic that it is often overlooked. The feedback loop that stabilizes Earth's climate on short time scales is the relationship between surface temperature and the flux of outgoing IR radiation (Fig. 3-22). (We have already hinted in Chapter 1 that there is another feedback loop that stabilizes Earth's climate on long time scales, but that is not what we are talking about here.) If

FIGURE 3-22

Systems diagram illustrating the negative feedback loop between surface temperature and the outgoing flux of infrared radiation. This feedback is the fundamental reason that Earth's climate is stable.

Earth's surface temperature were to increase for some reason, the outgoing IR flux from the top of the atmosphere would also increase. But if the outgoing IR flux were to increase, the surface temperature would tend to decrease, because more energy would be lost from the Earth system. This feedback loop might appear to be trivial, but it is not; there are situations in which it can fail. In particular, the positive correlation between surface temperature and the outgoing IR flux can break down if the atmosphere contains a very large amount of water vapor. This, we think, is what happened to our sister planet, Venus, and it led to what is sometimes called a *runaway greenhouse*. But we shall save that story for Chapter 16.

The Uncertain Feedback Caused by Clouds

Some feedback loops in the climate system are much more complex than the feedbacks we have discussed thus far. Arguably, the most important such feedback involves changes in clouds. A simple thought experiment demonstrates why even the sign of this cloud feedback loop is not easy to predict. Suppose that a forcing (a doubling of the CO_2 concentration, for example) is applied to the climate system and that this forcing would cause Earth's surface temperature to increase. By the argument given earlier in this section, such a change would cause an increase in evaporation rates and a corresponding increase in the amount of water vapor in the atmosphere. Because clouds form by the condensation of water vapor, we might expect that either cloud cover or cloud thickness, or perhaps both, would increase.

However, this conclusion may not be correct. Most clouds form in regions where the air is ascending; the reason is that air cools as it ascends, and cooling causes water vapor in the air to condense. Regions where the air is descending are generally cloud-free. Because what goes up must come down, roughly half the air in the troposphere is ascending at any given time, and the other half is descending. By this argument, clouds should always cover about 50% of Earth's surface regardless of how the climate system is forced. This estimate is not exact, as air tends to rise in columns of rapid convection, whereas it tends to sink more slowly and over a larger area. However, this argument does show that the possible response of clouds to a forcing is not at all straightforward.

The fact that cloud feedback is difficult to predict should not come as a surprise. We saw earlier that the effect of clouds on climate depends on both their altitude and their thickness. Furthermore, cloud cover and cloud type vary over different parts of the Earth. Because of these complexities, the only way that we can predict how they might change in response to forcing is by using a GCM. Even then, accurate prediction is difficult, because clouds are hard to simulate realistically. The fundamental problem is one of scale: A typical cloud is much smaller than the size of a grid square on even the highest-resolution GCM. Nevertheless, climate modelers do try to *parameterize*—that is, approximate in terms of simpler, known quantities—the effects of clouds by using knowledge gained from small-scale models or from observations. Somewhat surprisingly, some of the best, highest-resolution GCMs predict that cloud feedback should be positive for CO_2 doubling: Warmer ocean temperatures lead to deeper, more-vigorous convection in the troposphere, which in turn leads to an increase in high cirrus clouds. We have already seen that high clouds tend to promote surface warming. As a consequence of this feedback, the expected temperature increase for CO_2 doubling is often quoted as 2 to 5°C—considerably greater than what would be predicted in the absence of cloud feedback. This prediction, however, is rather uncertain, for the reasons just discussed. Clouds are, and will probably remain, one of the greatest obstacles in our understanding of how Earth's climate might change in the future.

In summary, we have now examined Earth's climate system in enough detail to understand how the atmospheric greenhouse effect warms the planet and how the planet's average surface temperature may respond to a human-induced increase in greenhouse gases. But Earth's climate cannot be described by just its average surface temperature. The term "climate" includes many other related factors, such as latitudinal and seasonal temperature gradients, winds, and precipitation. To study these phenomena, we need to broaden our spatial perspective and consider the Earth system from a three-dimensional perspective. The next two chapters describe how the transport of heat from one location to another by the atmosphere and oceans determines these other important features of Earth's global climate.

Chapter Summary

1. Earth is warmed by the absorption of visible radiation from the Sun and is cooled by the emission of infrared radiation to space.
   a. Much of the infrared radiation emitted by Earth's surface is absorbed and re-emitted by atmospheric gases.
   b. The result is a greenhouse effect that warms the surface by about 33°C. Without this natural greenhouse effect, Earth would be too cold to support life.
2. Only certain atmospheric gases, most importantly H_2O and CO_2, contribute to the greenhouse effect. These gases absorb infrared radiation by changing the rate at which individual molecules rotate or vibrate. Other trace gases, such as Freons, can contribute substantially to the greenhouse effect by absorbing radiation at different wavelengths than do H_2O and CO_2.
3. Clouds affect the atmospheric radiation budget both by reflecting incident sunlight and by contributing to the greenhouse effect. Low, thick clouds tend to cool the surface; high, thin clouds tend to warm it.

4. Earth's climate system contains several well-understood feedbacks that play important roles in regulating climate change.
   a. The climate system is stabilized by a strong negative feedback loop between surface temperature and the outgoing infrared flux.
   b. The system is destabilized by a positive feedback loop involving atmospheric water vapor. Because it acts on short time scales, this feedback is likely to play an important role in contemporary global warming. Climate models predict a surface temperature response to CO_2 doubling that is twice that of models in which this feedback is neglected.
   c. The system is also destabilized by a positive feedback loop involving the extent of snow and ice cover, due to the effect of albedo.
   d. Clouds may also contribute to climate feedback, but their effect is not well understood.

Key Terms

barometric law
blackbody radiation
conduction
convection
effective radiating temperature
electromagnetic radiation
electromagnetic spectrum
15-μm CO_2 band
flux
frequency
general circulation model

H_2O rotation band
infrared radiation
inverse-square law
Kelvin temperature scale
latent heat
mesosphere
photon
photosphere
radiation
radiative-convective model

relative humidity
Stefan–Boltzmann law
stratosphere
thermosphere
troposphere
ultraviolet radiation
visible radiation
visible spectrum
wavelength
Wien's law

Review Questions

1. How are the wavelength and frequency of an electromagnetic wave related?
2. What is a photon?
3. What physical law describes the manner in which the intensity of sunlight changes as the observer moves away from the Sun?
4. a. Name two physical laws that apply to blackbody radiation.
   b. What do these laws tell us about the nature of the emitted radiation?
5. What is the major contributor to Earth's albedo?
6. What are the three most abundant gases in Earth's atmosphere?
7. a. List the four layers of Earth's atmosphere.
   b. How are they defined?
8. Name three mechanisms by which heat energy can be transferred. Which two are important in Earth's global energy budget?
9. a. Identify two physical processes by which gases can absorb infrared radiation.
   b. Give examples of each process.
10. Why are O_2 and N_2 not greenhouse gases?
11. Describe the different ways in which climate is affected by high and low clouds.
12. Identify two positive feedback loops in Earth's climate system. Why is Earth's climate stable despite these destabilizing, positive feedbacks?

Critical-Thinking Problems

1. a. Given that a 300-K blackbody radiates its peak energy at a wavelength of about 10 mm, at what wavelength would a 600-K blackbody radiate its peak energy?
   b. If the two bodies in part (a) were the same size, what would be the ratio of the heat emitted by the hotter object to the heat emitted by the colder one?
2. a. Venus and Mars orbit the Sun at average distances of 0.72 AU and 1.52 AU, respectively. What is the solar flux at each planet?
   b. Venus has a planetary albedo of 0.8, and Mars has an albedo of 0.22. Using the answer to part (a), determine the effective radiating temperatures of these planets.
   c. How do the effective radiating temperatures determined in part (b) compare with the value for Earth, and why is this result surprising?
   d. The mean surface temperatures of Venus and Mars are 730 K and 218 K, respectively. Using the answer to part (b), determine the magnitude of the greenhouse effect on each planet.
   e. How do the results of (d) compare with the magnitude of the greenhouse effect on Earth?
3. a. The Sun radiates at an effective temperature of 5780 K and has a radius of about 696,000 km. Remembering that 1 AU = 149,600,000 km, derive the approximate value of the solar flux at Earth's orbit.
   b. Compare your answer with the value given in the text.
4. The tropospheric *lapse rate* (the rate at which temperature decreases with altitude) is approximately 6°C (11°F) per kilometer. Given that the mean surface temperature of Earth is 288 K and the effective radiating temperature is 255 K, from what altitude does most of the emitted radiation derive?
5. Solar luminosity is estimated to have been 30% lower than today at the time when the Solar System formed, 4.6 billion years ago.

a. If Earth's albedo was the same as it is now ($A = 0.3$), what would have been its effective radiating temperature at that time?
   b. If the magnitude of the greenhouse effect had also remained unchanged ($\Delta T_g = 33$ K), what would Earth's average surface temperature have been? How does this compare with today's value?
6. For atmospheric CO_2 concentrations not too different from the present value, the radiative forcing of CO_2 can be expressed by the formula

$$\Delta F = -6.3 \ln\left(\frac{C}{C_0}\right),$$

where $C_0 = 300$ ppm is the CO_2 concentration near the turn of the 20th century, C is the CO_2 concentration at some other time, and ΔF is the change (in watts per square meter) in the outgoing infrared flux caused by the change in CO_2 concentration. The function $\ln(x)$ denotes the natural logarithm of a given number x. Any scientific calculator has this function key.

a. By how much would the outgoing infrared flux decrease if the atmospheric CO_2 concentration were increased from 300 ppm to 600 ppm (that is, if $C = 600$ ppm)?
   b. By how much would surface temperature have to increase in order to bring the radiation budget back into balance in part (a), assuming that the planetary albedo and the amount of water vapor in the atmosphere do not change? (*Hint:* Use the planetary energy balance equation to calculate how much T_e would have to change to balance the radiation budget. Remember that the left-hand side of this equation represents the outgoing infrared flux. The quantity T_s will change by the same amount as T_e if the amount of water vapor is held constant.)

Chapter 4
The Atmospheric Circulation System

Convective towers in the ITCZ. Solar heating evaporates large amounts of water from the tropical oceans. The air cools and condenses as it rises, releasing the energy used for evaporation as latent heat. The release of latent heat in these convective towers is the pump that drives the Hadley circulation. (From NASA Science Source, Photo Researchers, Inc.)

Key Questions

- Why does air move?
- Are the movements of the winds random across the surface of the Earth, or do they follow regular patterns?
- What implications do these circulatory systems have for global climate?
- What other factors govern the geographic and seasonal distributions of temperature and rainfall?

Chapter Overview

Earth's climate is a central theme of ours. We focus on the role climate plays in the Earth system and explain how Earth's climate works, how climate has changed through time, and how it may change in the future. An important element of Earth's climate is the atmospheric circulation. In Chapter 3 we described the global energy budget and showed that if we average the radiation fluxes around the globe and over a few years there is a balance. Earth emits as much energy as it receives—aside from the issue of anthropogenic increases in the greenhouse effect. If we look at regions smaller than the globe and over time periods of less than a year, however, the situation is very different. There is a significant imbalance in the distribution of energy at various latitudes: The tropics receive a surplus of radiative energy, whereas the poles run a deficit. This imbalance causes an equator-to-pole temperature gradient that results in density and pressure differences in the atmosphere. The density and pressure differences cause air to move in a global-scale pattern of wind belts, which are modified by Earth's rotation and by the distribution of land and water. The net effect is to restore the latitudinal energy balance by moving surplus energy away from the tropics to cancel out the deficit at the poles. In the process, energy is used to evaporate water from the land and ocean surfaces, water vapor is carried by wind, and energy is released when the vapor condenses to form clouds. Thus, there are close interactions between the transport of energy and of water by means of circulating air. In other words, Earth's atmospheric circulation has a direct impact on the global distributions of temperature and precipitation.

The Global Circulatory SubSystems

Anyone who has felt the wind blow, watched the clouds move, and seen the rain fall is aware that large parts of the Earth system are in constant motion. Even the continents and oceans, despite their apparent permanence, are continuously moving. The island of Iceland in the North Atlantic, for example, is spreading, and its two sides are moving away from each other fast enough to be measured by today's instruments. Although these movements may sometimes appear random, they form part of a well-ordered circulation of energy and matter throughout the Earth system.

Like the circulatory system of humans (part of the cardiovascular system), Earth's circulatory subsystems work to maintain the planet in a thermal and chemical balance. The human circulatory system transports dissolved gases, nutrients, and hormones throughout the body; carries away waste products; helps regulate the acidity of body fluids; and is a vital part of the body's thermoregulatory system, carrying warm blood from one area to another. Although the human circulatory system is not an exact analogy to Earth's circulatory subsystems, these systems do have much in common. Essential gases and nutrients are transported throughout the Earth system, and waste products are removed from their area of production. All of Earth's circulatory subsystems act in some way to help regulate the global temperature: The winds and ocean currents redistribute the energy received from the Sun, and the motions of the solid Earth redistribute carbon and help regulate the CO_2 level of the atmosphere. The circulations within the solid Earth are discussed in Chapter 6; here and in Chapter 5 we examine those circulations that occur within the fluid part of the Earth system: the atmosphere and oceans.

The purpose of this chapter is to describe the major characteristics of the atmospheric circulation, to explain why they occur, and to illustrate the way in which they affect the transport of energy and materials around the globe. In Chapter 3 we described the energy input and output from the Earth system as a whole; now we take that system apart and examine some of its internal workings—specifically those related to climate. In doing so, we have two primary objectives. The first is to explain why weather and climate vary across the globe. The second is to emphasize that because of the internal workings of Earth's climate system, the response to global-scale processes and changes may not be uniform around the globe. This regional variability of our present climate is illustrated in Figure 4-1, which shows the annual and seasonal distributions of temperature, and Figure 4-2, which shows the same for precipitation. The remainder of this chapter is devoted to explaining why these particular distributions occur.

Organized movements of the atmosphere occur over many different time and space scales. These movements range from centimeter-scale swirls, or *eddies,* to global-scale motions of the wind belts. All of these are important in one way or another, but we will limit our discussion to processes that are global in extent and that have the greatest influence on the transport of energy and mass through the Earth system. One of Earth's most important constituents is water. Cycling continuously among the atmosphere, the oceans, and the land surface, water carries with it energy, dissolved nutrients, and other matter—all vital for maintaining an environment suitable for life.

In the same way that the functioning of the cardiovascular system in the human body ultimately depends on the ability of the heart to keep pumping, the functioning of Earth's circulatory subsystems rely on several different pumps. Each of these pumps drives a different circulatory mechanism, and each works at a different speed. Over shorter time scales (years to decades), the most important pump is found in the tropical oceans. This pump is responsible for the movements of the air and the surface ocean over most of the globe. The energy source that drives this pump is radiation from the Sun. Over longer time scales (about 1000 years), a second pump drives the deep-ocean circulation (Chapter 5). The ultimate energy source is again the Sun. The pump operating over the longest time scales (millions of years) is radioactive decay and the production of heat in Earth's interior. This pump causes the movements of the continents that we shall discuss in Chapter 6.

All these circulation subsystems play a vital role in the operation of the Earth system. We know that in humans the cardiovascular system can maintain stability only if the blood keeps moving. Similarly, the Earth system can maintain stability only as long as its circulatory subsystems continue to function. The long-term pump (that is, the processes of internal heat production and plate tectonics) ceased to function on Mars. As we shall see in a later chapter, the planet's inability to support an environment suitable for life may be at least partly a result of the failure of this circulation mechanism.

The Atmospheric Circulation

Recall from Chapter 3 that the troposphere is the lowermost layer of the atmosphere. Most of the processes we are interested in take place in the troposphere, so we limit our discussion here to that layer. Although the circulation of the stratosphere plays a role in the depletion of stratospheric ozone, we will save that discussion for Chapter 14.

The Movement of Air

Air moves over Earth's surface because there are horizontal differences in pressure. Air also moves vertically either because it is forced to rise mechanically (for example, when it encounters a mountain range) or because there are changes in *buoyancy.* **Buoyancy** is the tendency of an object to float in a fluid. Buoyancy is controlled by differences in

(a)

(b)

F°	5	9	18	27	36	45	54	63	72	81	90	99	108	F°
C°	3	5	10	15	20	25	30	35	40	45	50	55	60	C°

(c)

FIGURE 4-1

Global temperature distributions in degrees centigrade for (a) January, (b) July, (From T. McKnight: *Physical Geography: A Landscape Appreciation,* 6/e, 1999. Reprinted by permission of Prentice Hall, Upper Saddle River, N.J.), and (c) the annual range (difference between summer and winter). (From R.W. Christopherson, *Geosystems: An Introduction to Physical Geography,* 3/e, 1997. Reprinted by permission of Prentice Hall, Upper Saddle River, N.J.) [See color section]

density between the object and the fluid, where density is given by the mass of a substance within a unit volume. (The greater the mass within a given volume, the greater the density.) Ultimately, all of these horizontal and vertical movements (except those due to mechanical forces) can be attributed to differences in temperatures across the globe. To explain how these movements occur, we need to understand how pressure and density are related to temperature.

Vertical Movement. It is easiest to picture these relationships by thinking of vertical and horizontal movements separately. Imagine the situation with a hot-air balloon. Remember from Chapter 3 that the addition of energy in the form of heat causes molecules to move faster. In this case, the faster the air molecules move, the more they collide with each other and with the interior of the balloon. These collisions exert a force (that is, air pressure) on the interior surface. If the balloon was a fixed container (one that could not expand), there would be an increase in the air pressure within the container. Thus we see a connection between the temperature and pressure of a gas: As the temperature increases, the pressure increases. But the balloon *is* expandable. As the pressure starts to increase, the air pushes outward on the interior of the balloon, causing it to expand. So in a balloon, it is the volume rather than the pressure that increases.

If we begin with the balloon partially inflated, it will contain a certain number of air molecules. As these are heated, the number of molecules does not change, but they move faster, increasing the pressure on the interior of the balloon; this causes the balloon to expand. We now have the same number of air molecules as before, but they occupy a greater volume. This means that the density of the air must decrease. Because the air in the balloon is less dense than the air surrounding it, the balloon becomes *positively buoyant,* and it rises. The balloon will continue to rise until the density of the air outside the balloon matches that inside (*neutral buoyancy*). If the air in the balloon is more dense than the surrounding air, the balloon would have *negative buoyancy,* and it would sink. Exactly the same processes occur in the atmosphere when we heat the surface below a parcel (column) of air. The surface heats the parcel of air at the bottom of the column; the air parcel expands, its density decreases, and the parcel rises through the air column. Cooling a parcel of air higher in the column causes it to become more dense than the surrounding air, and the parcel sinks.

Horizonal Movement. How do horizontal movements occur? We saw that heating the surface below a parcel of air causes the parcel to expand. With fewer molecules in a given area, the number of collisions between molecules decreases and so the air pressure decreases. This appears

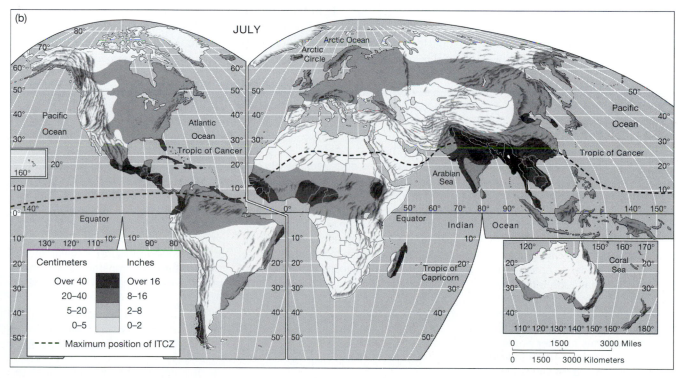

FIGURE 4-2

Global distributions of precipitation over land in (a) January and (b) July. (From T. McKnight: *Physical Geography: A Landscape Appreciation*, 6/e, 1999. Reprinted by permission of Prentice Hall, Upper Saddle River, N.J.) [See color section]

to contradict what we said above when we stated that as temperature increases, pressure increases—but remember, that was for a fixed container in which the volume could not change. In this case, volume increases and we now have a localized reduction in air pressure near the surface. If we consider a second parcel of air that is being cooled at the surface, the reverse will occur. The cooling reduces the movement of the air molecules, and the parcel's volume contracts; there are now more air molecules per unit volume, so the local air pressure increases. If these two air columns are adjacent to each other, the high-pressure air will push sideways against the low-pressure air, and the air will move sideways from high to low pressure.

From this discussion we can establish two important points that will help explain why air moves:

1. Air tends to move from an area of higher pressure to an area of lower pressure until the two pressures are equalized. In other words, air (wind) will move horizontally in the lower troposphere from higher to lower pressure. Pressure differences among air masses are typically related to the distribution of surface temperatures.

2. If an air mass is heated until its density is lower than that of its surroundings, the lower-density air will rise. This phenomenon is a form of convection. (We discussed this phenomenon in a more general sense in Chapter 3 when we demonstrated the convection of a fluid that is heated from below.) Conversely, if an air mass is cooled until its density is higher than that of the underlying air, it will sink. This phenomenon is referred to as **subsidence.**

The Driving Force: The Global Energy Distribution

We learned in Chapter 3 that the average global temperature is determined by the balance between the solar energy ab-

sorbed by Earth and the infrared radiation emitted to space. However, neither the radiation received from the Sun (our primary energy source) nor the infrared emission from Earth is distributed uniformly across Earth's surface. The incoming solar energy varies with latitude and with season, whereas the outgoing terrestrial radiation depends on the temperature of the surface and atmosphere at each location.

The distribution of the incoming solar radiation changes with latitude as a result of the change in surface area presented to the Sun's rays as Earth's surface curves (Fig. 4-3). The energy from the Sun radiates outward in all directions; however, by the time the Sun's rays reach Earth, they are essentially parallel to each other. This means that the flux of solar energy passing perpendicularly through the plane *A-B* in Figure 4-3 will be the same at any point. For example, the three "beams" in the diagram are equal in solar flux when they pass through the plane. Because of the curvature of Earth, however, when these beams reach the top of Earth's atmosphere, the same amount of light is spread over a much larger area at the poles than at the equator. Consequently, each square meter of surface receives proportionally less energy at the higher latitudes, and the incoming solar flux thus decreases from the equator toward the poles. (Recall the light bulb and sheet of paper experiment in Chapter 3.)

The solar radiation absorbed at the surface follows the same general pattern, although the actual amount absorbed varies with cloud cover and atmospheric absorption. This

The Relationship among Temperature, Pressure, and Volume: The General Gas Law

We can see from the text discussion that we can define specific relationships among the temperature (T), pressure (P), and volume (V) of a gas. (In this case, volume refers to the space occupied by a fixed quantity of gas molecules; density is the inverse of volume, so we need not consider it separately here.) First, we know that if the temperature is held constant, then an increase in gas pressure results in a decrease in volume, and a decrease in pressure results in an increase in volume. In other words, pressure is inversely proportional to volume: The product of pressure and volume is a constant. Mathematically, we can write

$$P_{\text{initial}}V_{\text{initial}} = P_{\text{final}}V_{\text{final}}, \quad \textbf{(Boyle's law)}$$

where $P_{\text{initial}}V_{\text{initial}}$ is the product of the initial pressure and volume, and $P_{\text{final}}V_{\text{final}}$ is the product of the final (new) pressure and volume. This relationship, known as

Boyle's law, was discovered in 1662 by the British chemist Robert Boyle.

If, instead, gas pressure is held constant, then we know that an increase in temperature results in an increase in volume and that a decrease in temperature results in a decrease in volume. In other words, volume is directly proportional to temperature: The quotient of volume and temperature is a constant. Mathematically, we can write

$$\frac{V_{\text{initial}}}{T_{\text{initial}}} = \frac{V_{\text{final}}}{T_{\text{final}}}, \quad \textbf{(Charles' law)}$$

where $V_{\text{initial}}/T_{\text{initial}}$ is the quotient of the initial volume and temperature, $V_{\text{final}}/T_{\text{final}}$ is the quotient of the final volume and temperature, and temperature is in degrees Kelvin. This relationship, known as **Charles' law,** was discovered in 1787 by the French physicist Jacques Charles.

Boyle's law and Charles' law can be combined. The relationship obtained from them is called the **general gas law.** According to this law, the product of the pressure and volume of a gas is directly proportional to temperature: The quotient of the product (pressure) × (volume) and temperature is a constant. Mathematically, the general gas law is as follows

$$\frac{P_{\text{initial}}V_{\text{initial}}}{T_{\text{initial}}} = \frac{P_{\text{final}}V_{\text{final}}}{T_{\text{final}}}.$$

(general gas law)

It is important to note that these relationships apply to idealized gases, that is, gases in which there are no attractive forces between molecules. In reality, gases may not respond to environmental changes exactly as described here. Nevertheless, these relationships do represent a very close approximation of how gases behave.

FIGURE 4-3

Variation of incoming solar energy with latitude. The radiation reaching Earth is spread over larger and larger areas as we move from the equator to the poles. Each square meter of the surface receives proportionately less energy as we move to higher latitudes. (From R.W. Christopherson, *Geosystems: An Introduction to Physical Geography,* 3/e, 1997. Reprinted by permission of Prentice Hall, Upper Saddle River, N.J.)

equator-to-pole gradient in the energy absorbed at the surface exerts a primary control on Earth's climate. Figure 4-4 shows this gradient (solid curve) as a function of latitude (that is, the amount averaged around each latitude band). As we might expect from the previous discussion, the maximum absorbed solar energy is found in the tropics, and the available solar energy decreases rapidly as we move toward the poles. This gradient in absorbed solar energy is the single most important control on temperature. In general more energy is available at the equator than at the poles, so we can assume that temperatures should be highest in the tropics and lowest at high latitudes. Indeed, this broad latitudinal distribution of temperature is the dominant feature of Figure 4-1. Figure 4-4 also shows the latitu-

dinal distribution of infrared radiation emitted from Earth to space (dashed curve). The higher emissions in the tropics are a result of the high surface temperatures there and the correspondingly high temperature in the middle troposphere, from whence the outgoing radiation is emitted.

The difference between the incoming solar radiation and the outgoing terrestrial radiation is referred to as *net radiation.* Referring again to Figure 4-4, note that the energy absorbed exceeds the energy emitted in the tropics (net radiation is positive); near the poles, the reverse is true (net radiation is negative). This distribution of available energy is a permanent feature of Earth's climate system. The gradient seems to imply that the tropics should get warmer while the poles get progressively colder. Clearly this does not happen; other processes must be operating to ensure an energy balance at each latitude. In reality, the latitudinal energy gradient produces atmospheric temperature and density differences that force the atmosphere to circulate, carrying warmer air toward the poles and colder air toward the equator. These circulations move energy from regions where there is a surplus to regions where there is a deficit. Most of what we experience as weather and climate is this response of the atmosphere to the unequal latitudinal distributions of energy.

The General Circulation of the Atmosphere

The major components of the large-scale tropospheric circulation are outlined schematically in Figure 4-5. Each of the links in Figure 4-5 represents a set of complex interactions that are difficult to quantify. Hence, although this diagram looks similar to the systems diagrams used in previous chapters, we do not attempt to indicate positive or negative interactions or feedbacks. However, as a whole, the diagram represents a negative feedback loop

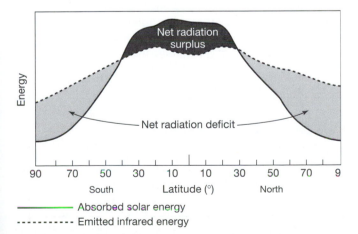

FIGURE 4-4

The distribution of absorbed solar and emitted infrared radiation with latitude. There is a surplus of energy in the tropics, where incoming radiation is greater than outgoing, and a deficit at high latitudes, where more radiation is emitted than is received.

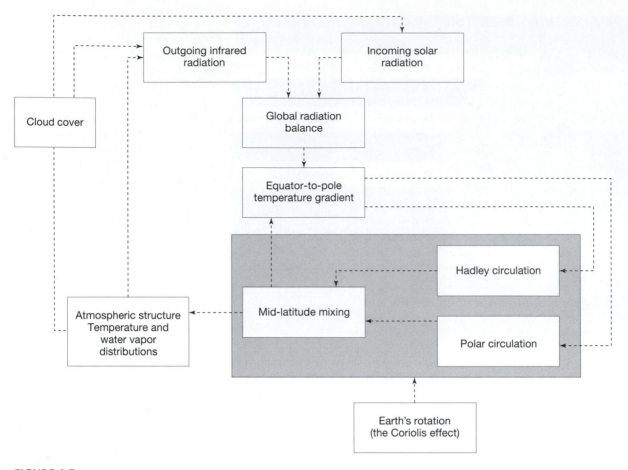

FIGURE 4-5

Schematic diagram of the tropospheric circulation.

as the atmosphere responds to the temperature gradient by transferring energy latitudinally to reduce the gradient and restore an energy balance.

Convergence. As we proceed through this chapter, we will take each link of Figure 4-5 in turn and explain the role of that link in the general circulation of the atmosphere and oceans. We will begin with the heating in the tropics. The large solar input to the tropics heats the sur-

face (primarily ocean), which in turn heats the overlying air. As we saw earlier, when heated from below, air will rise by convection. The tropical air near the surface rises, creating a low-pressure region there. But we saw that air tends to move horizontally from an area of higher pressure to an area of lower pressure. Thus, the rising air is replaced by surface air moving equatorward into the region of low pressure from regions of higher pressure (Fig. 4-6). The merging of air masses that are moving inward

FIGURE 4-6

Convergence, divergence, and the Hadley circulation in the tropics. There is a Hadley cell on either side of the intertropical convergence zone (ITCZ), located over the equator. Rising air in the ITCZ is replaced by inflowing air (convergence) at the surface. Outflowing air (divergence) in the upper troposphere sinks at about 30° N and 30° S, completing the circulations in the two cells.

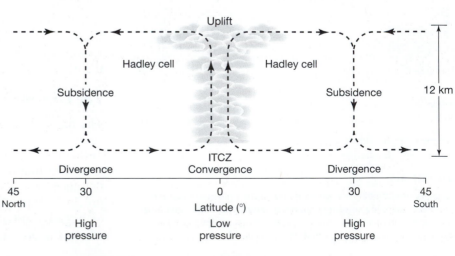

toward a low-pressure region is called **convergence.** The converging air masses that meet at the tropics and rise make up the **intertropical convergence zone (ITCZ).**

The surface heating produces evaporation in addition to convection. As the convecting air rises, it cools, and the evaporated water (water vapor) in the convecting column condenses to form clouds. As a consequence, the ITCZ is characterized by extensive areas of cloud cover and heavy precipitation. We will talk more about evaporation, condensation, and rainfall later in the chapter.

Divergence. The top of the troposphere, located at about 12–15 km in the tropics, forms a barrier to further uplift. Remember that temperatures generally increase in the stratosphere and that the higher temperatures produce a stable structure that limits convection from below. The air that rises in the ITCZ, on reaching this barrier, is forced to diverge poleward. **Divergence,** in this case, refers to the movement of air outward from a region in the atmosphere. This poleward-moving air subsides at about 30° N and 30° S latitude, replacing the air that is moving equatorward at the surface (Fig. 4-6). The air warms as it sinks, which prevents condensation from occurring and clouds from forming. As a result these regions are characterized by clear skies and low rainfall amounts. If you check an atlas, you will find that such areas coincide with some of the world's largest deserts (for example, the Sahara and Arabian deserts and the Great Australian Desert). The subsiding air also leads to an area of high pressure and divergence at the surface.

Hadley Circulation. This pattern of air movement, with convergence occurring in the tropics and divergence and subsidence some 30° away in one large convection cell, is called **Hadley circulation.** This circulation pattern was named for George Hadley, the British meteorologist who first explained the phenomenon. The convection cells on either side of the equator, referred to as *Hadley cells,* represent the dominant north–south mode of circulation between 30° N and 30° S latitude. Note, however, that the Hadley cells—and the ITCZ—are not continuous around the globe. The circulation takes place in individual cells of rising and subsiding air, and the pattern is further broken up by land–ocean contrasts. These large-scale circulation features can be observed in satellite images such as that shown in Figure 4-7.

These cells result directly from surface heating in the tropical oceans. In fact, although solar heating provides the fuel for tropospheric circulation, the actual pump that drives the circulation is the release of latent heat during convection. (We discussed latent-heat release in Chapter 3.) The energy of solar radiation, used to evaporate water from the ocean surface, is converted to latent heat, and the latent heat is released to the atmosphere in huge towers of convective cloud clusters within the ITCZ (Fig. 4-8). It is this release of latent heat that pumps the air around each Hadley cell.

Midlatitude and High-latitude Circulation. Thus far we have discussed the atmospheric circulation only between the equator and 30° N or S. What about atmospheric circulation from there to the poles? The very low temperatures at

FIGURE 4-7

Satellite image of the eastern Pacific and Central America. These images are obtained from geostationary satellites, which orbit over the equator at an altitude of about 35,000 km and at an orbital speed that keeps pace with Earth's rotation; thus the satellite appears to remain stationary over the same spot on the equator. This image was obtained by the National Oceanographic and Atmospheric Administration (NOAA) during Northern Hemisphere summer. A line of convective clouds marks the ITCZ just north of the equator. The clear areas to the north and south of the ITCZ mark the descending arms of the Hadley cells. (From Earth Satellite Corporation/Science Photo Library, Photo Researchers, Inc.)

FIGURE 4-8

Convective towers in the ITCZ. Solar heating evaporates large amounts of water from the tropical oceans. The air cools and condenses as it rises, releasing the energy used for evaporation as latent heat. The release of latent heat in these convective towers is the pump that drives the Hadley circulation. (From NASA Science Source, Photo Researchers, Inc.)

FIGURE 4-9

Mixing of air in the midlatitudes. The lower-density warm air from the tropics rises above the colder air moving equatorward from high latitudes. These contrasting air masses do not mix very easily. This zone is characterized by large temperature contrasts over very short distances.

FIGURE 4-10

The north–south (meridional) circulation of the troposphere. The tropical circulation is dominated by the Hadley circulation, whereas midlatitude circulation and weather are controlled by the location of the polar front zone and the mixing of cold polar air with warm air from the tropics.

the poles, particularly in winter, result in increased air density near the surface and, thus, in higher pressures than occur in the tropics. The higher density and pressure lead to divergence and a general movement of cold air outward at the surface—that is, toward the equator. The divergence is accompanied by subsidence from above. The equatorward-moving cold air meets the warm air moving poleward from the subtropics, producing a zone of steep temperature gradients called the **polar front zone** at approximately 60° N and S latitude. The two air masses do not mix easily: The warm air is less dense than the cold air, which therefore sinks below the warm air when the two air masses meet (Fig. 4-9). The polar front zone, therefore, slopes poleward with increasing altitude in the atmosphere. Note that, because of dynamic processes that come into play when air moves over a curved surface, this frontal zone forms a wavelike structure around the hemisphere. The actual latitude at which the front is located, therefore, varies from place to place.

When we put Figures 4-6 and 4-9 together, we see an alternating pattern of northward and southward moving air at the surface (Fig. 4-10). Such north–south movement is called *meridional* circulation. If we look at Figure 4-10 from above, we might expect to see a general pattern of surface winds such as those depicted in Figure 4-11. We would ex-

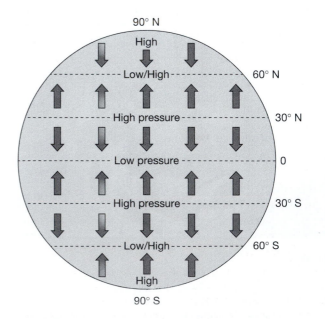

FIGURE 4-11

A possible model of the surface winds obtained by plotting, on a globe, the pattern of surface winds that would be deduced from Figure 4-10. Surface winds blow out of the high-pressure zones at the poles and at 30° N and 30° S and blow toward the low-pressure zones at the equator and in the midlatitudes.

(a)

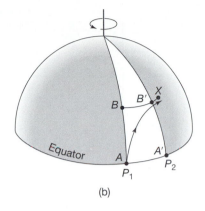

(b)

FIGURE 4-12

The Coriolis effect. (a) As Earth rotates, the speed of the surface is greatest at the equator and is zero at the poles. (b) At the equator, Earth has rotated from *A* to *A'*. Points *A* and *B* have moved to *A'* and *B'*. If the effects of Earth's rotation are ignored, air moving from *A* to *B* would actually curve to the right and arrive at point *X*.

pect surface winds to blow out of the high-pressure zones at the poles and at about 30° N and S, and to blow toward the low-pressure zones at the equator and at about 60° N and S. The actual pattern, however, is more complicated because winds tend to blow in east–west directions as well. Indeed, the east–west motions are considerably greater than the north–south motions. We know that differences in solar heating cause the equator-to-pole movement we have been discussing. What causes the east–west movements?

The Coriolis Effect

East–west movements of surface winds are the result of the Coriolis effect. The **Coriolis effect** is the apparent tendency for a fluid (air or water) moving across Earth's surface to be deflected from its straight-line path. (Some scientists refer to a *Coriolis force* in relation to this effect. This force, however, is only an apparent force due to the observer's frame of reference, not a real force due to an identifiable source, such as the gravitational pull of a planet.) Viewed from Earth, a north–south moving object appears to be deflected to the east or west. Viewed from space, the same object is in fact seen to move in a straight line. The apparent curve that we see is the result of our frame of reference—we normally view the object's movement from *within* the system. This phenomenon is named for Gaspard Gustav de Coriolis, the French mathematician who in 1835 proposed that the concept applies to surface winds.

The Coriolis effect applies to any object moving on a rotating body. To visualize this, let us first consider Earth rotating on its axis. The two vertical lines in Figure 4-12a represent the distance moved in a given time interval, and the arrows represent the rotation speed of Earth's surface at different latitudes over that interval. The speed of rotation is greatest at the equator (approximately 464 m/sec), and it decreases as we move north—or south— until it becomes zero at the poles.

Now imagine an object, such as an air mass, that is apparently stationary at a point on Earth's surface. Although this air mass is not moving relative to the surface, it is traveling eastward at Earth's rotation rate for that location. (For example, an object that is stationary at any one of the points marked by the left-hand edges of the rotation arrows in Fig-

ure 4-12a would, to an observer in space, appear to move along the arrow's path.) Because the rotation rate changes with latitude, an air mass moving northward from the equator—from point *A* to point *B* in Figure 4-12b—will appear to curve off to the right of its straight-line path, arriving at *X* rather than at *B'*. Why does this happen?

In the time it would take the air mass to travel from *A* to *B*, the Earth rotates from *A* to *A'* (and *B* moves to *B'*). Remember that the air mass at *A* is moving not only northward but also eastward at Earth's speed of rotation (represented by the distance *A*-*A'*). As long as it is between the equator and point *B*, the air mass is moving from west to east faster than Earth is rotating at *B*. Thus, the air mass will "gain" on the ground below it and will arrive at point *X* instead of at *B'*. Although it is difficult to visualize, the air does in fact move in a straight line; if we were watching from space rather than from Earth, that is what we would see.

We have seen how an air mass (or any object) moving northward in the Northern Hemisphere is deflected to its right. Following the same reasoning, we can see that an air mass moving southward in the Northern Hemisphere also curves to its right (relative to the direction of initial movement), because now the air mass is moving eastward more slowly than Earth's surface immediately underneath. The easiest way to keep track is to think of the deflection direction relative to the direction of initial motion of the object—it is always to the right of the direction of initial motion in the Northern Hemisphere and to the left in the Southern Hemisphere.

What happens when the initial direction of movement is due east or due west? As it happens, the Coriolis effect still comes into play, but for a different reason. When an object is set in motion along a circular path, *centrifugal force*—another apparent force—tends to push the object away from the center of rotation. (This is the same phenomenon that forces your car off the road if you try to turn a corner too fast.) If an air mass in the Northern Hemisphere is moving eastward faster than Earth is rotating at that latitude, that air mass will experience an apparent centrifugal force that pushes it directly away from Earth's spin axis.

We can break this apparent force down into two components: one component that is acting perpendicular to the surface, and one that is horizontal (parallel) to the surface

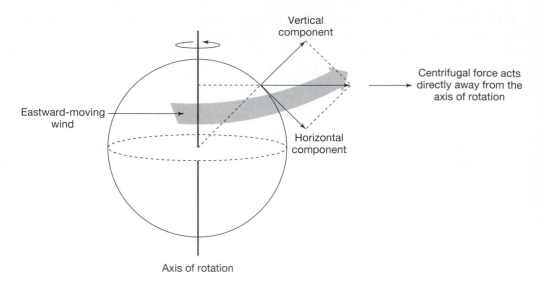

FIGURE 4-13

The Coriolis effect produced by the centrifugal force acting on eastward- or westward-moving winds.

(Fig. 4-13). For an eastward-moving wind in the Northern Hemisphere, the horizontal component is to the south; the wind would curve to the south, or to the right. For a westward-moving wind, the horizontal component is to the north; the wind still curves to the right. In the Southern Hemisphere, the effects would be opposite: Both eastward-moving and westward-moving winds would curve to the left.

Although these descriptions of the deflection effect for north–south and east–west moving objects appear to be very different, mathematically the deflections are identical, and both are referred to simply as the Coriolis effect. The Coriolis effect increases as the speed of the object increases. And whereas the speed at which Earth's surface is moving due to its rotation is zero at the poles and a maximum at the equator, the Coriolis effect is zero at the equator and increases with latitude. The only place on Earth's surface where the Coriolis effect does not come into play is at the equator. An air mass moving eastward or westward around the equator is not deflected from its original path. Such an air mass is not changing latitude, so there is no Coriolis effect due to the difference in rotation rate with latitude. Nor is there a horizontal component to the centrifugal force, so again there is no Coriolis effect.

Distribution of Surface Winds

We can now modify the simplistic pattern of northward-moving and southward-moving surface winds in Figure 4-11 to obtain the more realistic pattern of surface winds shown in Figure 4-14. There the winds are deflected to the right and left of the paths of initial motion in the Northern and Southern hemispheres, respectively. This deflection of the winds due to the Coriolis effect gives rise to easterly winds at high latitudes. Meteorologists refer to winds in terms of the direction from which they blow.

In other words, an "easterly" wind is a wind that blows from east to west. The midlatitudes are characterized by westerly flow, and the tropics by easterly winds called the *northeast* and *southeast trade winds*. Winds at the equator tend to be highly variable in direction. This region where winds are light and frequently change direction is referred to as the *doldrums*.

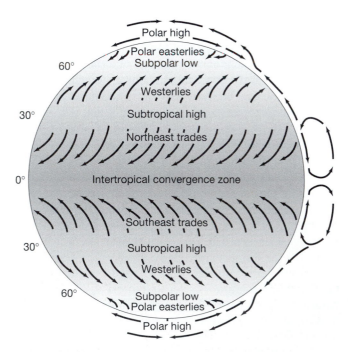

FIGURE 4-14

The pattern of surface winds. This shows the same general pattern of winds as Figure 4-11, but the wind directions have been changed to include the deflection due to the Coriolis effect. (From T. McKnight: *Physical Geography: A Landscape Appreciation*, 6/e, 1999. Reprint by permission of Prentice Hall, Upper Saddle River, N.J.)

Considerations of pressure differences, buoyancy, and the Coriolis effect have led us to a good first approximation of the general circulation of the troposphere. This pattern, however, is still a little too simplistic. In reality the indicated winds, for example, do not blow continuously, and they are not continuous around the globe. As we noted earlier, uplift in the ITCZ takes place in clusters of convective cells rather than in two giant cells, one on each side of the equator. The rising air moves poleward and, under the influence of the Coriolis effect, turns to the right in the Northern Hemisphere (to the left in the Southern Hemisphere). It thus becomes a westerly flow in the upper troposphere. Some of this air subsides near 30° N or S latitude to form the subtropical high-pressure belt (Fig. 4-14), but the subsidence too is concentrated in localized areas. The locations of these high-pressure systems vary with season, although they are always found in these approximate locations. The trade winds blow from the equatorward side of these semipermanent high-pressure systems. Similarly, poleward-moving air from the subtropical high-pressure zone curves due to the Coriolis effect, producing a generally westerly flow in the Northern Hemisphere midlatitudes (and easterly in the Southern Hemisphere midlatitudes). The actual flow pattern, however, is highly variable from day to day.

The pressure and wind patterns in the midlatitudes depend on the location of the subtropical highs as well as on the distribution and movement of temporary areas of high or low pressure that form in association with the steep temperature gradients in the polar front zone. Small areas of low pressure (on the order of 1000 km wide) form in this zone in part due to the surface-temperature gradient, but also because of dynamic processes occurring higher in the troposphere. As air blows into these regions, it curves to the right (in the northern hemisphere), producing a localized circular flow pattern referred to as *cyclonic flow.* Air flowing out of a high-pressure region (referred to as an *anticyclone*) will also curve to the right in the Northern Hemisphere, creating an *anticyclonic,* or clockwise, flow. (The direction of air flow around cyclones and anticyclones is reversed in the Southern Hemisphere.)

Low-pressure systems that form outside the tropics are referred to as *extratropical cyclones.* The circular flow mixes warm air from the equatorward side of the system with cooler air from the high latitudes. As we noted earlier, the warm and cool air masses do not mix easily; hence these systems are characterized by extensive uplift as the warmer, less dense air rises above the cooler and denser air mass. As we will see later in the chapter, this rising air results in the formation of rain or snowfall. These circulation features move along the polar front, bringing low pressure as they move over a region, which is then replaced by higher pressure as they move past. These transient high-pressure and low-pressure systems are characteristic features of midlatitude climates and account for much of the day-to-day variability in weather in these regions. At high latitudes, the polar easterlies are most clearly developed in winter (when the coldest surface temperatures occur and when this area of low temperatures is most extensive). However, low-pressure systems from the midlatitudes often migrate into the polar regions in summer, breaking down the surface high-pressure systems and disrupting the easterly flow.

Seasonal Variability. The simple pattern of global winds shown in Figure 4-14 is also modified by seasonal variations. As well as changing with latitude, the distribution of solar energy varies with the seasons. Figure 4-15 shows the seasonal pattern of Earth's orbit around the Sun. The time when Earth is closest to the Sun is referred to as *perihelion;* Earth is farthest from the Sun at *aphelion.* This difference in distance from the Sun affects the seasonal distribution of temperature. More important to seasonality is the *tilt,* or **obliquity,** of the Earth. Obliquity refers to the angle of the Earth's spin axis relative to a line drawn perpendicular to the plane of the planet's orbit around the Sun. Each planet has a different angle of tilt. Earth's axis is tilted 23.5° from the perpendicular. On human time scales, the obliquity remains constant as the Earth revolves around the Sun (Fig. 4-15).

For six months of each year, the Northern Hemisphere faces the Sun and the Southern Hemisphere faces away; for the other six months, it is the Southern Hemisphere that faces the Sun while the Northern Hemisphere faces away. The hemisphere that faces the Sun receives much more solar energy than does the other hemisphere. It is this factor that determines the seasons. Consider the June 21 solstice (the day with the longest period of sunlight in the Northern Hemisphere) in Figure 4-15. Imagine Earth spinning around its axis. The North Pole will remain in sunlight the whole time, while the South Pole will remain in darkness. The opposite is true at the solstice on December 21 (the day of shortest sunlight duration in the Northern Hemisphere). The result is six months of sunshine and six months of darkness at the poles.

The greatest heating occurs where the Sun is directly overhead. Due to Earth's obliquity, the latitude at which this occurs varies continuously throughout the year: from 23.5° S (called the Tropic of Capricorn) on December 21 to 23.5° N (called the Tropic of Cancer) on June 21. There are also two days on which the Sun is directly overhead at the equator—the *vernal equinox,* March 20, and the *autumnal equinox,* September 22. On these dates, which correspond to the first days of spring and fall, respectively, in the Northern Hemisphere, daytime and nighttime are equal in length. (These dates vary by a day or so because the calendar is adjusted every four years to reconcile it with Earth's revolution around the Sun.) Notice, though, that the Sun is always almost overhead within a narrow band near the equator. Hence there is a large input of solar radiation to the tropics at all times. The difference between daytime and nighttime temperatures in this region, in fact, is usually much greater than the seasonal difference there.

The seasonal variability in incoming energy shifts the atmospheric circulation patterns northward and

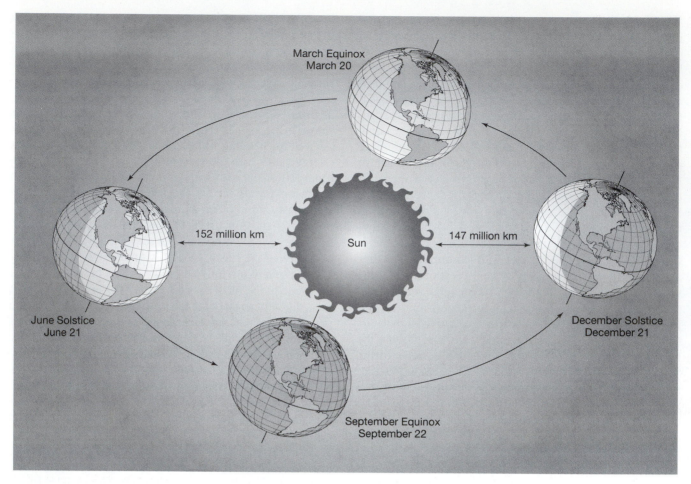

FIGURE 4-15

The seasons. The seasons are controlled by Earth's obliquity and Earth's orbit around the Sun. The hemisphere that is "tilted" toward the Sun experiences summer while it is winter for the hemisphere that is "tilted" away from the Sun. The equinoxes (when the Sun is directly overhead at the equator) mark the transition seasons: fall and spring. (From T. McKnight: *Physical Geography: A Landscape Appreciation*, 6/e, 1999. Reprinted by permission of Prentice Hall, Upper Saddle River, N.J.)

southward as the seasons change (Fig. 4-16). The hemisphere experiencing summer has less of a temperature gradient between the tropics and the pole than does the opposite hemisphere. The fact that the Sun shines continuously for six months at each pole compensates for the fact that the poles do not receive as much solar energy per unit area as do the tropics. This reduced temperature gradient weakens the strength of the atmospheric circulation. At the same time, because the Sun is directly overhead somewhere away from the equator, the maximum solar energy is directed somewhere poleward. Consequently, the steepest temperature gradients are shifted toward the poles, and the circulation patterns are also shifted poleward. In the winter hemisphere (with six months of darkness at the pole), the equator-to-pole temperature gradient is much stronger and the steepest gradients are shifted equatorward. As a result, the atmospheric circulation is more intense and the circulation patterns are shifted toward the equator.

Global Distributions of Temperature and Rainfall

In the first part of this chapter, we described some of the main features of the global-scale atmospheric circulation. For the remainder of the chapter, we will look at the effect this circulation has on other parts of the Earth system—specifically, the global temperature and rainfall distributions.

As we have learned, the ultimate cause of the atmospheric circulation is the distribution of available energy. More interesting for our purposes is the fact that the interaction between temperature and the circulation is not a one-way process. As we indicated earlier, the circulation itself is an important component of Earth's thermoregulatory system, transporting energy (heat) from areas where there is a surplus to areas where there is a deficit.

The transport of water is also strongly affected by the atmospheric circulation. The distribution of water about the globe is important in that organisms require a suffi-

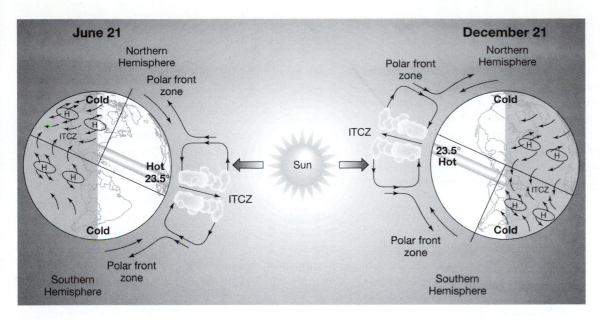

FIGURE 4-16

Seasonal migration of the atmospheric circulation patterns. The ITCZ is found in the summer hemisphere, where the circulation is weaker and the patterns are shifted toward the pole. The subtropical high-pressure cells that mark the descending arms of the Hadley circulations are denoted by H.

cient supply of water to maintain life. That distribution is also important for the transport of dissolved materials. As we will soon see, evaporation and precipitation are strongly influenced by temperature and, therefore, by the distribution of energy. Furthermore, the transport of water in its various forms (liquid, water vapor, and ice) also modifies the temperature distribution by affecting the radiation budget (Chapter 3) and thus feeds back to affect the circulation. Hence, we see that temperature, precipitation, and the atmospheric circulation are all closely linked and that interactions and feedbacks exist among all three of these components of Earth's climate.

Land–Ocean Contrasts

Beyond the latitudinal distribution of energy, global temperature patterns are also strongly influenced by the distribution of land and ocean. Recall from Chapter 2 that the albedo of the ocean surface is considerably lower than the albedo of most land surfaces. In consequence, oceans absorb more of the available solar energy than do land surfaces at the same latitude.

Land and ocean surfaces also behave very differently in what they do with that energy. An ocean surface rapidly transfers heat downward by turbulent mixing and to the atmosphere above by convection. Part of the contrast between land and ocean is due to differences in their thermal properties. The land surface rapidly loses heat to the atmosphere by convection, but it transfers heat downward relatively slowly by conduction. How easily this transfer occurs depends on the physical and chemical properties of the material; the rate at which this occurs is described by its *thermal conductivity.* More formally, thermal conductivity is the rate at which heat

energy passes through a column of material that has a temperature gradient along the column of 1 K, or 1°C, per meter. Water has a high thermal conductivity, whereas land surfaces have low thermal conductivities. Furthermore, we can consider the *heat capacity* of the two types of surfaces. Heat capacity at constant volume is the heat energy required to raise the temperature of a unit mass of a substance by 1 K or 1°C without changing its volume. In other words, heat capacity is a measure of how much heat must be added to an object to change its temperature. The heat capacity of water is about three to four times that of dry soil. Thus, the input of a given amount of energy will raise land temperatures much more than it will raise sea-surface temperatures.

More important is that the ocean surface transfers heat rapidly downward by turbulent mixing. As we will see in Chapter 5, the surface layers of the ocean are well mixed; when the ocean surface is heated, the heat is mixed downward within the surface layers. The amount of material that must warm (or cool) is very large, so the temperature change is very slow. A further factor is differential absorption of the two surfaces. Whereas all the solar radiation falling on the land surface is reflected or absorbed right at the surface, some of the solar radiation falling on the ocean penetrates and is absorbed below the surface. Hence, energy is transferred downward even more rapidly in water than on land.

The Sea Breeze. Putting all this together, we see that with equal amounts of incoming energy, land surfaces will heat up much more rapidly than do ocean surfaces but will also cool down much more rapidly once the input of energy is reduced. Land surfaces heat up quickly during the day and cool quickly at night, whereas ocean surfaces warm slowly in the day, and the temperature drops very little at night.

The sea breeze that occurs near coastlines is a direct result of this diurnal variability. The heating of the land surface during the day warms the overlying air and gives rise to small areas of low pressure and uplift; cooler temperatures over the ocean result in relatively higher pressures and subsidence of the cooled air above (Fig. 4-17). Air flows down the pressure gradient (from the area of higher pressure to the area of lower pressure), creating on-shore flow from ocean to land. At night this temperature structure breaks down, and the atmospheric circulation weakens. If the land cools sufficiently, the circulation pattern may reverse.

Continentality.   We can see the same effect on a longer scale in terms of seasonal climate variability. As we noted earlier, the seasonal variation over midlatitudes and high latitudes is much greater than in the tropics. This variability is also much greater over land surfaces than over the oceans because of their different thermal characteristics. This property is referred to as *continentality.* The more continental the climate, the more it is characterized by seasonal temperature extremes. Land surfaces are much warmer than ocean surfaces in summer and much colder in winter. The effect of continentality on global temperatures is visible in Figure 4-1. The greatest seasonal variability is found in the interior of large continental masses, and the lowest variability is over the tropical oceans. The oceans provide a moderating effect in coastal regions that reduces the temperature extremes. The temperature difference between ocean and land surfaces also affects the mean sea-level pressure distribution (again feeding back to affect the circulation). Figure 4-18 shows the average atmospheric pressure that would be found over land if the land surfaces were at sea level. In Northern Hemisphere winter (Fig. 4-18a), the very cold surface temperatures of interior North America and Asia cool the lower layers of the atmosphere, producing high surface pressures over those land masses. The North Atlantic and North Pacific are characterized by low-pressure zones produced by the low-pressure systems forming along the steep temperature gradient in the polar front zone. In summer this gradient decreases (Fig. 4-18b). The low-pressure zones are less well developed and are displaced poleward, the subtropical highs expand, and the continental regions of high pressure are replaced by regions of low pressure.

Monsoons.   The most extreme consequence of the seasonal variability due to differential heating of land and ocean surfaces is the monsoon regime of Southeast Asia. The **monsoon** is a seasonal reversal in the surface winds. In summer the large Asian land mass, with its high elevations in the Tibetan Plateau of central Asia, causes high surface temperatures, low atmospheric pressures, and intense convection of air above the surface. The rising air is replaced by air moving in from the high-pressure region over the Indian Ocean to the south (Fig. 4-19a). The moist air drawn in from the Indian Ocean cools as it rises above the mountains of southwest India and over the Himalayas. In both instances the rising air produces clouds and heavy rainfall (the monsoon rains). In winter the pattern reverses: high elevations and persistent snow cover enhance the continentality, producing even lower temperatures. This results in high atmospheric pressure and subsidence of air over the continent and a southward flow of air (Fig. 4-19b). A similar feature, but on a much smaller scale, is found in the southwestern United States, where a "monsoonal" flow from July through to mid-September brings moist air in from the Gulf of California and the eastern Pacific.

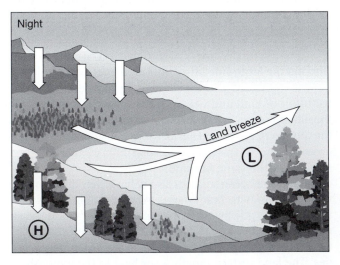

FIGURE 4-17

The sea breeze. The heating of the land during the day causes localized convection with low pressure at the surface. This convection establishes a pressure gradient from the ocean toward the land that results in onshore wind. At night, the rapid cooling of the land (relative to the ocean) causes this circulation to break down and may even reverse the flow, causing the wind to blow from the land toward the water. (From T. McKnight: *Physical Geography: A Landscape Appreciation,* 6/e, 1999. Reprinted by permission of Prentice Hall, Upper Saddle River, N.J.)

(a)

(b)

FIGURE 4-18

Average sea-level pressure patterns: January (a) and July (b). Units are in millibars (mb), where 1 bar (1000 mb) equals 1.013 atmospheres (atm). Low-pressure areas are designated by L and high-pressure areas by H. (From R.W. Christopherson, *Geosystems: An Introduction to Physical Geography,* 3/e, 1997. Reprinted by permission of Prentice Hall, Upper Saddle River, N.J.)

The Southern Hemisphere (Fig. 4-18c and d), with its huge expanse of ocean and very little land area, presents a less complicated picture. The midlatitude air flow is much more *zonal*—that is, the air circulation more closely follows lines of latitude—than it is in the Northern Hemisphere. The seasonal temperature change causes the air flow pat-terns to shift north and south slightly and causes pressure gradients to change, but the distribution of land and oceans produces much less variability of winds around the Southern Hemisphere than around the Northern Hemisphere.

We see, therefore, that the broad pattern of global temperatures is determined by the latitudinal distribution

FIGURE 4-19

The monsoon flow over southeast Asia. (a) Summer heating of the Tibetan Plateau produces intense convection and low surface pressures, drawing in moist air from the Indian Ocean to the south. (b) The reverse occurs in winter, when low temperatures and extensive snow cover on the plateau produce high surface pressures, subsidence, and outflowing air. (From R.W. Christopherson, *Geosystems: An Introduction to Physical Geography,* 3/e, 1997. Reprinted by permission of Prentice Hall, Upper Saddle River, N.J.)

of net radiation, so higher temperatures occur in the tropics and lower temperatures at the poles. This distribution varies with season such that the seasonal range of temperatures is slight in the tropics and increases poleward. Beyond this, we see from the preceding discussion that the seasonal variability is strongly modified by land–ocean contrasts, the interior of continents having a much greater seasonal range than do coastal locations.

Global Precipitation Patterns

In addition to transporting energy, the circulation of the troposphere also involves the movement of material across Earth's surface. We can imagine how effective the atmosphere is at transporting material from the fact that pollution from midlatitude industrial sources has been found on both polar ice caps.

The most important substance transported in the atmosphere is water, in the form of water vapor and clouds. Both water vapor and clouds are important for several reasons: They play a dominant role in the global energy balance; they are a significant factor in determining the distribution of fresh water around the globe; and they are highly variable in time and space (making them difficult to predict).

The Global Hydrologic Cycle. Water is not only the most important chemical compound in the atmospheric circula-

tion but also in the entire Earth system. The human body is 60% water by weight, and all organisms require some water in order to live. From space, we see that Earth is a planet dominated by water (Fig. 4-20). Seventy percent of Earth's surface is covered by oceans. The poles are encased in extensive sheets of ice that either float on the surface of the ocean (*sea ice*) or form glaciers many kilometers thick over land. Clouds, which are made of condensed water vapor, swirl across the surface—continuously changing in size, shape, and location but always covering about 50% of the globe at any time. The water that exists in gaseous form (water vapor) also varies in amount across the globe—from near zero over the ice caps to about 7% in the tropics.

Water is unique in the Earth system for a variety of reasons, not the least of which is that it is the only naturally occurring substance that can exist in all three phases (solid, liquid, and gas) at the temperatures found on Earth's surface. Because water changes so readily from one phase to another, it cycles easily among all the system components. In so doing, water plays a vital role in many Earth system processes. Changing from a solid to a liquid or from a liquid to a gas requires a large addition of energy. That energy is stored in the water molecule in the form of latent heat. The **latent heat of vaporization,** which we described in Chapter 3 as the energy needed to convert liquid water to water vapor, is 2260 kJ/kg at 100°C. (This much energy

FIGURE 4-20

Earth, viewed from space at about 37,000 km (23,000 mi), is dominated by water. (From NASA Headquarters) [See color section]

must be added to each kilogram of boiling water to convert it to water vapor.) When the process is reversed and the water changes from a gas back to a liquid, this same amount of energy is released to the environment (Fig. 4-21). The **latent heat of fusion,** or the energy needed to convert ice to liquid water, equals 335 kJ/kg at 0°C. When the process is reversed and the water changes from a liquid back to a solid, this amount of energy is again released to the environment. To raise the temperature of liquid water from 0°C to 100°C requires 419 kJ/kg. To convert ice to water vapor thus takes 3014 kJ/kg ($= 2260 + 419 + 335$). These values apply to water at sea level, and they will vary slightly as atmospheric pressure changes. If these conversion processes occur in different locations, then there is a net transfer of energy from one place to another. Therefore, the distribution and movement of water in its various

phases has important consequences for the transfer of energy and the global pattern of surface temperatures.

We saw in Chapter 3 that water vapor and carbon dioxide are the most important of the present-day greenhouse gases. Without them, much of Earth would be too cold to support life. We saw also that clouds have a major impact both on Earth's albedo and on the emission of terrestrial radiation to space. And we shall see in Chapter 6, water plays a vital role in breaking down rocks (weathering) and in transporting essential nutrients throughout the Earth system. *Water, in all its phases, is the primary medium by which energy and matter are circulated among the Earth system components.*

Water in the Earth system is concentrated in several major reservoirs:

1. The oceans, where the water exists in the form of seawater;

2. The land surface, in the forms of ice sheets, glaciers, snow, lakes, and rivers, and the land subsurface, in the form of groundwater; and

3. The atmosphere, in the form of water vapor.

These reservoirs and the pattern of water storage and movement throughout the system comprise the global **hydrologic cycle** (Fig. 4-22a). Most of Earth's water—about 97%—is stored in the first reservoir, the oceans (Fig. 4-22b). Almost 3% is on or in the second reservoir, land. Of this amount, three-quarters is trapped in the polar ice sheets (Greenland and Antarctica). Should the Greenland ice sheet melt, it would raise the global sea level by about 6 m; should all of Antarctica melt, sea level would rise by about 60 m. A small amount of water exists in mountain glaciers as well. Most of the remainder occurs as **groundwater,** or water that penetrates through soil and rock and collects below the surface. Water stored in rivers, lakes, and the soil accounts for less than 1% of all the water found on land. Almost two-thirds of this amount is stored in lakes and reservoirs; about one-third occurs in the soil, and a tiny fraction occurs in rivers. The third reservoir, the atmosphere, contains less than

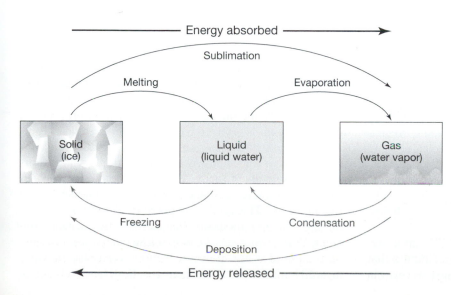

FIGURE 4-21

Schematic diagram of the different phases of water. Energy is absorbed as water changes from a solid to a gas (moving left to right in the diagram) and is released as water changes from a gas to a solid (from right to left).

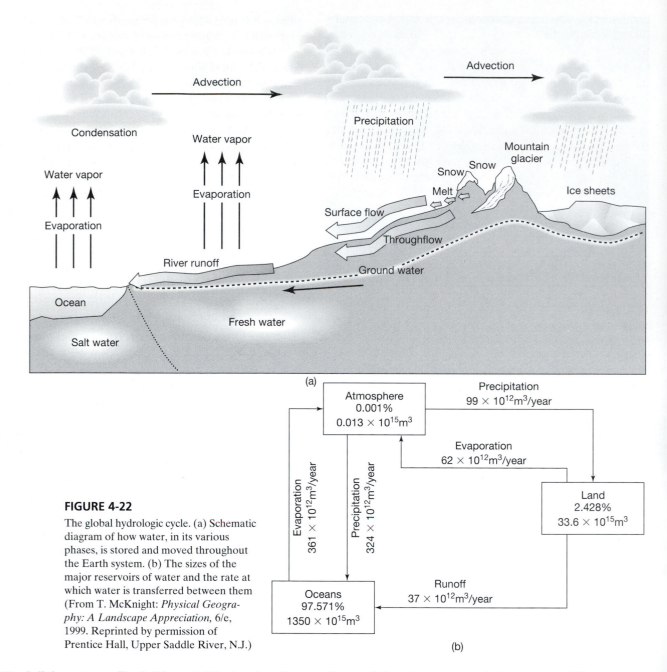

(a)

(b)

FIGURE 4-22

The global hydrologic cycle. (a) Schematic diagram of how water, in its various phases, is stored and moved throughout the Earth system. (b) The sizes of the major reservoirs of water and the rate at which water is transferred between them (From T. McKnight: *Physical Geography: A Landscape Appreciation,* 6/e, 1999. Reprinted by permission of Prentice Hall, Upper Saddle River, N.J.)

0.001% of all the water on Earth. Figure 4-22b also gives the annual exchange of water among the three major reservoirs.

Precipitation and Saturation Vapor Pressure. The transfer of water between the land/ocean surface and the atmosphere takes place through evaporation and *precipitation.* Precipitation occurs when atmospheric water vapor condenses to form small droplets of liquid water. When the water droplets reach sufficient size, they fall because of gravity. If they do not evaporate before they reach Earth's surface, we experience them as rain. If atmospheric temperatures are below freezing, the droplets fall as snow or sleet instead.

One way of expressing the amount of water vapor present in the atmosphere is to measure the contribution that water vapor makes to the atmospheric pressure. We saw in

Chapter 3 that air is composed of numerous different gases. Each gas exerts its own pressure. What is measured as the atmospheric pressure is the sum of all the **partial pressures** of the individual gases—that is, the pressure each gas would exert if it were the only gas present. The pressure exerted by water vapor is referred to as the *vapor pressure.*

Imagine a body of water. Water molecules at the surface that have a little more energy than do their neighbors can overcome the attractive forces that hold the molecules together and thereby escape as water vapor molecules into the air above. This is the process of **evaporation.** Some of these water vapor molecules that subsequently come in contact with the water surface would lose energy, be "caught" by the liquid water molecules, and become liquid water again. This is the process of **condensation.** Once the rate of condensation equals the evaporation rate—that is, as many molecules leave the gas as are added to it—the gas is at equilibrium. At this point, the vapor pressure of water is referred to as the *saturation vapor pressure*. In this scenario, the saturation vapor pressure depends only on the rate at which molecules are transferred from liquid to gas and back again. This rate depends on the energy of the molecules, which means it depends on temperature. (Recall that the higher the energy, the higher the temperature.) Therefore, as temperature increases, the saturation vapor pressure increases.

Figure 4-23 is a graph of saturation vapor pressure versus temperature for water. In general, we can think of clouds as forming when the air is at the saturation vapor pressure for water. Further evaporation adds water vapor molecules to the air, where they condense to form water droplets in clouds. When these droplets become large enough to overcome the upward motion of the air, they fall as precipitation. Assume that an air mass is at the temperature and vapor pressure indicated by point *P* in Figure 4-23. Point *P* is not on the curve; the air is not at the saturation vapor pressure for that temperature, hence clouds would not form and precipitation could not occur. We can bring that air mass to the *saturation point*—that is, to equilibrium at a point on the temperature versus saturation vapor pressure curve—in two ways. First, we could add more water vapor through increased evaporation from the surface and thereby increase the saturation vapor pressure. That action would move the air mass from point *P* up along the vertical dashed line toward the curve. Second, we could reduce the air temperature, which would move the air mass from point *P* to the left along the horizontal dashed line toward the curve.

Because the saturation vapor pressure varies with temperature, knowing just the vapor pressure on a given day does not give a good indication of when clouds will form. Consequently, we normally think in terms of *relative humidity*—the ratio of the actual vapor pressure to the saturation vapor pressure at that temperature. (We are using vapor pressure as a measure of the amount of water present; hence this definition is equivalent to that presented in Chapter 3.) The relative humidity is usually expressed as a percentage; a relative humidity of 100% represents air at the saturation vapor pressure. In general, water vapor will condense to form water droplets and clouds when the air is fully saturated. But in fact, very clean air may have greater than 100% relative humidity (that is, it can be *supersaturated*) without condensation taking place. Condensation is facilitated by impurities in the air—microscopic particles (solid or liquid) that are small enough to remain in suspension in the air. Such particles are known as *cloud condensation nuclei* (CCNs) when they are used in cloud formation. These nuclei can come from many sources, both natural and anthropogenic. It is likely that many of the clouds that form from CCNs over land surfaces derive from human-produced sources, such as sulfates.

We said earlier that the vapor pressure can be brought to the saturation point either by increasing the vapor pressure or by cooling the air. We can visualize both processes taking place as air moves over different surfaces: evaporation increasing as unsaturated air moves over lakes or the ocean, and temperatures decreasing as the air moves over cooler surfaces. The largest and most rapid changes take place, however, when air rises in the troposphere. As we saw in Chapter 3, temperature generally decreases with altitude in the troposphere, where most clouds exist. Thus most rainfall situations occur with some form of **uplift,** or rising of air masses. Uplift is a general term denoting any process by which air at a given level in the atmosphere is lifted to a higher altitude.

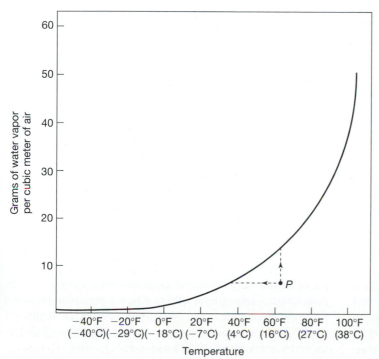

FIGURE 4-23

Saturation vapor pressure versus temperature for water. (From T. McKnight: *Physical Geography: A Landscape Appreciation*, 6/e, 1999. Reprinted by permission of Prentice Hall, Upper Saddle River, N.J.)

(a) Sonoran

(b) Taklimakan

(c) Atacama

(d) Kalahari

FIGURE 4-24

The distribution of the world's major deserts. (From R.W. Christopherson, *Geosystems: An Introduction to Physical Geography,* 3/e, 1997. Reprinted by permission of Prentice Hall, Upper Saddle River, N.J.)

We can get some appreciation for how precipitation is distributed around the globe, therefore, by recognizing that most precipitation takes place as air cools when it is forced to rise. We have already mentioned two of the processes that result in uplift: first, large-scale uplift that occurs with the mixing of air masses of different densities; second, uplift due to convection. Consequently, there is heavy precipitation along the polar front zone in the midlatitudes and in the vicinity of the ITCZ. Convection, however, occurs not only in the tropics, but wherever there is intense surface heating. Therefore, although convection does not always produce rain, it is the dominant rainfall-producing process over warm land masses in summer. A third process that forces air to rise is the confrontation between a moving air mass and a mountain range. Such encounters cause *orographic* precipitation on the windward (upwind) slopes of mountains. (Orography is the branch of geography that involves mountains and mountain systems.) For example, orographic precipitation commonly occurs on the western slopes of the Sierra Nevada mountain range and on the southern slopes of the Himalayas.

Under these circumstances, precipitation is enhanced due to the atmospheric circulation. Under other circumstances, precipitation is inhibited in certain areas. We call such areas **deserts,** and we can examine why they are located where they are (Fig. 4-24). Remembering that condensation results from uplift (which cools the air) or from increasing the amount of available moisture, we would expect to find deserts in areas where uplift is suppressed or where there is an inadequate moisture supply.

In general, precipitation is low in the interior of large land masses, simply due to the distance from moisture supplies. Deserts are located in the vicinity of the descending arms of the Hadley cells, as we noted earlier, and on the leeward (downwind) slopes of mountains. They occur also, perhaps unexpectedly, on the west coasts of large continents in areas that lie equatorward of the midlatitude low-pressure systems. For reasons that we will discuss in Chapter 5, these regions are characterized by cold offshore ocean currents. The cold ocean currents reduce evaporation and cool the air that moves over them—a combination that inhibits convection and precipitation over the adjacent coastline. The deserts that form in this way are called littoral (alongshore) deserts. In fact, one of the driest deserts in the world is the Namib Desert along the coast of southwest Africa. The desert of Baja, California (actually in Mexico), is another example. Although you generally think of deserts as being hot, some deserts are also located in the cold polar regions. The low temperatures inhibit uplift, and, where precipitation does occur, quantities are small because of the low saturation vapor pressures. The central part of Antarctica is, in fact, a desert. There is enough ice on Antarctica to raise global sea levels by 60 m, but the average annual snowfall on the plateau is the equivalent of less than 51 mm (2 in.) of liquid water

Chapter Summary

1. The driving force for the atmospheric circulation is the global distribution of energy.
   a. The angle at which the Sun's rays strike the Earth changes from the equator toward the poles. The result is that incoming solar radiation decreases with latitude. More solar radiation is received in the tropics than at the poles, resulting in an equator-to-pole temperature gradient.
   b. This temperature gradient drives the atmospheric circulation because of the inverse relationship between the temperature and the density of a gas: Higher temperatures produce lower densities. Differences in the distribution of global temperatures cause differences in air density and, therefore, pressure.
   c. Air tends to move from areas of high pressure to areas of low pressure. These large-scale movements of air produce the global windbelts.
   d. These windbelts are significantly modified by the Coriolis effect, which is caused by the Earth's rotation.
2. The net effect of these atmospheric movements is to redistribute available thermal energy. There is a negative feedback between the energy gradient and the circulation. Energy is moved from low latitudes, where Earth is hot, toward the poles, where it is cold.
3. The atmospheric circulation exerts a major control on global temperature patterns. The movement of the air carries water vapor from one region to another. Because evaporation and condensation are largely a function of temperature, the redistribution of water around the globe is also strongly tied to temperature distributions and to the atmospheric circulation.
   a. Precipitation is enhanced wherever the circulation promotes uplift and is inhibited in areas dominated by subsidence.
   b. Precipitation amounts are also affected by continentality and the distance from moisture sources.
   c. The distribution of land and ocean affects the distribution and variability of surface temperatures. Variability increases as the distance from the ocean increases.
   d. The circulation, temperature, and precipitation distributions are modified by seasonal variations in incoming solar energy caused by Earth's obliquity and Earth's orbit around the Sun.

Key Terms

Boyle's law
buoyancy
Charles's law
condensation
convergence
Coriolis effect
deserts
divergence

evaporation
general gas law
groundwater
Hadley circulation
hydrologic cycle
intertropical convergence zone
latent heat of fusion

latent heat of vaporization
monsoon
obliquity
partial pressure
polar front zone
subsidence
uplift

Review Questions

1. What are the functions of the global circulatory system?
2. Explain why the distribution of solar energy varies with latitude.
3. a. Draw a graph showing the variation of incoming solar energy and outgoing infrared radiation with latitude.
   b. Indicate the regions of energy surplus and energy deficit.
   c. Explain why this distribution is important for the atmospheric circulation.
4. Explain why heating an air mass causes it to rise.
5. a. Use a diagram to describe Hadley cells.
   b. Why does the Hadley circulation change seasonally?
6. a. What is the Coriolis effect?
   b. How does the Coriolis effect help determine the global pattern of winds?
7. a. Explain why Earth experiences different seasons throughout the year.
   b. Which parts of Earth experience the greatest seasonal variability, and which parts experience the least? Explain why.

8. Contrast the different roles of turbulent heat transfers and conduction in modifying the thermal response of a land surface and an ocean surface.
9. Use map sketches to explain the processes that drive the Southeast Asian monsoon.
10. a. What is latent heat?
    b. Explain why latent heat is important for the redistribution of energy.
11. a. What is meant by saturation vapor pressure?
    b. Draw a graph that plots saturation vapor pressure as a function of vapor pressure and temperature.
    c. Explain why the information shown in the plot is useful for understanding the relationships between atmospheric circulation and precipitation.
12. Describe three processes that produce uplift in the atmosphere and are important in causing precipitation.

Critical-Thinking Problems

1. Sketch a map of India. Locate the major mountain ranges. Show which areas you think would have high rainfall and which areas you think would have low rainfall, and explain why.
2. In this chapter, we discussed several types of desert, including polar deserts. The center of the Antarctic ice sheet, for example, receives very little precipitation each year and is regarded as a desert, although it does not match the customary idea of what a desert is. From this chapter we saw that
   - Precipitation generally decreases as temperature decreases (because saturation vapor pressure is much lower in cold air than in warm air).
   - Much midlatitude and high-latitude precipitation occurs in extratropical storm systems that move along the polar front zone.
   - The polar front zone is located in the latitudes where the temperature gradient is greatest. (This will be equator-

ward of the ice margin, where cold air draining off the ice cap moves equatorward to meet the warm air that is blowing poleward from the subtropical highs.)
   a. Put this information together in a systems diagram that has two feedback loops: one that links ice extent, albedo, temperature, and snowfall; and one that links ice extent, temperature, the location of the polar front zone, and snowfall.
   b. Are these feedback loops positive or negative?
   c. What implications do the feedback loops in part (a) have for the long-term growth of an ice sheet?
3. Indicate on two world maps the areas where you would expect to find relatively high rainfall and where you would expect to find relatively low rainfall, or even deserts, in (a) July and in (b) January. (c) Explain these distributions.

Further Reading

General

Henderson-Sellers, A., and P. J. Robinson. 1986. *Contemporary Climatology.* London: Longman Scientific and Technical; New York: John Wiley and Sons.

Advanced

Hartmann, D. L. 1994. *Global Physical Climatology.* San Diego: Academic Press.

Chapter 5
The Circulation of the Oceans

An idealized map of the deep-water flow (solid lines) and the returning surface circulation (dashed lines). This circulation has been described as a global conveyor belt. The deep water flows out of the North Atlantic, mixing with warmer water to the south. It is recooled by mixing with the cold surface water that subsides around Antarctica. Joining with the Antarctic Bottom Water, it flows around Antarctica in the Antarctic Circumpolar Current. Branches then flow back into the Atlantic as well as the Pacific and Indian Oceans, where upwelling brings the cold waters to the surface. The water eventually returns via the surface currents to the North Atlantic to complete the circulation. (From W.K. Hamblin and E.H. Christiansen: *Earth's Dynamic Systems,* 8/e, 1998. Reprinted by permission of Prentice Hall, Upper Saddle River, N.J.)

Key Questions

• Why do ocean currents form?

• How can the circulations of both the surface ocean and the deep ocean basins be driven by solar radiation and be closely linked, yet operate at very different time scales?

• What role does ocean circulation play in the global climate system?

Chapter Overview

We continue our discussion of Earth's circulatory subsystems by describing the processes that drive the circulation of the world's oceans. The movement and circulation of the oceans is tied very closely to the circulation of the atmosphere: Both are ultimately driven by the distribution of available solar energy, and their motions are linked by friction at the sea surface. In Chapter 4, we described an imbalance in the latitudinal distribution of energy that produces an equator-to-pole temperature gradient at the surface—the driving force for the pattern of Earth's surface wind. These wind patterns are responsible for the circulation of the ocean surface and the formation of the world's major ocean currents. As with the atmosphere, once the ocean starts to move, it comes under the influence of the Coriolis effect, which plays a significant role in the resulting circulation patterns. The oceans are vertically stratified, with more dense water at the bottoms of the major ocean basins and less-dense water near the surface. The density is controlled by the temperature and by the salt content (*salinity*) of the water. The deep-ocean water is separated from the surface layer of the ocean by a transition zone with sharply defined density, temperature, and salinity gradients. This deep-ocean water moves as a response to small changes in density that occur over wide areas, and the movement is largely independent of the surface-ocean circulation. Together, however, both types of ocean circulation contribute to the redistribution of available energy in the Earth system, albeit over very different time scales. And both play a major role in the distribution of nutrient supplies in the oceans.

The Ocean Circulation

Like the circulation of the atmosphere, the circulation of the oceans is ultimately driven by solar energy. Figure 5-1 is a systems diagram of ocean circulation. The distribution of solar energy over space and time results in the formation of the global wind belts. These roughly latitudinal wind patterns in turn produce the ocean currents that determine the circulation patterns of the upper ocean. The distribution of surface-ocean temperatures, which is partly a result of these circulation patterns, strongly influences the density structure of the ocean. It is this density structure that drives the circulation within the deep ocean. As with the atmosphere, the feedback loops in Figure 5-1 are negative—surface-temperature gradients drive the circulations, but the net effect is to move warmer water poleward and cooler water toward the tropics. Both the surface-ocean and deep-ocean circulations also play a vital role in climate. The surface circulation is one of the primary factors controlling climate variability on the order of years to decades, but even the deep circulation, which generally operates on the time scale of hundreds of years, is being implicated in short-term climate change. The two different circulation systems act together in controlling the distribution of the nutrients that are essential to marine life.

Winds and Surface Currents

Chapter 4 showed that the circulation in the troposphere is caused by atmospheric pressure gradients that result from vertical or horizontal temperature differences. We saw that from a global perspective, these temperature variations are caused by latitudinal differences in solar heating. But ocean surfaces are also heated by incoming

solar radiation. Do the oceans, therefore, circulate for the same reason as the atmosphere? The answer is no, because the solar heating of the ocean takes place at the *upper* surface of the fluid, whereas the solar heating of the atmosphere occurs largely at the *lower* surface of the fluid-near Earth's surface, where clouds and the greenhouse effect warm the atmosphere. Solar heating results in warmer water at the surface of most of the world's oceans. But the Sun's rays warm only the top few hundred meters of the ocean; 90% of the radiation that penetrates the surface is absorbed in the first 100 m. The warmer water is less dense than the cooler water below, which is not affected by the surface heating. This situation is inherently stable, so there is very little vertical movement. It is similar to the situation in the stratosphere. Recall from Chapter 3 that the atmosphere at this level is stable because the maximum solar heating occurs high in the stratosphere, the site of peak absorption of ultraviolet radiation by ozone. Where temperature increases with height, there is no density imbalance, and convection cannot take place. The fluid—water or air—remains well stratified. The true situation in the ocean is actually more complicated than this, as we shall see, because the density of seawater is also affected by its salt content. It remains true, however, that the ocean overturns very slowly.

At the same time, temperature changes in the ocean occur slowly. Remember from Chapter 4 that the oceans have a high heat capacity—it takes a considerable amount of heat to produce just small changes in temperature. Slight differences in incoming solar radiation from place to place thus have little impact on the surface temperature of the ocean, so lateral temperature and density differences are slight over large areas. Unlike the troposphere, therefore, the surface ocean does not circulate as a *direct* response to the surface heating. Instead, surface temperature plays a more indirect role: The surface temperature influences the atmospheric circulation, and the resulting pattern of global winds determines the circulation of the upper ocean.

The movement of the wind over the ocean causes friction at the surface. As a result of friction, the wind drags the ocean surface with it as it blows, thus setting up a pattern of surface-ocean *wind-drift* currents. The force of the wind acting on the surface is referred to as *wind stress*. The water movement is usually confined to the top 50 to 100 m of the ocean, although well-developed currents such as the Gulf Stream in the North Atlantic and the Kuroshio Current in the North Pacific may extend as much as 1.0 to 2.0 km below the surface. The Coriolis effect influences ocean currents just as it does winds, so the water is deflected to the right of the path of the wind in the Northern Hemisphere (and to the left of the wind's path in the Southern Hemisphere). Observations show that this deflection tends to be approximately 20 to 25° from the wind direction. Thus, as a first approximation of what the ocean circulation should look like, we can take the surface winds of Figure 4-14 and add a large

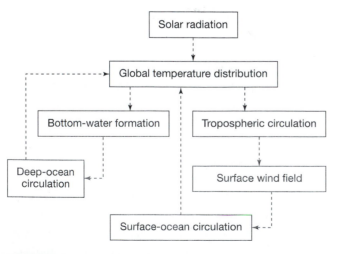

FIGURE 5-1

Systems diagram of the ocean circulation.

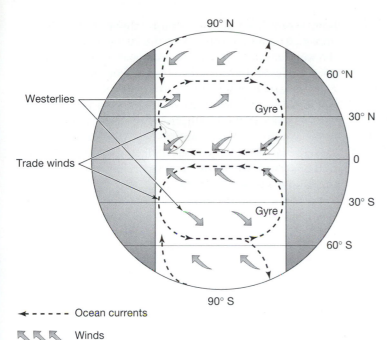

- ◄ - - - - Ocean currents

🍂🍂🍂 Winds

FIGURE 5-2

A simplified view of the surface-ocean circulation.

ocean bounded by continents on the east and west. In doing so we obtain the surface-ocean circulation pattern shown in Figure 5-2. The trade winds produce westward-flowing currents in the tropics. When these currents reach the western continental boundary, they are deflected northward and southward. They then come under the influence of the westerlies, which cause the currents to flow eastward in the midlatitudes. When these currents reach the eastern land mass, some water is deflected to the pole and some toward the equator. The waters that flow to-

ward the poles are replaced by equatorward flow along the western land mass. The waters that flow toward the equator come back under the influence of the trade winds and are blown westward again. The currents complete a large circular circulation pattern (called a **gyre**) in the subtropical oceans. The circulation of these gyres is clockwise in the Northern Hemisphere and counterclockwise in the Southern Hemisphere.

Compare the simplified model in Figure 5-2 with the observed distribution of ocean currents in Figure 5-3. The

FIGURE 5-3

The major surface-ocean currents. (From R.W. Christopherson, *Geosystems: An Introduction to Physical Geography,* 3/e, 1997. Reprinted by permission of Prentice Hall, Upper Saddle River, N.J.)

same general circulation features are apparent. Figure 5-3 shows counterclockwise gyres in the Southern Hemisphere and clockwise gyres in the Northern Hemisphere. The figure identifies the world's major ocean currents and designates them as "warm" or "cold," labels that we will explain later in this chapter. The pattern in the real world is more complicated, because the distribution of land and water is not as simple as it is in Figure 5-2. In the Southern Hemisphere, the westerlies result in an eastward-flowing current—the *West-Wind Drift*—that extends around the globe because, in the real world, there is very little land in the middle and higher latitudes to deflect the water back toward the west.

Convergence

There are further differences between Figure 5-2 and the actual circulation of the ocean surface that are not apparent from the diagram. To begin with, although our predicted gyres are present, the explanation of their occurrence is a little more complicated than we suggest above. In order to explain why they form, first we need to describe a few more processes that take place at the ocean surface. If we consider the circulation pattern shown in Figure 5-2, we might expect that water would pile up as it reached the coasts. In the Northern Hemisphere, therefore, we would expect to find water piling up in the northeast and southwest portions of the gyre. This does not happen; rather, water piles up (or converges) in the *middle* of the gyre. This *convergence* results from the combined effects of the wind-driven surface-ocean currents, Earth's rotation, and, ultimately, friction.

The Norwegian explorer Fidtjof Nansen made a key observation that led to a better understanding of convergence during an expedition across the Arctic Ocean in the 1890s. His ship, the *Fram,* was frozen into the ice at the beginning of winter and drifted with the ice for over a year. It had long been thought that the surface-ocean currents were produced by the winds. Among many observations made during the expedition, however, Nansen noted that the ice (and the ship) did not drift *with* the wind but at 20 to 40° to the right of the surface wind path.

Walfrid Ekman, a Swedish physicist, first made the connection between wind-driven currents and Earth's rotation and derived a mathematical explanation of Nansen's observations. Due to friction between wind and the water surface, some of the kinetic energy of the air is transferred to the top layer of the water. As that layer moves, it drags along the water just below it, which in turn drags along the water just below that, and so on. The water appears to move as many thin, independent layers, and kinetic energy is transferred down the water column. However, as the energy is transferred downward, friction causes some of the energy to be dissipated in the form of

heat, so each level moves more slowly than the level above. At some depth below the surface, the effects of the wind-induced movement disappear. However, as each layer moves, it is again subject to the Coriolis effect. Once a layer starts to move, the water is deflected to the right of the path of the layer above (or the wind path, for the surface layer) in the Northern Hemisphere and to the left in the Southern Hemisphere. The deeper below the surface, the farther each layer is deflected to the right or left of the layer above, producing a spiraling effect known as the **Ekman spiral** (Fig. 5-4a).

Ekman's theory predicts, under strong and persistent winds in the open ocean, (1) that the surface current will flow at 45° to the surface-wind path, (2) that the flow

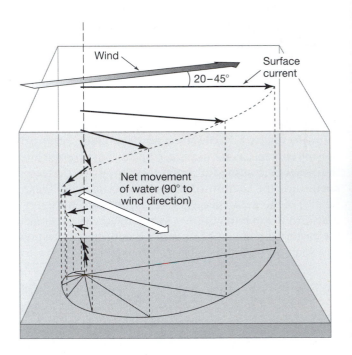

FIGURE 5-4a
The Ekman spiral.

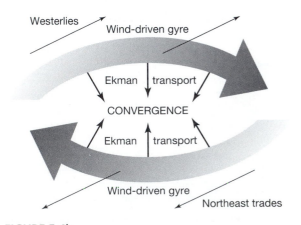

FIGURE 5-4b
Convergence in the center of a subtropical gyre due to Ekman transport.

will be reversed at approximately 100 m below the surface (that is, the current at 100 m will flow in a direction opposite to the surface current), and (3) that it will also be considerably reduced in speed. In practice, there are few observations of a well-developed Ekman spiral, but observations do show that the surface flow is to the right of the surface-wind path (although usually at an angle less than 45°). The observations also bear out a further prediction from the theory—that when the movements of all the individual layers of water in the spiral are added, the net direction of transport within the water column is at a right angle (90°) to the wind direction. This net movement of water is referred to as **Ekman transport.** In a clockwise gyre in the Northern Hemisphere, the effect of Ekman transport is to push water into the center of the gyre (Fig. 5-4b). Note that the counterclockwise gyres in the Southern Hemisphere will produce exactly the same result, because the Coriolis effect deflects the water to the left.

Divergence

Just as there are parts of the ocean where convergence occurs, there are also parts of the ocean where divergence occurs. In the equatorial Atlantic of the Northern Hemisphere, for example, the Northeast trades (which, remember, blow from the northeast toward the southwest) result in a westward-flowing surface current, the North Equatorial Current. The net Ekman transport is 90° to the right of this flow, which means that the bulk of water transport is directed due north. Conversely, the Southeast trades (in the Southern Hemisphere) produce the westward-flowing South Equatorial Current, and the net Ekman transport is to the left of the current flow, toward the south. Hence divergence occurs near the equator. Both divergence and convergence can also be found along coastlines where the Ekman transport may push water toward or away from the coast, depending on the direction of wind movement and the surface current. Important areas of divergence occur off the southwest coast of North America and the west coast of North Africa, due to the easterly winds and southward-moving currents in these regions. Similar areas of divergence are found off the west coasts of South America and southern Africa, where northward-moving currents have the same effect.

Upwelling and Downwelling

In areas of convergence, the surface water piles up, the sea surface rises, and the surface layer of water thickens (Fig. 5-5a). In areas of divergence, the surface water moves away, the sea surface drops, and the surface layer thins. Where convergence occurs, the accumulation of water causes it to sink in a process known as **downwelling.** Conversely, where divergence occurs at the surface, water must rise from below to replace it (Fig. 5-5b). Water at depth is cooler than water at the surface. The rising of cooler water to the surface to replace warm, divergent surface water is referred to as **upwelling.** As we will see

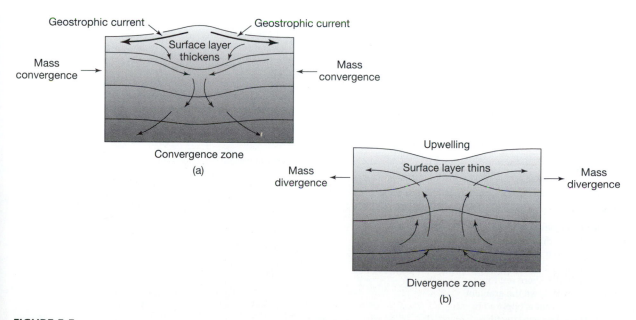

FIGURE 5-5

Schematic representation of zones of convergence and divergence. (a) Surface water accumulates in convergence zones, increasing the surface elevation (very exaggerated in the diagram) and thickening the surface layer. (b) The opposite happens in divergence zones—there is a decrease in surface elevation, and the surface layer thins.

later in this chapter, these deeper waters also tend to be rich in nutrients. Upwelling, therefore, brings these nutrient-rich waters to the surface.

Geostrophic Flow

Having explained Ekman transport and convergence, we now come to the real reason why the circulation in the subtropical oceans takes the form of very distinct oceanic gyres. Areas of convergence and divergence produce slight variations in sea-surface elevation across the ocean basins, so the sea surface actually slopes from one point to another. This difference in elevation is very slight—on the order of a few meters over 10^2 to 10^5 km (that is, slopes of 1 in 10^5 to 1 in 10^8). Yet these slight elevation gradients are sufficient to cause a downslope force on the water due to gravity. If we consider the subtropical ocean in the Northern Hemisphere, for example, we have already seen that the northeast trade winds produce a westward-flowing ocean current near the equator, whereas the prevailing westerly winds in the midlatitudes result in an eastward flowing current. The circulation is completed by the deflection of water along the coastlines at the ocean margins. Ekman transport in the surface layers causes convergence and the pile-up of water in the middle of the ocean (Figs. 5-4b and 5-6a).

The sea surface is only about 50 cm higher in the center of the gyre than at the edges, but gravity acting on this pile of water results in a force (referred to as the *pressure-gradient force*) that pushes outward, down the gradient, from the center. As the water flows, however, it is deflected by the Coriolis effect until that effect balances the pressure-gradient force acting down the slope. The result of the two forces acting in opposition is to cause a flow of water off to the side—to the right in the Northern Hemisphere and to the left in the Southern Hemisphere (Fig. 5-6b). Thus we end up with a circular flow of water around the gyre that is approximately parallel to the ocean slope (Fig. 5-6c). The resulting current is called a **geostrophic current,** which flows around the gyre clockwise in the Northern Hemisphere (and counterclockwise in the Southern Hemisphere), in the same direction as the original wind-driven flow. In practice, the flow is a little less than 90° to the slope, so in fact the water tends to spiral inward as it moves around the gyre and the convergence in the gyre results in downwelling.

Boundary Currents

Ocean gyres are a prominent feature of the surface circulation. Figure 5-3, however, gives the impression that the flow around these gyres is not symmetric. Indeed, the flow in the western part of the gyre is confined to a narrow path with a fast-flowing current (a *western boundary current*); that in the east is more diffuse, spread over a much larger area and with much-reduced current speeds (an *eastern boundary current*). Eastern boundary currents

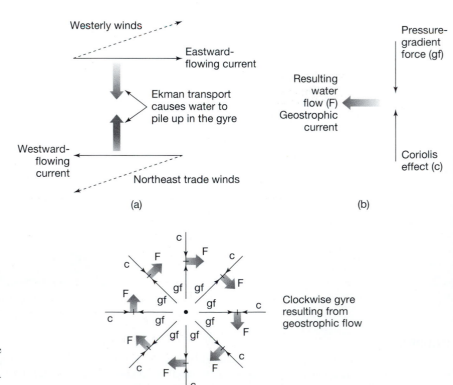

FIGURE 5-6

(a) The subtropical gyres are formed by geostrophic currents that occur when Ekman transport from the wind-driven currents causes water to pile up in the center of the gyre. (b) There is a force due to gravity, acting down the gradient of the surface slope, that is opposed by the Coriolis effect. The net effect is a flow of water at approximately 90° to the slope. (c) The result is a geostrophic current that flows approximately perpendicular to the slope of the sea surface around the gyre.

also tend to be divergent; the Ekman transport is away from the continent, thinning the surface layer along the coastline. The thinner surface layer and the divergent flow promote upwelling in these regions.

The most-studied western boundary current is the Gulf Stream in the western North Atlantic. The Gulf Stream begins as a narrow (50–75 kilometers wide), fast-flowing stream of warm water (20°C or higher) in the Florida Current, flowing northward between Bermuda and Cuba. The current can reach depths of more than a kilometer, with surface speeds between 3 and 10 km/hr. This current follows the coast northeastward to Cape Hatteras, North Carolina, where it continues across the North Atlantic as the Gulf Stream (see Fig. 5-3). Moving northeastward across the Atlantic, the Gulf Stream decreases in speed and the flow broadens into the North Atlantic Drift, which eventually flows into the Arctic Basin north of Norway. On reaching Europe, the North Atlantic Drift splits north and south; the southward-flowing component becomes an eastern boundary current—the Canary Current. The Canary Current moves much more slowly than the Gulf Stream, is shallower (reaching depths of only about 500 m), and is much broader—up to 1000 km across. As the water flows back toward the tropics, it comes under the influence of the northeast trade winds that push it to the west in the Northern Equatorial Current to complete the gyre. This elliptical flow of water essentially isolates the area in the center of the gyre—the Sargasso Sea. The Sargasso Sea is named for the extensive cover of seaweed often found in this area. Low current speeds and light, variable winds made this region difficult to traverse back in the days of sailing ships. The ancient mariners were also afraid of being entangled by the huge mats of seaweed that covered the surface.

Our discussion of wind-driven currents illustrates how wind stress, the Coriolis effect, and the pressure-gradient force serve to produce convergence, geostrophic flow, and gyres in the subtropical oceans. However, our discussion still does not account for the asymmetric nature of the gyres and the very different modes of flow in the eastern and western boundary currents. This pattern is caused by dynamic forces that operate when fluids tend to move in a rotary motion. How this happens is explained in the box "Vorticity."

The Circulation of the Surface Ocean

We can now add a few more details to the portions of the systems diagram in Figure 5-1 that relate to surface-ocean circulation. The large-scale surface-wind pattern produces gyres in the surface layer of the midlatitude oceans. In the Northern and Southern Hemispheres, the Coriolis effect and Ekman transport cause a net movement of water into the center of the gyres. The higher surface elevation in the center of the gyres causes a geostrophic current to flow around the gyres in the same direction as the wind-driven flow, thus reinforcing the surface circulation (Fig. 5-7). The shape of the gyre and the nature of the eastern and western boundary currents are then determined by the need to balance the forces that produce a tendency for water to rotate differently in different parts of the gyre.

The resulting circulation pattern has a significant impact on the redistribution of energy around the globe and on regional temperature (see Fig. 5-3). The equatorial currents are warmed by the large input of solar radiation at low latitudes. When these currents are deflected poleward, they carry warmer water to middle and high latitudes. However, these currents lose heat as they travel poleward. When they are deflected northward and southward by the eastern land mass, the water moving poleward is warmer than the polar ocean, whereas the water moving equatorward is colder than the tropical ocean. At the same time, the surface water that originates in the polar oceans and moves equatorward is also colder than the midlatitude oceans. Ocean currents thereby aid in the latitudinal redistribution of energy: They move warmer water toward the poles and cooler water toward the equator.

FIGURE 5-7

Systems diagram of the surface-ocean circulation.

Vorticity

The text explains why large-scale gyres form in the subtropical oceans. But to explain the asymmetric pattern of the gyres and the differences between eastern and western boundary currents, we need to introduce an additional concept—that of vorticity. **Vorticity** describes the tendency of a fluid to rotate. A tendency to rotate in a counterclockwise direction is referred to as positive vorticity, whereas a tendency for clockwise rotation is negative vorticity. We refer to vorticity as being the *tendency* to rotate (rather than the actual rotary motion of the fluid) because different forces could impose both a positive and a negative vorticity on the same mass of water at the same time, and the actual amount and direction of the rotation would then depend on the net effect of the different forces.

So what produces vorticity? Imagine yourself standing exactly on the North Pole. You would be spinning around a vertical axis at the rate of one rotation every 24 hours (actually, 23 hours 56 minutes). In other words, you would be experiencing a counterclockwise rotation about your vertical axis. At the equator you would experience no angular rotation, because you would be standing at exactly 90° to Earth's axis of rotation (Box Fig. 5-1). Anywhere between the equator and the pole you would experience some fraction of the pole's angular rotation. This angular rotation about a vertical axis at Earth's surface, brought about because of Earth's rotation, produces vorticity that is referred to as **planetary vorticity.** Mathematically, the planetary vorticity is identical to the Coriolis effect. Like the Coriolis effect, planetary vorticity acts in the opposite direction in the Southern Hemisphere.

A tendency for rotary motion can be created by a number of factors other than planetary rotation. Surface waters being driven by cyclonic or anticyclonic circulations in the atmosphere (that is, low-pressure or high-pressure systems) will produce positive or negative vorticity. Similarly, *current shear,* in which the speed of the current changes across the current, will also produce vorticity. Representing the speed of the current by the length of the arrows in Box Figure 5-2a, for example, would produce negative (clockwise) vorticity; the faster-moving water tends to curl in toward the slower part of the current. The current shown in Box Figure 5-2b, conversely, would produce positive (counterclockwise) vorticity. Current shear can be quite dramatic where friction with a coastal boundary slows the edge of currents that flow parallel to the coastline. The vorticity produced by the pattern of surface winds and by current shear is referred to as **relative vorticity.** The **absolute vorticity** experienced by a body of water is then simply the sum of the planetary and relative vorticities.

How does this help us explain the difference between western and eastern boundary currents? The answer lies in the fact that, like mass and energy, absolute vorticity is a conserved property. In Box Figure 5-3, we can see the vari-

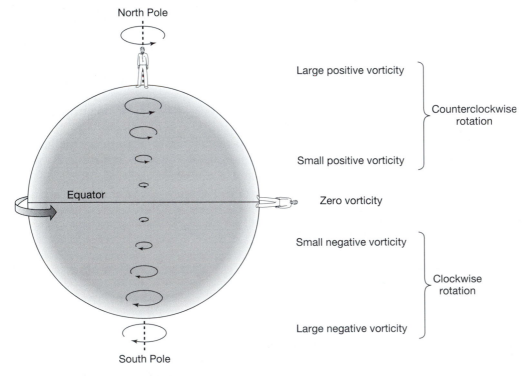

BOX FIGURE 5-1
Planetary vorticity, increasing from zero at the equator to a maximum at the poles. (After Open University, Ocean Circulation, Pegamon Press, 1989)

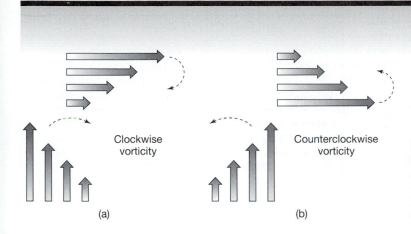

BOX FIGURE 5-2

Schematic diagram of current shear producing negative (clockwise) and positive (counterclockwise) relative vorticity. The lengths of the arrows represent relative speeds in the current.

ous factors that contribute to vorticity around a gyre. To begin with, the anticyclonic surface-wind pattern produces *negative relative vorticity* all around the gyre. At the eastern boundary current, this wind-supplied negative relative vorticity is balanced:

• The negative relative vorticity is balanced in part by a small increase in positive relative vorticity due to friction (current shear) along the coast. This factor is not large, because of the current only a small portion interacts with the coastline.

• Positive relative vorticity also arises from the southward flow of the water. The water is moving into an area of lower positive planetary vorticity; for vorticity to be conserved, there has to be a decrease in negative relative vorticity (or, looked at another way, a gain in positive relative vorticity).

The situation is very different on the western boundary:

• Current shear again produces positive relative vorticity.
• The change in latitude, however, leads to a gain in *negative relative vorticity*

(the water is moving into regions of large positive planetary vorticity and so must acquire negative relative vorticity). This gain in negative relative vorticity reinforces rather than offsets the negative relative vorticity imparted by the surface-wind pattern.

The only way balance can be achieved in the western boundary current is by increasing the friction along the boundary, which is achieved with a narrower, deeper, and faster-flowing current. Hence the asymmetric nature of the gyre.

BOX FIGURE 5-3

Contributions to the changes in relative vorticity around an asymmetric subtropical gyre. (After Open University, Ocean Circulation, Pegamon Press, 1989)

Consider the warm North Atlantic Drift as an example. This northwest-flowing current brings fairly warm waters to northern Europe (see Fig. 5-3). Due to the predominantly westerly flow of air in the midlatitude troposphere, most of northern Europe is warmed by the waters from the west and ultimately from the south. Contrast the seasonal variability and much milder conditions in southern Scandinavia with the more extreme conditions on the Labrador coast at the same latitude (see Fig. 4-1). Southern Scandinavia benefits from the warmth of the North Atlantic Drift. The air passing over Labrador, however, comes from the cold interior parts of Canada, and a cold offshore current (the Labrador Current) brings sea ice down to this area in winter. The result is lower temperatures and a much greater seasonal temperature range in that area. Notice that we now also have the explanation for the cold offshore currents, such as the Benguela and Humboldt Currents, that are responsible for the Namib and Atacama deserts on the west coasts of South Africa and South America, respectively.

The Circulation of the Deep Ocean

Salinity

In contrast to the surface-ocean circulation, which is driven by atmospheric winds, the deep-ocean circulation is driven by differences in water density. These differences are caused by variations in temperature and in **salinity,** or the salt content of a water mass. Salinity is measured in terms of the proportion of dissolved salt to pure water. The salinity of the ocean is a measure of the quantity of different salts—sodium chloride (NaCl) being the most common—dissolved in a given mass of seawater (g salt/*kg* seawater). Oceanographers express salinity in parts per thousand (%), or *per mil.* (Do not confuse per mil with parts per million.) A salinity of 1%, or 1 out of a possible 100 parts salt, is equivalent to 10%, or 10 out of a possible 1000 parts salt. The average salinity of the world's oceans is approximately 35%, but there is some variability from ocean to ocean.

The primary constituents of sea salt are the ions chlorine (Cl^-), sodium (Na^+), sulfate (SO_4^{2-}), magnesium (Mg^{2+}), calcium (Ca^{2+}), and potassium (K^+) (Table 5-1). Except for calcium, which shows some variability from place to place, these elements are found in nearly constant proportions around the globe. Many of the minor constituents—but not all—show a similar uniformity. Most of the variability occurs in those constituents that are utilized by marine organisms.

The salts contained in seawater are largely the result of the breakdown of crustal rocks, or *weathering*. Weathering occurs when rocks are altered by physical or chemical processes. When water flows over or through rocks, it removes soluble materials (ions). Rivers eventually

TABLE 5-1

Salt content of the Earth's oceans		
Salt ion	*Grams per kilogram (g/kg) of ion in seawater*	*Ion by weight (%)*
Chloride (Cl^-)	18.980	55.04
Sodium (Na^+)	10.556	30.61
Sulfate (SO_4^{2-})	2.649	7.68
Magnesium (Mg^{2+})	1.272	3.69
Calcium (Ca^{2+})	0.400	1.16
Potassium (K^+)	0.380	1.10
Bicarbonate (HCO_3^-)	0.140	0.41
Bromide (Br^-)	0.065	0.19
Boric acid (H_3BO_3)	0.026	0.07
Strontium (Sr^{2+})	0.013	0.04
Flouride (F^-)	0.001	0.00
Total	34.482	99.99

Source: PINET, P. R., *Oceanography* (St. Paul, MN: West Publishing Co., 1992).

carry the soluble ions to the ocean. In fact, it is estimated that rivers deliver between 2.5×10^{15} and 4×10^{15} g (about 4 billion tons) of dissolved salts to the oceans each year. The oceans are much saltier than the river water because, when ocean water evaporates, the salt is left behind (increasing the salt concentration) in the ocean. Some of that evaporated fresh water falls on the land and eventually runs back to the ocean. It weathers rocks along the way and thus delivers a new supply of salts that accumulate over time to increase the saltiness of the ocean.

If such great volumes of salts reach the oceans each year, are the oceans still getting saltier with time? The answer is no, because many processes also remove salts from seawater. These processes include the following:

1. Evaporation of seawater from shallow seas. The remaining salts are concentrated and precipitate from solution as **evaporite deposits,** such as halite (table salt, NaCl) and anhydrite ($CaSO_4$).

2. Biological processes. For example, some marine microorganisms remove the elements calcium or silicon from seawater to form their shells, some of which are eventually deposited in ocean sediments.

3. Chemical reactions between seawater and newly formed volcanic rocks on the sea floor.

4. The formation of sea spray. As small droplets of seawater become airborne, salts, especially sodium and chlorine, are removed when the spray is deposited on land. These salts are eventually returned to the oceans via rivers.

Overall, salts are removed from seawater at a rate that essentially equals the rate of input. In other words, the present salt content of the oceans does not represent the

The Salt Content of the Oceans and the Age of Earth

Following ideas first expressed by the British astronomer Sir Edmund Halley in 1715, the Irish scientist John Joly attempted to calculate the age of Earth on the basis of estimates of the salt content of the ocean and the rate of delivery of salts to the ocean. Two hundred years after Halley, Joly calculated Earth to be 80 to 89 million years old. However, we now know that Earth is approximately 4.6 billion years old—so where did Joly go wrong?

Joly assumed that the ocean had simply been accumulating all the salts delivered to it by rivers at a constant rate since Earth first formed. Joly neglected the various processes that remove salts from seawater (see the accompanying text). Repeating Joly's calculations but using current estimates of ocean volumes and salinities, we obtain the following:

- The total amount of salt in the oceans is approximately 5×10^{22} g.
- The rate at which rivers deliver salt is 4×10^{15} g/yr.

- Therefore, the "age" of Earth is $5 \times 10^{22}/4 \times 10^{15} = 13 \times 10^6$ yr.

Thirteen million years is somewhat less than Joly calculated with his knowledge of the world's river discharge, chemical composition, and ocean volume, and salt content. The "age" that we have calculated is, in fact, the average length of time salt remains in the ocean. As we shall see in Chapter 7, the length of time a substance remains in a given reservoir is called the *residence time*.

---

result of continuous accumulation but simply a balance between the rates of input and output of salts.

Variations in salinity are caused by regional differences in evaporation, precipitation, sea-ice formation and melt, and river runoff. Surface salinities increase where evaporation exceeds precipitation. We see this effect in such areas as the Mediterranean Sea, the Red Sea, and the Arabian Gulf. In contrast, the Gulf of Bothnia in the Baltic Sea, which also has little exchange of water with the open ocean but experiences much greater precipitation, has relatively low salinities. A similar effect is seen in the Chesapeake Bay on the Atlantic coast of the United States.

Thermohaline Circulation

Because deep-ocean circulation depends on temperature and salinity, this circulation is referred to as **thermohaline circulation** (*thermo* is Greek for "heat," and *haline* comes from the Greek *hals,* for "salt"). In discussing atmospheric circulation, we showed that large horizontal (in particular, latitudinal) changes in temperature and pressure lead to steep pressure gradients and a relatively rapid air circulation. In the deep oceans, horizontal changes in density are small, whereas vertical changes can be larger. But the densest water is at the bottom, so the structure is very stable. Consequently, the movement of water through the deep ocean is relatively slow. Although density-driven movements are much slower than the surface currents, they are no less important in shaping Earth's climate on time scales of hundreds to thousands of years.

The Vertical Structure of the Oceans.
The vertical structure of the deep oceans is determined by water density: the highest densities tend to occur in the deepest layers, while the lowest densities are typically found near the surface. Water density, in turn, is controlled by temperature and salinity: Usually, density increases as salinity increases or as temperature decreases; density decreases as salinity decreases

or as temperature increases. However, water is an unusual fluid in that the density of fresh water *increases* as temperature increases from 0°C to 4°C. (The temperature at which maximum density occurs varies as salinity changes.) After that point, water acts more like other fluids in that its density *decreases* as temperatures continue to increase.

The lower-density zone, which occurs in the top 60 to 100 m of the ocean, is called the *surface zone,* a layer that interacts with the overlying atmosphere. This interaction takes place through evaporation, precipitation, exchanges of kinetic energy (the effect of winds and friction), radiative exchanges (the absorption of solar radiation and the emission of long-wavelength radiation), and the exchange of heat. This zone is well mixed by wind action, and so the surface zone is often referred to as the **mixed layer.**

The transition zone between the surface zone and the deep is on the order of a kilometer in thickness and is characterized by a rapid increase in density with increasing water depth. The very sharp increase in density is called the **pycnocline;** the transition zone is referred to as the *pycnocline zone* (Fig. 5-8). In some regions this density gradient is dominated by salinity changes, and salinity rises rapidly with increasing depth. In this case, the salinity gradient is specifically referred to as the **halocline.** In most other regions, temperature changes dominate the density gradient, and temperature drops rapidly with increasing depth. There the transition is called the **thermocline.** In either case, the steep density gradient makes this layer very stable. This stability limits vertical movements and insulates the deep ocean from seasonal changes in temperature and salinity. The deep ocean below the pycnocline (typically 1–5 km depth) contains about 80% of the volume of all oceans. Deep-ocean water is further stratified, with the highest densities at the sea floor. The water column within the deep ocean, therefore, is also stable, and little vertical movement takes place. The movement that does occur is subhorizontal, along sloping layers of equal density.

Figures 5-9 and 5-10 show temperature and salinity profiles, respectively, from the Atlantic and Pacific ocean

FIGURE 5-8

Vertical profiles of temperature, salinity, and density in the ocean. These diagrams show the ocean to be divided into three layers: the surface zone, where there is little change in temperature, pressure, and density with depth; the pycnocline zone, where density increases rapidly (the pycnocline) and where there is an increase in salinity (the halocline) and a rapid decrease in temperature (the thermocline); and the deep ocean, where salinity generally increases slowly with depth, temperatures gradually decrease with depth, and there is little change in salinity.

FIGURE 5-9

Vertical distribution of temperature in the (a) Atlantic and (b) Pacific oceans. The thermocline separates deep water from the surface layer in the tropics, but deep water extends to the surface at high latitudes. Temperature values are given in degrees Celsius.

basins. We see that the simple vertical structure outlined above applies to most of the world's oceans, with the exception of those in high latitudes. The high-latitude seas are characterized by low temperatures and relatively low salinities at the surface, similar to the waters of the deep oceans. We will see next that there is a connection between these high-latitude surface waters and the deep oceans and that the formation of very dense surface water near the poles is, in fact, the primary driving force for the deep-ocean circulation.

FIGURE 5-10

Vertical distribution of salinity in the (a) Atlantic and (b) Pacific oceans. The salinity profiles are more complicated than the temperature distributions in Figure 5-9. In the deep ocean, salinity tends to increase in the deeper waters. There is a salinity maximum at the surface in the tropics, however, due to evaporation. When water evaporates, the salts are left behind. Where evaporation exceeds precipitation, there is a net loss of water from the surface layer, and the remaining water has a higher salt concentration.

Bottom-Water Formation. Deep-ocean circulation begins with the production of dense (cold and/or salty) water at high latitudes. This very dense water can be produced by several processes. For example, cooling and increased salinity may result from a large difference between evaporation and precipitation or from the formation of sea ice. Forming along the sea-ice margin in just a few regions in the polar oceans, **bottom water** constitutes the densest water produced in the oceans. Near the poles, the surface waters are cooled below freezing ($-1.9°C$ in some areas) by contact with the cold overlying atmosphere. (The freezing point is lower than that of pure water because of the presence of salt.) When that water freezes, it forms a layer of sea ice several meters thick that floats on the surface of the polar oceans. When the ocean surface freezes, most of the sea salt is excluded, because the salt does not fit into the crystal structure of the ice. As a result, the water just beneath the sea ice becomes saltier, and an underlayer of very cold, highly saline water forms. The combination of low temperatures and high salinity results in very dense water that sinks and flows down the slope of the basin and spreads toward the equator as the bottom layer of water

in the deep ocean basins. Figure 5-11 shows the primary components of the deep-ocean circulation system.

FIGURE 5-11

Systems diagram of the deep-ocean circulation.

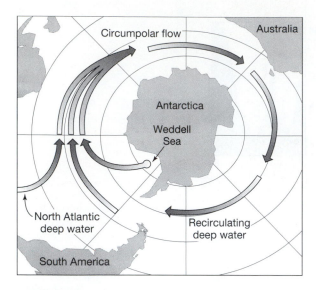

FIGURE 5-12

The Weddell Sea. Part of the Weddell Sea is occupied by an ice shelf (a mass of ice several hundred meters thick that flows from the West Antarctic ice cap). The remainder of the sea is covered by sea ice in winter. The ice forms near the coast and is pushed northward by persistent winds blowing off the ice cap. As the ice is pushed away from the coast, open water is exposed that freezes rapidly in the very cold temperatures. This ice, in turn, is pushed northward, allowing even more ice to form in a continuous process throughout the winter. This region is thus an ice-making factory. The result is the formation of very cold, highly saline water at the surface, which sinks to produce Antarctic Bottom Water. (After Broecker, W. S., and Peng, T.-S., *Tracers in the Sea,* New York: Eldigio Press, 1982, Figure 7-17.)

One of the major sites of bottom-water formation is the Weddell Sea off Antarctica (Fig. 5-12). The water formed there, called **Antarctic Bottom Water (AABW)** circles Antarctica and flows northward as the deepest layer in all three major ocean basins (Atlantic, Pacific, and Indian). Although few direct measurements of deep-ocean circulations have been made, reconstructions using observations of the distributions of temperature and salinity have identified Antarctic Bottom Water as far north as 45° N in the North Atlantic and 50° N (at the Aleutian Islands) in the North Pacific. Typical speeds for deep-ocean currents are only 0.03 to 0.06 km/hour, yet Antarctic Bottom Water has traveled more than 10,000 km from its formation site in the Weddell Sea, a trip that has taken some 250 years.

Similar masses of cold, dense water form in the Arctic Ocean—off the coast of Greenland—and flow south at depth into the western North Atlantic. These water masses comprise **North Atlantic Deep Water (NADW).** North Atlantic Deep Water, produced by winter cooling and evaporation, provides approximately half the input of deep water to the world's oceans, and the remainder comes from the Weddell Sea. The NADW that forms to the west of Greenland in the Labrador Sea sinks directly into the western Atlantic; the NADW that forms in the Norwegian Basin subsides and is dammed behind the *sills*

(undersea ridges) that connect Greenland to Iceland and Iceland to the British Isles. This water periodically flows over the sills and cascades into the deep basins of the North Atlantic.

North Atlantic Deep Water flows southward through the Atlantic Ocean and joins the Antarctic Circumpolar Current, which flows around Antarctica. There the NADW and the AABW combine and circle the continent. They then proceed to branch off into the Indian and Pacific Oceans (Fig. 5-13). Some of the water completes the circle, reentering the Atlantic or continuing around for another circuit.

The Thermohaline Conveyor Belt

We have now described how surface waters in the polar regions sink and spread at depth throughout the world's oceans. Ultimately these waters must return to the surface to complete the circulation (Fig. 5-14). Water is not compressible (we cannot cram more and more of it into the same space), so if water is sinking at high latitudes, it must be rising somewhere else. There must also be some flow of surface water into the high latitudes to replace the water that is subsiding and moving equatorward. This return flow is even more difficult to monitor than the flow of bottom water. It takes place very slowly through the pycnocline over the whole ocean and more rapidly through upwelling in currents along the eastern boundary of ocean basins and other regions of upwelling. Once the former deep water has reached the surface, the surface circulation that we discussed earlier returns the water to the polar regions. According to Wallace Broecker, a geochemist at Columbia University, this complete cycling of ocean water that is driven by thermohaline circulation can be likened to a giant conveyor belt.

The thermohaline conveyor belt is a significant feature of the Earth system in several respects: It plays a dominant role in the recycling of ocean nutrients, and it has a major impact on Earth's climate. Much of the life that exists in the oceans can be found in the near-surface layers, utilizing sunlight for photosynthesis—phytoplankton, for example—or living off the animals that feed on phytoplankton. These plants and animals use the nutrients in ocean water, so the surface layers become relatively depleted in nutrients. When these organisms die, they sink through the water column, decompose, and release the nutrients back into the water. The deeper ocean, therefore, is relatively rich in nutrients. The thermohaline circulation transports these nutrient-rich waters around the globe, returning the nutrients to the surface in areas of upwelling—primarily along the continental margins. Consequently, the concentrations of marine life are greatest in these upwelling regions.

Our description of the ocean circulation depicts a complex system of surface-wind-driven currents overlying a deep ocean with a relatively simple circulation

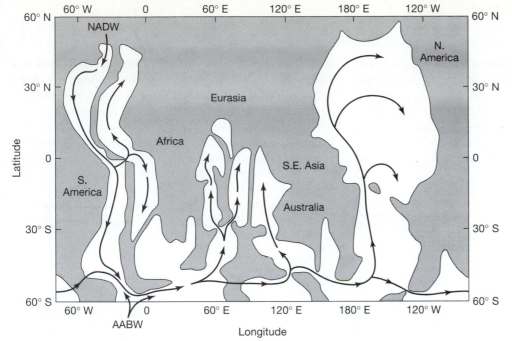

FIGURE 5-13

Flow pattern of the North Atlantic Deep Water and the Antarctic Bottom Water. This diagram represents the flow at a depth of 4000 m; the strange-looking continent/ocean configuration is what we would obtain if the oceans were drained to this depth. (After Broecker, W. S., and Peng, T.-S., *Tracers in the Sea,* New York: Eldigio Press, 1982, Fig. 1-12.)

Warm shallow current

Cold salty deep current

FIGURE 5-14

An idealized map of the deep-water flow (solid lines) and the returning surface circulation (dashed lines). This circulation has been described as a global conveyor belt. The deep water flows out of the North Atlantic, mixing with warmer water to the south. It is recooled by mixing with the cold surface water that subsides around Antarctica. Joining with the Antarctic Bottom Water, it flows around Antarctica in the Antarctic Circumpolar Current. Branches then flow back into the Atlantic as well as the Pacific and Indian Oceans, where upwelling brings the cold waters to the surface. The water eventually returns via the surface currents to the North Atlantic to complete the circulation. (From W.K. Hamblin and E.H. Christiansen: *Earth's Dynamic Systems,* 8/e, 1998. Reprinted by permission of Prentice Hall, Upper Saddle River, N.J.)

driven by bottom-water formation and surface divergence. In reality, the deep oceans are much more complex. Clearly distinguishable water masses can be identified at different depths and in different geographic locations where variations in temperature and salinity impart different characteristics to the water bodies. For example, high evaporation rates and low rainfall (to-

gether with little river runoff) produce relatively warm, highly saline water in the Mediterranean Sea. This water flows out of the Mediterranean at depth through the Straits of Gibraltar and is clearly recognizable as a plume of warm, saline water spreading out into the mid-Atlantic at a depth of about 1000 m (Fig. 5-15). We need not concern ourselves with these added complexities,

FIGURE 5-15

The distribution of (a) temperature (in degrees Celsius) and (b) salinity (per mil) at 1000 m depth in the North Atlantic, showing the spread of Mediterranean Sea water. (After Open University, Ocean Circulation, Pegamon Press, 1989)

but it is worth noting that our knowledge of the deep-ocean circulation is limited and that much scientific investigation remains to be done in order for us to fully understand what is going on. This understanding is particularly important because the oceans play such a significant role in global climate.

Ocean Circulation and Climate

As we discussed earlier, the ocean circulation has a strong influence on global temperatures. The transport of warm surface water toward the poles, to replace the bottom water that forms near the sea-ice margin, is one mechanism by which excess solar energy is transferred poleward. Figure 5-16 shows the Northern Hemisphere poleward heat transport in the atmosphere and ocean. The total heat transport data are derived by calculating the heat transfer necessary to balance the radiation budget at each latitude (see Fig. 4-4) and estimating how much of this transfer can be accomplished by the atmosphere. The ocean heat transport is then obtained by subtracting the atmospheric transport from the total. The estimates indicate that the ocean (1) provides almost as much poleward heat transport as does the atmosphere and (2) transports more heat than does the atmosphere at low latitudes, whereas the atmospheric transport dominates at middle to high latitudes.

At the same time, the oceans represent a vast reservoir of heat, absorbing heat from the atmosphere in some areas and releasing it in others. Because water heats up and cools down relatively slowly, pools of water that are cooler than normal, or warmer than normal will cool or warm the atmosphere on time periods of months to seasons or years—the time needed for the pools of water to heat up or cool down. On much longer time periods, however, the average effect of the oceans on the

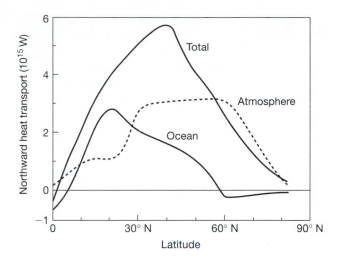

FIGURE 5-16

Poleward heat transport in the Northern Hemisphere. (After Open University, *Ocean Circulation*, Pegamon Press, 1989)

atmosphere is determined by the overall temperature of the oceans. Most of the water in the oceans lies in the deep oceans, and its temperature is largely determined by the process of bottom-water formation and by the transport of bottom water around the ocean basins. If the process of bottom-water formation changes, the ocean temperatures will change—and so will climate. Estimates of the rate of bottom-water formation under the present climate, together with measurements of ocean volume, indicate that it would take about 1000 years to recycle all of the deep water in the oceans. Hence, the thermohaline circulation can moderate climate over time periods of about 1000 years. However, brief interruptions or changes in the thermohaline circulation can also have rapid and large impacts on regional climates, as we will see in Chapter 12.

Chapter Summary

1. As with the atmosphere, the driving force for the oceanic circulation is the global distribution of energy. Unlike the atmosphere, however, the oceanic circulation is driven indirectly by temperature differences: The surface-ocean circulation is, in fact, driven by the circulation of the atmosphere.
   a. Due to friction, wind blowing over the ocean surface drags the surface waters along, producing ocean currents.
   b. The pattern of surface-ocean currents is modified by the Coriolis effect, a consequence of Earth's rotation, and by the distribution of land and oceans.
2. The thermohaline circulation of the deep oceans results from temperature and salinity variations, which control the density of ocean waters.
   a. Cold, saline water is formed in the North Atlantic and in the Weddell Sea off Antarctica.

   b. The combination of low temperatures and high salinities produces very dense water that sinks to the ocean floor and flows as bottom water throughout the world's oceans.
3. Bottom water eventually rises to the surface in zones of upwelling and returns in surface currents back to the high-latitude source regions, completing a vast oceanic conveyor belt.
   a. In combination with the atmospheric circulation, the net effect of these oceanic circulations is to redistribute thermal energy from low latitudes, where Earth is hot, toward the poles, where it is cold.
   b. The thermohaline conveyor belt and the associated zones of upwelling and downwelling play a significant role in climate and in the distribution of nutrients in the oceans.

Key Terms

absolute vorticity
Antarctic Bottom Water
bottom water
downwelling
Ekman spiral
Ekman transport
evaporite deposit

geostrophic current
gyre
halocline
mixed layer
North Atlantic Deep Water
planetary vorticity
pycnocline

relative vorticity
salinity
thermocline
thermohaline circulation
upwelling
vorticity

Review Questions

1. What effects does the surface-wind pattern have on the circulation of the oceans?
2. Why do ocean currents not move in exactly the same direction as the wind?
3. a. What is the Ekman spiral?
   b. Explain why Ekman transport occurs.
4. a. What is upwelling?
   b. Where does upwelling occur?
5. What is meant by a geostrophic current?
6. Explain the different characteristics of western and eastern boundary currents.
7. a. Where does the salt in the oceans originate?
   b. Are the oceans getting saltier and saltier with time? If not, then why not?
8. a. Define thermohaline circulation.
   b. What are the processes that drive the circulation of the deep oceans?
9. Explain the differences among the pycnocline, the halocline, and the thermocline.
10. a. What is bottom water?
    b. Where and how does bottom water form?
11. What is meant by the term *thermohaline conveyor belt*?
12. Explain what effects the ocean has on modifying the global temperature distribution.

Critical-Thinking Problems

1. Explain what is meant by Ekman transport and what role it plays in producing oceanic gyres in the surface waters of the subtropical oceans.
2. Use a rough map sketch to help explain the role that the oceans play in determining the climates of southern South America and southern Africa, poleward of 20° S.
3. If bottom-water production around Antarctica suddenly stopped, how long would it take before the effects could be found in the deep-ocean waters of the North Pacific basin? Explain why you would, or would not, expect these changes to affect the surface-ocean temperatures in the North Pacific at that time.

Further Reading

General

Open University. 1989. *Ocean Circulation.* Oxford: Pergamon Press.
Perry, A. H., and J. M. Walker, 1977. *The Ocean-Atmosphere System.* New York: Longman.

Advanced

Broecker, W. S., and T.-S. Peng. 1982. *Tracers in the Sea.* New York: Eldigio Press.

Chapter 6
Circulation of the Solid Earth: Plate Tectonics

Tomographic image of the mantle's (a) P-wave and (b) S-wave velocity variations underneath North America, along the transect line shown in the insert. (From W.K. Hamblin and E.H. Christiansen: *Earth's Dynamic Systems*, 8/e, 1998. Reprinted by permission of Prentice Hall, Upper Saddle River, N.J.)

Key Questions

- How do the physical and chemical characteristics of Earth change with depth toward its center?
- What is plate tectonics?
- What provides the energy that drives plate tectonics?
- How can we relate the surface features of Earth to plate tectonics?
- What is the rock cycle?
- How have the geographical positions of the continents changed through time as a result of plate tectonic activity?

Chapter Overview

Our understanding of solid Earth processes has taken a great leap forward since the mid-1960s with the development of the theory of *plate tectonics*. Yet there is still much to learn about the composition and dynamics of the solid Earth. Earthquakes and volcanoes demonstrate that the interior of Earth is not a static place. Rather, like the oceans and atmosphere clinging to its surface, the solid Earth is in motion. The energy that drives the circulation of the solid Earth derives not from the Sun but from Earth's interior. Convective currents in the interior are coupled to the rigid rocks that form the continents and sea floor, putting the continents in motion. Where continents collide, huge mountain belts form; where oceanic blocks collide with each other or with continents, deep-sea trenches and volcanoes form. These plate tectonic forces join with the surface processes of rock weathering and erosion to generate landscapes and recycle elements from solid Earth reservoirs into the soils, hydrosphere, and atmosphere, making those elements available to the biota once again. Thus, plate tectonic activity is critical to the maintenance of a biologically active planet.

We will study the circulation of the solid Earth in the way we studied atmospheric and oceanic circulations. First we will explore the anatomy of the planet, from its exterior to its greatest depths. The tools used to reveal Earth's internal structure give us clues about the temperature and compositional variations in the interior, but very little direct information exists. Then we will discuss how the heat flux from the interior produces the motions within the solid Earth and how these motions form and modify Earth's major surface features: mid-ocean ridges and mountains, deep-sea trenches, and transform faults. We will then trace the history of these motions over the past 3 billion years, during which time the continents have drifted apart and joined together again and again in a global tectonic cycle.

Introduction

The German meteorologist Alfred Wegener is largely credited with establishing the fundamental underpinnings of the theory that we now call **plate tectonics.** According to this theory, Earth's surface is divided into rigid plates of continent and ocean floor that move relative to each other through time. (*Tectonics* is the study of Earth's crust and the processes that deform it.) Wegener was fascinated by the near-perfect fit between the coastlines of Africa and South America and by the correspondence among the geological features, fossils, and evidence of glaciers on these two separate continents. Could all the continents once have been assembled into

a *supercontinent?* Wegener believed so, as had others before him. He, however, was the first to put together all the diverse evidence in support of that concept. He named the proposed landmass **Pangea,** meaning "all Earth." He proposed that Pangea began to break apart just after the beginning of the Mesozoic era, about 200 million years ago, and that the continents then slowly drifted into their current positions. This theory is called **continental drift.** Wegener's maps, produced in 1924, are remarkably similar to the best global paleogeographic reconstructions available today (Fig. 6-1).

Although it was welcomed by paleontologists because of its consistency with the terrestrial fossil record, Wegener's theory of continental drift was not well received by the geophysicists of his day. The British scientist Sir Harold Jeffreys presented calculations in 1925 demonstrating that the continents could not possibly plow through the rigid sea floor, as seemed to be required by the theory. Other scientists were unconvinced because Wegener could not propose a physical mechanism for driving the motion of the continents. Indeed, many of Wegener's own calculations and proposed mechanisms were found to be in error and untenable.

The acceptance of Wegener's theory of continental drift awaited a better understanding of the structure and operation of the solid Earth. That understanding has come about largely as the result of geophysical exploration, which has revealed the complex anatomy of the Earth's interior.

FIGURE 6-1

Wegener's reconstructions of the positions of the continents in the geological past. (From W.K. Hamblin and E.H. Christiansen: *Earth's Dynamic Systems,* 8/e, 1998. Reprinted by permission of Prentice Hall, Upper Saddle River, N.J.)

Upper Carboniferous (about 300 million years ago)

Eocene (about 50 million years ago)

Older Quaternary (about 1 million years ago)

Anatomy of Earth

Seismic Probing of Earth's Interior

For centuries geologists have been probing Earth's surface, cataloguing the variety of rocks exposed and studying the processes that have led to their formation. The study of liquid rock that has risen to the surface has given some insight into the chemical and *mineralogical* composition of the shallow interior, but virtually everything we know about Earth's deep interior has been derived by indirect methods. Paramount among these methods is the science of *seismology,* the study of earthquakes and related phenomena.

Earthquakes. An **earthquake** is the sudden release of stored energy as a result of rapid movement between two blocks of rock. This energy radiates away from the earthquake in the form of vibrations. The site of energy release, known as the *focus,* can be anywhere from very near the Earth's surface to as deep as 700 km below. This area, Earth's uppermost shell, is rigid; when it deforms, it does so *elastically.* This means that the material recovers its shape after the force that is tending to deform it is removed, unless it is deformed to the point of fracture and the original shape cannot be recovered. If there is differential movement on either side of the break, the fracture is called a *fault.* The *epicenter* of an earthquake is the position on the Earth's surface directly above the focus (Fig. 6-2).

Seismic Waves. Just as a person jumping into a swimming pool produces waves and ripples in the water, earthquakes create vibrations called **seismic waves** that ripple through Earth's interior, away from the earthquake's

focus, as a result of that deformation. Two types of seismic waves are generated: body waves and surface waves. Both types spread outward from the focus. As we might expect, **body waves** travel through Earth's interior, whereas **surface waves** travel only across the surface. Surface waves transmit earthquake energy along Earth's surface, where movement is unconstrained vertically. The motion is much like that of a water wave, easily seen by watching a cork bobbing up and down: Particles are displaced upward, backward, downward, and then forward in a circular motion. There is no net movement of the particles, but energy is transmitted away from the center. Body waves are categorized as either P waves or S waves on the basis of their mode of propagation through Earth. **P waves,** or primary waves, result from the compression of material in Earth's interior. The material is alternately compressed and, as the wave travels away, stretched. Thus, a P wave travels as a series of compressions and expansions in the overall direction of wave movement, similar to the way sound travels or to the response of a spring or a Slinky® (Fig. 6-3a). **S waves,** which

(a)

(b)

FIGURE 6-3

The movement of seismic body waves. (a) P waves alternately compress and expand materials, like the transfer of a compressive force imposed on a spring. (b) S waves move material from side to side, perpendicular to the overall direction of wave motion, just as a spring responds to repeated side-to-side shaking. (From J.P. Davidson, W.E. Reed and P.M. Davis: *Exploring Earth: An Introduction to Physical Geology,* 1997. Reprinted by permission of Prentice Hall, Upper Saddle River, N.J.)

FIGURE 6-2

An earthquake's focus is located at depth. The point on the surface that directly overlies the focus is the epicenter. (From J.P. Davidson, W.E. Reed and P.M. Davis: *Exploring Earth: An Introduction to Physical Geology,* 1997. Reprinted by permission of Prentice Hall, Upper Saddle River, N.J.)

The Principle of the Seismograph

In the simplest sense, seismographs consist of a recording drum, which is anchored to bedrock and situated below a suspended weight with a pen attached. The weight has sufficient inertia to remain vertical as the ground (and the recording drum) vibrate beneath it (Box Figure 6-1). During an earthquake, the recording drum vibrates back and forth, allowing the seismologist to determine the amount of displacement that occurs as the seismic wave passes through the ground below.

(a)

BOX FIGURE 6-2

(a) Aerial view of collapsed section of the Cypress viaduct on Interstate 880, Oakland, California, after the Loma Prieta earthquake. (b) Map view of the Loma Prieta earthquake surface waves as recorded in October 1989 at various sites, shown as triangles. The star marks the epicenter, solid lines are faults, and dashed lines are highways. All seismograph plots appear at the same scale. (From H.G. Wilshire, U.S. Geological Survey, U.S. Department of Interior.)

BOX FIGURE 6-1

A seismograph. (From S. Judson and S.M. Richardson: *Earth: An Introduction to Geologic Change,* 1995. Reprinted by permission of Prentice Hall, Upper Saddle River, N.J.)

The *amplitude,* or size, of seismic waves is directly related to the amount of energy released during an earthquake. Although seismologists now have more-sophisticated ways of assessing earthquake amplitude, the public has grown accustomed to the use of the *Richter scale.* On the Richter scale, the lower limit of detection is arbitrarily assigned a value of 0. For every tenfold increase in seismic-wave amplitude (and en-

are also called secondary or shear waves, are transmitted as displacements perpendicular to the overall direction of wave travel. An analogy is the movement of a spring swung from side to side (Fig. 6-3b). Although both solids and fluids can transmit P waves, only materials with structural rigidity (solids) can transmit S waves. This fact has proved to be important in the characterization of Earth's interior, allowing us to identify particular regions as fluids rather than solids, as we will see later.

Eventually the path of all body waves intersects the Earth's surface. There they can be detected and record-

ed by a *seismograph,* which is a sensitive instrument that detects slight vertical and horizontal displacements of Earth's surface (see the box "The Principle of the Seismograph"). The rate at which seismic body waves travel through Earth depends on the properties of the material in Earth's interior. If we know how much time it takes waves to travel from the earthquake source to a site where they are detected at the surface by a seismograph, and if we can determine the path a particular seismic wave has taken, then we can calculate an average wave speed along that path. For a single earthquake event, a seismograph near the wave source records waves that

(b)

ergy), the Richter scale increases by one unit. Thus, an earthquake with a magnitude of 5 on the Richter scale releases 10 times as much energy as one with a magnitude of 4.

Seismographs from around the world recorded the October 17, 1989, Loma Prieta earthquake with an epicenter 16 km northeast of Santa Cruz, California (Box Fig. 6-2b). The earthquake registered 7.1 on the Richter scale and created the first major

rupture along the San Andreas Fault since the famous 1906 San Francisco earthquake. Sixty-eight people died as a result of the Loma Prieta event, and nearly four thousand were injured. Thousands of homes and businesses were damaged or destroyed, with an estimated dollar loss on the order of $7 billion. One of the lasting images of the earthquake is the collapsed sections of the Cypress viaduct on Interstate 880 (Box Fig. 6-2a).

traveled very shallowly through Earth, whereas a seismograph far from the source of the earthquake receives seismic waves that may have traveled through Earth's center. Thus, by comparing several seismographic records from various places around the world for a particular event, we can construct a fairly detailed three-dimensional view of the paths along which seismic waves travel through Earth. This process is called *seismic tomography* (Fig. 6-4).

Generalized Structure of the Earth. The gross picture revealed by seismic imaging is of a layered Earth com-

prising a *crust,* a *mantle* (consisting of an upper mantle and a lower mantle), an *outer core,* and an *inner core* (Fig. 6-5), defined on the basis of contrasts in seismic wave velocities. The shallowest of these transitions is the crust/mantle boundary, first discovered by the Croatian seismologist Andrija Mohorovičić who was investigating shock wave traveling through the Earth from Zagreb (the former Yugoslavia) in the early 20th century. The boundary is now known as the *Mohorovičić discontinuity,* or the **Moho,** in his honor. This boundary is defined by a sharp increase in seismic wave velocities, from P-wave velocities in the crust of around 5 to

FIGURE 6-4

Tomographic image of the mantle's (a) P-wave and (b) S-wave velocity variations underneath North America, along the transect line shown in the insert. Blue colors indicate regions of fast seismic velocities, while reds indicate slow seismic velocities. The blue region cutting across the center of the diagram is the downgoing Farallon slab, which has been subducting under North America for 100 million years. (From W.K. Hamblin and E.H. Christiansen: *Earth's Dynamic Systems,* 8/e, 1998. Reprinted by permission of Prentice Hall, Upper Saddle River, N.J.) [See color section]

6 km/sec to uppermost mantle velocities of about 8 km/sec. Beneath the continental crust, the depth to the Moho ranges from as much as 75 km in young mountain belts to 20 km in areas undergoing extension and crustal thinning. Beneath the oceanic crust, the Moho is at a nearly constant depth of around 7 km below the ocean floor.

Below the Moho, the velocities of both P waves and S waves generally increase with depth through the mantle, although a *low-velocity zone* (or LVZ) exists at a

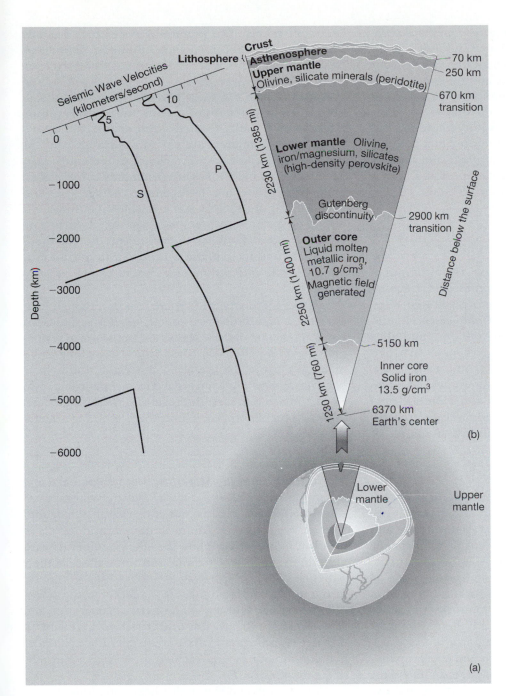

Seismic Wave Velocities (kilometers/second)

Crust
Lithosphere {
Asthenosphere
Upper mantle
Olivine, silicate minerals (peridotite)

70 km
250 km
670 km transition

Lower mantle Olivine, iron/magnesium, silicates (high-density perovskite)

Gutenberg discontinuity

2900 km transition

Outer core
Liquid molten metallic iron, 10.7 g/cm³
Magnetic field generated

5150 km

Inner core
Solid iron
13.5 g/cm³

6370 km
Earth's center

2230 km (1385 mi)

2250 km (1400 mi)

1230 km (760 mi)

Distance below the surface

Depth (km)

Lower mantle

Upper mantle

(a)

(b)

FIGURE 6-5

Internal structure of the Earth, showing the distribution of seismic wave speeds. The speed changes with depth, defining the boundaries between the crust, mantle, outer core, and inner core. (From R.W. Christopherson, *Geosystems: An Introduction to Physical Geography,* 3/e, 1997. Reprinted by permission of Prentice Hall, Upper Saddle River, N.J.)

depth of between 80 and 300 km (see Fig. 6-5). Velocities then increase again through the transition zone between the upper and lower mantles, but the increase is not smooth: It occurs in a stepwise fashion, which indicates some sort of incremental change in mantle properties. This change is thought to be related to a transformation of the minerals present to more compact, denser forms. Seismic velocities then increase more gradually with depth through the lower mantle.

The boundary between the lower mantle and the outer core, at a depth of 2900 km, is distinguished by a significant drop in P-wave velocities and by the disappearance of S waves. Because S waves travel only through solids, geophysicists infer that the outer core is a metallic fluid. At a depth of 5150 km, P-wave velocities increase markedly. Moreover, S waves travel through this region, indicating that the inner core is solid. Tremendous pressure at Earth's center is thought to be responsible for converting fluids to solids there.

This description of Earth's interior structure, although essentially correct, is greatly simplified. Seismic tomography reveals a more heterogeneous distribution of seismic wave velocities (see Fig. 6-4), perhaps related to the exchange of materials between Earth's interior and its surface. Clearly Earth's interior is a dynamic place. Plate tectonic activity has altered the chemical and thermal structure of the mantle and may have operated in response to influences conveyed through the mantle from the core. Let us explore these heterogeneities in more detail, utilizing information not only from seismology, but also from *petrology,* the study of the origin and evolution of the chemical and mineralogical compositions of rocks.

The Crust

Earth's uppermost layer, the crust, is not homogeneous; rather, it varies in both thickness and composition. The most pronounced differences are between the continental and oceanic crust. *Continental crust* underlies the continents, whereas *oceanic crust* underlies the ocean basins. As delineated by variations in Moho depth, continental crust is thicker than oceanic crust. It is also less dense and on average older. The two types of crust differ in chemical and mineralogical compositions as well. To understand these differences, we must first introduce a genetic classification of crustal rocks.

Igneous, Sedimentary, and Metamorphic Rocks. All rocks are composed of **minerals,** defined as naturally occurring inorganic solids of definite crystal structure and chemical composition. Geologists recognize three major types of rocks: *igneous, sedimentary,* and *metamorphic.* **Igneous rocks** form by the cooling and solidification of **magma,** which is molten, or liquid, rock. If the magma solidifies beneath Earth's surface, the rocks are called *intrusive* igneous rocks. **Granite** is a well-known intrusive rock. If the magma is carried to Earth's surface at a volcano and erupts, it is called *lava* and cools rapidly, forming *extrusive* igneous rocks. **Basalt** is an abundant type of extrusive igneous rock. Both intrusive and extrusive igneous rocks vary in composition, especially in the amount of the mineral quartz (SiO_2) they contain. *Felsic* igneous rocks (for example, granite or *rhyolite,* its extrusive analogue) are quartz-rich, light-colored, and less dense than *mafic* igneous rocks (such as basalt or *gabbro,* its intrusive analogue).

Rocks (of any type) that are exposed at Earth's surface tend to decompose, or *weather,* into finer materials called **sediments**—layers of unconsolidated mineral matter that is transported by water, wind, or gravity. As new sediments are deposited on top of existing sediments, the underlying sediments become compacted, expelling water from the pores between

sediment grains. The remaining pores may become filled with mineral cements precipitated from subsurface fluids. In this way, sediments ultimately form thick, cohesive, layered deposits known as **sedimentary rocks** at Earth's surface. This overall process of conversion from sediment to rock is called **lithification.** Sedimentary rocks formed from sand-sized grains (>63 μm in diameter) are called *sandstone,* whereas rocks composed of finer grains are called *mudstones.* Finely layered mudstones are commonly called *shales.* Sedimentary rocks may instead form chemically or biochemically. For example, some marine organisms form a shell or skeleton by precipitating calcium carbonate ($CaCO_3$) minerals. When the organisms die, the shells and skeletons accumulate on the sea floor and ultimately lithify into sedimentary rocks known as *limestones.*

Rocks (of any type) that are exposed to high temperatures, high pressures, chemically active fluids, or any combination of these agents are transformed in mineralogical and chemical compositions. As long as no melting is involved, the altered material is said to have been *metamorphosed* and is called a **metamorphic rock.** (When melting occurs, the resulting rock is igneous.) *Marble* is metamorphosed limestone, and *slate* is metamorphosed shale.

Major Rock-Forming Minerals. Both continental and oceanic crust are composed primarily of rocks made of **silicate minerals**—that is, minerals rich in silicon and oxygen. *Feldspars* are the most abundant minerals in the continental crust. They are silicate minerals with aluminum, sodium, calcium, and potassium in their structures. Quartz also is an abundant silicate mineral in the upper continental crust. Together with certain other minerals, feldspars and quartz form granite and *granodiorite* (a slightly less quartz-rich rock), the hard, erosion-resistant rocks that create the high peaks of many mountain ranges, including the Sierra Nevadas of California.

Magnesium and iron-rich silicate minerals such as *olivine* and *pyroxene* characterize the basalts of the oceanic crust. The Hawaiian Islands are composed of basaltic materials. The high relative abundance of these dense minerals in the oceanic crust accounts for the higher density of mafic oceanic crust in relation to felsic continental crust.

Sedimentary Cover. The basalts and granites/diorites of the oceanic and continental crust, respectively, are overlaid by sediments and sedimentary rocks. The source of these materials is easy to identify in the ocean: Sediments settle through the water column and accumulate on the sea floor as a sequence of relatively flat layers. Sedimentary rocks are also abundant in the continental crust, however. Some sedimentary rocks accumulated in basins on the continents themselves, but most were originally deposited as sediments on the sea floor and may have been deeply buried. Subsequent tectonic activity transported these sediments onto the continents. There they became exposed in mountain belts where the once-flat layering has typically become highly deformed through uplift. The oldest parts of the continents were once sedimentary rocks that have become significantly deformed and altered through many cycles of tectonic activity and metamorphism.

The Mantle

Beneath the crust is the mantle, which extends from the Moho to the top of the fluid outer core. The exact structure and composition of the mantle is a hotly debated topic among geologists, largely because the mantle is very difficult to observe. The samples of deeper mantle material that are available at the surface were brought up during rare geological events, such as the formation of kimberlites. *Kimberlites* are long, pipe-shaped igneous bodies that were emplaced after having passed rapidly from the upper mantle to the near-surface. They are remarkable in that they contain diamonds that formed under the high-pressure conditions of the mantle.

Most of what we know about the mantle is inferred from seismology. The velocity structure so determined indicates that the mantle is relatively uniform in composition and formed of silicate minerals. However, as depth increases so too do pressures and temperatures, causing changes in the structural and mineralogical composition of these silicates. These changes affect seismic wave velocities.

The recognition of the seismic low-velocity zone from depths of 80 to 300 km in the upper mantle proves to be an important link in plate tectonic theory. Most geologists accept that the low seismic wave velocities are the result of the presence of some molten rock at this depth. There need not be much; the data can be explained if only 1% or less of the rock is molten. Yet the small amount of melt present is critical, because it allows the crust and upper mantle to move relative to the underlying mantle—a basic tenet of plate tectonics.

A transition zone of rapidly increasing seismic wave velocities from depths of 370 to 650 km separates the upper and lower mantles. Geologists disagree on the reason for the transition zone. Many conclude from seismic evidence and theory that the transition zone is the consequence of mineralogical changes, whereas others conclude that differences in elemental compositional are the cause. We will return to this controversy later in our discussion of mantle convection.

The Core

The abundance of most elements in the crust and mantle can be explained by using meteorite compositions as a basis for comparison. Particular types of meteorites, the *chondrites,* are thought to be fragments of larger planetary bodies that now reside in the asteroid belt of the solar system. Their parent bodies formed at the beginning of the solar system from the solar nebula, and thus chondrites are thought to be representative of the average abundances of elements in the solar system, including silicon, sodium, magnesium, calcium, and oxygen, and in Earth as a whole. By comparison with chondritic meteorites, the mantle and crust are significantly depleted in iron. This deficiency is made up for in the core, which is believed to be dominated by iron, along with small amounts (about 6%) of nickel, and trace amounts of a light element, most likely oxygen or sulfur. The iron–nickel core is much denser than the overlying mantle—which, together with changes in seismic wave velocity, explains why seismic waves reflect off the core–mantle boundary.

Although the core is far removed from Earth's surface, it affects surface conditions because it is the source of Earth's magnetic field. Like a simple bar magnet, Earth has a magnetic field with north and south poles. Unlike a bar magnet, though, Earth's magnetic field is thought to be generated by a *magnetic dynamo* (Fig. 6-6). A **magnetic dynamo** is a mechanism that transforms energy from fluid motions (convection) into electrical currents that create a magnetic field. In a dynamo, the convecting fluid (liquid iron, in Earth's case) must be a conductor of electricity. The outer core convects because a temperature gradient is established across it by heat loss from the solid, inner core.

Why does the liquid outer core convect? As we discussed in Chapters 3 and 4, thermal convection occurs when a fluid is heated from below. Our example was the troposphere, which is heated by the warm Earth's surface. The ultimate source of this energy is sunlight. No sunlight makes it down to Earth's core, of course, so we must look for a different energy source there. One possibility is radioactive decay (which we shall cover later in this chapter), but the elements responsible for this heat source are preferentially concentrated in the crust and mantle rather than in the core.

FIGURE 6-6

Earth's magnetic field is like that of a bar magnet, except that this field is generated electromagnetically by convection in the outer core. Dashed arrows indicate lines of force of the magnetic field. (From W.K. Hamblin and E.H. Christiansen: *Earth's Dynamic Systems,* 8/e, 1998. Reprinted by permission of Prentice Hall, Upper Saddle River, N.J.)

The Theory of Plate Tectonics

Sea-floor Spreading

As we discussed in the introduction to this chapter, the theory of continental drift lay more or less dormant for several decades after the publication of Wegener's ideas, largely because of the lack of acceptance by the geophysicists. It is ironic that the resurgence of interest in Wegener's theory was the result of information obtained in the 1960s by geophysicists investigating the topographic and magnetic features of the sea floor. (*Topography* refers to the difference in elevations from the high point to the low point.) During and just after World War II, an intensive period of mapping took place that revealed intriguing details of the sea floor

(Fig. 6-7). This work gave the world the first evidence of chains of subsea volcanic mountains running down the centers of the ocean basins; we now call these features **mid-ocean ridges.** A *rift,* or narrow valley, runs down the center of such ridges. In the early 1960s, scientists proposed that these linear volcanic chains represent new sea floor that is extruded along the mid-ocean ridges. Once it forms, the new sea floor spreads to the sides of the ridges, generating the central rift, and is replaced at the *ridge axis*—the middle of the rift—by even younger new sea floor. This process was named **sea-floor spreading.** Having just crystallized from the magma, newly formed sea floor at sea-floor spreading centers is hot and expanded. As it spreads to either side of the plate boundary at the ridge axis, the material cools and contracts, and the sea floor subsides. This process occurs symmetrically across the axis of spreading and so creates symmetrical undersea mountain belts some 1000 to 4000 km wide that rise 2 to 3 km from the sea floor.

Additional clues to the origin of these features came from a better understanding of the magnetic characteristics of the sea floor. From studies of volcanic rocks extruded on land, scientists knew that Earth's magnetic **polarity,** the geographic orientation of the North and South Poles, has flipped numerous times in Earth's history. As the lava that formed these volcanic rocks cooled beyond a critical temperature (called the *Curie point*), the rocks became magnetized in the direction of Earth's magnetic field at the time of cooling. More-recent flows record switches in the magnetic field, with the North magnetic pole roughly coincident with the South geographic pole, and vice versa (Fig. 6-8).

Because the basaltic rocks of the sea floor were known to be of volcanic origin, in the late 1950s oceanographic expeditions were designed to map the magnetic character of the sea floor. As a result, a startling observation was made: The sea floor has a striped magnetic pattern, with the stripes running essentially parallel to the mid-ocean ridges. The stripes are alternating bands of polarity. Today's magnetic polarity is considered to be *normal,* and the opposite polarity is considered to be *reversed* (Fig. 6-8). Stripes on one side of a mid-ocean ridge were matched to others of similar width and polarity on the opposite side of the ridge (Fig. 6-9).

Geophysicists soon concluded that this pattern of magnetic stripes must be caused by sea-floor spreading. Accordingly, at the time of its formation, the new sea floor locks in the contemporaneous magnetic direction. It is then transported in opposite directions away from the ridge axis as new, molten rock is extruded from the volcano. Each reversal of the magnetic field produces a magnetic stripe on the sea floor. New material forming at a mid-ocean ridge is thereby

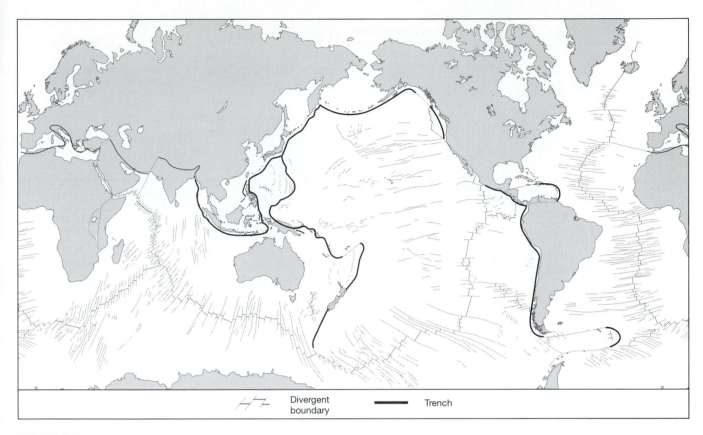

FIGURE 6-7

Distribution of oceanic trenches and mid-ocean ridges. (From W.K. Hamblin and E.H. Christiansen: *Earth's Dynamic Systems*, 8/e, 1998. Reprinted by permission of Prentice Hall, Upper Saddle River, N.J.)

differentiated from the older sea-floor material that was produced during a previous magnetic interval at the same ridge and has subsequently drifted away from the ridge axis.

Sea-floor spreading provided the solution to the problem plaguing geologists interested in continental drift since the days of Wegener: How could the continents drift through the rigid sea floor? The answer was that the con-

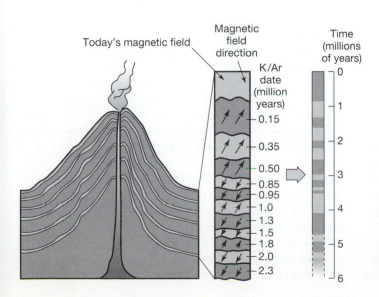

FIGURE 6-8

The geologic record of magnetic reversals from lava flows on land, calibrated to an absolute time scale (shown on the right). (From J.P. Davidson, W.E. Reed and P.M. Davis: *Exploring Earth: An Introduction to Physical Geology,* 1997. Reprinted by permission of Prentice Hall, Upper Saddle River, N.J.)

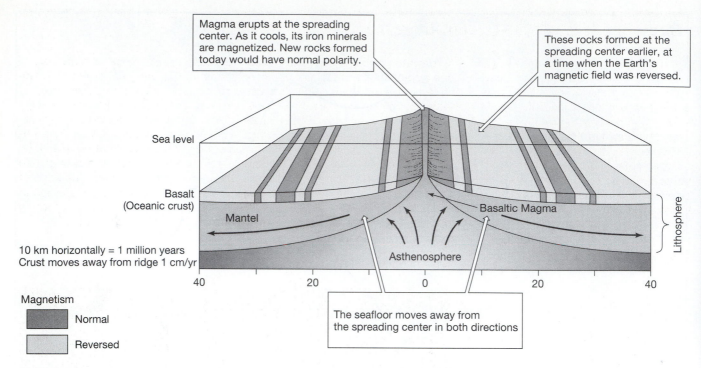

FIGURE 6-9

Magnetic stripes develop as new crust is added to the ocean floor at mid-ocean ridges and cools, becoming magnetized according to the magnetic field that exists at the time. As this material moves away from the axis, new sea floor is created, and its magnetization may be reversed if Earth's magnetization has reversed polarity in the intervening time. (From S. Judson and S.M. Richardson: *Earth: An Introduction to Geologic Change,* 1995. Reprinted by permission of Prentice Hall, Upper Saddle River, N.J.)

tinents do *not* plow through the sea floor. Rather, continents and segments of ocean floor are connected into plates that continuously move away from one another at mid-ocean ridges.

New Structural Categories— Lithosphere and Asthenosphere

As Wegener learned long ago, the conventional separation of the solid Earth into the core, the mantle, and the crust on the basis of seismic wave velocities is inadequate in view of plate tectonic theory. To explain the drift of continents, the mobile plates need to be distinguished from the lubricating layer below. To do so, the mantle and crust are best recategorized according to material strength (Fig. 6-10). The plates extend through the crust and into the uppermost mantle; we call this outermost sphere the **lithosphere.** The upper, crustal part of the lithosphere is *brittle,* that is, it fractures in response to stress. Below the lithosphere is the **asthenosphere,** a region of the upper mantle that reacts more like a fluid than a solid. The mantle lithosphere

and asthenosphere are *ductile*—they flow plastically, or deform easily, in response to stress. The top of the asthenosphere is coincident with the mantle's low-velocity zone. Recall that the preferred explanation for low seismic wave velocities there is the presence of a small amount of molten rock. The asthenosphere extends through the mantle's transition zone to a depth of around 700 km. Below this depth, the lower mantle is thought to be much less ductile because of the effects of very high pressure.

Plates and Plate Boundaries

According to the theory of plate tectonics, the lithosphere is divided into about 20 rigid plates (Fig. 6-11). The crustal portions of some plates are entirely oceanic, whereas other plates include both oceanic and continental crusts. "Oceanic lithosphere" describes a plate that is topped by oceanic crust; "continental lithosphere" refers to a portion of a plate topped by continental crust.

Deep-Sea Life at Mid-Ocean Ridge Vents

The axial portion of a mid-ocean ridge is marked by volcanic activity, earthquakes with a shallow focus, and the venting of hot fluids rich in dissolved metals and hydrogen sulfide. Seawater drawn into the ridge along its flanks flows through cracks in the oceanic crust and is expelled through vents in the axis. Along the way, the seawater is heated, and chemical reactions with basalt alter its composition: Magnesium is removed and sulfate is reduced to sulfide, and calcium and trace metals are added. The circulation of seawater through the mid-ocean ridge also alters the chemical composition of the oceans. While exiting through the vents, iron sulfide minerals precipitate from the altered seawater and give the plumes a black coloration. For this reason, the vents through which the fluids exit are called *black smokers* (Box Fig. 6-3a). They are also known as *hydrothermal vents,* because they release hot, watery solutions.

The fluids released by black smokers sustain a unique community of organisms. These organisms synthesize organic matter with the help of bacteria that carry out **chemosynthesis** rather than photosynthesis. In other words, these bacteria utilize energy from inorganic matter—energy released during chemical reactions between seawater and hydrogen sulfide. Chemosynthetic bacteria do not use the energy of sunlight. Feeding off these bacteria are unusual species of clams, crabs, and giant red and white tube worms (Box Fig. 6-3b).

(b)

(a)

BOX FIGURE 6-3

Abundant and bizarre life thriving under the harsh conditions of the deep-sea floor, in the vicinity of hydrothermal venting. (a) A black smoker chimney is shown spewing out sulfide-rich solutions that provide the energy source for this food chain. (From Dudley Foster, Woods Hole Oceanographic Institution.) (b) Tube worms, crabs, and other organisms can be seen. (From Richard A. Lutz and Michael J. Kennish, Reviews of Geophysics, vol. 31, p. 210, August 1993. Copyright © 1993 by the American Geophysical Union.) [See color section]

We now know that tectonic activity (such as earthquakes or volcanism) is concentrated at plate boundaries; there is little activity within a plate. This activity is the result of plate motion: The plates move relative to each other at average speeds of a few centimeters per year. As a result of friction between the plates, there are alternating periods of stasis (during which stresses build) and periods of movement (when they are released) both at the plate boundary and near the surface. (Seismic and satellite measurements indicate that at greater depths or farther from the plate boundary, the motions are more continuous.) After a period of stasis, pent-up energy is released suddenly as the plates jump past each other, causing earthquakes. As predicted, the distribution of earthquakes at Earth's surface follows plate boundaries quite closely (compare Figs. 6-11 and 6-12).

There are three types of plate boundaries (or margins): divergent, convergent, and transform (Fig. 6-13). At *divergent margins,* lithospheric plates are moving away from each other. At *convergent margins,* plates are moving toward each other. At *transform margins,* plates are slipping past each other. Each boundary type is represented differently at Earth's surface. In other words, each type of plate margin is reflected in distinctive surface features: mid-ocean ridges, deep-sea trenches, and transform faults, respectively.

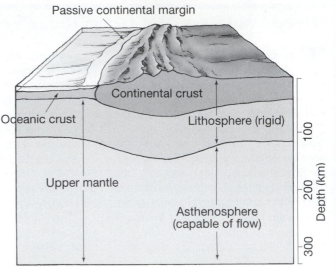

FIGURE 6-10

(a) Internal structure of the Earth, comparing the traditional classification by seismic wave velocities with the plate tectonic classification by material strength. (From R.W. Christopherson, *Geosystems: An Introduction to Physical Geography,* 3/e, 1997. Reprinted by permission of Prentice Hall, Upper Saddle River, N.J.) (b) A cross section of the upper mantle and crust showing the relative positions of the lithosphere (crust plus uppermost mantle) and asthenosphere. (From J.P. Davidson, W.E. Reed and P.M. Davis: *Exploring Earth: An Introduction to Physical Geology,* 1997. Reprinted by permission of Prentice Hall, Upper Saddle River, N.J.)

Divergent Margins

Divergent margins are regions where stresses are pulling apart the lithosphere. The mid-ocean ridges, described above, represent most of the divergent plate boundaries on Earth. Divergent boundaries that occur on land represent sites of continental fragmentation, or *rifting,* where the continental crust stretches. Tensional forces pull the continent apart. In the process, faulting occurs and flat-bottomed valleys called *rift valleys* form. During the breakup of Pangea some 200 million years ago, these types of plate boundaries were much more common than they are today. Today the rift valleys of East Africa are our best example of a continental divergent boundary in the making (Fig. 6-14).

Convergent Margins

Convergent margins are regions where two lithospheric plates are forced together. Although there has been some spirited controversy in the past, most geologists now are convinced that Earth is not increasing in size. But if new sea floor is produced at mid-ocean ridges, then what happens to the old sea floor? The mapping

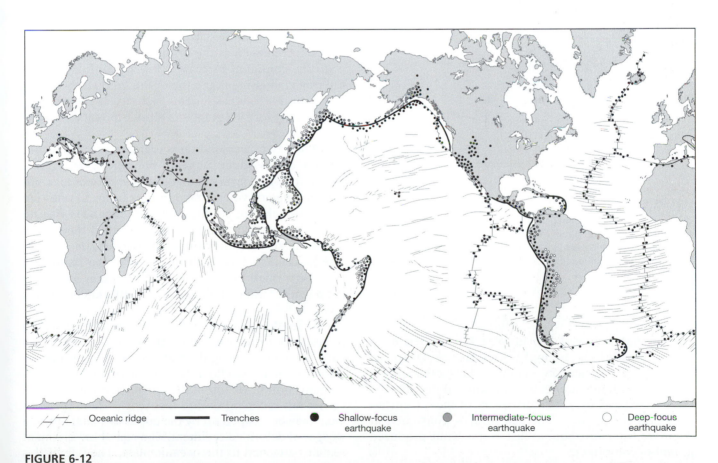

Legend for Figure 6-11:
— Ridge axis — Transform ⊥⊥ Subduction zone --- Zones of extension within continents ···· Uncertain plate boundary

FIGURE 6-11

The lithosphere is divided into rigid plates. (From S. Judson and S.M. Richardson: *Earth: An Introduction to Geologic Change*, 1995. Reprinted by permission of Prentice Hall, Upper Saddle River, N.J.)

Legend for Figure 6-12:
Oceanic ridge — Trenches ● Shallow-focus earthquake ◐ Intermediate-focus earthquake ○ Deep-focus earthquake

FIGURE 6-12

Distribution of earthquakes of shallow, intermediate, or deep focus. Deep-focus earthquakes occur only at subduction zones. (From W.K. Hamblin and E.H. Christiansen: *Earth's Dynamic Systems*, 8/e, 1998. Reprinted by permission of Prentice Hall, Upper Saddle River, N.J.) [See color section]

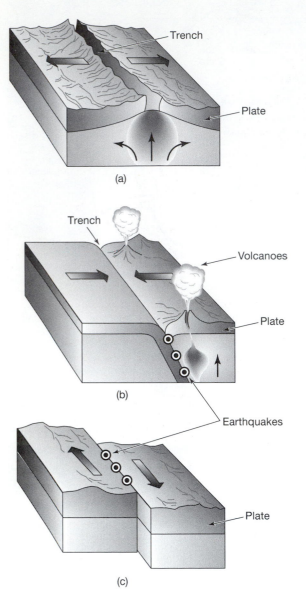

FIGURE 6-13
The three types of plate boundaries: (a) divergent; (b) convergent; and (c) transform fault. (From J.P. Davidson, W.E. Reed and P.M. Davis: *Exploring Earth: An Introduction to Physical Geology,* 1997. Reprinted by permission of Prentice Hall, Upper Saddle River, N.J.)

efforts that followed World War II revealed deep basins in addition to mid-ocean ridges. Deep-sea trenches are long, narrow, very deep basins that are especially common along the margins of the Pacific Ocean (see Fig. 6-7). The discovery of these trenches provided the answer: The sea floor is consumed at deep-sea trenches about as fast as it is being produced at mid-ocean ridges.

Deep-sea trenches form at two of the three types of convergent plate boundaries: those that involve two oceanic plates and those that involve an oceanic and a continental plate. A third type of convergent margin forms the world's highest mountains rather than the deepest trenches; these involve the collision of two continental plates. Let us explore each of these cases.

Oceanic/Continental and Oceanic/Oceanic Convergent Margins. Recall that the upper portion of a lithospheric plate (the crustal section) is brittle. When oceanic plates collide, the leading edge of one plate sinks entirely beneath the other. When the leading edge of one of the plates at a convergent margin is oceanic lithosphere and the leading edge of the other plate is continental lithosphere, the denser oceanic lithosphere sinks beneath the less-dense continental plate (Fig. 6-15a). The sinking of an oceanic plate at a convergent margin is called **subduction,** and the entire region is called a *subduction zone.* The downgoing plate, called a **slab,** subsides into the mantle. In so doing, the plate bends, creating deep, linear depressions at the surface—deep-sea trenches. These trenches are the deepest parts of the oceans. Friction between the downgoing plate and the overriding plate generates a substantial amount of seismic activity near the surface (in the upper 60–100 km); other forces generate earthquakes within the subducting slab at greater depths. The earthquake foci deepen as the distance from the trench toward the continent increases (Fig. 6-12). Inland from the trench, melting at the top of the subducting slab of lithosphere produces igneous activity at the surface. This activity forms a range of volcanic mountains called a *volcanic arc.*

When two oceanic plates collide at a convergent margin, one subducts beneath the other (Fig. 6-15b). Similar to what occurs in an oceanic–continental collision, a range of volcanic mountains forms to one side of the trench, but in this case the volcanoes rise up along the sea floor rather than on land. If they reach the ocean surface, they produce volcanic *islands arcs.* The Marianas Islands, off the coast of the Philippine Islands, formed in this way. The Marianas Trench is the deepest trench of all, more than 10.5 km below sea level.

As sea floor spreads from its place of origin at a mid-ocean ridge to its place of destruction at a subduction zone, sediment settling through the overlying water accumulates on the sea floor. Like a conveyor belt, the convergent motion of the plates in the subduction zone carries this sediment toward the trench. There the sediment may be scraped off by the opposing plate, forming wedges of deformed sediment. The rest of the sediment remains attached to the oceanic plate. The fate of this sediment is an area of active research. Part of it appears to become *underplated*—that is, attached to the base of

FIGURE 6-14

The rift valleys of East Africa. (From J.P. Davidson, W.E. Reed and P.M. Davis: *Exploring Earth: An Introduction to Physical Geology,* 1997. Reprinted by permission of Prentice Hall, Upper Saddle River, N.J.)

the overlying plate—whereas some of it is carried into the asthenosphere. In the asthenosphere, it undergoes *dehydration* (loss of water) and *decarbonation* (loss of carbon) to the surrounding mantle, as well as a host of mineralogical transformations. These reactions prove to be very important to the global recycling of elements such as carbon, because volcanoes that form in such sub-

duction zones derive carbon dioxide gas from this sedimentary source.

Continental/Continental Convergent Margins. When two continental plates meet at a convergent margin, the continents collide abruptly. Because continental

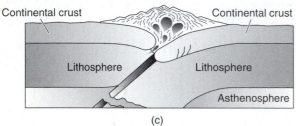

FIGURE 6-15

Three types of convergent plate boundaries: (a) oceanic–continental; (b) oceanic–oceanic; and (c) continental–continental. (From W.K. Hamblin and E.H. Christiansen: *Earth's Dynamic Systems,* 8/e, 1998. Reprinted by permission of Prentice Hall, Upper Saddle River, N.J.)

crust is too buoyant to be subducted, continental collision results in the separation of the crustal portion of the lithospheric plate from the mantle portion below. Subduction of the mantle portion of one plate may occur while the continental crust on both plates becomes compressed and crumpled. As a consequence, tall mountain belts and high plateaus form (Fig. 6-15c). Around 50 million years ago, India collided with Asia. The collision between these two segments of continental lithosphere led to massive deformation and uplift of the continents. The tall peaks of the Himalayas and the uniformly high Tibetan Plateau bear firm testament to the awesome energetics of this collision. Older mountain belts (such as the Appalachians) are the products of collisional tectonics that occurred hundreds of millions of years ago. Subsequent erosion has reduced what were once majestic mountains into the more modest ridges we observe today.

Transform Margins

When the relative motion along a plate boundary is parallel to the boundary, lithosphere is neither created (extruded) nor destroyed (subducted); the plates merely slip past one another at a fault. Faults that form boundary-parallel margins are known as **transform faults.** The San Andreas Fault of California, which marks a segment of the boundary between the North American and Pacific plates, is a transform fault (Fig. 6-16). The jagged shape of parts of the mid-ocean ridge system is caused by offsets between ridge segments created by transform faulting (Fig. 6-17).

FIGURE 6-16

Transform faults. Sea floor generated at the Juan de Fuca ridge moves southeastward, past the Pacific plate and beneath the North American plate, at the Mendocino transform fault. This fault connects a divergent boundary to a subduction zone. The San Andreas fault, another transform fault, forms the connection between two spreading centers: the Juan de Fuca ridge and a divergent zone in the Gulf of California. (From J.P. Davidson, W.E. Reed and P.M. Davis: *Exploring Earth: An Introduction to Physical Geology,* 1997. Reprinted by permission of Prentice Hall, Upper Saddle River, N.J.)

FIGURE 6-17

Ocean sea-floor topography inferred from high-resolution satellite measurements of anomalies in the gravitational field. Note transform faults extending across the Atlantic, especially near the equator and across the eastern Pacific. (From the National Oceanic and Atmospheric Administration [NOAA].)

Overview of Plate Interactions

Figure 6-18 provides an overview of the types of plate interactions and the surface features they generate. The production of new oceanic lithosphere at mid-ocean ridges is matched by the destruction of older oceanic lithosphere at subduction zones, which are manifested at the surface by deep-sea trenches. Part of the sedimentary layer riding atop the oceanic plate is incorporated into the overriding plate (oceanic or continental), and the rest is subducted into the mantle.

As Figure 6-18 shows, some lithospheric plates consist of oceanic lithosphere welded to continental lithosphere. The ocean–continent boundary is not a plate boundary at all. The North American plate, for example, is made up of the North American continent and oceanic lithosphere to the east. The North American plate continues beyond the continent–ocean lithospheric

boundary to the axis of the Mid-Atlantic Ridge (the mid-ocean ridge down the center of the Atlantic Ocean). In such situations, the ocean widens as the continents drift away from the mid-ocean ridge; the continental margin is referred to as passive. *Passive continental margins* consist of a broad, gently seaward-dipping continental shelf that gives way to the more steeply dipping continental slope. The continental lithosphere is welded to the oceanic lithosphere beneath the continental shelf. In other cases, such as along much of the Pacific Ocean, the ocean–continent boundary is a convergent margin and is therefore a site of subduction; such margins are said to be active. On *active continental margins,* including that off the Pacific coast of the northwestern United States, the continental shelf is narrow.

Where the relative motion of two adjacent plates is such that lithosphere is neither created nor destroyed,

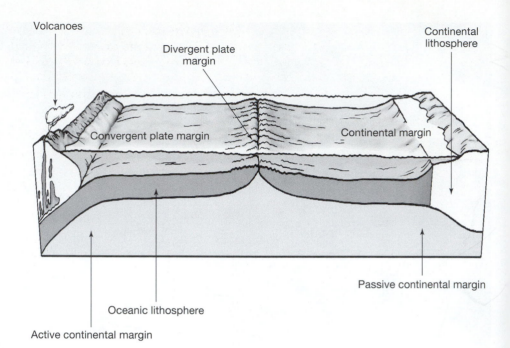

FIGURE 6-18

Schematic view of the relationships among the types of plate boundaries. (From J.P. Davidson, W.E. Reed and P.M. Davis: *Exploring Earth: An Introduction to Physical Geology,* 1997. Reprinted by permission of Prentice Hall, Upper Saddle River, N.J.)

the plates simply slide past each other along a transform fault. The oceanic ridge is offset in numerous places by transform faults. The relative motion across these faults is parallel to the plate boundary. Transform faults generate considerable amounts of earthquake activity. Around the globe, convergent and divergent margins are connected by transform faults.

The Physiology of the Solid Earth: What Drives Plate Tectonics?

Heat from the Deep

In the simplest sense, plate tectonics is the surface expression of the mechanism by which heat escapes from Earth's interior. Although there are spatial variations, temperatures generally increase through the mantle. This heat is transported to the surface, where it escapes to the atmosphere. (The average *geothermal heat flux,* or heat transported to the surface, is 0.1 W/m², which is trivially small compared to the net absorbed solar flux of about 240 W/m², the energy budget discussed in Chapter 3.) Some heat is brought to the surface conductively through the oceanic and continental lithosphere. But the conduction of heat through the solid Earth is very slow; hence, geothermal heat finds other ways to escape, namely, by mantle convection.

What is the origin of this heat in Earth's interior? There are two major sources: (1) radioactive decay and (2) residual heat from Earth's formation.

Radioactive Decay. Where does heat originate in Earth's interior? One of the major heat sources is the radioactive decay of potassium, uranium, and thorium. **Radioactive decay,** or *radioactivity,* is the spontaneous disintegration of an unstable nucleus of one element, creating a different, generally more stable nucleus of a different element. This process also releases particles (such as α particles, which are helium nuclei, or β particles, which are electrons) and radiation (such as gamma rays, which are high-energy photons). We learned in Chapter 1 that, for certain elements, some isotopes of that element are stable but others are unstable. It is these unstable isotopes that undergo radioactive decay. Besides converting one element to another, radioactive decay releases heat.

The rate of heat release by radioactive decay depends on the amount of radioactive material present and on the half-life of the radioactive isotope that is decaying. The **half-life** is the time it takes for half the initial quantity of radioactive isotope (or *radioisotope*) to decay (Fig. 6-19). For example, the half-life of the isotope ${}^{40}K$ (pronounced potassium-40) is 1.25 billion years. In other words, the amount of radioactivity of a ${}^{40}K$-containing mineral is halved every 1.25 billion years.

The half-lives of the important isotopes of potassium, uranium, and thorium are on the order of hundreds of millions to billions of years. Thus, the decay rates of these isotopes are quite low. However, the crust and mantle contain significant concentrations of these elements, so their radioactive decay generates a considerable amount of heat. Because radioactive decay leads to a continuous loss of radioactive materials from Earth's interior, the

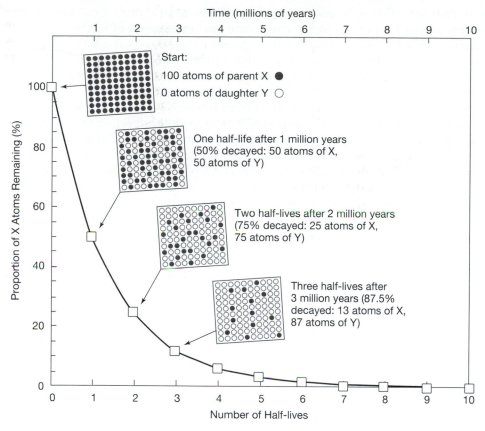

FIGURE 6-19

The graph of radioactive decay is exponential. In other words, half of the radioactive parent is left after one half-life. After a second half-life, a quarter of the parent is left, and so on. (From J.P. Davidson, W.E. Reed and P.M. Davis: *Exploring Earth: An Introduction to Physical Geology*, 1997. Reprinted by permission of Prentice Hall, Upper Saddle River, N.J.)

abundance of these materials must have been greater in the past than it is now. Similarly, the rate of heat production (and of heat loss) must have been much higher in the past. On the basis of the abundance of potassium, uranium, and thorium in the crust and mantle, we can calculate that the amount of radioactive heat production has decreased by about a factor of 5 since Earth formed 4.6 billion years ago.

Radiometric Dating of Geological Materials

uppose that we have a rock sample that contains both a radioisotope and its decay product, and we know exactly how much of that isotope existed when the material formed. If we also know the half-life of that isotope, then we can calculate the age of the sample. This method, called **radiometric dating,** has proved extremely useful in providing absolute dates for the geologic time scale and for other specific events in Earth's history. The accuracy of radiometric dat-

ing drops rapidly after eight or nine half-lives. For example, the radioactive isotope of carbon, ^{14}C or *radiocarbon,* has a half-life of just 5730 years. Thus, *radiocarbon dating* is accurately applied only to samples less than a few tens of thousand of years old. The long-lived radioisotopes of potassium and uranium, however, are useful for determining the ages of the oldest rocks on Earth, of lunar material, and of meteorites. Especially useful are the radioactive isotopes of uranium, ^{238}U

(half-life of 4.5 billion years) and ^{235}U (half-life of 0.713 billion years), which decay through a series of intermediate steps to stable lead isotopes ^{206}Pb and ^{207}Pb, respectively. The radioactivity of the uranium in these ancient rocks is not high enough to be measured directly, but we can use an instrument called a *mass spectrometer* to determine the relative amounts of the lead isotopes. From these ratios and the known half-lives, we can accurately date the rocks.

Other Heat Sources. Other sources of heat are residual; they are associated with heating events during Earth's formation. Earth (and the other planets) formed by the accretion of larger and larger clumps of matter into moon-sized objects called *planetesimals.* The planetesimals collided and merged, forming a large, primitive planet. A tremendous amount of energy was transferred to Earth during the accretion of the planet by collisions with planetesimals. The larger collisions probably caused widespread melting of Earth's upper mantle. (One such collision, with a Mars-sized object, is believed to have melted the entire surface of Earth and sent bits and pieces of material into orbit that eventually coalesced into the Moon.) The segregation of Earth into a less-dense mantle and crust and a denser core released gravitational energy in the form of heat. Convection of the outer core and mantle has been transferring this heat to Earth's surface ever since.

Convection in the Mantle

But how can a solid convect? Convection generally is thought of as a process that affects fluids. Yet solids need not be rigid; witness the flow of glaciers or the ductility of plastics. Rocks are ductile at the temperatures and pressures that occur within the mantle. When heated locally, these materials expand, become less dense, and rise buoyantly, although very slowly. Cooler, denser material sinks to replace the buoyant material. In this way, mantle rocks can convect. On rising to the base of the lithosphere, the material cools as heat is transferred conductively to the lithosphere. As the material continues to cool, it travels laterally. It cools so much that it eventually becomes denser than the underlying lithosphere and descends. Thus, the material sinks back into the asthenosphere. The cycle continues as the material is again heated and becomes buoyant.

The size of mantle convection cells is unknown. Smaller cells may be generated separately within the upper mantle and within the lower mantle, or the whole mantle below the lithosphere may be involved. The nature of the mantle transition zone is the distinguishing factor between the two convection mechanisms. If this zone marks a change in chemical composition, then convection cells do not cross it because if they did, the compositional distinctions would be lost. In this case, there are likely to be separate convective cells in the upper and lower mantle (Fig. 6-20a). Conversely, if the transition zone is the result of mineralogical rather than chemical changes, and if these changes take place quickly relative to the rate of convection, whole-mantle convection is possible (Fig. 6-20b).

The lithosphere is an integral part of the mantle convection system. In a sense, the lithosphere is the cool upper boundary of the convective cell. However, the subduction of cool oceanic lithosphere undoubt-edly perturbs the internal thermal structure of the upper mantle and the distribution of convective cells. Oceanic lithosphere is so dense that a slab of oceanic lithosphere at a subduction zone may sink to great depths within the mantle, become detached from the surface portion of the plate, and actually cool regions of the mantle from below. Such regions would become thermally stable, preventing convection locally. Lateral movements of the plates might also induce lateral movements of the underlying asthenosphere. And finally, the separation of the lithosphere at sea-floor spreading centers might drive the mantle to upwell. As the asthenosphere rises, it expands and melts, which further enhances the upwelling of more magma. Thus there are a number of lithosphere–asthenosphere interactions that affect the nature of convection in the mantle.

FIGURE 6-20

Mantle convection may (a) separate into upper and lower mantle convective cells or (b) involve the whole mantle. (From S. Judson and S.M. Richardson: *Earth: An Introduction to Geologic Change,* 1995. Reprinted by permission of Prentice Hall, Upper Saddle River, N.J.)

Forces Acting on Plates

The most important lithosphere–asthenosphere interaction is that which drives the motion of the plates. When plate tectonics was first proposed, *mantle drag,* or friction between the convecting asthenosphere and the overlying rigid lithosphere, was considered to be the cause of plate motions (force F_1 in Fig. 6-21). Now geologists recognize a number of other forces that act on plates. These forces include the push generated by the high topography of a mid-ocean ridge on the rest of the oceanic plate (F_2); the increasing density of the oceanic lithosphere as it cools, which pulls the opposite end of the plate into a subduction zone (F_3); the elastic resistance of the oceanic plate to being bent into a subduction zone (F_4); the tendency for the overriding plate to be drawn toward a subduction zone as the subducting slab bends (which otherwise would move the trench away from the overriding plate; F_5); friction between the subducting slab and the overlying lithosphere (F_6); and a tendency for the oceanic plate to sink as it cools and becomes denser (F_7). The overall motion of a given plate is the result of the balance of all these forces. Analysis of plate motions today argues for a predominant role played by the push of the ridges and the pull of the subduction zone, but in the past, mantle drag may have played a more important role in the rifting apart of supercontinents (see below).

Recycling of the Lithosphere: The Rock Cycle

All rocks of the lithosphere ultimately derive from igneous rocks. Igneous rocks of the oceanic lithosphere are born when extruded volcanically at mid-ocean ridges, and they die on average some 80 million years later when subducted and incorporated into the asthenosphere. (The oldest oceanic crust is about 200 million years old.) In contrast, the oldest-known continental rocks formed nearly 4 billion years ago. These record-setters reside in the old, previously active but now tectonically dormant regions of the continental interiors known as **cratons.** These very old continental blocks form the nucleus on which subsequent plate collisions have plastered on new material over the past 3 to 4 billion years, leading to the growth of the continents (Fig. 6-22).

Weathering and Erosion

Once formed, igneous rocks are subject to a variety of processes that can alter their chemical composition and weaken their structural integrity, typically leading to complete disintegration or *dissolution* (dissolving away) of the original rock. On the sea floor, oceanic lithosphere is altered as hydrothermal solutions circulate the oceanic crust (see above), but these processes tend not to destroy the rocks. In contrast, igneous rocks that are exposed on land are subject to a variety of physical, biological, and chemical forces that tend to degrade the rock. These are referred to as *weathering* processes. Weathering transforms solid rock into small particles (sediments) and dissolved material.

A number of processes contribute to physical weathering. Rocks expand and fracture as the weight of overlying material is removed through erosion. In temperate latitudes, water seeps into these fractures in the spring, summer, and fall. It then expands when it freezes during the winter, cracking the rock. Rocks exposed at the sur-

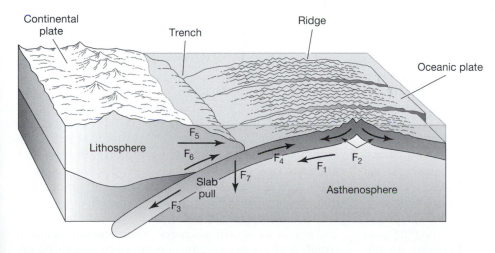

FIGURE 6-21

The various forces (labeled F) acting on plates at their leading and trailing edges. The motion of the plates responds to the sum of these forces. See the text for a discussion of the origin of each force. (From W.K. Hamblin and E.H. Christiansen: *Earth's Dynamic Systems*, 8/e, 1998. Reprinted by permission of Prentice Hall, Upper Saddle River, N.J.)

Age of Continental Crust
(billions of years)

	> 2.5
	1.9 – 1.8
	1.8 – 1.7
	1.7 – 1.6
	1.2 – 1.0

FIGURE 6-22

The ages of the components of the North American continent reveal that the continent has grown by the amalgamation of very old cratons, followed by accretion of younger material onto the periphery of the craton during plate collisions. (From J.P. Davidson, W.E. Reed and P.M. Davis: *Exploring Earth: An Introduction to Physical Geology,* 1997. Reprinted by permission of Prentice Hall, Upper Saddle River, N.J.)

and deltas and deeper basins in the ocean. In the process, landscapes are created: Valleys form as erodable material is removed, leaving peaks and ridges of material that is resistant to erosive forces.

Sediment Accumulation

The accumulation of sediments depends on two factors: the rate of supply of sediment and the amount of space available to accumulate the sediment. The great depth of water in deep marine basins allows for the thick accumulation of sediments. As the sediments accumulate, however, the sea floor rises toward sea level, and the amount of available space diminishes. Sedimentary deposits also form in shallow-water settings, though, such as the margins of the Gulf of Mexico, where subsidence of the sea floor (resulting from tectonic forces that stretch and thin the lithosphere, or cooling and contraction) allows for continued accumulation of sediment.

As sediments continue to accumulate in these basins, the material is compacted by the increasing weight of overlying sediments. That weight can become so great that fluids trapped between the sediment grains are expelled. Eventually, as the burial process continues, sediments may become buried to a depth of several kilometers below the sea floor. At these depths, temperatures can exceed 200°C, and the pressures can be hundreds of times the atmospheric pressure. In addition, the fluids that circulate through these buried sediments are quite distinct chemically from surface waters. As a result of these environmental changes, sediments undergo further compaction, and the small voids that remain between sediment grains become filled with mineral cements precipitated from the subsurface fluids. In other words, the sediments lithify and become sedimentary rocks.

Uplift

If the continents were at sea level, there would be virtually no driving force for weathering and erosion, and thus no rock cycle. However, continental collisions crumple the crust, producing mountains and high plateaus, and oceanic collisions produce volcanic islands. These plate collisions generate the topography that is then destroyed by erosion. Part of the process involves the recycling of sedimentary rocks before metamorphism or melting occurs: Sedimentary deposits situated along active continental margins can be entrained into the convergent plate motions and uplifted onto the continents. As soon as the sedimentary rocks become exposed at the surface, weathering and erosion commence.

face in high latitudes or at high altitudes are ground up as glaciers advance and retreat. Finally, there are biophysical mechanisms of weathering, including the action of plant roots that penetrate along these fractures, wedging the rock apart.

Chemical weathering results from the tendency for minerals to dissolve when exposed to rainwater and acidic soil waters generated by bacteria, fungi, or plant root discharges. The products of these chemical reactions include dissolved materials and relatively insoluble clays that form in the soils.

The transport of the products of weathering to basins where sediment accumulates is called **erosion.** In this process, crustal materials, decomposed and loosened by weathering, and the clays that form in the soils are transported by winds, landslides, and streams to sites of deposition. These sites include lakes and flood plains on land

Metamorphism and Melting

If instead sedimentary (or igneous) rocks are subjected to high temperatures and pressures in Earth's interior, they metamorphose. If they ultimately melt, they form magma that can itself ascend and become igneous rock. Metamorphism occurs deep in sedimentary basins along passive margins, in the deformed regions of active margins, and in other regions of the crust where igneous activity has injected hot igneous rock into what was cooler crustal rock (igneous or sedimentary). The generation of magma from crustal rocks typically occurs only where those rocks have been carried deep within Earth's interior—for example, in subduction zones.

The Rock Cycle

There are a number of alternative pathways in this process, but overall the complete regeneration of rock is called the **rock cycle** (Fig. 6-23). The rock cycle is a consequence of plate tectonics. One complete cycle takes about 100 million years. However, the average lifetime of continental lithosphere as a whole is actually much longer (a few hundred million years), because the interiors of the continents are well insulated from the tectonic activity that occurs along their margins.

It is important to recognize that the rock cycle is not completely closed: New crustal material is produced through the emplacement of magmas derived from the mantle, and older crustal materials are taken back to the mantle at subduction zones. This slow exchange between the mantle and crust replenishes the crust, on average, every 2 to 3 billion years. In other words, a large proportion of the geologic record of the early Earth's crust has not only been through the rock cycle many times, but has been ingested (perhaps once and for all) into Earth's interior. Nevertheless, averages can be deceptive: The interior parts of the continents (the cratons) can be billions of years old and represent crustal material that has never been recycled, whereas the continental margins are recycled on a time scale much shorter than a billion years.

Plate Tectonics through Earth History

Evolution of the Driving Force

Earth has been losing heat throughout its 4.6-billion-year history. Although early in this history other mechanisms may have dominated, for at least the past 3 billion years heat loss has occurred by mantle convection. The rate of heat loss on the early Earth was many times the present value, presumably fueling higher relative rates of sea-floor production and subduction. But the style of subduction, and the balance of forces acting on the plates, may have been somewhat different on the early Earth than they are today. Presuming that plate velocities of a few centimeters per year have prevailed over the past 3 billion years or so, the continents have moved great distances during that time. Geologists have used a variety of types of evidence to reconstruct this drift of the continents throughout Earth's history, largely confirming the notion of Pangea proposed by Wegener decades earlier.

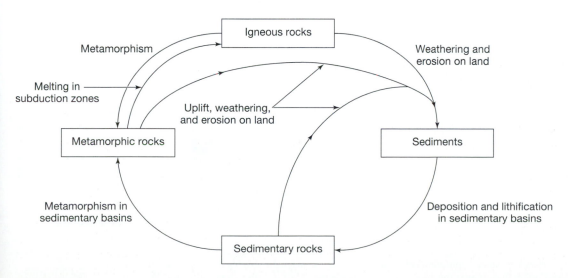

FIGURE 6-23

The rock cycle.

Continental Drift and Paleogeographic Reconstructions

The symmetrical magnetic stripes on the sea floor increase in age to either side of the Mid-Atlantic Ridge, recording the opening of the Atlantic Ocean (Fig. 6-24). In a sense, we can reverse time by rolling the sea floor back into the Mid-Atlantic Ridge, bringing together once again stripes of equal age at the ridge axis. In this technique, the Atlantic Ocean closes first in the Southern Hemisphere and then in the Northern Hemisphere, as South America slips into place alongside Africa. The close fit between the two continents, especially at the edges of their continental shelves, was part of what convinced Wegener in the early 1900s that the continents have drifted apart over geological time.

Sea-floor magnetic stripes provide the best tool for *paleogeography,* the reconstruction of the positions of the continents in the past. However, subduction has destroyed much of the sea-floor record of the past 200 million years and all of the record older than that. So paleogeographers turn to other sorts of evidence to determine ancient continental positions.

Sedimentary rocks prove very useful in this regard. We know today that glacial deposits generally form at high latitudes (poleward of about 45°), so we assume that glacial sediments indicate high *paleolatitudes*—that is, the latitudes at which the rocks formed. Similarly, because coral reefs are located in the tropics today, we assume that reef limestones indicate tropical paleolatitudes. Salt deposits indicate subtropical paleolatitudes because they form preferentially in arid regions underlying the descending branches of the tropical Hadley cells. In addition, similar fossils on two continents indicate that the continents were in close proximity, or joined, at the time the organisms lived, allowing their migration.

m.y. = millions of years

0–2 m.y.	58–66 m.y.
2–5 m.y.	66–84 m.y.
5–24 m.y.	84–117 m.y.
24–37 m.y.	117–144 m.y.
37–58 m.y.	144–208 m.y.

FIGURE 6-24

The age of the ocean floor is shown as bands of different color on the basis of the magnetic striping developed during sea-floor spreading. The youngest ocean floor is near the mid-ocean ridge, while the oldest is furthest away. (From R.W. Christopherson, *Geosystems: An Introduction to Physical Geography,* 3/e, 1997. Reprinted by permission of Prentice Hall, Upper Saddle River, N.J.) [See color section]

The least ambiguous paleolatitudinal indicator comes from magnetism in rocks. By measuring the angle of the preserved magnetic field with respect to the sedimentary layering of the rocks (which is presumed to have originally been horizontal), geologists are able to estimate the angle of the magnetic field with respect to the horizontal. This angle gives the latitude at which the sedimentary deposit formed (Fig. 6-25). Rocks in which the magnetic field is nearly parallel to the bedding plane must have formed near the equator; rocks in which the magnetic field is perpendicular to the bedding plane must have formed near the poles. These inferences, of course, are based on the assumption that Earth's magnetic field has always had two poles (North and South) and that the magnetic poles have always been approximately aligned with the geographic poles.

Unfortunately, most tools paleogeographers use (paleomagnetism, rock types) constrain only the latitude, not the longitude, of the continents. Longitudinal positioning is much more difficult. Similarity of fossil assemblages on adjacent continents is taken to indicate proximity, whereas large differences between the types of fossils found on two continents imply that the continents were separated by wide oceans that prevented the dispersal of organisms. However, features other than oceans, such as mountains and deserts, can also prevent dispersal. Thus, there are large uncertainties in the longitudinal positions of the continents prior to 200 million years ago.

The sequence of maps shown in Figure 6-26 (pp. 124–125) shows how the continents are thought to have drifted over the past 500 million years. In the Cambrian period (from 540 to 500 million years ago), the continents were widely dispersed along the equator. Over the next 300 million years, the continents drifted together and collided. The collisions created large, Himalayan-style mountain ranges, including the Appalachians of the eastern United States. By 300 million years ago, the continents were assembled into the gigantic supercontinent Pangea, centered on the equator. Pangea eventually began to disassemble about 200 million years ago. The Atlantic Ocean formed as a rifting apart of continents first between North America and Africa, next between South America and Africa, and finally between North America and Europe. By 120 million years ago, Africa, Antarctica, India, and Australia had begun their separate paths. North America rotated and drifted from the low latitudes into its present position. Most impressive was the long journey of India as it detached from Antarctica and ultimately collided with Asia some 50 million years ago.

The paleomagnetic and geological evidence of continental positions before 600 million years ago is sparse. Available paleomagnetic data seem to indicate that between about 900 and 600 million years ago the continents were assembled into a Pangea-like supercontinent. Prior to this time the data are so insufficient that paleogeographic reconstructions are currently not feasible.

Wilson Cycles

A pattern seems to be emerging: Continents assemble into a supercontinent, which then breaks apart; these smaller continents eventually disperse and then reassemble. This plate tectonic cycle has been dubbed the **Wilson cycle,** in honor of one of the pioneers of plate tectonics, the Canadian geologist J. Tuzo Wilson. From paleogeographic reconstructions it appears that the cycle of supercontinent assembly and destruction takes about 500 million years. We can also approximate the duration of a Wilson cycle from the time it takes a plate to make its way halfway around the Earth. Taking a typical plate speed of 4 cm/year (or 40 km/million years) and a half-circumference (at the equator) of around 20,000 km, we confirm that two continents rifting apart would meet each other again, on the far side of the planet, in around 500 million years. Accordingly, given that Pangea formed about 300 million years ago, the next supercontinent should be formed in about 200 million years as the Pacific Ocean closes, swallowed up by the subduction zones that surround it.

Why do all the continents come together into a supercontinent rather than displaying a less organized, more random pattern of collision and rifting apart? A

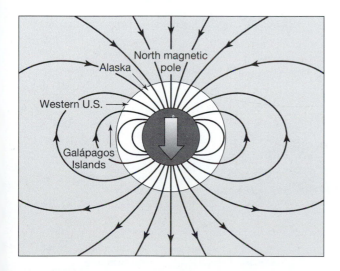

FIGURE 6-25

Earth's magnetic field, indicating that the angle of the field lines with respect to Earth's surface varies from horizontal at the equator to vertical at the poles. Paleogeographers use this feature to determine paleolatitudes of rocks, on the basis of the magnetic field orientation they acquired at the time of their formation. (From W.K. Hamblin and E.H. Christiansen: *Earth's Dynamic Systems,* 8/e, 1998. Reprinted by permission of Prentice Hall, Upper Saddle River, N.J.)

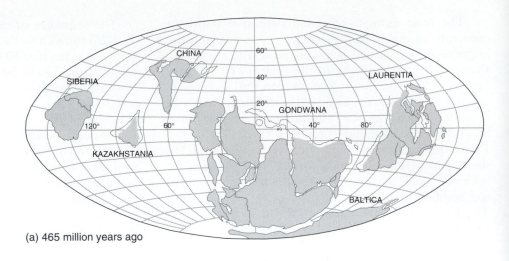

(a) 465 million years ago

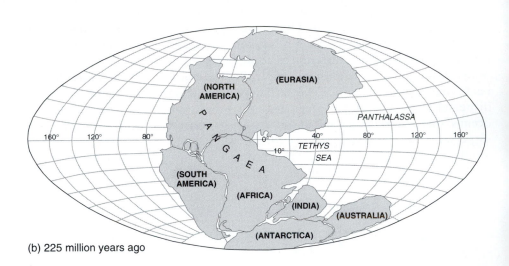

(b) 225 million years ago

FIGURE 6-26

Paleogeographic reconstructions from the Cambrian to the Eocene. Also shown are the regions' active and inactive mountain belts, shallow seas, and ice caps for each time slice. (From R.W. Christopherson, *Geosystems: An Introduction to Physical Geography,* 3/e, 1997. Reprinted by permission of Prentice Hall, Upper Saddle River, N.J.)

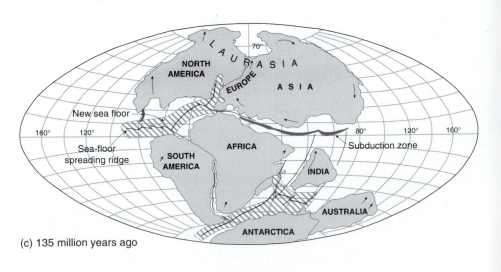

(c) 135 million years ago

(d) 65 million years ago

somewhat controversial hypothesis argues that the continents are drawn toward cold regions of the asthenosphere (Fig. 6-27). Once assembled, the thick supercontinent acts as an insulator, slowing the release of heat from the mantle. Mantle temperatures rise beneath the supercontinent, modifying the pattern of convection. The resulting tension at the surface eventually rips the supercontinent apart. The continents begin to move from this region of hot upwelling mantle to a site thousands of kilometers away, where the mantle has cooled and downwelling has commenced. This hypothesis is consistent with present-day plate speeds and seismic tomography, which together suggest that the continents, with the exception of Africa, appear to be moving toward regions of cold mantle. Africa is situated above hot mantle, as evidenced by the East African Rift zone, a place where tension within the continental lithosphere is creating a rift in the continent. Africa has apparently moved little since the breakup of Pangea, and the underlying mantle still retains the heat built up during Pangea's existence.

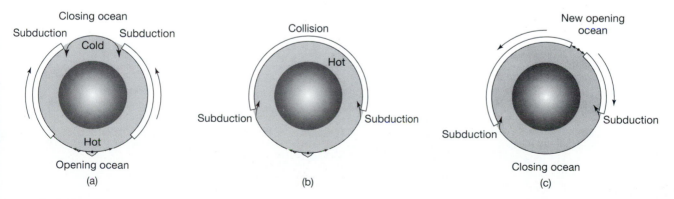

FIGURE 6-27

The Wilson cycle of supercontinent assembly and fragmentation. (a) The continents are drifting toward a region of cold asthenosphere. The closing ocean is lined by subduction zones and is contracting. The other ocean is opening, and the oceanic lithosphere is connected to the continental lithosphere at both margins. (b) The continental fragments have collided, forming a supercontinent. Subduction has begun along the margins of the formerly opening ocean. The insulating effects of the thick continental lithosphere leads to the buildup of heat and the initiation of rifting. (c) What once was an opening ocean has become a closing ocean, with cool asthenosphere beneath. One Wilson cycle is now complete. (Alter P. Kearey and F.J. Vine: Global Tectonics, 1990. Blackwell Scientific Publishing, Oxford, U.K.)

Chapter Summary

1. The solid Earth is dynamic, not static. Wegener's idea of drifting continents proposed in the early 20th century has largely been substantiated.
   a. New sea floor is created at mid-ocean ridges and moves outward as ocean basins grow.
   b. Old sea floor is destroyed at deep-sea trenches in subduction zones. Earthquakes outline the surface of the slab of oceanic lithosphere being subducted beneath the continent; their foci along a continental margin become ever deeper away from the trench.
   c. Deeper probing of Earth's interior has revealed heterogeneity in composition and temperature that can be the result only of large-scale circulation in the mantle and outer core.
      i. This circulation is fueled by residual heat from the formation of the planet 4.6 billion years ago and by heat that continues to be produced as the result of the radioactive decay of potassium, uranium, and thorium in the mantle and crust.
      ii. Mantle circulation is the result of convection, not unlike that of the troposphere. Scientists continue to debate whether convective cells extend throughout the mantle or whether a dual system of upper and lower mantle convection is in operation.
2. The signature of plate tectonics is best seen at the surface, along the margins of lithospheric plates.
   a. At plate margins, divergent, convergent, and boundary-parallel plate motions generate impressive topographic features: mid-ocean ridges and continental mountain belts and volcanoes, deep-sea trenches, and transform faults, respectively. These features themselves change with time.
      i. Mountains grow through plate collision but shrink through weathering and erosion.
   ii. Sediments are transported to the oceans, where they fill in deep-sea trenches as well as basins generated through subsidence.
   iii. Burial converts sediments to sedimentary rocks, but in time these rocks are likely to become re-exposed as the result of plate convergence and uplift, or metamorphosed and perhaps ultimately melted.
   b. This cycle of rock production and destruction (the rock cycle), driven largely by plate tectonic processes, continuously resurfaces the planet and recycles material between the crust and mantle.
3. The movement of lithospheric plates is driven by forces at plate margins and, at the base of the lithosphere, by friction with the convecting asthenosphere.
   a. Magnetic stripes displaying mirror-image patterns across the mid-ocean ridges document the creation of sea floor at the ridges.
   b. Plate motions are slow on human time scales (on the order of centimeters per year), but over geological time they can lead to the complete redistribution of continents on the globe.
      i. These movements appear to be organized, following a pattern called Wilson cycles. These cycles consist of the alternating assembly of supercontinents (perhaps at the position of mantle downwelling) and their subsequent breakup (as the insulating effects of the supercontinent lead to sublithospheric heating and mantle upwelling).
      ii. The most recent supercontinent, Pangea, formed more than 300 million years ago. Its breakup, beginning some 200 million years ago, led to the creation of the Atlantic Ocean and the separation of North America and South America from Europe and Africa.
      iii. The Pacific Ocean is currently shrinking, and it is estimated that in another 200 million years or so the continents will again reassemble into another supercontinent.

Key Terms

asthenosphere	magma	rock cycle
basalt	magnetic dynamo	sea-floor spreading
body waves	metamorphic rocks	sediment
chemosynthesis	mid-ocean ridge	sedimentary rock
continental drift	mineral	seismic wave
craton	Moho	silicate mineral
earthquake	Pangea	slab
erosion	plate tectonics	subduction
granite	polarity	surface waves
half-life	P wave	S wave
igneous rock	radioactive decay	transform fault
lithification	radiometric dating	Wilson cycle
lithosphere		

Review Questions

1. Why was the theory of continental drift not immediately embraced by the scientific community in the 1920s?
2. What is the Moho?
3. What are the bases for the two major divisions of Earth's interior—one that distinguishes crust, mantle, and core and the other that distinguishes lithosphere and asthenosphere?

4. Compare and contrast P and S seismic waves.
5. Why are earthquakes focused along plate margins?
6. What are the sources of heat in the Earth's interior?
7. What is magnetic polarity? What role did it play in the generation of ideas regarding sea-floor spreading?
8. What are the three types of plate boundaries, and what surface features are characteristic of each?

9. What is erosion?
10. How can radioactivity be used to determine the age of a rock?
11. What are the driving forces for plate movement?
12. What is hypothesized to drive the Wilson cycle of plate fragmentation and reassembly?

Critical-Thinking Problems

1. We have seen that cooling of the oceanic lithosphere causes contraction, leading to subsidence of the sea floor away from the axis of spreading. The depth *d* of the ocean floor, measured in meters, increases with age *t*, measured in millions of years from the present, according to the following equation (valid for sea floor younger than 80 million years old):

$$d = 2500 + 350\sqrt{t}.$$

   Graph a cross section of a mid-ocean ridge that is spreading symmetrically in both directions at a rate of 1 cm/yr (10 km/million yr). The age of the oldest sea floor shown should be 80 million years.

2. Duplicate Figure 6.24 and answer the following questions:
   a. Draw a line from the tip of Florida horizontally across the Atlantic to northwest Africa, a distance of about 6400 km. Now, graph the age of the sea floor (on the y-axis) against the distance from the ridge axis (on the x-axis). From this graph, determine the spreading rate for each geologic interval represented, averaging the two values determined for eastward and westward spreading. Graph these values (y-axis) as a function of time (in million years, on the x-axis).
   b. How has the Atlantic spreading rate varied over the last 200 million years?
   c. Now, duplicate your map four more times. Trim the Atlantic margins at the boundaries between the regions of sea floor of various ages (as identified in the legend of the figure). For example, for the first map, cut out that part of the Atlantic that is younger than 38 million years, and then try to match the two halves. On the next sheet, cut out all the sea floor younger than 65 million years, and so on. Then compare your reconstructions of the relative positions of North and South America, Europe and Africa to the paleographic maps in Figure 6-26.

Further Reading

General

Tarbuck, Edward J., and Frederick K. Lutgens, 1998. *Earth: An Introduction to Physical Geology,* 6/e. Upper Saddle River, NJ: Prentice Hall.
"The Dynamic Earth," 1983, *Scientific American.*

Wegener, Alfred, 1924. *The Origins of the Continents and Ocean Basins.* London: Metheune and Co.

Advanced

Kearey, Philip, and Frederick J. Vine. 1996. *Global Tectonics,* 2nd ed. Oxford, England: Blackwell Scientific Publications.

Chapter 7
Recycling of the Elements: The Carbon Cycle

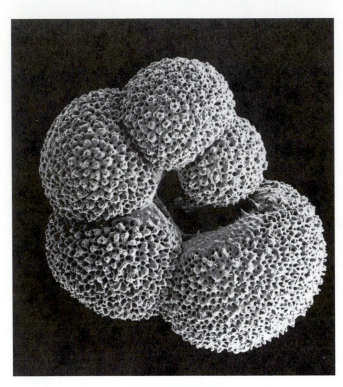

Foraminifer. (Courtesy of
R. Bernstein, Univesity of South Florida.)

Key Questions

- What determines how reservoirs (such as the atmospheric CO_2 reservoir) respond to imbalances in the flow of material to and from them?
- Which reservoirs and processes are important to the recycling of carbon and other essential nutrients in the Earth system?
- Do feedback mechanisms regulate the amount of atmospheric CO_2?

Chapter Overview

The recycling of the elements among the components of the Earth system is key to the continued functioning of Earth as a living planet. Although many elements are critical, none is more central to the workings of the Earth system than carbon. All life is based on carbon; gaseous carbon dioxide is an important greenhouse gas; the acidity of the ocean is regulated by carbon compounds; and the maintenance of an oxygen-rich atmosphere depends on the transfer of carbon to sedimentary rocks. To perform these functions, the *carbon cycle* involves a hierarchy of subcycles that operate on different time scales, ranging from decades (for the recycling of CO_2 through plants) to hundreds of millions of years (for the recycling of carbon through sedimentary rocks and for the exchange of carbon within Earth's interior). Both biological and physical processes are involved in the recycling of carbon, and they are so closely intertwined that it becomes difficult to separate the two. In this chapter we will trace the movement of carbon as it cycles through the Earth system; we will develop additional systems theory notions of *steady state* and *residence time* along the way. We will also discuss the *nutrient elements,* because the rate of carbon recycling depends strongly on the rate of nutrient recycling.

Systems Approach to the Carbon Cycle

Why is Earth the only planet in our solar system that supports life? The direct answer is that Earth is the only planet that has liquid water at its surface. (Jupiter's moon

Europa may have water only a few kilometers beneath its icy surface.) But part of the reason Earth is able to maintain liquid water is that our planet has natural recycling systems for the elements essential for life, including carbon, nitrogen, phosphorus, and sulfur. These recycling systems are ultimately linked with the global process of plate tectonics, discussed in Chapter 6. The link between tectonic activity and the carbon cycle is important to the regulation of atmospheric CO_2 concentrations and thus to climate as well.

The winds and ocean currents discussed in Chapters 4 and 5 and the moving lithospheric plates discussed in Chapter 6 make up Earth's circulatory system: They transport energy and material to different parts of the Earth system where they are utilized in biological and physical processes. This mixing of Earth's fluid and solid parts also helps accomplish an important task: the recycling of the elements. Essential elements are released to the **biosphere** (the part of Earth that supports life, including the oceans, atmosphere, land surface, and soils) as rocks weather, volcanoes erupt, and nitrogen is made available from the atmosphere by chemical transformations stimulated by lightning discharge. Compared with the rates of utilization by the biota, these releases are very slow; they would support only very low rates of biological activity were there not highly efficient nutrient recycling mechanisms. **Nutrients** are substances, normally obtained in the diet, that are essential to organisms. Nutrient elements are incorporated into living tissue during

growth and rapidly returned to the soil or ocean on death. This cycle is repeated many times before the elements are lost from the biosphere, mostly as constituents of sedimentary rocks. The situation is much like our recycling of aluminum cans: Recycling substantially reduces our dependence on the extraction of aluminum from Earth and allows us to produce aluminum products much more rapidly. Similarly, element recycling within the biosphere allows for much higher rates of biological productivity.

A number of important recycling systems operate on Earth. We've already discussed the water cycle in Chapter 4. The cycles of the nutrient elements nitrogen and phosphorus are discussed later in this chapter. The recycling of carbon is especially important: As a major constituent of the greenhouse gases carbon dioxide and methane, it affects not only biological productivity, but Earth's climate as well. We focus on the carbon cycle in this chapter because of its overarching importance to the Earth system.

A Journey through the Terrestrial Organic Carbon Cycle

As an introduction to just one part of the global carbon cycle, imagine that we could follow the carbon atom of a CO_2 molecule as it cycled through the terrestrial (land-based) part of the cycle (Fig. 7-1). The carbon in CO_2 is **inorganic carbon**—it is not associated with compounds

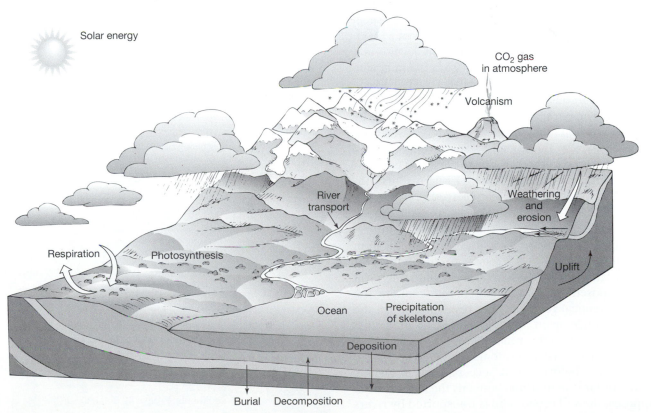

FIGURE 7-1

The global carbon cycle. (From J.P. Davidson, W.E. Reed and P.M. Davis: *Exploring Earth: An Introduction to Physical Geology,* 1997. Reprinted by permission of Prentice Hall, Upper Saddle River, N.J.)

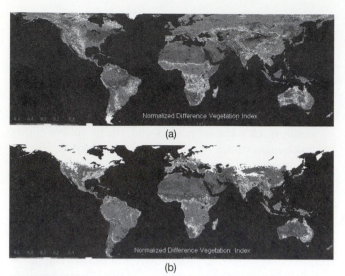

FIGURE 7-2

Satellite image of the vegetation coverage of the land surface in Northern Hemisphere in (a) summer and (b) winter. (From Felix Kogan, NOAA/NESDIS/ORA, Climate Research and Application Division.) [See color section]

formed by living organisms and it does not contain carbon-carbon or carbon-hydrogen bonds. After spending nearly a decade moving with the winds in the troposphere, the gaseous CO_2 molecule will have visited both the Northern and Southern Hemispheres several times. Then one spring, during the annual greening of the Northern Hemisphere (Fig. 7-2), the CO_2 molecule passes through a small opening in a leaf, the photosynthetic apparatus of a plant. Through a frenzy of collisions with other molecules and atoms, the oxygen atoms are ripped from the molecule while hydrogen, nitrogen, and other carbon atoms become attached. Our carbon atom, as part of the leaf, is now **organic carbon.** Some leaves are consumed and digested by animals. The carbon in these leaves is then released back to the atmosphere by the animals' respiration as CO_2.

Summer passes, fall arrives, and the leaf that contains the carbon atom has not been eaten. The nourishing substances and water that the leaf has received from the tree have ceased to flow. The leaf is released from the branch and settles to the ground. Other leaves fall on top, burying it in a thick mat of decaying matter. The carbon atom is part of the soil, where it will remain for about the next 50 years. By the end of that time, bacteria and fungi will have decomposed the organic matter that contains the atom. The chemical reactions that result transform the carbon atom once again into a gaseous CO_2 molecule, which escapes back to the atmosphere.

This life cycle of a carbon atom is repeated nearly 500 times on average before a "leak" occurs. Once in a while, before the organic matter that contains the carbon atom decomposes, the soil erodes and is transported by rivers to the oceans. There it settles with the other particles to the sea floor and is buried by subsequent sediments. The carbon atom then lies deep within the sedimentary layer.

The carbon atom may spend millions of years in this reservoir, as mountain belts form, thrusting deeply buried sedimentary rocks to Earth's surface and beyond to great elevations. Eventually our carbon atom will be transferred from its burial place within Earth's interior to the surface. Here environmental forces, both biological and physical, will cause the sedimentary rock containing the carbon atom to disintegrate during the process of weathering. In this process, the organic carbon reacts with oxygen from the atmosphere and forms (inorganic) CO_2, which escapes as a gas to the atmosphere. The weathering process, then, is the connecting link in the long path this particular carbon atom has taken—from the atmosphere, to the plant, to the soil, to the sediment, to the sedimentary rock, and back to the atmosphere.

The path the carbon atom has taken encompasses the *terrestrial organic carbon cycle,* operating on time scales that are short (years to decades) and those that are long (centuries and millennia to multi-million-years). If instead of being incorporated into a plant leaf the carbon atom had entered the ocean and been converted to organic carbon by marine algae, it would have become part of the *marine organic carbon cycle.* There are also a host of processes not involving organic carbon that compose the *inorganic carbon cycle.* These various parts of the carbon cycle are discussed below.

Carbon Reservoir Dynamics

Carbon resides in many reservoirs at or near Earth's surface (Fig. 7-3), ranging in size from the relatively tiny amount of carbon in atmospheric methane to the tremendous amount of carbon stored in sedimentary rocks. One of our ultimate goals is to understand how this system of

FIGURE 7-3

Reservoirs of carbon at or near Earth's surface (circa 1997).

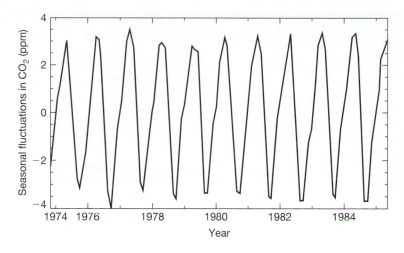

FIGURE 7-4

Seasonal fluctuations in atmospheric CO_2 from the Mauna Loa Observatory for 1974–1986. The persistent increase due to fossil-fuel burning and deforestation has been subtracted out to emphasize seasonal fluctuations. The tick marks are at January of the year.

reservoirs responds to perturbations. We will use the response to the release of carbon dioxide from the burning of fossil fuels as an example of the dynamics of the carbon cycle and of material recycling systems in general.

Figure 7-4 shows the seasonal fluctuations in the atmospheric CO_2 level over several years (1974–1985) measured from atop Mauna Loa, Hawaii. We saw a similar graph in Figure 1-2. In Figure 7-4, however, the gradual increase in atmospheric CO_2 caused by the burning of fossil fuels and by deforestation has been subtracted out to emphasize the natural, seasonal cycle itself. The CO_2 content falls during the Northern Hemisphere summer, when photosynthesis (and the growth of leaves) surpasses respiration and decomposition. It then rises during the late fall to early spring, when respiration and decomposition of the previous season's crop of leaves exceeds photosynthesis. Because Hawaii is in the Northern Hemisphere (17° N), measurements made at Mauna Loa reflect the annual cycle in that hemisphere (see Chapter 1 for further discussion). Let us cast these observations in terms of systems theory.

Reservoirs. From the systems point of view, the atmosphere is a *reservoir* of carbon in the form of CO_2 (Fig. 7-5). Reservoirs are typically characterized in terms of the amount of material they are holding at any particular time. Their sizes are commonly expressed either in mass units or volume units. (The units are typically *moles;* refer to the box "The Concept of the Mole" if you need a refresher.) In Figure 7-5, the amount of carbon is expressed in gigatons (Gton) of carbon, or Gton(C). A gigaton is 1 billion

metric tons, and 1 metric ton is 1000 kg. With the notation Gton(C), we are keeping track of the mass of only the carbon atoms, not the other atoms to which they are attached.

Reservoirs are temporary repositories for material that flows through them, and their sizes vary in response to imbalances between inflow and outflow, typically expressed in unit mass, unit volume, or moles per unit time. The *inflow* to the atmospheric CO_2 reservoir is the combination of the processes of respiration and decomposition. This inflow can be expressed in units of gigatons of carbon per year. The *outflow* from this reservoir is photosynthesis, and the rate of outflow is also expressed as gigatons of carbon per year.

Steady State. If the rates of inflow and outflow were such that the atmospheric CO_2 level remained at a constant value with time, we would say that *steady state* had been achieved. **Steady state** is a condition in which the state of a system component is constant with time. Steady state could be achieved if no inflow and no outflow existed— that is, if both processes ceased. A constant level could also be maintained if the rate of inflow of CO_2 into the atmosphere equaled the rate of outflow.

Any imbalance in these rates leads to a change in the level of atmospheric CO_2. When the inflow exceeds the outflow, the atmospheric CO_2 level rises. (This situation is analogous to the Northern Hemisphere winter condition.) When the outflow exceeds the inflow, the level falls (analogous to the Northern Hemisphere summer condition).

In the record of atmospheric CO_2 variations in Figure 7-4, we see that steady state is achieved only twice per year, when CO_2 levels reach maximums and minimums. These states are fleeting, however; for an instant the rate of input equals the rate of output. Averaged over longer times, however, the *natural* cycle of CO_2 is thought to be close to steady state, despite seasonal imbalances. Because of anthropogenic disturbances, the atmospheric CO_2 level is not currently at steady state, as demonstrated by the persistent rise in CO_2 over the past several decades.

Steady state can be maintained over time only if the rates of inflow and/or outflow are sensitive to changes in

FIGURE 7-5

The atmospheric carbon (that is, CO_2) reservoir, showing inflows and outflows.

The Concept of the Mole

Atoms and molecules are typically measured in units called moles. A *mole* (abbreviated mol) of a substance is defined as the amount of that substance that contains the same number of atoms or molecules (or any other particle) as the number of atoms in 12 g of ^{12}C. There are 6.022×10^{23} atoms in 12 g of ^{12}C; this number is called *Avogadro's number,* after the Italian chemist Amedeo Avogadro. A mole of any type of particle contains Avogadro's number of these particles. In concept, a mole is no different than, say, a dozen; it simply converts a number that would be cumbersome into one that is more practical. Even when we use moles, global-scale reservoirs such as that of atmospheric CO_2 are huge. There are presently about

6×10^{16} moles of CO_2 in the atmosphere.

One mole of ^{12}C weighs exactly 12 g. However, 1 mol of H weighs only approximately 1 g, and 1 mol of ^{16}O weighs approximately 16 g. These weights are defined as the mass numbers of the given isotopes. Recall from Chapter 1 that the mass number of an isotope is the total number of protons plus neutrons in the nucleus. Equivalently, *atomic weight* is the weight of 1 mol of a particular isotope. For an element with more than one stable isotope, the precise atomic weight (as opposed to the mass number) of the element is determined from the relative amounts of the various isotopes.

Expressing quantities of substances in moles rather than in mass units (for ex-

ample, grams) can be useful when we are studying chemical reactions. Consider the chemical reaction for the formation of salt (sodium chloride, NaCl):

$$Na + Cl \longrightarrow NaCl.$$

This equation shows that one atom of sodium (Na) will react with one atom of chlorine (Cl) to form one molecule of NaCl. Therefore, 1 mol of Na will react with 1 mol of Cl to form 1 mol of NaCl. To express this equation in mass units, we would need to use the atomic weights of sodium and chlorine. If we did, we would find that 22.99 g of Na reacts with 35.45 g of Cl to produce 58.44 g of NaCl—a more cumbersome calculation.

the size of the reservoir. In systems terminology, this means that there must be couplings that link the reservoir size to the processes that govern inflow and outflow (see Chapter 2). Consider what would happen if a reservoir in steady state were perturbed by an addition of material. If the coupling governing inflow was negative or if that governing outflow was positive (that is, inflow decreased or outflow increased), the reservoir would return to its original state. One such negative feedback loop exists between the photosynthetic rate of plants and atmospheric CO_2. As CO_2 levels go up, plants photosynthesize more rapidly; this effect has been called *CO_2 fertilization* (Fig. 7-6). But as they do, CO_2 levels tend to fall, because CO_2 is consumed by plants during photosynthesis. Thus, the terrestrial biota tend to stabilize atmospheric CO_2 levels.

Residence Time. To help us monitor the recycling of elements through the Earth system, the concept of residence time can be useful. **Residence time** is defined as the average length of time a substance spends in a given reservoir that is at steady state. We can calculate the residence time by dividing the reservoir size at steady state by the inflow or outflow rate:

$$\text{residence time} = \frac{\text{reservoir size at steady state}}{\text{inflow rate or outflow rate}}.$$

Reservoir size has units of mass, and the inflow or outflow rate is in mass per unit time. Thus the quotient has units of time.

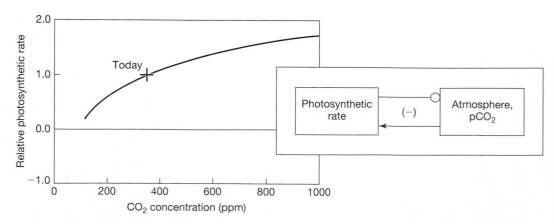

FIGURE 7-6

Effect of changes in CO_2 concentration on the photosynthetic rate of typical plants. Photosynthetic rates are relative to the value for today's atmospheric CO_2 level. Inset represents the negative feedback loop that results from this dependence.

We can determine the residence time of the atmospheric carbon reservoir from the size of that reservoir, which Figure 7-3 gives as 760 Gton(C), and from the rate of respiration and decomposition (the inflow rate) or the rate of photosynthesis (the outflow rate), both of which are given as 60 Gton(C)/yr. Thus, if we assume steady state, the residence time of carbon in the atmosphere with respect to these processes is 760 Gton(C)/60 Gton(C)/yr = 12.7 yr. This means that carbon in the atmosphere is replenished about once per decade.

We can also think about residence time as an indicator of how long a reservoir takes to respond measurably to large imbalances in inflow or outflow. In our atmospheric carbon example, if photosynthesis were to cease but respiration and decomposition were to continue at their current rate, the atmospheric CO_2 level would double in about a decade. Thus, the residence time, defined at steady state, becomes the **characteristic response time** when a system is *not* at steady state.

Oxidized and Reduced Carbon

The many identities that carbon assumes in the Earth system can be lumped into two general categories: oxidized carbon and reduced carbon. **Oxidized carbon** is carbon that is combined with oxygen. Examples of oxidized carbon include the carbon in the skeletons of some organisms and in atmospheric CO_2. **Reduced carbon** is carbon that is combined mainly with other carbon atoms, hydrogen, or nitrogen. Organic carbon is a form of reduced carbon. Perhaps a more familiar pair of reduced and oxidized substances is metallic iron and its oxidation product, rust (iron oxide). In the presence of oxygen gas at Earth's surface, reduced substances, such as organic carbon and metallic iron, tend to be highly chemically reactive. Oxidized substances, such as CO_2 and rust, tend to be more inert.

In the next few sections, we will explore the organic and inorganic carbon cycles. We begin with the organic carbon cycle as it operates on land—the terrestrial organic carbon cycle, which we already introduced in our journey through the carbon cycle. We then move to the oceans, where both oxidized and reduced carbon recycling are important on short time scales. Finally, we consider the longer-time-scale cycles, which involve geological processes.

The Short-Term Organic Carbon Cycle

The short-term organic carbon cycle involves processes ranging from those we can observe and appreciate on a daily to seasonal timescale (see Fig. 7-2), such as the processes of photosynthesis and respiration, to processes of decomposition that are somewhat slower (Fig. 7-7). The key step in this cycle is the conversion of inorganic carbon (atmospheric CO_2) to organic carbon by the process of **photosynthesis.** We are not so much interested in the process itself but rather in its impact on the global cycle, which is generally expressed as primary productivity. **Primary productivity** is the amount of organic matter produced by photosynthesis in a unit time over a unit area of Earth's surface. That amount depends on the population size of **primary producers**—that is, plants (or other types of photosynthesizers or even chemosynthesizers) that provide energy other organisms can use. The relationship is not simple, however, because some primary producer species are much more productive than others. In its simplest representation, primary production involves a chemical reaction between CO_2 and water to form organic matter and oxygen:

$$\text{Photosynthesis:} \quad CO_2 + H_2O \longrightarrow CH_2O + O_2.$$
**(Primary carbon water carbohydrate oxygen
Production) dioxide gas**

Here, organic matter is represented by CH_2O, the simplest *carbohydrate,* or compound of carbon, hydrogen, and oxygen. In reality the molecules making up organic matter are much larger than this simple carbohydrate, and they contain small amounts of many other elements, including nitrogen and phosphorus. For our purposes, however, this simpler representation suffices.

Photosynthesis does not occur spontaneously at Earth's surface but instead requires an input of energy from the Sun. Plants, algae, and bacteria have evolved pigments that are able to capture the energy of sunlight and convert it to chemical energy, part of which is stored in living tissues. This chemical energy is then utilized by other organisms that cannot utilize solar energy directly. Such organisms, including animals, are called **consumers.**

Most of the photosynthesis that occurs each year leads to the formation of tissue that is recycled rapidly, including the leaves of trees. This recycling is the cause of the seasonal variations observed in the CO_2 record of Figure 7-4. However, most of the organic carbon that makes up plant tissues has a residence time of many decades. That is because the bulk of the organic carbon in terrestrial plants is contained not in the leaves but in the roots and trunks of slow-growing trees. In other words, most of the *biomass* of primary producers on Earth is contained in tree roots and trunks. **Biomass** is the total mass of organic matter in living organisms in a particular reservoir. In terms of carbon, the total living biomass—the combined biomasses of all primary producers and consumers—is about equal to the atmospheric carbon reservoir (see Fig. 7-3).

Consumer biomass is a small percentage (only about 1%) of the biomass of the producers. Consumers derive their metabolic energy from the chemical energy stored in plant tissues by ingesting the tissues and respiring. Respiration is the reverse of photosynthesis: It is the chemical reaction between oxygen and organic tissue that yields CO_2 and water:

$$\text{Respiration:} \quad CH_2O + O_2 \longrightarrow CO_2 + H_2O.$$
**carbohydrate oxygen carbon water
gas dioxide**

FIGURE 7-7

The short-term, terrestrial organic carbon cycle, showing reservoir sizes, inflows, and outflows. Dark shaded area represents the long-term cycle (see Fig. 7-12).

Unlike photosynthesis, respiration would proceed spontaneously (without intervention by consumers) because it releases rather than requires energy. However, it would do so very slowly. Organisms are able to accelerate this chemical reaction by the use of *enzymes,* chemical compounds (specifically, proteins) that they synthesize for this purpose.

On land, about half the organic material produced by photosynthesis is respired by animals and by plants themselves. The remaining material is added to the organic-rich upper layers of soil. A host of microscopic bacteria and fungi live in soil. Their metabolic requirements are satisfied through the decomposition of the large store of organic matter that is buried there, fueled by the supply of oxygen from the overlying layers. A biological process that uses oxygen is said to be *aerobic,* and an organism that undergoes aerobic metabolism is an *aerobe.* The chemical reaction for this aerobic decomposition is identical to that for respiration.

Because the only source of oxygen is the air, the microorganisms that live well below the surface must be adapted to environments that are devoid of oxygen. A biological process that occurs in the absence of oxygen is said to be *anaerobic,* and an organism that undergoes anaero-

bic metabolism is an *anaerobe.* In the O_2-free environment of deep soil live anaerobic bacteria that decompose organic matter by fermentation. **Fermentation** is an anaerobic form of metabolism in which both oxidized carbon (in CO_2) and reduced carbon (often in methane) are produced:

$$\text{Fermentation:} \quad 2CH_2O \longrightarrow CO_2 + CH_4.$$
$$\text{carbohydrate} \qquad \text{carbon} \quad \text{methane}$$
$$\text{dioxide}$$

(Other, more-complex compounds can also be used in and produced by fermentation, but for the purposes of studying the carbon cycle, this representation is adequate.) The gases CO_2 and CH_4 can escape to the atmosphere. Once there, the CO_2 continues the short-term cycling path described earlier. Methane, however, is chemically unstable in our O_2-rich atmosphere and is destroyed by a series of oxidation reactions. The carbon contained in CH_4 combines with O_2 to form CO_2. With an atmospheric reservoir size of 5 Gton(C) and a supply rate from fermentation of 0.5 Gton(C)/yr (see Fig. 7-7), the residence time of CH_4 in the atmosphere is approximately 10 years.

The land surface is continuously stripped of its soil cover by the action of winds and water. On average, about 5 cm of soil is eroded from the land surface every 1000 years and transported by rivers to the oceans. Although river systems have a substantial capacity for storing sediment in flood plains and deltas, eventually most of the sediment makes its way to the oceans and is deposited on the sea floor. These sediments contain whatever organic matter has survived the trip from land to sea. Thus, there is a transfer of organic carbon from the terrestrial to the marine realm, one that amounts to about 0.1 Gton(C) per year. However, this transfer is small compared with the flux through the oceanic water column of organic matter produced in the ocean.

The Marine Organic Carbon Cycle on Short Time Scales

Producers and Consumers. The dominant primary producers in the ocean are the free-floating, photosynthetic marine microorganisms referred to as *phytoplankton,* which we introduced in Chapter 1. (In more general terms, *plankton* are organisms with any type of metabolism that float freely in aquatic environments.) These organisms—primarily *diatoms* (Fig. 7-8a) and other algae,

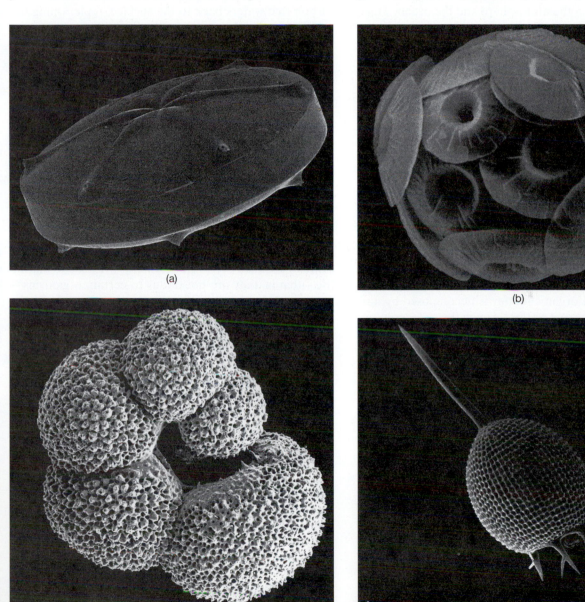

(a)

(b)

(c)

(d)

FIGURE 7-8

Typical phytoplankton: (a) diatom and (b) coccolithophorid. Typical zooplankton: (c) foraminifer and (d) radiolarian. (Courtesy of R. Bernstein, University of South Florida.) [See color section] such as *coccolithophorids* (Fig. 7-8b)—live in the photic zone. The **photic zone** is the uppermost part of the oceanic water column where there is sufficient light for photosynthesis: about the upper 100 m of the water column in the open ocean, and in shallower waters near shore, where water clarity is reduced. It roughly corresponds to the "surface ocean," which, as we saw in Chapter 5, is the upper part of the ocean mixed by the winds.

Phytoplankton consume CO_2 and produce O_2 through photosynthesis in much the same way as do land-based plants. Although the gases phytoplankton use and produce are dissolved in seawater, there is continuous gas exchange between the atmosphere and the ocean. Thus, the activities of phytoplankton affect the atmosphere as well as the ocean.

Much of the organic matter produced in the surface ocean by phytoplankton is consumed by zooplankton. **Zooplankton** are free-floating marine consumers, including small invertebrates and microorganisms such as *foraminifera* (Fig. 7-8c) and *radiolarians* (Fig. 7-8d), that cannot photosynthesize. Zooplankton produce fecal pellets that, together with other large particles of organic matter, settle through the water column to great depths. In contrast to the flux of organic matter from treetops to the ground, though, only about 1% of this material survives the trip to the sea floor. Even then, the material is subject to efficient recycling by aerobic and anaerobic decomposers that live on the sea floor or in the uppermost layers of sediment. As a result, only about 0.1% of the organic matter that settles from the surface ocean is preserved in marine sediments.

Most marine organic matter is decomposed by animals and microbes as it settles through the water column. This decomposition releases CO_2 (the product of both oxygen-breathing animals and microbial respiration) and nutrients to the oceanic deep waters. For marine organisms, nutrients lead to high rates of primary productivity if they are available in the appropriate concentrations. Thus, it is critical to the productivity of the marine biota to get these nutrients back to the surface.

The Biological Pump.   The overall effect of photosynthesis in shallow waters, of the settling of organic matter, and of decomposition in deep waters is the transfer of CO_2 and nutrients from the surface waters to the deep ocean. This process is known as the **biological pump** (Fig. 7-9). It is balanced by the conveyor-belt-like thermohaline circulation of the ocean (Chapter 5), which brings nutrients and carbon-rich waters back to the surface, replenishing the nutrients and carbon removed by the biological pump.

The biological pump has a profound effect on ocean chemistry. As a result of its operation, surface waters are measurably depleted in carbon and severely depleted in phosphate and nitrate (the major nutrient elements) relative to deep waters. If the biological pump were to cease—for example, as the result of a mass extinction—the ocean would assume a more uniform composition in a few thousand years, as the thermohaline circulation homogenized the ocean.

Nutrient Limitation.   Some elements are classified as nutrients for marine phytoplankton because they are essential for growth and exist at suboptimal concentrations; an increase in their concentration leads to higher rates of primary productivity. Other elements are *toxic* to marine life—that is, they are poisonous to certain organisms—because their concentrations in seawater are above the optimum for growth. (Even nutrient elements can be toxic if their concentrations become too high.) There is great variability in the concentration of many elements throughout the world's oceans. Each element, depending on its concentration, can be either a nutrient or a toxic substance. The situation is analogous to the parabolic growth curves for daisies as discussed in Chapter 2 (Fig. 2-9a), except that element concentration substitutes for temperature. For each element, there is some optimum concentration that favors biological productivity.

Marine phytoplankton incorporate many nutrient elements into their tissues in ratios that appear to be nearly identical in all species. These ratios are called **Redfield ratios,** in honor of Alfred C. Redfield, the oceanographer who first described this phenomenon. Even more remarkably, the ratios of many of these elements in seawater is nearly identical to that in phytoplankton. For example, the elemental ratio of carbon:nitrogen:phosphorus is very nearly $106:16:1$ (Table 7-1). A chicken-or-egg question arises: Does the composition of seawater determine the composition of organisms that live in the sea, or does the composition of marine organisms determine the composition of seawater? To answer this question, we must consider the distribution of primary productivity in the oceans.

FIGURE 7-9

The marine biological pump.

Oxygen Minimum Zone

The decomposition of organic matter settling through the water column consumes oxygen, leading to oxygen depletion. Decomposition also releases nutrients to the water, so nutrient concentrations increase. The lowest oxygen concentrations are achieved at intermediate depths, about 1 km below the surface. This region is known as the *oxygen minimum zone* (Box Fig. 7-1). In this zone, dissolved oxygen concentrations reach a minimum as a result of high oxygen demand by aerobic decomposers and low oxygen supply from the surface ocean or from below.

In the waters below the oxygen minimum zone, oxygen content increases with depth. Why is this so, if the source of oxygen is exchange with the atmosphere and production by phytoplankton in the surface ocean?

Recall from Chapter 5 that the circulation of the ocean resembles a giant conveyor belt that brings waters from the surface in the high-latitude North Atlantic and Antarctic oceans to the deep sea. This water is enriched in oxygen as a consequence of having originated at the surface in the high latitudes. Like most gases, oxygen is more soluble in cold water than in warm water. So, gas exchange with the atmosphere leads to higher gas concentrations in high-latitude surface waters than in low-latitude surface waters. High-latitude surface waters are injected beneath the intermediate waters of the world's oceans by thermohaline circulation. Deep water therefore ends up with more dissolved O_2 than does the water above it. Thus, the oxygen minimum zone owes its existence to the combined operation of the biological pump and the global conveyor-belt circulation.

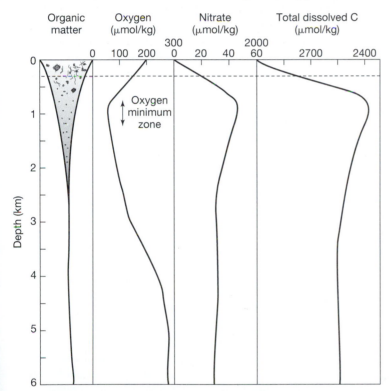

BOX FIGURE 7-1

The effect of the biological pump and thermohaline circulation on the chemical composition of the ocean. Typical vertical profiles of the amount of organic matter, dissolved oxygen, nitrate (a nutrient) and inorganic carbon are shown. Concentrations are in micromoles (10^{-6} mol) per kilogram of seawater.

TABLE 7-1

Redfield Ratios	
Element	*Relative number of atoms in living phytoplankton*
Carbon	106
Nitrogen	16
Phosphorus	1
Iron	0.01

Where are rates of primary productivity greatest in the oceans? We now have a sophisticated way of answering this question, by satellite. With a color scanner, researchers can quantify the color of seawater from space (Figure 7-10). The color of the ocean surface is strongly influenced by the density of phytoplankton, which contain photosynthetic pigments. In the early 1960s and 1970s, oceanographers began to study the relationship between the concentration of these pigments and the abundance and productivity of near-surface-dwelling phytoplankton. The researchers found that regions of the ocean that have low concentrations of *chlorophyll* (a green pigment that is the dominant pigment in algae) are biological deserts, with low abundances of organisms and low nutrient concentrations. In

FIGURE 7-10

The concentration of photosynthetic pigments as determined by the Coastal Zone Color Scanner (CZCS) on the *Nimbus 7* satellite. Pigment concentrations are indirect indicators of rates of primary production. (From Gene Feldman, NASA/GSFC/Science Photo Library, Photo Researchers, Inc.) [See color section]

contrast, more productive waters tend to have high chlorophyll concentrations.

Although we might expect that the warm, sunny low-latitude oceans would be most conducive to phytoplankton growth, the satellite patterns reveal that the waters with the highest productivities are the cold waters of the high-latitude Atlantic, Pacific, and Southern oceans. The reason is that the thermocline is weak or nonexistent at these latitudes because surface waters are just as cold as the deep waters there. Thus, wind-driven mixing of nutrient-rich deep waters up to the surface occurs much more readily at high latitudes than at low latitudes, where mixing is inhibited by the strong temperature and density gradients (Chapter 5). Apparently, the inhibitory effects of cold temperatures and low light availability experienced by organisms that live at high latitudes is more than compensated for by an enhanced nutrient supply.

The satellite images also indicate high productivity in regions of upwelling. As we saw in Chapter 5, many of these regions occur along the western continental margins. There the wind-driven surface currents cause off-shore transport, which allows nutrient-rich waters from intermediate depths to well up to the surface. Other up-welling regions occur where surface currents diverge (for example, along the equator, due to the action of the trade winds and the Coriolis effect). Surface divergence also allows intermediate-depth waters to rise and replace the water transported away at the surface. Finally, the coastal regions of the continental shelves tend to be highly productive because of nutrient inputs from rivers (often strongly enhanced by anthropogenic inputs of nitrogen and phosphorus) and localized upwelling driven by the topography of the sea floor.

Nutrient supply therefore seems to be a major limitation on the productivity of the surface ocean. Except in upwelling regions, phosphate and nitrate concentrations in surface waters are driven essentially to zero as a result of intense uptake by phytoplankton. Deep waters begin as surface waters, depleted in nutrients, at high latitudes. These waters increase in nutrient concentration as nutrients are released by decomposing organic matter that rains down from the overlying water column (the biological pump). In the North Atlantic, the effect of this "rain" on the composition of the water is small, because the water mass is young and has not had time to build up a large supply of organic matter. Farther along the conveyor belt's path, the aging water mass begins to show its increasing influence of the biological pump. Accordingly, the nutrient contents of deep North Pacific waters (the oldest waters in the ocean) are much greater (Fig. 7-11a), and the O_2 contents much lower (Fig. 7-11b), than those of the North Atlantic. Thus, the explanation for the observed similarity in nutrient elemental ratios between seawater and marine organisms— the Redfield ratios—is that the nutrient composition of the world's oceans is dominated by the production and decomposition of organic matter. In other words, the composition of marine organisms determines the composition of seawater.

The Long-Term Organic Carbon Cycle

The processes we have discussed thus far affect the atmospheric CO_2 balance on time scales shorter than a century. On longer time scales, these processes must be very closely in balance: Because the fluxes involved are so large, persistent imbalances would lead to intolerable fluctuations in atmospheric CO_2. Geological processes become the important controls on atmospheric CO_2 on longer time scales (Fig. 7-12). The fluxes of carbon involved in these processes are small, and the reservoirs involved are large. Together these two adjustments in scale mean that the importance of the geological processes affecting the sedimentary reservoir and the atmosphere are influential only on long time scales (on the order of 1000 years to 1 million years).

Carbon Burial in Sedimentary Rocks

The flux of land-derived and marine sediments to the sea floor fills sedimentary basins, many of which flank the margins of the continents. The continuous supply of sediment to these basins leads to the burial of previously deposited material. Eventually, as the process continues, sediments become buried to a depth of a few kilometers below the sea floor and become lithified. The organic carbon associated with these sediments is then entombed in sedimentary rock until weathering liberates the material to the biosphere.

(a)

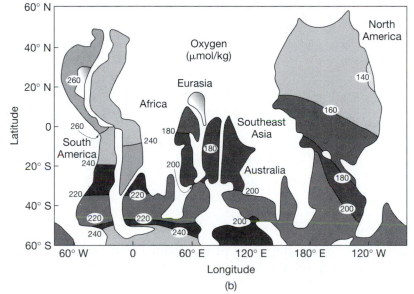

(b)

FIGURE 7-11

Concentrations of dissolved (a) nitrate and (b) O_2 in the deep ocean, in micromoles per kilogram of seawater. Maps are drawn for a water depth of 4000 m. At this depth the mid-ocean ridges appear as mountain ranges, and the continental outlines are substantially modified. (After W. S. Broecker and T.-H. Peng, 1982. *Tracers in the Sea.* New York: Eldigio Press, Columbia University, p. 31.) [See color section]

Carbon Leaks and Oxygen Replenishment. This organic carbon burial represents a leak of material from the short-term organic carbon cycle (see Fig. 7-12). It is this leak, rather than photosynthesis alone (which is nearly balanced by respiration and decay), that maintains the O_2 content of the atmosphere. Oxygen is continuously removed from the atmosphere by chemical reactions with reduced materials (especially organic matter) that are preserved in rocks exposed at Earth's surface and with reduced volcanic gases such as hydrogen, sulfur dioxide, and carbon monoxide. The loss of O_2 is very slow, but the O_2 concentration would reach zero in a few million years if that gas were not supplied from other sources. Oxygen is replenished by the leak of organic matter into the sedimentary rock reservoir. For every carbon atom that enters this reservoir as organic carbon, one oxygen molecule is left behind. This is because the O_2 liberated during the photosynthesis of that carbon was not utilized during respiration or decomposition, and thus the gas remains in the atmosphere.

Formation of Fossil Fuels. During burial, the organic material in the sediment also undergoes significant changes in its structure and chemistry, and fossil fuels may form. Sediments derived from land plants (especially *peat,* which forms in swamps and bogs) accumulate on land or in basins near shore. If the concentrations of terrestrial organic matter are high in these basins, burial processes under high pressures and temperatures can lead to the formation of **coal.** Similarly, high concentrations of organic matter in marine sediments can produce sedimentary rocks that, under high-pressure and high-temperature conditions of burial, serve as sources of **petroleum.** This material is fluid and tends to migrate through the basin

FIGURE 7-12

The combined short-term and long-term organic carbon cycles, showing the geological processes of sedimentation, burial, and weathering.

until it becomes trapped and accumulates. If the accumulation is sufficient, the petroleum will be an economical source of fossil fuel. More commonly, however, the concentration of organic matter is 1% or less of the total sediment material. As a result, most sedimentary rocks do not represent economically viable energy sources.

The Sedimentary Organic Carbon Reservoir. Sedimentary rocks contain by far the greatest quantity of organic carbon on Earth: approximately 1×10^7 Gton(C) (see Fig. 7-3). Most of the organic carbon is found in *shales,* which are fine-grained sedimentary rocks formed by the lithification of muds. The residence time for the sedimentary organic carbon reservoir is about 200 million years.

Weathering of Organic Carbon in Sedimentary Rocks

Weathering of the organic carbon in sedimentary rocks is an oxidation process requiring atmospheric O_2, either by direct exposure to the atmosphere or by exposure to groundwaters containing dissolved O_2. The oxidation of

this material can be represented by the same schematic chemical reaction that we used previously for respiration and aerobic decomposition. The organic matter reacts with oxygen, releasing carbon dioxide to the atmosphere or groundwater.

In this sense, the mining, pumping, and combustion of fossil fuels represent merely an acceleration of the weathering process. The rocks from which humans have removed these fuels would likely have become exposed at the surface and undergone oxidation to form CO_2 some time in the distant future. Human intervention, however, has speeded up this process by a factor of a million or more for these fossil-fuel deposits. The release of organic matter from sedimentary rocks is occurring much faster than it can be replaced. Hence fossil fuels represent only a short-term energy source.

Summary of the Organic Carbon Cycle

We have now explored the entire global organic carbon cycle. Pathways exist to recycle carbon from all reservoirs (see Fig. 7-12). Every reservoir in this cycle is directly connected to the atmosphere. Thus, the CO_2 concentration of the atmosphere changes continuously in response to changes in the flux of carbon to and from these reservoirs. The responses to these changes are rapid for the large fluxes associated with the biota, soils, and marine sediments but slow for the small fluxes from the sedimentary rock reservoir. This observation will prove to be important in later chapters when we consider the fate of CO_2 added to the atmosphere from the burning of fossil fuels.

The Inorganic Carbon Cycle

The photosynthesis of CO_2 to reduced carbon in organic matter, and its subsequent reoxidation to CO_2 through respiration, decomposition, and weathering, is central to the organic carbon cycle. But there are other sources and sinks for atmospheric CO_2. Carbon dioxide readily dissolves in rainwater and seawater and then undergoes rapid chemical reactions to other ionic forms of inorganic carbon. The oxidized carbon in these waters is chemically reactive and becomes involved in a number of chemical processes. Because they do not involve organic carbon directly, these processes are together referred to as the ***inorganic carbon*** cycle.

The important reservoirs of inorganic carbon are the atmosphere, which we have discussed at length; the ocean; sediments; and sedimentary rocks (see Fig. 7-3). The sediment and sedimentary-rock carbon reservoirs consist primarily of limestone. **Limestone** is a rock composed largely of *calcium carbonate* ($CaCO_3$)—generally in the form of the mineral *calcite.* The magnesium-rich carbonate mineral *dolomite,* $CaMg(CO_3)_2$, is abundant in older sedimentary rocks.

Carbon Exchange between Ocean and Atmosphere

Carbon dioxide is continuously exchanged between the atmosphere and ocean. The distribution of sources and sinks of CO_2 is tied to the circulation and productivity patterns of the oceans (Fig. 7-13). In regions of the ocean where high rates of primary productivity have created surface waters with low CO_2, CO_2 *diffuses* from the atmosphere to the ocean. In other words, the net flow of CO_2 is down the concentration gradient—from regions of higher CO_2 concentration (in this case, the atmosphere) to regions of lower concentration (the ocean). Conversely, upwelling regions, such as the equatorial Pacific surface waters, have high CO_2 concentrations because deep waters rich in CO_2 (as the result of decomposition associated with the biological pump) have risen to the surface there. In such regions, CO_2 flows from the ocean to the atmosphere. Thus, the oceans serve as both a source and a sink for atmospheric CO_2.

Before the carbon cycle was disturbed by human activity, the flux of CO_2 from oceanic CO_2 source areas was probably closely balanced by the flux to oceanic CO_2 sinks. This pattern is changing in response to the burning of fossil fuels. As we have seen, the atmospheric CO_2 concentration has been increasing. Regions of the ocean that were previously weak sources have now become sinks, and the ocean as a whole has become a sink for CO_2.

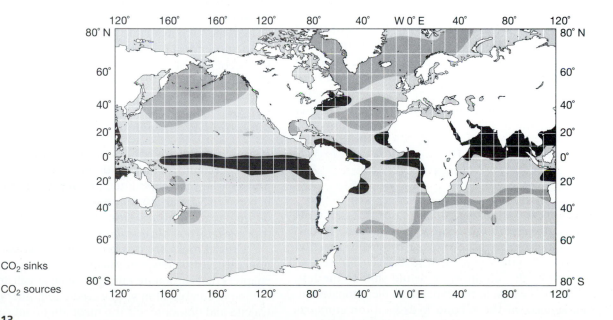

FIGURE 7-13

Oceanic sources (darker shading) and sinks (lighter shading) of atmospheric CO_2. Sources have CO_2 concentrations larger than those in equilibrium with the atmosphere, whereas sinks have lower-than-equilibrium CO_2 concentrations. (After T. Takahashi, 1989. *Oceanus,* vol. 32, pp. 22–29.)

The Chemistry of Inorganic Carbon in Water

In the oceans, inorganic carbon exists in a number of dissolved forms. When CO_2 dissolves in water (whether it be freshwater or seawater), carbonic acid is generated:

$$CO_2 + H_2O \longleftrightarrow H_2CO_3. \qquad (1)$$
$$\text{carbon} \quad \text{water} \quad \text{carbonic}$$
$$\text{dixoide} \qquad\qquad \text{acid}$$

The double arrows in this reaction indicate that the reaction can proceed both forward and backward. The rates of the forward and reverse reactions depend on the concentrations of the *reactants* (on the left-hand side) and of the *products* (on the right-hand side), respectively. If the concentration of the reactants is high, the forward reaction proceeds fast, depleting the concentration of the reactants and enhancing that of the products until the forward rate matches the reverse rate. Conversely, if the concentration of the products is high, the reverse reaction proceeds more rapidly, and the products become depleted (and reactants enriched) until the rates balance. In other words, *chemical equilibrium* is rapidly achieved in such reactions.

Like many substances that dissolve in water, carbonic acid molecules break apart, or *dissociate,* into ions. Recall that ions are charged atoms or molecules. Ions with negative charges are referred to as **anions.** Ions with positive charges are referred to as **cations.** When carbonic acid dissociates, carbon-bearing anions and hydrogen cations are formed. The relative abundance (concentration) of the carbon anions is thus linked to the *pH* of seawater.

Carbonic Acid, Bicarbonate, and Carbonate Ion Equilibrium. Returning to the discussion of the forms of inorganic carbon in water, we see that the dissociation of carbonic acid involves the release of one or both of its hydrogen atoms to yield carbon-oxygen anions. When the first hydrogen atom is lost, *bicarbonate ion* (HCO_3^-) is formed:

$$H_2CO_3 \longleftrightarrow H^+ + HCO_3^-. \qquad (2)$$
$$\text{carbonic} \qquad \text{hydrogen} \quad \text{bicarbonate}$$
$$\text{acid} \qquad\quad \text{ion} \qquad\quad \text{ion}$$

If the H^+ concentration were to decrease (that is, the pH increased), more carbonic acid would dissociate to balance the equilibrium between the two forms (the reaction would proceed to the right). If, instead, seawater were to become acidic, this equilibrium would shift to the left, forming carbonic acid at the expense of bicarbonate ion.

The release of the second hydrogen ion converts bicarbonate ion in the previous reaction to *carbonate ion* (CO_3^{2-}):

$$HCO_3^- \longleftrightarrow H^+ + CO_3^{2-}. \qquad (3)$$
$$\text{bicarbonate} \qquad \text{hydrogen} \quad \text{carbonate}$$
$$\text{ion} \qquad\qquad \text{ion} \qquad\quad \text{ion}$$

(Note that we have previously used the term "carbonate" to refer to calcium and magnesium carbonate minerals.) For a given hydrogen ion concentration (pH), the relative amounts of bicarbonate and carbonate ions are adjusted until equilibrium is achieved. A pH above neutral favors the carbonate ion, whereas a pH below neutral favors the bicarbonate ion.

Perturbation of this equilibrium (for example, by the diffusion of anthropogenic CO_2 from the atmosphere to the ocean) changes the pH of seawater. These changes in pH affect the relative concentrations of carbonic acid, bicarbonate ion, and carbonate ion in the following way: CO_2 dissolves, forming carbonic acid (reaction 1); the carbonic acid dissociates, forming bicarbonate and hydrogen ions (reaction 2), which causes the pH to drop. Hydrogen ion then reacts with carbonate ion, forming another bicarbonate ion (the reverse of reaction 3). The overall chemical reaction describing the uptake of anthropogenic CO_2 is the sum of these three reactions (reactions 1, 2, and the reverse of 3):

$$CO_2 + CO_3^{2-} + H_2O \longleftrightarrow 2HCO_3^-.$$

Thus, the ocean's capacity to take up CO_2 derived from fossil fuels is enhanced relative to what it would be if it simply equilibrated with the atmosphere, because it is converted to other forms of inorganic carbon. There is a limitation on the amount that can be taken up, however: the amount of carbonate ion in the ocean, which is smaller than the total available fossil fuel. Thus, if we continue to utilize fossil fuels, the oceans' capacity to take it up will become depleted, and a larger fraction will remain in the atmosphere.

Chemical Weathering

A similar chemistry applies when atmospheric CO_2 dissolves in raindrops, making them naturally acidic. As we noted earlier, the unpolluted pH of rainwater is generally between 5 and 6. Rocks exposed at Earth's surface undergo chemical attack from this rain of dilute acid, a process known as *chemical weathering.*

Crustal rocks are composed mainly of two types of minerals: carbonates and silicates. Recall from previous chapters that carbonate minerals, such as calcite, contain carbon in combination with oxygen and other elements. Calcite and dolomite are the most abundant carbonates at Earth's surface, occurring as the dominant minerals in limestones and dolostones and as minor minerals in a host

pH

The hydrogen ion H$^+$ is the smallest of all cations. The small size and ionic charge make hydrogen ions extremely reactive: They tend to infiltrate solids, breaking bonds and causing the molecules that make up the solids to dissolve. *Solutions* (liquids with dissolved material) with high concentrations of hydrogen ions are called **acids.** Vinegar and hydrochloric acid are common acids. Solutions with low concentrations of hydrogen ions are called **bases** (or *alkalis*). Baking soda and lye dissolved in water are common bases.

Strong acids are solutions that completely dissociate: When dissolved in water, the anions separate completely from the hydrogen cations. This dissociation leads to high hydrogen ion concentrations. For example, the strong acid hydrochloric acid (HCl) dissociates to form hydrogen ions and chloride ions:

$$HCl \longrightarrow H^+ + Cl^-.$$

hydrochloric	hydrogen	chloride
acid	ion	ion

Other strong acids include nitric acid (HNO$_3$) and sulfuric acid (H$_2$SO$_4$). Weak acids, such as boric acid (H$_3$BO$_3$) and carbonic acid (H$_2$CO$_3$), only partially dissociate when dissolved in water.

The concentration of dissociated hydrogen ions (expressed in moles per liter of solution) determines the *acidity* of the solution. Acidity is commonly measured by the pH scale, which is a logarithmic scale. The **pH** of a solution is a close approximation of the negative of the logarithm (to the base 10) of the hydrogen ion concentration, [H$^+$] (in moles per liter):

$$pH = -\log[H^+].$$

(We say "approximation" because chemists define pH in terms of the *activ-*ity of hydrogen, which is the concentration of only the fraction of hydrogen ions available for chemical reaction. The rest are involved in electrostatic interactions with other ions.) For example, a solution with a hydrogen ion concentration of 10^{-4} mol per liter would have a pH of 4, because the logarithm of 10^{-4} is −4. At room temperature, pure water is defined as having a *neutral* pH, or a pH of exactly 7. Solutions with a pH less than 7 are acids, whereas solutions with a pH greater than 7 are bases (Box Fig. 7-2). The pH of the surface and deep oceans is slightly basic—about 8 and 7.5, respectively. Rainwater that is in equilibrium with atmospheric CO$_2$ has a slightly acidic pH, between 5 and 6. Lakes, rivers, and streams range in pH from about 6 to 9—that is, from slightly acidic to mildly basic.

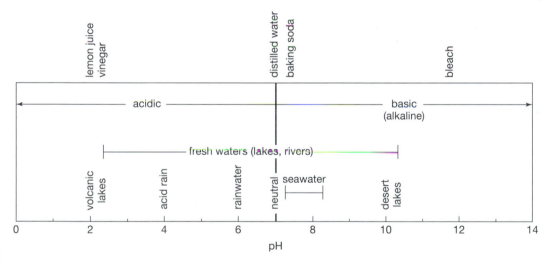

BOX FIGURE 7-2

The pH scale.

---

of other rock types. These minerals have fairly simple chemical formulas: CaCO$_3$ and CaMg(CO$_3$)$_2$, respectively. Silicate minerals contain compounds of silicon and oxygen. They tend to have rather complicated compositions. They are most abundant in igneous rocks but are also common minerals in sedimentary and metamorphic rocks. (See Chapter 6 for a further discussion of rocks and minerals.)

In discussing the carbon cycle, we are most interested in the weathering of the calcium-bearing minerals, because calcium ions released by weathering are used by organisms in the construction of calcium carbonate shells and skeletons. Important calcium-bearing silicates include *anorthite*, CaAl$_2$Si$_2$O$_8$, and *hornblende*, NaCa$_2$(Mg,Fe,Al)$_5$(Si,Al)$_8$O$_{22}$(OH)$_2$. However, we will use *wollastonite* (CaSiO$_3$), which has a much simpler formula, to represent calcium silicates. It is the relative amounts of calcium and silicate that matter to our discussion of chemical weathering, not the detailed compositions of the weathered minerals.

When exposed to rain, both carbonates and silicates weather chemically, although the carbonates dissolve

(a)

(b)

FIGURE 7-14
Differential weathering of (a) granite (From Bill Aron, Photo Researchers, Inc.) and (b) limestone (From Mark Burnett, Photo Researchers, Inc.) The granite headstone is a few years older than the limestone tombstone.

much more rapidly (Fig. 7-14). Chemical weathering neutralizes the acidity of carbonic acid in much the same way that an antacid neutralizes the acidity of your stomach:

Carbonate weathering:
$$CaCO_3 + H_2CO_3 \longrightarrow Ca^{2+} + 2HCO_3^-,$$

calcium   carbonic   calcium   bicarbonate
carbonate   acid   ion   ion

Silicate weathering:
$$CaSiO_3 + 2H_2CO_3 \longrightarrow$$

wollastonite   carbonic
acid

$$Ca^{2+} + 2HCO_3^- + SiO_2 + H_2O.$$

calcium   bicarbonate   silica   water
ion   ion

In both weathering reactions, the products include Ca^{2+} and HCO_3^-. Silicate weathering also yields dissolved silica (SiO_2), which, together with Ca^{2+} and HCO_3^-, is transported to rivers and ultimately to the oceans. Silicate weathering consumes twice as much dissolved CO_2 (in the form of carbonic acid) as does carbonate weathering.

This fact will prove important later in this chapter when we attempt to balance the carbon cycle.

Carbonate Mineral Deposition

The oceans eventually receive the products—soil particles and dissolved materials—of the chemical and physical weathering and erosion of the land surface. Most of the particulate material is deposited near the mouths of rivers in deltas, beaches, and other deposits near shore. In contrast, the dissolved materials mix with seawater and, if their residence times are longer than the ocean's mixing time, are distributed throughout the world oceans. The constant flux of dissolved materials from land to sea would gradually increase the saltiness of seawater by dilution were it not for processes that continuously remove material from the sea (see Chapter 5).

Some organisms, such as diatoms (Fig. 7-8a), radiolarians (Fig. 7-8d), and sponges, remove dissolved silica from seawater. They convert it into solid (opaline) silica as the structural part of their skeletons. Other organisms, such as foraminifera (Fig. 7-8c), coccolithophorids (Fig. 7-8b), corals, and shellfish, produce solid $CaCO_3$ in forming their shells and skeletons. Although these minerals can form *abiotically* (without the aid of organisms), most of the $CaCO_3$ precipitated from the ocean today is formed by such organisms.

These carbonate-producing marine organisms remove Ca^{2+} and HCO_3^- from seawater and precipitate $CaCO_3$ as a shell or skeleton. The overall chemical reaction is essentially the reverse of the carbonate weathering reaction:

Carbonate precipitation:
$$Ca^{2+} + 2HCO_3^- \longrightarrow CaCO_3 + H_2CO_3.$$

calcium   bicarbonate   calcium   carbonic
ion   ion   carbonate   acid

Calcium carbonate producers cause a shift in the ocean's carbon chemistry. By precipitating $CaCO_3$ shells or skeletons, they enhance H_2CO_3 (and, because of the equilibrium in reaction 1, dissolved CO_2) concentrations and reduce HCO_3^- concentrations—as well as pH. By increasing the concentration of dissolved CO_2 in surface waters, carbonate-producing organisms produce a CO_2 gradient between the oceans and the atmosphere. This gradient promotes the diffusion of CO_2 from the oceans to the atmosphere.

Some carbonate producers (for example, coccolithophorids) are also phytoplankton. They photosynthesize organic matter, which tends to drive the dissolved carbon system in the opposite direction, toward lower CO_2 concentrations; hence these microorganisms have opposing effects on atmospheric CO_2 concentrations. The majority of phytoplankton do not produce carbonate skeletons, however. Thus, the overall effect of biological production in the surface ocean is in favor of reduced CO_2 concentrations: On average, the ratio of organic matter to carbonate mineral production for plankton is about 4:1.

When plankton die, their shells or skeletons sink through the water column but are less subject to destruction during the trip to the sea floor than is their organic debris. In regions where the total water depth is less than 4 km or so, carbonate particles accumulate more or less intact on the sea floor. These waters are said to be *saturated* with respect to $CaCO_3$. This shallower-water deposition of planktonic debris is part of the material that eventually becomes limestone. Other limestones form in very shallow tropical waters, where reefs and other carbonate-producing organisms live on the sea floor. Deeper waters, however, have higher concentrations of dissolved CO_2 (carbonic acid), due to the decomposition of organic matter. As a result, these waters are corrosive to $CaCO_3$; they are said to be *undersaturated* with respect to $CaCO_3$. The level at which the rate of dissolution of carbonate sediment balances the flux of carbonate settling through the water column is called the *carbonate compensation depth* (CCD). Below this depth, carbonate materials, in the form of shells or skeletons, dissolve in transit as they settle through the water column and do not accumulate on the sea floor. Thus, the deep-ocean basins are devoid of the cover of plankton-derived carbonate sediments that occupy the shallower parts of the ocean—for example, along the mid-ocean ridges (Fig. 7-15). Interestingly, carbonate sediments deposited on mid-ocean ridges are carried slowly to depths below the CCD as the sea floor spreads away from the ridge, cools, and subsides (see Chapter 6). Carried below the CCD, the sediments ultimately are subjected to the action of corrosive deep waters. In this way, plate tectonics aids in the slow digestion of carbonate sediments.

Summary of the Inorganic Carbon Cycle

In the past few sections, we have discussed several processes that affect the transfer of inorganic carbon. These processes can be combined into a mass-flow diagram for the inorganic carbon cycle (Fig. 7-16). The free exchange of CO_2 between the atmosphere and the oceans tends to keep the atmosphere near equilibrium with surface waters. During chemical weathering, CO_2 dissolved in rain and soil waters becomes converted to HCO_3^- (neutralized), which is carried to the oceans. There organisms combine it with Ca^{2+} in precipitating $CaCO_3$ shells or skeletons. Some of the $CaCO_3$ dissolves in the water column and on the sea floor. The rest is buried beneath the sea floor, becoming part of the limestone reservoir. When tectonic uplift occurs after millions of years, the limestones become exposed at Earth's surface. There they undergo chemical weathering, and the cycle continues.

Net Removal of CO_2 from the Ocean and Atmosphere

We can combine the equations that we presented in previous sections for carbonate and silicate weathering and for $CaCO_3$ precipitation to determine the net effects of these processes on the chemistry of the oceans and on the composition of the atmosphere. To do this, we simply add the equations by adding the left-hand

= Areas with >75% $CaCO_3$

FIGURE 7-15

The distribution of $CaCO_3$-rich sediments on the sea floor. The pattern closely matches the areas of higher topography (shallower water) along the mid-ocean ridge system. (After W. S. Broecker and T.-H. Peng, 1982. *Tracers in the Sea*. New York: Eldigio Press, Columbia University, p. 59.)

FIGURE 7-16

The global inorganic carbon cycle.

sides of both equations, adding the right-hand sides of both, and then canceling out terms that appear on both sides of the resulting equation—exactly as we would in adding two algebraic expressions. The net effect of $CaCO_3$ weathering on land and $CaCO_3$ precipitation in the ocean is zero, because the two processes balance each other:

Carbonate weathering:
$$CaCO_3 + H_2CO_3 \longrightarrow Ca^{2+} + 2HCO_3^-$$

Carbonate precipitation:
$$Ca^2 + 2HCO_3^- \longrightarrow CaCO_3 + H_2CO_3 \ .$$

Net result: 0

The weathering equation is the reverse of the precipitation equation.

The situation is different for the weathering of calcium silicate minerals on land. First of all, calcium-silicate minerals do not re-form under Earth surface conditions. Thus, the calcium liberated during silicate weathering ultimately leaves the ocean as $CaCO_3$. Secondly, recall

from the silicate weathering equation that two molecules of H_2CO_3 are needed to dissolve silicate minerals, whereas only one H_2CO_3 molecule is liberated when carbonate-bearing organisms precipitate their shells or skeletons. Thus:

Silicate weathering:
$$CaSiO_3 + 2H_2CO_3 \longrightarrow Ca^{2+} + 2HCO_3^- + SiO_2 + H_2O$$
Carbonate precipitation:
$$Ca^{2+} + 2HCO_3^- \longrightarrow CaCO_3 + H_2CO_3$$
Ocean/atmosphere CO_2 exchange:
$$CO_2 + H_2O \longrightarrow H_2CO_3$$
.
Net result:
$$CaSiO_3 + CO_2 \longrightarrow CaCO_3 + SiO_2$$

(The SiO_2 liberated during silicate weathering is utilized by diatoms and radiolaria in skeleton production.) So, the combined processes of silicate weathering on land and carbonate precipitation in the sea lead to a net conversion of atmospheric CO_2 to solid $CaCO_3$. This process serves as a net outflow of CO_2 from the atmospheric reservoir, analogous to the net outflow of photosynthetically produced organic carbon that leaks out of the organic carbon cycle due to sedimentary burial.

Like the leak in the organic carbon cycle, the rate of conversion of atmospheric CO_2 to limestone by silicate weathering and $CaCO_3$ deposition is quite small: about 0.03 Gton(C)/yr. Nevertheless, if left unbalanced, this flux would apparently deplete the atmosphere of carbon in about 20,000 years. Actually, this number is misleading, because we have not accounted for the fact that the various oceanic reservoirs hold more than 50 times more carbon than the atmospheric reservoirs hold (see Fig. 7-3). A reduction in atmospheric CO_2 creates a concentration gradient between the atmosphere and the oceans. As a consequence, CO_2 diffuses from the oceans to the atmosphere until the gradient is removed. Thus, the depletion time would be more like a million years, the time it would take for all the oceans' inorganic carbon to diffuse into the atmosphere and react with silicate rocks. Although this characteristic response time is significantly longer than the calculated 20,000 years, it is still short in terms of changes in atmospheric composition and climate over geologic time scales of many millions of years. On such time scales, then, there must be a return flux of CO_2 to the atmosphere and oceans to offset this outflow. This balance, which is in essence the long-term inorganic carbon cycle, has come to be known as the *carbonate-silicate geochemical cycle.*

The Carbonate-Silicate Geochemical Cycle

After inorganic carbon is involved in chemical weathering and carbonate mineral precipitation and is removed by sedimentary burial, plate tectonics provides the needed return flux of CO_2 in the form of metamorphic and volcanic CO_2 inputs to the atmosphere. Mantle-derived CO_2 is released to the ocean-atmosphere system at mid-ocean ridges and along convergent margins (Fig. 7-17). This carbon is derived from the mantle and so is in a sense "new."

At convergent plate margins (deep-sea trenches), some of the sediments resting on the downgoing slab are subducted along with the plate (see Chapter 6). The plate and its sediment cover are carried to depths as great as hundreds of kilometers within the mantle, where high temperatures and pressures promote chemical reactions that transform the sediments into metamorphic rock. Among these reactions is the reaction between sedimentary carbonate minerals and silica-rich sediments that forms silicate minerals and releases CO_2:

Carbonate metamorphism:
$$CaCO_3 + SiO_2 \longrightarrow CaSiO_3 + CO_2.$$

calcite   silica   wollastonite   carbon dioxide

This process is termed **carbonate metamorphism.** As before, we use the mineral wollastonite to represent the more-complex silicate minerals that are typically generated by this process.

If sufficiently high temperatures are reached at depth during carbonate metamorphism, magmas are generated. These magmas may erupt in volcanoes at the surface, releasing CO_2 to the atmosphere. The CO_2 in these volcanoes probably includes some mantle-derived CO_2 and some CO_2 from the subducted crust and sediments, but scientists do not yet know the relative proportions of these two sources. Under metamorphic conditions the CO_2 produced can migrate as a fluid toward the surface. Although a substantial fraction of it reacts with minerals along the way, some CO_2 is released through springs and seeps to the atmosphere.

Together, silicate weathering, carbonate precipitation, and ocean–atmosphere exchange are the reverse of carbonate metamorphism. (Compare the equation for the net result of these processes, given in the previous section, with the carbonate metamorphism equation.) Without a fairly close balance between the inflows and outflows of carbon, the supply of this important greenhouse gas to the atmosphere and ocean would, on geological timescales, quickly be depleted. Earth would soon become a frozen ball of ice.

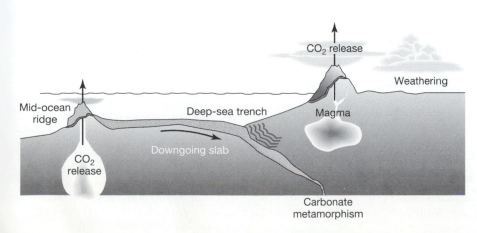

FIGURE 7-17
Pictorial representation of the carbonate–silicate geochemical cycle.

That may have been the fate of Mars, which appears to once have had flowing water on its surface and, perhaps, a CO_2-rich atmosphere with a stronger greenhouse effect. In contrast, on Earth the release of CO_2 after carbonate metamorphism and volcanism has essentially balanced the consumption of CO_2 during silicate weathering over the history of the planet. What has ensured this balance? We cannot call on simple chemical equilibrium, because the reactions involved are representative of a whole host of processes rather than a single chemical reaction. Rather, we must look for feedback loops that, according to the amount of CO_2 in the atmosphere, adjust the rates of CO_2 input by volcanism or of CO_2 removal by silicate weathering and thereby keep the reservoir at steady state.

Long-Term Feedbacks in the Carbonate–Silicate Cycle

Because it is driven largely by heat flow from the Earth's interior, the rate of volcanism is probably not very sensitive either to the amount of CO_2 in the atmosphere or to the climate of Earth's surface. In contrast, many climatic factors affect the rate of chemical weathering. The regulation of atmospheric CO_2 on long time scales (millions of years) likely is the consequence of the feedback between climate and rates of silicate weathering. The climatic factors that help regulate the chemical weathering rates of silicate rocks include the following:

• temperature: Rates of reactions, including chemical weathering, tend to increase as temperature increases; and

• net rainfall: Weathering requires water as a medium both for the dissolution of minerals and for the transport of the dissolved material to the oceans, and thus weathering rates rise as precipitation increases.

These environmental factors are responsive to atmospheric CO_2 levels. Recall from Chapter 3 that, as a result of the greenhouse effect, global temperatures rise as the atmospheric content of CO_2 increases and that rates of evaporation increase with increasing temperature. We know from Chapter 4 that water that evaporates from the ocean must fall as precipitation. For these reasons, we would expect that a warmer world would be a wetter world: Net precipitation should increase as temperature increases. Thus, the silicate weathering rate should increase as the atmospheric CO_2 level rises. Figure 7-18 shows a feedback loop for these processes. On the other side of the feedback loop, increased silicate weathering rates tend to reduce atmospheric CO_2 levels because silicate weathering uses up carbonic acid.

The overall feedback loop, as shown in Figure 7-18, is negative. The feedback tends to stabilize Earth's climate against perturbations. In Chapter 8 we shall examine the operation of this feedback loop through Earth's history.

Links between the Organic and Inorganic Carbon Cycle

Although we have differentiated between the terrestrial and marine organic carbon cycles and between the organic and inorganic carbon cycles, all these cycles are inextricably linked as parts of the global carbon cycle. Changes that occur on land rapidly affect the oceans through the transport of carbon and nutrients by rivers and through variations in atmospheric CO_2. Changes in the recycling of organic matter affect atmospheric and oceanic CO_2 and in turn the whole of the carbon cycle. Thus, the cycle divisions we have made are artificial, but they help us represent a complicated system in simple terms.

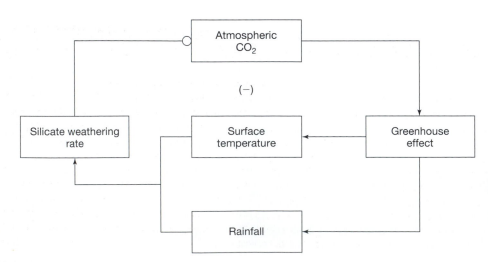

FIGURE 7-18
Systems diagram showing the negative feedback loop that results from the climate dependence of silicate–mineral chemical weathering and its effect on atmospheric CO_2. This feedback loop is thought to be the major factor regulating atmospheric CO_2 concentrations and climate on long time scales.

Biological Enhancement of Chemical Weathering

A particularly active and controversial area of geological research centers on the extent to which terrestrial biota may enhance rates of chemical dissolution of minerals exposed at Earth's surface. Several factors are involved, including the following:

- Microbial decomposition and respiration by plant roots in soils generate large amounts of carbon dioxide. The CO_2 concentration in soils is generally 10 to 100 times that in the atmosphere.
- Microbial decomposition also releases organic acids that cause mineral dissolution. By increasing weathering rates, these biological processes enhance the release rate of essential mineral nutrients to the biota.
- Plant root development leads to the stabilization of soils on steep slopes, where erosion would otherwise strip soils away. This process creates a more stable weathering environment where mineral dissolution can take place for longer times before erosion takes place.
- Roots penetrating through fractured rock tend to enlarge the fractures, causing further disintegration of the rock and allowing water and soil acids to penetrate deeper into the rock.

Field experiments to test the hypothesis of biological enhancement of chemical weathering span environments that have huge contrasts in climate (Iceland and Hawaii) but similar types of exposed rocks (basalts). In Iceland, scientists have found that streams draining areas of plant cover have significantly higher dissolved ion concentrations than do those draining areas of bare rock.

In Hawaii, scientists are looking at rock surfaces covered with lichens (Box Fig. 7-3) and comparing them with bare surface exposures. Underneath the lichen there generally is a thin layer of material that is leached of the more soluble elements;

BOX FIGURE 7-3

A lichen-covered surface in Hawaii. (From Douglas Faulkner, Photo Researchers, Inc.)

bare surfaces have no such layer. These leached layers appear to be the result of mineral dissolution aided by the production of organic acids, although some scientists argue that the leached layers actually are wind-blown dust trapped by the lichens. Further research is needed to resolve this controversy.

Calculations of the extent of biological enhancement from various field studies are somewhat wide-ranging but indicate that biological activity accelerates rates of chemical weathering by at least a factor of 2. Some lichens studies have concluded that that factor is several hundred times. In any case, biological processes are very important components of the feedback loop that regulates global chemical weathering rates, and thus atmospheric CO_2 levels, on geological time scales.

Chapter Summary

1. The global carbon cycle involves processes that occur on land and in the oceans and involve both biological and non-biological chemical reactions.
2. The terrestrial and marine organic carbon cycles operate on a variety of time scales. On the short time scale (tens to hundreds of years),
   a. Carbon dioxide is removed from the atmosphere during photosynthesis on land and returned during respiration and decomposition. Methane is released to the atmosphere from soils, where anaerobic fermentation is taking place.
   b. A small amount of terrestrial organic carbon survives respiration and decomposition and is buried in sedimentary basins on land or is transported to the sea.
   c. In the oceans, phytoplankton produce organic matter that is consumed by zooplankton and decomposed by aerobic and anaerobic bacteria.
   d. A small fraction of the organic matter settling through the water column is not decomposed and is instead buried in marine sediments.
3. On longer time scales (millions of years), the organic matter buried in sediments undergoes lithification with the sediments. Most of these sediments are muds, and the rocks formed are shales.
   a. When concentrations of organic matter are very high in the sediment, fossil fuels may form during burial and lithification.
   b. The sedimentary rocks may eventually undergo uplift through tectonic processes, exposure, and weathering. During weathering, organic matter undergoes oxidation, producing CO_2, which escapes to the atmosphere.
4. An inorganic carbon cycle, involving oxidized forms of carbon, is important on both short and long time scales (millions of years).

a. Atmospheric CO_2 exchanges with CO_2 dissolved in the surface ocean on a time scale of decades. The uptake of CO_2 is enhanced by reactions among the forms of dissolved inorganic carbon in seawater.

b. Atmospheric CO_2 dissolves into rainwater, creating an acidic solution. When the rain falls on the land surface, reactions with carbonate and silicate minerals convert carbonic acid to bicarbonate ion. The bicarbonate ion is carried by rivers to the oceans.

c. In the oceans, carbonate-secreting organisms use the bicarbonate ion in the construction of their shells or skeletons. This material may dissolve in transit or it may become part of the sediment that covers the sea floor.

5. The regulation of atmospheric CO_2 on long time scales (millions of years) is the consequence of the feedback between climatic factors and rates of chemical weathering of silicate rocks, as part of the long-term inorganic carbon cycle, referred to as the carbonate–silicate geochemical cycle.

a. Any disturbance in the amount of atmospheric CO_2 affects climate through the greenhouse effect. Changes in climate affect silicate weathering rates and thus the rates of CO_2 consumption.

b. The overall feedback loop is negative, implying that on long time scales the climate system is stable against a wide range of perturbations.

6. The terrestrial and marine organic carbon cycles, the organic and inorganic carbon cycles, and the carbonate–silicate geochemical cycle are inextricably linked as parts of the global carbon cycle.

a. Changes on land are "communicated" rapidly to the oceans by riverine transport of carbon and nutrients and through variations in atmospheric CO_2.

b. Changes in the recycling of organic matter affect atmospheric and oceanic CO_2 and in turn the whole of the carbon cycle.

Key Terms

acid	coal	pH
anion	consumers	photic zone
base	fermentation	photosynthesis
biological pump	inorganic carbon	primary producer
biomass	limestone	primary productivity
biosphere	nutrient	Redfield ratio
carbonate metamorphism	organic carbon	reduced carbon
cation	oxidized carbon	residence time
characteristic response time	petroleum	zooplankton

Review Questions

1. Which of the following carbon reservoirs has the longest residence time: plants, the oceans, or sedimentary limestone?

2. One or more of the following processes involves organic carbon; identify it (them): the precipitation of a calcite skeleton, the exchange of carbon between the oceans and the atmosphere, dissolution at the sea floor, or oxidation during weathering?

3. Describe the biological pump.

4. Why is plate tectonics critical to the maintenance of an atmosphere–ocean reservoir rich in carbon?

5. Limestone (carbonate) weathering does not lead to the net removal of carbon dioxide from the atmosphere. Why not?

Critical-Thinking Problems

1. The key to stability is feedback between the reservoir and the fluxes into and/or out of the reservoir. Assume that the rate of outflow from a reservoir depends on the size of the reservoir according to the following relationship: outflow rate = $k \cdot$ (size of reservoir), where k is a constant.

a. A reservoir of water has a volume of 5000 liters, and the rate of outflow at steady state is 25 liters per minute. What is k? (Give both the numerical value and its units.) What is the residence time? What is the relationship between k and the residence time?

b. The inflow rate is 25 liters per minute. Describe graphically and in words how the reservoir size would change with time, beginning with a reservoir size of zero and continuing until the reservoir reaches steady state.

2. Use Figure 7-4 to answer the following questions:

a. During which season is the rate of photosynthesis greatest, relative to the combined rate of respiration and decomposition, and during which season is it smallest? Explain your reasoning.

b. At what times during the year is the atmosphere at steady state with respect to CO_2?

c. On the basis of your answers to parts (a) and (b), sketch a graph depicting how net photosynthesis and net respiration–decomposition vary throughout the seasons. (The graphs should extend over several years.)

d. Using Figure 7-4 and the graphs that you sketched in part (c) and averaging over several seasons, estimate the maximum net rates of CO_2 removal and addition during the seasons when these changes are greatest.

3. A giant meteor crashes into Earth, causing devastating environmental changes that kill off all life in the oceans.
   a. Describe how the vertical distribution of dissolved oxygen, carbon, and nutrients would respond.
   b. Would the temperature and salinity of the ocean be affected by the loss of the biological pump? Why or why not?
   c. If the meteor caused, instead of the complete loss of marine life, the sudden cessation of thermohaline circulation in the oceans, what would be the effect on the vertical and spatial distribution of dissolved nutrients, carbon, and oxygen in the world's oceans?

4. a. The atmosphere consists of 78% N_2, 21% O_2, 1% Ar, and about 0.036% (360 ppm) CO_2. What is the mean molecular weight of air? Round your answer to three significant figures, and use the following table of atomic weights:

Element	Atomic Weight
C (Carbon)	12.011
N (Nitrogen)	14.0067
O (Oxygen)	15.9994
Ar (Argon)	39.948

   b. The total mass of the atmosphere is about 5×10^{18} kg. How many moles each of air, O_2, and CO_2 are present in the atmosphere? (*Note:* Calculate the latter two answers from the first one rather than by computing the masses of O_2 and CO_2. The values listed in part (a) for the various gases are abundances by volume, not by mass. This fact, and the fact that a mole of any gas takes up the same volume at a given pressure and temperature, mean that you need to work in moles.)
   c. Forests contain about 600 Gton(C) in the form of wood and leaves. Suppose that all the world's forests were to burn down instantaneously. By how much would atmospheric CO_2 increase? By how much would O_2 decrease? Express your answers in percentages. Assume that the equation for burning is the same as that for respiration (given earlier in this chapter).

5. Explain why lakes and rivers have slightly basic pH values, whereas rainwater (the ultimate source of water for lakes and rivers) is slightly acidic.

Further Reading

General

Berner, R. A., and A. C. Lasaga. 1989. "Modeling the Geochemical Carbon Cycle." *Scientific American* 260(3):74.
Mackenzie, F. T. 1995. "Biogeochemistry." In *Encyclopedia of Environmental Biology*. New York: Academic Press, vol. 1, p. 249.

Advanced

Post, W. M., T.-H. Peng, and W. R. Emanuel. 1990. "The Global Carbon Cycle." *American Scientist* 78(4):310.

Chapter 8
Long-Term Climate Regulation

(a)

(b)

(c)

Geological indicators of glaciation: (a) a tillite from the 2.4-b.y.-old Gowganda formation in Canada; (b) glacial striations from the 0.65-b.y.-old Smalfjord tillite in Norway; (c) a dropstone from the Gowganda formation. (Schopf, J.W.; Earth's Earliest Biosphere: It's Origin and Evolution. Copyright© 1983 by Princeton University Press. Reprinted by permission of Princeton University Press.)

Key Questions

- How old is Earth?
- How did the atmosphere and ocean form?
- Why did Earth's climate remain warm despite reduced solar luminosity in the distant past?
- Has Earth's climate generally been warmer or colder than today's climate?
- Why was the climate warm during the time of the dinosaurs, and why has it cooled over the past few tens of millions of years?

Chapter Overview

Earth formed some 4.6 billion years ago by the accretion of solid particles from a cloud of gas and dust surrounding the young Sun. As Earth formed, it developed a thick, CO_2-rich atmosphere and a deep, globe-encircling ocean. The greenhouse effect of this dense atmosphere helped keep Earth's climate warm despite lower solar luminosity at that time. As the Sun gradually brightened, atmospheric CO_2 concentrations diminished, maintaining the climate within the limits favorable to life. This decrease in CO_2 was not accidental; rather, it was a natural consequence of a negative feedback in the carbonate-silicate cycle. Over the past few hundred million years, Earth's climate has fluctuated between warm and cold conditions, primarily as a result of changes in atmospheric CO_2 induced by plate tectonics and the carbonate–silicate cycle.

Introduction

Earth has been in existence for billions of years and has been inhabited by organisms for most of that time. All known organisms require water during at least part of their life cycles (although some can do without water for extended periods). This implies that Earth's surface temperature has remained warm enough to support liquid water throughout this entire time. But we have already seen that the Sun was approximately 30% less bright early in Solar System's history. Why did the climate remain relatively warm despite such a large change in solar heating?

Based on what we learned in Chapter 3, the solution to the "faint young Sun paradox" probably involves either a stronger atmospheric greenhouse effect or a lower planetary albedo. That is what is needed to balance the planetary energy budget. When we last looked at this problem, we had no good reason to believe that either of these factors should have changed in the needed direction. In Chapter 7, however, we studied another component of the Earth system, the carbon cycle, in some detail. In particular, we found that a strong negative feedback exists between atmospheric CO_2 concentrations and the rate of silicate weathering. This feedback has a tendency to stabilize climate over long time scales because it causes atmospheric CO_2 levels to increase when the climate gets too cold or to decrease when it becomes too warm.

This negative feedback in the carbonate–silicate cycle is the most plausible solution to the faint young Sun paradox. But it cannot explain all the details of Earth's climate evolution. If we examine Earth's climate history more closely, we find that its surface temperature has experienced substantial swings: there have been periods (like the present) when the polar regions have been covered with ice, and there have been other, longer periods when polar ice appears to have been completely absent. There have even been periods, like the Late Precambrian Era (about 650 million years ago), when continental ice sheets may have existed in the tropics! In this chapter, we will examine the question of just how stable Earth's climate has been, and we will look at how plate tectonics and biological evolution may have caused large climatic shifts.

Introduction to Geologic Time

One of the most important points that any geology professor makes to an introductory class is the immense amount of time represented in the geologic record. For reasons outlined below, scientists believe that Earth and the rest of the Solar System formed about 4.6 billion years (b.y.) ago. The universe itself has existed for roughly 12 billion years, on the basis of estimates of its current rate of expansion. Both the age of the Earth and the age of the universe are almost inconceivably longer than a typical human lifetime of about 80 years, or even the total amount of time that humanlike species have been in existence, about 4 million years (m.y.). An analogy that is sometimes made is to scale Earth's age down to a single calendar year beginning at 12:01 A.M. on January 1. In that case humans would have first appeared at about 4:40 P.M. on December 31, and the oldest human (about 120 years) would have been born less than one second before midnight later that evening.

In Chapter 1, we introduced the geologic time scale and discussed two events, the Pleistocene glaciations and the K/T mass extinctions, that occurred relatively recently in Earth history. In this chapter, we will focus on events that occurred hundreds of millions to billions of years ago. The major **eons** into which geologic time is divided are the Hadean (4.6–3.8 b.y. ago), Archean (3.8–2.5 b.y. ago), Proterozoic (2.5–0.6 b.y. ago), and Phanerozoic (0.6 b.y. ago–present). The first three of these time intervals are often collectively termed the Precambrian, because they come before the Cambrian Period, when the first **shelly fossils** (fossilized remains of shelled organisms) appeared in the rock record. Figure 8-1 shows some of the other major events that have occurred during Earth's history.

The Age of the Earth

The problem of determining Earth's age is somewhat complicated even though the basic principles of radiometric dating (Chapter 6) are straightforward. The reason is that standard parent–daughter age-dating techniques, like the uranium–lead system, yield only the crystallization ages of the minerals to which they are applied, not the age of the material itself. But, as far as geologists know, none of the rocks and minerals that composed Earth's original crust have been preserved. The oldest minerals that can be dated by standard techniques are a handful of **zircons** (a zirconium silicate mineral) that yield ages of about 4.1 b.y. How do we deduce that Earth is actually half a billion years older than this?

To find Earth's age, one must first answer a related question: How old are meteorites? **Meteorites** are pieces of rock and/or metal that are thought to have drifted around the Solar System for billions of years before hitting Earth. They have been collected from all over the planet, but are found most readily in Antarctica, where they are easily spotted on the ice. The gradual flow and melting of the ice concentrates meteorites in a few specific localities, making them especially easy to find. The most primitive meteorites (because they have not been altered by melting) are called **chondrites.** Dating of chondrites provides an upper limit on Earth's age because these objects are thought to have formed at the same time as the Solar System as a whole.

Meteorites can be dated using standard parent-daughter techniques, but the most accurate method—**lead–lead dating**—involves the use of multiple lead isotopes. The reason this is useful is that the U/Pb ratio differs from one meteorite to another and even among different minerals within the same meteorite, so it is difficult to determine what the initial U/Pb ratio must have been. By using multiple isotopes, we can avoid this problem. The three isotopes used are ^{204}Pb, ^{206}Pb, and

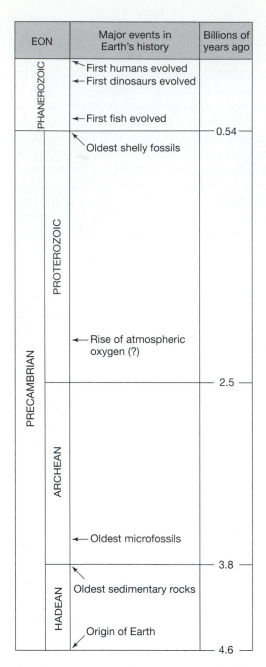

EON	Major events in Earth's history	Billions of years ago
PHANEROZOIC	← First humans evolved ← First dinosaurs evolved ← First fish evolved	
	↖ Oldest shelly fossils	— 0.54
PROTEROZOIC	← Rise of atmospheric oxygen (?)	
		— 2.5
ARCHEAN	← Oldest microfossils	
		— 3.8
HADEAN	↖ Oldest sedimentary rocks Origin of Earth	
		— 4.6

(PRECAMBRIAN spans PROTEROZOIC, ARCHEAN, and HADEAN)

FIGURE 8-1

The geologic time scale, showing major events in Earth's history. (From R.W. Christopherson, *Geosystems: An Introduction to Physical Geography*, 3/e, 1997. Reprinted by permission of Prentice Hall, Upper Saddle River, N.J.)

^{207}Pb. The lead isotopes ^{206}Pb and ^{207}Pb are radiogenic isotopes that derive, respectively, from the decay of ^{238}U and ^{235}U. The half-lives of these decay processes are 4.5 b.y. and 0.713 b.y. The third lead isotope, ^{204}Pb, is a nonradiogenic isotope that is used for comparison in the measurements. By measuring the abundances of these isotopes in different meteorites and then plotting the $^{206}Pb/^{204}Pb$ ratio on one axis and the $^{207}Pb/^{204}Pb$ ratio on

the other, we can construct what is known as an **isochron diagram** (Fig. 8-2). If all the samples being analyzed have the same age, as is true for chondritic meteorites, then the data should (and do) all fall on a straight line. The slope of this line tells us the age of the collection of meteorites, which is accurately determined as 4.55 b.y.

Moon rocks, which were brought back by the Apollo spacecraft missions of the late 1960s and early 1970s, can be dated in a similar manner. The oldest Moon rocks are about 4.44 b.y. old, suggesting that the Moon formed soon after the Solar System itself.

Earth's age is more difficult to obtain because the oldest rocks are all gone. They were probably recycled back into the interior by plate tectonics. However, Earth's age can be deduced indirectly by examining lead isotope ratios in rocks containing lead minerals, that is, minerals that initially contained lead, but little or no uranium. (U–Pb dating, in contrast, is performed on minerals that initially contained uranium, but little or no lead.) The isotopic composition of lead minerals is the same as that of the magma from which they formed. Because Earth's mantle contains uranium as well as lead, the

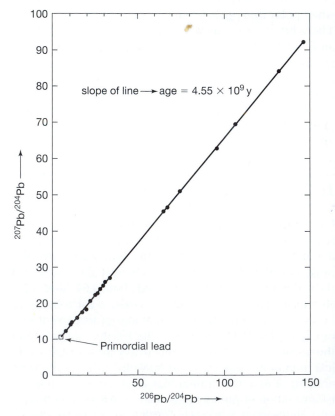

FIGURE 8-2

An isochron diagram, showing lead isotope ratios from a collection of chondritic meteorites. The fact that all the data fall on a straight line shows that all the meteorites have the same age. The age of the meteorites, 4.55 b.y., is determined from the slope of the line.

abundances of ^{206}Pb and ^{207}Pb in the mantle have increased with time, as have those of the magmas derived from it. If we plot the lead isotope ratios from rocks of different ages and then analyze the resulting curve mathematically, we can show that, 4.5 to 4.6 b.y. ago, Earth's mantle should have had the same ^{206}Pb/^{207}Pb ratio as meteorites. This, in turn, implies that Earth formed at the same time as the rest of the Solar System, about 4.5 b.y. ago. This radiometric age scale is the fundamental underpinning to most of our theories about how Earth formed and evolved.

Formation of the Solar System

How did Earth and the rest of the Solar System form? This question has fascinated astronomers for hundreds of years. It is interesting to Earth system scientists as well because it sets the boundary conditions for the rest of Earth's subsequent evolution.

Formation of the Solar Nebula

The Sun is thought to have formed from a collapsing cloud of interstellar dust and gas. Such **interstellar clouds** are observed today by both optical and radio telescopes (Fig. 8-3). They are concentrated in the spiral arms of our Milky Way galaxy, where the density of material is highest. If such a cloud is dense enough and cold enough (about 10 K), it will collapse under its own self-gravity and the process of star formation can begin. It may seem counterintuitive that a cold cloud is required to form a hot star, but this is indeed the case. The reason is that, according to the general gas law (Chapter 4), a cold gas exerts less pressure than a warm gas of the same density. A cold interstellar cloud has less internal pressure to counteract the force of gravity and is therefore more likely to collapse. Once it does so, the cloud immediately warms up because the infall of material releases gravitational energy, which is converted into heat. The innermost part of the cloud becomes hot enough for thermonuclear fusion reactions to begin, and a new star is born.

The clouds that are observed in interstellar space are typically many thousands of times the mass of the Sun. As they contract, however, they produce smaller fragments that can themselves contract to form one or more stars. Whether the collapse results in a single star, or a multiple star system, depends largely on how fast the cloud fragment is rotating: the faster it rotates, the more likely it is to form two or more stars. In the case of our own Sun, a single star formed. This is fortunate for us, because a multiple star system would probably

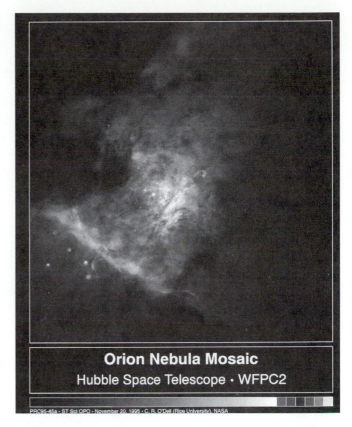

FIGURE 8-3

A star-formation region in the constellation Orion. (From NASA Headquarters.) [See color section]

be a very difficult place to form a habitable planet like Earth. The cloud fragment did have a certain amount of rotation, however, and this caused some of its material to spread out into a disk. The gas and dust that made up the disk are referred to as the **solar nebula** (Fig. 8-4).

Formation of Planets

Once the solar nebula was in place, the process of planetary formation would have begun. The nebula itself would have been heated by the emerging Sun—the **proto-Sun**—so its interior would have warmed. At the same time, small particles of solid material would have begun to condense from the gas. In the hot, inner parts of the nebula, these grains consisted mainly of rocky materials such as iron and silicate minerals that can condense at temperatures as high as 2000 K. That is why Earth and the other **terrestrial planets** (Mercury, Venus, and Mars) are composed principally of rock. In the cooler, outer parts of the nebula, icy materials such as water (H_2O), methane (CH_4), and ammonia (NH_3) would also have condensed. Thus, the giant planets

FIGURE 8-4
The solar nebula: The rotating cloud of dust and gas from which the Sun and the planets formed.

(Jupiter, Saturn, Uranus, and Neptune) contain large amounts of these more volatile compounds. **Volatile compounds** are substances that have low boiling points. The particles formed by condensation were gravitationally attracted to the midplane of the nebula, where they clumped together to form **planetesimals,** small proto-planets. These planetesimals collided with each other, often sticking together to form larger bodies in a process called **accretion.** This process is shown schematically in Figure 8-5. Over tens to hundreds of million of years, the planetesimals grew to form the planets that we see today.

While Earth was growing by accretion, its core should have started to form. Recall that the innermost parts of Earth are its solid inner core and liquid outer core, both composed mainly of iron and nickel. Core formation was once thought to have been triggered by radioactive heating after Earth was fully formed. But it is now believed that core formation occurred as the planet itself was forming. Some of the planetesimals that collided with Earth during accretion were so large that they melted large portions of the crust and upper mantle. This allowed the iron and nickel to separate out and flow down to form the core.

Formation of Jupiter

Two specific events that occurred during the process of planetary accretion have special significance for surface conditions on Earth. The first was the formation of the giant planet Jupiter. Jupiter has over 300 times the mass of Earth and more than three times the mass of the next largest planet, Saturn. The reason Jupiter is so large, astronomers believe, is that its core accreted early enough to capture large amounts of hydrogen and helium from the solar nebula before the nebula dissipated. Accretion would have been rapid at Jupiter's orbit because volatiles could condense in addition to metal and silicates. To capture hydrogen efficiently, Jupiter's core must have grown very rapidly to a mass several times that of Earth. Observations of young stars indicate that nebular gas and dust persist for at most a few million years, after which time they are either incorporated into planets or they spiral back into the star.

Jupiter affects surface conditions on Earth by perturbing asteroids from the asteroid belt into Earth-crossing orbits and by preventing most comets from reaching the inner Solar System. The first effect makes Earth a more dangerous place to live, whereas the second one tends to make it safer. As mentioned in Chapter 1, impacts of comets and asteroids are thought to have played a major role in the evolution of life. A large asteroid impact may have caused the extinction of the dinosaurs, and this may in turn have paved the way for the rise of mammals. Thus, biological evolution might have taken an entirely different course had Jupiter not attained the size that it did.

Formation of the Moon

Another celestial event that had a profound influence on Earth's subsequent evolution was the formation of the Moon (Fig. 8-6). Most of us know that the Moon's gravitational pull affects ocean tides. However, few

(a) A slowly rotating portion of a large nebula becomes a distinct globule as a mostly gaseous cloud collapses by gravitational attraction.

(b) Rotation of the cloud prevents collapse of the equatorial disk while a dense central mass forms.

(c) A protostar "ignites" and warms the inner part of the nebula, possibly vaporizating preexisting dust. As the nebula cools, condensation produces solid grains that settle to the central plane of the nebula.

(d) The dusty nebula clears by dust aggregation into planetesimals or by ejection during a T-Tauri stage of the star's evolution. A star and a system of cold bodies remains. Gravitational accretion of these small bodies leads to the development of a small number of major planets.

FIGURE 8-5

The process of planetary accretion. By colliding with each other, (a) small planetesimals (b) grow gradually into (c) large planets. (From W.K. Hamblin and E.H. Christiansen: *Earth's Dynamic Systems,* 8/e, 1998. Reprinted by permission of Prentice Hall, Upper Saddle River, N.J.)

people are aware that the Moon affects climate as well. It does so by stabilizing Earth's obliquity. Recall from Chapter 4 that it is Earth's *obliquity,* or tilt (currently 23.5°), that gives rise to the normal progression of the seasons at middle to high latitudes. We will see in Chapter 10 that Earth's obliquity varies slightly and that these variations have influenced the glacial–interglacial cycles of the past 3 million years. Computer modeling studies have shown that, without the Moon, Earth's obliquity would vary by much larger amounts, occasionally reaching values as high as 85°. This would wreak havoc with climate because the seasonal cycles would be extremely large over much of Earth's surface. Thus, from the standpoint of planetary habitability, the formation of the Moon may be one of the most important events to occur during the formation of the Solar System.

How exactly did it happen? Many theories have been advanced, including co-accretion (accreting in Earth's orbit at the same time as did Earth), fission (splitting apart of a rapidly rotating Earth), and capture (gravitational capture of a body that originated elsewhere in the Solar System). Most of these theories, however, involve steps that are either physically implausible

FIGURE 8-6

The front side of the Moon. The dark, circular regions, called *maria,* were formed by large impacts. (From NASA Headquarters.)

or that would produce a lunar composition different from that observed in Moon rocks. From the samples collected by the Apollo astronauts, we know that the Moon is depleted in volatile elements compared to the bulk Earth and that its oxygen isotopic composition is similar to that of Earth's mantle. Its density is substantially lower than that of Earth, indicating that the Moon is also depleted in iron. Furthermore, the Moon appears to have had a completely molten surface, or **magma ocean,** shortly after it formed.

One theory that is consistent with all of the available evidence is the **giant impact hypothesis.** According to this hypothesis, Earth received a glancing blow from a Mars-sized planetesimal during the latter stages of accretion, and the debris from the impact reassembled in orbit around Earth to form the Moon. This type of cosmic accident is statistically unlikely, but not so unlikely as to be implausible. We know that it is not a commonplace occurrence because none of the other terrestrial planets have large moons. (Mars has two tiny moons, Phobos and Deimos, but these are thought to be captured asteroids.) Current models of accretion suggest that giant impacts themselves should occur fairly frequently, but that most would not have the right geometry to form a large moon. Our own Moon resulted from one giant impact that did. That is fortunate for us, because the Moon has made Earth's climate much more stable than it would otherwise have been.

Formation of the Atmosphere and Ocean

Even after Earth had recovered from the effects of the Moon-forming impact, its surface would still have been an active and hostile environment. Continued bombardment of the surface by smaller planetesimals should have released large amounts of water and other volatile compounds directly into the atmosphere. This phenomenon is referred to as **impact degassing.** The process has been studied in the laboratory by firing high-speed bullets into targets of volatile-compound–rich material, such as carbonate rock. The shock from the bullet's impact causes the carbonate rock to release gaseous CO_2. Similar shock-induced degassing should occur when volatile-compound–rich planetesimals from the asteroid belt region or from the outer Solar System collided with Earth's surface.

The energy released by the impacts, combined with the greenhouse effect of the gases given off, may have kept Earth's surface so hot that all the water would have remained in the atmosphere as steam. Alternatively, Earth's oceans may have periodically condensed and been re-evaporated many times by impacts. Any in-

coming object larger than about 450 km in diameter would have had sufficient energy to evaporate today's oceans. In either case, both the atmosphere and the ocean should have begun forming as the planet itself formed. This modern conception of planetary formation contrasts with older theories in which Earth was thought to have accreted as a cold, airless body, and in which the atmosphere was thought to have formed later from gases given off by volcanos. Volcanos undoubtedly contributed material to the surface, but the bulk of the atmosphere and ocean were probably formed directly by impacts.

The main period of accretion is believed to have lasted for only about 100 million years. After this time, Earth's surface would have become somewhat more quiescent. Big impacts probably continued to occur sporadically, however, until about 3.8 b.y. ago. The evidence for this **heavy bombardment period** comes from the Moon, Mars, and Mercury. All three bodies appear to have been heavily cratered during their early histories. (Venus does not provide a record of this period because, like Earth, during its history it has been resurfaced many times by volcanism.) The lunar cratering record is the best understood because some of the craters have been dated by using rocks collected near their rims. If the Moon and the other terrestrial planets were being pelted by large objects, it is almost certain that Earth was getting hit, too. The bombardment may have made it difficult for life to have originated before about 4 to 4.2 b.y. ago (see Chapter 9). It may also have brought additional water and other volatile compounds to Earth, particularly if the impactors were comets or volatile-compound–rich asteroids from the outer asteroid belt. So, the atmosphere and oceans may have continued to grow throughout this period.

Composition of the Early Atmosphere

What would have been the composition and surface pressure of the atmosphere during the heavy bombardment period? No one knows for sure, but we can make some educated guesses (Fig. 8-7). Free oxygen, O_2, which makes up about 21% of today's atmosphere, should have been virtually nonexistent, because life—and therefore photosynthesis, the source of free oxygen—had probably not yet arisen. Nitrogen, N_2, does not participate very actively in geochemical cycles; hence, most of it should have been in the atmosphere, as it is today, at about 78% of the total. The present partial pressure of N_2 is about 0.8 bar. Nitrogen would have formed from nitrogen-rich organic compounds and ammonia ice (NH_3) in incoming planetesimals. The shock of their impact on Earth's surface should have converted much of this nitrogen to N_2.

Carbon dioxide, in contrast, could have been much more abundant than the trace amounts present today.

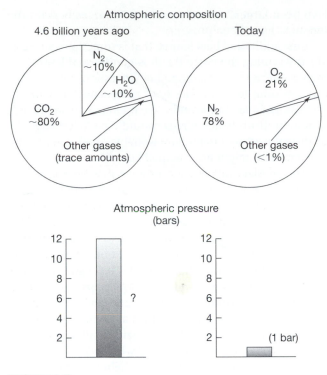

Atmospheric composition

4.6 billion years ago

Today

Atmospheric pressure
(bars)

FIGURE 8-7

Possible atmospheric composition and pressure during the heavy bombardment period, compared with today.

One reason is that the continents must originally have been much smaller. Today, about 30% of Earth's surface is covered by continents. These continents are thought to have grown with time as a consequence of plate tectonics. Geologists do not agree about how fast this process occurred. Many, though, imagine that Earth was initially encircled by a global ocean dotted with chains of volcanic islands. If this was the case, the rate of silicate weathering on land would have been very slow. Weathering reactions may have taken place at the mid-ocean ridges and on the *ejecta* (ejected material) from impacts; however, there would have been no stable place to store the carbonate rocks formed by these processes. The sea floor at that time was presumably being continuously recycled, as it is today. So, the CO_2 released by impact degassing and by volcanism would have tended to accumulate in the atmosphere.

Some estimates suggest that the atmospheric CO_2 partial pressure could have been as high as 10 bars, or 30,000 times the present value, during the first few hundred million years of Earth's history. Such an atmosphere should have produced a surface temperature of 80 to 90°C, according to climate model calculations. The oceans would not have been close to boiling, however, because the atmosphere would have acted like a giant pressure-cooker. Water boils at a higher temperature in such a vessel because the pressure is higher. A dense primitive atmosphere would have acted much the same way.

These ideas about early atmospheric composition are highly speculative. It is difficult to evaluate them, because no record of that time remains. We mention them mainly to show that extremely large changes in atmospheric composition during Earth's early history are possible, even probable. There is every reason to believe that the primitive atmosphere was quite different from today's atmosphere.

The Faint Young Sun Paradox Revisited

Let us return now to the problem that we considered briefly at the end of Chapter 1: The faint young Sun paradox. We have already mentioned that the Sun gets brighter as it ages. The luminosity increase is a direct consequence of the density change caused by the conversion of hydrogen into helium. As such, it is considered to be a *robust* prediction of solar evolution models, meaning that it does not depend sensitively on the details of the model. Thus, even if physicists are wrong about precisely how fusion occurs in the Sun's core, the faint young Sun problem is not likely to go away. Physicists have been bothered for years about the apparent underabundance of **neutrinos** emitted by the Sun. **Neutrinos** are massless particles emitted during nuclear reactions. Either we do not understand neutrinos very well, or our understanding of the processes taking place in the Sun's core is flawed.

There is one way in which the faint young Sun paradox could be avoided. If the Sun was a few percent more massive in the past, it would have been brighter then than it is now. But this would imply that it must have lost large quantities of material during its lifetime. The Sun does lose mass by way of the **solar wind,** an outflow of charged particles from the Sun's **corona** (the hot outermost layer that is visible during a solar eclipse). But the solar wind mass flux is about 10,000 times too small to account for a 1% change in the Sun's mass over geologic time. Rapid mass loss does occur in stars that are still in the process of forming. The outflow from young stars is often referred to as a *T-Tauri wind,* named after a particular star in the constellation Taurus that is behaving in this way. However, T-Tauri winds are expected to last only about 1 million years after the star begins to form, so this process should have no bearing on Earth's climate hundreds of millions of years later. (Earth itself is thought to have taken at least 10 million years to form.) Thus, the faint young Sun paradox appears to require an Earth-bound explanation.

Let us see if we are any closer to resolving this paradox now that we have examined the Earth system in more detail. If you worked out Critical-Thinking Problem 5 in Chapter 3, you should have determined that a 30% reduction in solar luminosity would have led to a 22° decrease in mean surface temperature, T_s, if Earth's albedo and greenhouse effect remained unchanged. As T_s is 15°C today, this would make the surface temperature at 4.5 b.y. ago equal to about −7°C, well below the freezing point of water. The actual surface temperature could have been even colder because of the positive feedback provided by water vapor and ice cover. A climate model calculation that includes the water vapor feedback (by assuming fixed relative humidity) is shown in Figure 8-8. The solar luminosity curve is the same as the one shown in Chapter 1. The lower dashed curve represents the effective radiating temperature, T_e, calculated using the principle of planetary energy balance from Chapter 3. The upper dashed curve is the mean surface temperature calculated by the climate model. The shaded region between the two curves represents the greenhouse effect, ΔT_g; this increases with time because of the water vapor feedback. A constant atmospheric CO_2 concentration of 340 ppm and a constant surface albedo are assumed.

Under these assumptions, T_s falls below the freezing point of water prior to 1.9 b.y. ago and reaches a chilly 255 K (−18°C) at 4.6 b.y. ago. This prediction is at odds with evidence from the geologic record, which shows that liquid water was present at 3.8 b.y. ago and that life has probably been present since that time as well (as we shall see in Chapter 9).

Possible Solutions to the Problem

How can the faint young Sun problem be resolved? According to the planetary energy balance equation, three types of solutions exist. Either the planetary albedo must have been lower in the past, the greenhouse effect must

have been larger, or additional heat sources besides the Sun must have been present.

One additional heat source that is sometimes suggested for warming the early Earth is **geothermal heat** from Earth's interior. As mentioned in Chapter 6, some of this heat is produced by the decay of radioactive elements in Earth's crust and mantle and some is left over from Earth's formation. Both heat sources should have been larger in the distant past, so it is not unreasonable to ask whether geothermal heat might have helped keep Earth warm.

The problem with this suggestion is that the geothermal heat flux is simply not big enough to supply the required energy. A deficit of 30% in the solar flux is equivalent to a loss of about 70 W/m^2 of heating, averaged over Earth's surface. The modern geothermal heat flux is only about 0.06 W/m^2. Theoretical models of Earth's interior evolution suggest that the geothermal heat flux at 4 b.y. ago was higher than the present value by a factor of perhaps 4 or 5, so the available heat flux should have been approximately 0.3 W/m^2. The release of this much heat at the sea floor would have prevented the oceans from freezing to the bottom, but they should still have been covered with an ice layer several hundred meters thick. Little or no light would have penetrated such an ice layer, so this would be hard to reconcile with the evidence for ancient photosynthetic life.

It is also difficult to solve the faint young Sun problem by simply changing Earth's albedo. The current planetary albedo is about 0.3, so the factor $(1 - A)$ in the energy balance relation is equal to 0.7. If the solar luminosity, S, were 30% lower, A would have to be near zero to keep T_e constant. It is difficult to imagine how a water-covered planet could have an albedo near zero, because clouds, which are highly reflective, would almost certainly have been present. Sea ice would likely have been present also, even if continents and continental ice sheets were not. So, although Earth's albedo may have changed with time, it is unlikely that this by itself could have kept the planet warm.

The most likely solution to the faint young Sun problem is that Earth's greenhouse effect was larger in the past. If so, though, which greenhouse gases would have been more abundant? As we have seen previously, water vapor is the strongest greenhouse gas in the modern atmosphere. However, water vapor cannot solve the faint young Sun problem by itself because it is close to saturation and, hence, acts as a feedback on climate rather than as a forcing. Sagan and Mullen, who pointed out the faint young Sun problem in the first place, suggested that ammonia, NH_3, might be the solution. Ammonia is a good absorber of infrared radiation and it is also a *reduced gas,* that is, one that combines with oxygen to form stable compounds. As discussed at greater length in Chapter 9, reduced gases should have been much more abundant prior to the rise of atmospheric oxygen around 2 b.y. ago. That is why Sagan and Mullen thought that ammonia might be a good candidate for warming early Earth. After their paper was published, however, other researchers demonstrated that

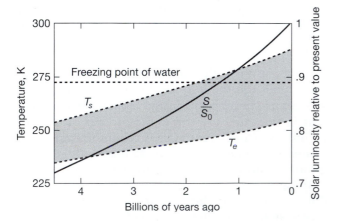

FIGURE 8-8

The faint young Sun paradox. The scale on the right applies to the solar luminosity curve, labeled S/S_0; the scale on the left applies to temperature curves. The shaded area represents the magnitude of the atmospheric greenhouse effect. (From J.F. Kasting et al., How Climate Evolved on the Terrestrial Planets, Scientific American 256(2): 90–97, 1988. Used with permission, © George V. Kelvin/Scientific American.

ammonia would have been rapidly destroyed by ultraviolet radiation. Hence, it is unlikely to have been abundant enough to have provided the necessary warming.

A Dense, CO_2-Rich Early Atmosphere?

The most likely greenhouse gas for keeping early Earth warm is carbon dioxide. There are several reasons for suspecting that atmospheric CO_2 levels were originally much higher. As discussed earlier, smaller continents would have reduced the amount of land available on which to weather silicate rocks and to store carbonate rocks; thus, the CO_2 sink should have been smaller. Impact degassing of late-arriving planetesimals, along with enhanced volcanism on the hot, young Earth, would have created a larger CO_2 source. All these changes would have favored higher atmospheric CO_2 concentrations.

As mentioned at the beginning of this chapter, there is an even more fundamental reason that atmospheric CO_2 levels should have been higher in the past. The carbonate-silicate cycle, which affects atmospheric CO_2 levels over long time scales, contains a strong negative feedback. If Earth's surface temperature were lower as a result of low solar luminosity, the rate of silicate weathering should have been slower, thereby lowering the CO_2 loss rate. CO_2 emitted from volcanos would have accumulated in the atmosphere until the global rate of silicate weathering balanced the volcanic outgassing rate. If Earth had ever become entirely ice-covered, silicate weathering would have ceased entirely and volcanic CO_2 should have accumulated in the atmosphere until the associated greenhouse effect became large enough to melt the ice. Thus, the Earth system has a natural way of avoiding global glaciation and solving the faint young Sun problem.

How much atmospheric CO_2 would have been required to keep the early oceans from freezing? If CO_2 and H_2O were the only important greenhouse gases, then the minimum CO_2 level needed to compensate for a 30%

reduction in solar luminosity is 0.3 bar—about 1000 times the amount of CO_2 in the atmosphere today (Fig. 8-9). Although this sounds like a lot, this amount of CO_2 is not large compared to the total amount of carbon available. The CO_2 stored in carbonate rocks today would produce a partial pressure of some 60 bars were it all present in the atmosphere. Only a small fraction (0.5%) of this CO_2 is needed to resolve the faint young Sun problem. Indeed, atmospheric CO_2 levels may initially have been much higher than this, as discussed earlier in this chapter. The upper limit of 10 bars shown in Figure 8-9 corresponds to the amount predicted on an ocean-covered early Earth.

The Long-Term Climate Record

Although our discussion to this point has been focused on the very early Earth, the negative feedback cycle described above should have operated throughout its history. As the Sun gradually brightened, atmospheric CO_2 concentrations should have fallen in such a way as to keep the carbonate–silicate cycle in balance. If the source for atmospheric CO_2 was higher in the past, as seems likely, the climate at that time should have been warmer than today, despite the faintness of the young Sun. As the source for atmospheric CO_2 gradually decreased toward its present value, Earth's climate should have cooled in a smooth and fairly predictable manner.

If one looks at the geologic record, however, this smooth variation in climate is not what is observed. A variety of different geologic indicators, discussed below, show that **paleoclimate** (past climate) has actually varied in a complex manner, with long periods of warmth separated by shorter periods of intense cold. This suggests that other factors that we have not yet talked about affect long-term climate as well. Determining what those other factors might be occupies the last two sections of this chapter. Before discussing them, though, let us examine some of the

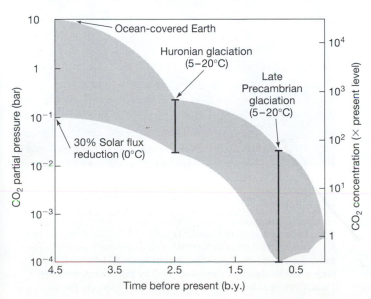

FIGURE 8-9

Atmospheric CO_2 concentrations needed to compensate for changing solar luminosity. The vertical bars at 2.5 b.y. ago and 0.65 b.y. ago show limits estimated from climate model calculations during glacial periods. (Reprinted with permission from J.F. Kasting, Science 25a: 920–926. Copyright 1993 American Association for the Advancement of Science.)

different types of geologic evidence that are used to study paleoclimate and see if we can piece them together to determine the broad outlines of Earth's climate history.

Geological Indicators of Paleoclimate

The types of geological indicators that are useful for determining paleoclimate depend on the time scale being considered. For recent periods of Earth history, it is possible to obtain a reasonably accurate estimate of ocean temperatures by measuring oxygen isotopes in carbonate sediments obtained from *deep-sea cores*. This technique will be described in Chapter 10, where it is used to identify the glacial–interglacial cycles of the past 3 m.y. This method works only for time periods more recent than about 200 m.y. because most of the sea floor older than this has been subducted.

During the past 540 m.y. (the Phanerozoic Era), we can learn about paleoclimate by examining the fossil record. Species of plants and animals that are known to live in certain climates can be used to estimate surface temperatures in the localities where their fossils are found. In doing so, we must account for the fact that the continents have drifted over Earth's surface as a consequence of plate tectonics. This technique is of limited use in the Precambrian Era, however, because the single-celled organisms that were the only extant forms of life during most of this time could have survived under a wide range of climatic conditions.

Evidence of Past Glaciations

The best available evidence for climate change on billion-year time scales comes from geologic deposits formed by glacial ice. Three such types of deposits are shown in Figure 8-10. **Tillites** (Fig. 8-10a) are mixtures of cobbles, pebbles, sand, and mud that have been packed together to form rocks. They are formed from debris produced when glaciers grind up surface rocks. The debris is carried along by the glaciers as they move and is deposited in piles of rubble called **moraines** along the margins of the ice sheets. (Moraines mark the terminal points or flanks of the glaciers.) In association with tillites, geologists sometimes find rocks with long, parallel scratches, or **glacial striations,** formed when moving glaciers drag other rocks across their surfaces (Fig. 8-10b).

A third signature of glaciation is **dropstones,** "misplaced" chunks of rock that occur in otherwise finely laminated marine sediments (Fig. 8-10c). They form when rocks trapped in glacial ice are carried out to sea by icebergs, a process termed **ice-rafting.** When the iceberg melts, the trapped debris falls to the bottom of the ocean and becomes incorporated into sediments.

The Long-Term Glacial Record

Geologists who have studied the long-term glacial record have concluded that Earth's climate history has been marked by five main periods of glaciation (Fig. 8-11). The first such period occurred at approximately 2.2 to 2.4 b.y. ago, the time from which the tillite and dropstone pic-

(a)

(b)

(c)

FIGURE 8-10

Geological indicators of glaciation: (a) a tillite from the 2.4-b.y.-old Gowganda formation in Canada; (b) glacial striations from the 0.65-b.y.-old Smalfjord tillite in Norway; (c) a dropstone from the Gowganda formation. (Schopf, J.W.; Earth's Earliest Biosphere: It's Origin and Evolution. Copyright© 1983 by Princeton University Press. Reprinted by permission of Princeton University Press.)

EON	GLACIATIONS	ERA	Duration in millions of years	Millions of years ago
PHANEROZOIC		CENOZOIC	65	— 65 —
PHANEROZOIC		MESOZOIC	186	— 251 —
PHANEROZOIC		PALEOZOIC	293	— 544 —
PRECAMBRIAN — PROTEROZOIC	Late Proterozoic glaciations	LATE	330	— 900 —
PRECAMBRIAN — PROTEROZOIC	WARM	MIDDLE	700	— 1600 —
PRECAMBRIAN — PROTEROZOIC	WARM	EARLY	900	— 2500 —
PRECAMBRIAN — PROTEROZOIC	Huronian glaciations			
PRECAMBRIAN — ARCHEAN	WARM (?)	LATE	500	— 3000 —
PRECAMBRIAN — ARCHEAN	WARM (?)	MIDDLE	400	— 3400 —
PRECAMBRIAN — ARCHEAN	WARM (?)	EARLY	400	— 3800 —
PRECAMBRIAN — HADEAN			800	— 4600 —

Era	Period	Epoch	Glaciations	Duration in millions of years	Millions of years ago
CENOZOIC	Quaternary	Holocene	Pleistocene glaciations	0.01	0.01
CENOZOIC	Quaternary	Pleistocene	Pleistocene glaciations	1.8	1.8
CENOZOIC	Tertiary	Pliocene		3.5	5.3
CENOZOIC	Tertiary	Miocene		18.5	23.8
CENOZOIC	Tertiary	Oligocene		9.9	33.7
CENOZOIC	Tertiary	Eocene		21.1	54.8
CENOZOIC	Tertiary	Paleocene		10.2	65
MESOZOIC	Cretaceous		WARM	79	144
MESOZOIC	Jurassic		WARM	62	206
MESOZOIC	Triassic		WARM	45	251
PALEOZOIC	Permian		Permo-Carboniferous glaciations	35	286
PALEOZOIC	Carboniferous	Pennsylvanian	Permo-Carboniferous glaciations	39	325
PALEOZOIC	Carboniferous	Mississippian		35	360
PALEOZOIC	Devonian		WARM	50	410
PALEOZOIC	Silurian		WARM	30	440
PALEOZOIC	Ordovician		Late Ordovician glaciations	65	505
PALEOZOIC	Cambrian		WARM	39	544
PRECAMBRIAN					

FIGURE 8-11

The major cold and warm periods during Earth's history. (From W.K. Hamblin and E.H. Christiansen: *Earth's Dynamic Systems,* 8/e, 1998. Reprinted by permission of Prentice Hall, Upper Saddle River, N.J.)

tured in Figure 8-10 derive. Because these deposits were first identified near the banks of Lake Huron in North America, this early cold period is sometimes called the "Huronian glaciation." The Huronian glaciation was followed by over 1 billion years of ice-free conditions during the Middle Proterozoic.

Eventually the climate became cool once again. Indeed, the Late Proterozoic, between 0.8 and 0.6. b.y. ago, was so cold that it is considered something of a mystery. Evidence for glaciation during this time interval is found on six of the seven present-day continents. (The only exception is Antarctica. Antarctica may have been glaciated, too, but it is impossible to tell because it is still largely buried in ice.) The reconstruction shown in Figure 8-12 suggests that the continents were at that time grouped into two supercontinents, one of which was centered near the equator. Alternatively, the continents may have been grouped into a single supercontinent centered on the equator but extending a significant distance to the north and south. The Late Proterozoic glaciation is particularly well documented in Australia, which was situated almost entirely within the tropics in both reconstructions. The geologic evidence suggests that continental ice sheets formed on Australia at latitudes only a few degrees away from the equator. Recent paleomagnetic evidence suggests that the Huronian glaciation may also have involved low-latitude ice sheets. The box "Low-Latitude Glaciations" explains why this is considered a major paleoclimate mystery.

The past three glacial periods have occurred during the Phanerozoic Period: a brief one during the Late Or-dovician Period (about 440 m.y. ago), a long series of glaciations near the boundary between the Permian and Carboniferous Periods (about 280 m.y. ago), and the most recent episode, which is called the Pleistocene glaciation. This name is something of a misnomer because Antarctica began to be glaciated some 15 to 30 million years ago, well before the Pleistocene epoch began, and both Antarctica and Greenland continue to be ice-covered today. To qualify as a long-term "glacial period," it is only necessary that ice be present over part of Earth's surface for an extended period of time. This is different from common usage, whereby the term "glacial period" refers to one of several episodes of maximum ice extent during the last 3 m.y. These shorter-time-scale climate fluctuations, which shall be discussed in Chapter 10, can be thought of as modulations of an overall climate which is basically "glacial."

Causes of Long-Term Climate Change

Changes in Continental Positions

How can one account for the alternating periods of climatic warmth and coolness observed in the geologic record? Geologists realized many years ago that part of the answer must lie in the positioning of the continents. Big ice sheets are most likely to occur when plate tectonics carries a continent near one of Earth's poles (assuming that Earth's obliquity is low and the poles are

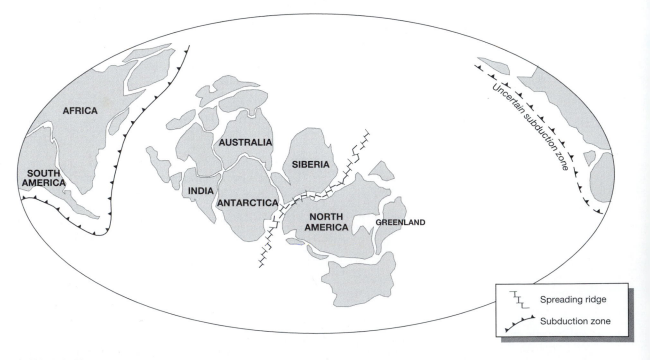

FIGURE 8-12

Possible continental reconstruction for the Late Proterozoic Period. All the continents appear to have been glaciated at that time. (After J.L. Kirshvink in the Proterozic Biosphere: A Multidisciplinary Study, J.W. Schopf and C. Klein, eds., Ch. 12.1, Cambridge University Press, Cambridge, 1992.)

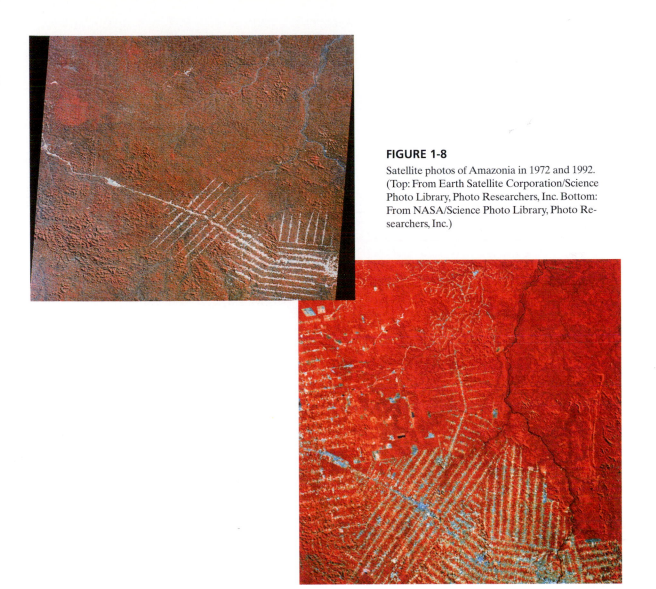

FIGURE 1-8

Satellite photos of Amazonia in 1972 and 1992. (Top: From Earth Satellite Corporation/Science Photo Library, Photo Researchers, Inc. Bottom: From NASA/Science Photo Library, Photo Researchers, Inc.)

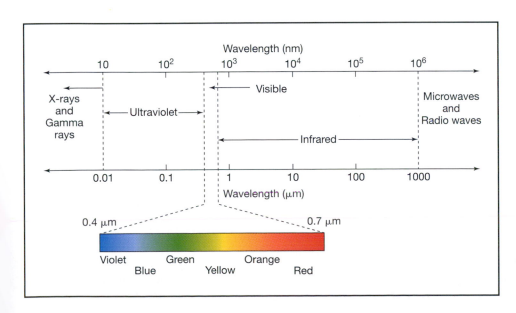

FIGURE 3-3

The electromagnetic spectrum.

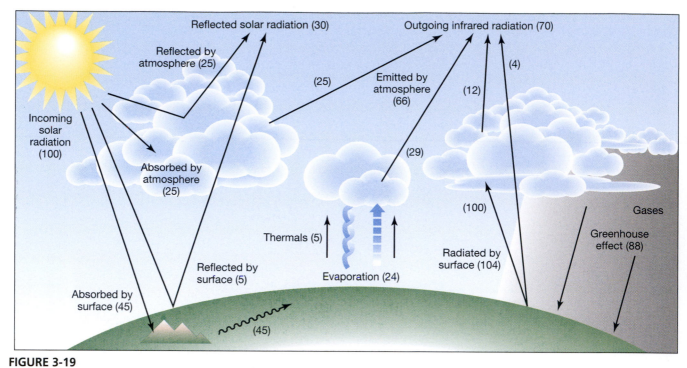

FIGURE 3-19

Earth's globally averaged atmospheric energy budget. All fluxes are normalized relative to 100 arbitrary units of incident radiation.

FIGURE 4-1

Global temperature distributions in degrees centigrade for (a) January, (b) July (From T. McKnight: *Physical Geography: A Landscape Appreciation*, 6/e, 1999. Reprinted by permission of Prentice Hall, Upper Saddle River, N.J.), and (c) the annual range (difference between summer and winter). (From R.W. Christopherson, *Geosystems: An Introduction to Physical Geography*, 3/e, 1997. Reprinted by permission of Prentice Hall, Upper Saddle River, N.J.)

JULY

	-50 -40 -30 -20 -10 0 10 20 30 40 50 60 70 80 90 100 110 120	
°F		°F
°C		°C
	-40 -30 -20 -10 0 10 20 30 40 50	

0 1500 3000 Miles

0 1500 3000 Kilometers

(b)

ANNUAL RANGE

0 1500 3000 Miles

0 1500 3000 Kilometers

F°	5	9	18	27	36	45	54	63	72	81	90	99	108	F°
C°	3	5	10	15	20	25	30	35	40	45	50	55	60	C°

(c)

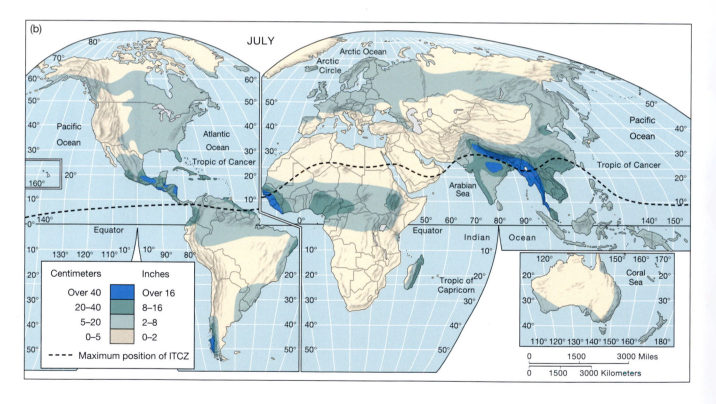

FIGURE 4-2

Global distributions of precipitation over land in (a) January and (b) July. (From T. McKnight: *Physical Geography: A Landscape Appreciation,* 6/e, 1999. Reprinted by permission of Prentice Hall, Upper Saddle River, N.J.)

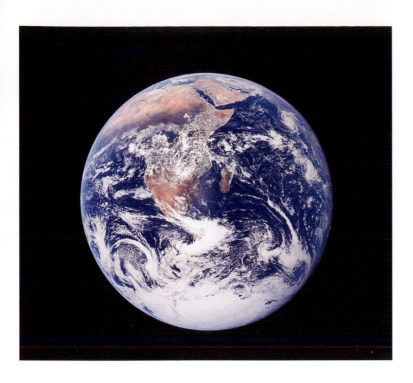

FIGURE 4-20
Earth, viewed from space at about 37,000 km (23,000 mi), is dominated by water. (From NASA Headquarters)

(a)

BOX FIGURE 6-3
Abundant and bizarre life thriving under the harsh conditions of the deep-sea floor, in the vicinity of hydrothermal venting. (a) A black smoker chimney is shown spewing out sulfide-rich solutions that provide the energy source for this food chain. (b) Tube worms, crabs, and other organisms can be seen. (From Richard A. Lutz and Michael J. Kennish, Reviews of Geosphysics, vol. 31, p. 210, August 1993. Copyright © 1993 by the American Geophysical Union.)

(b)

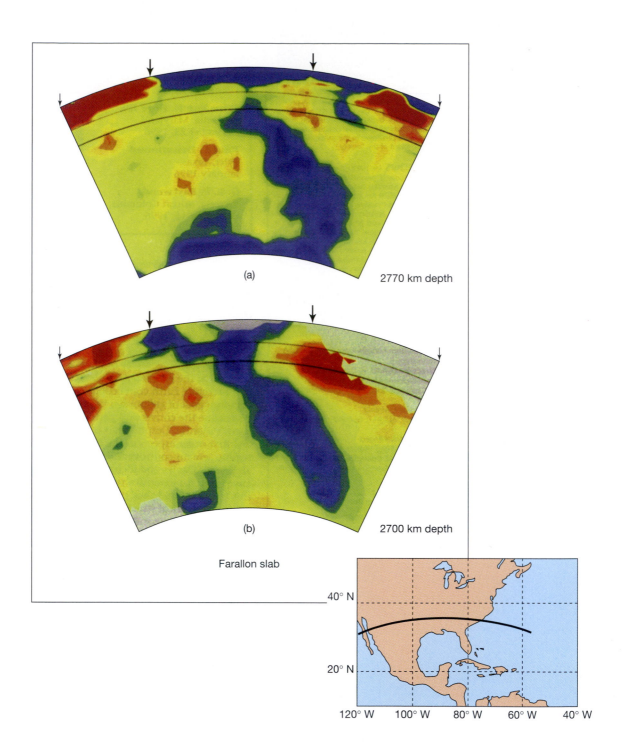

FIGURE 6-4

Tomographic image of the mantle's (a) P-wave and (b) S-wave velocity variations underneath North America, along the transect line shown in the insert. Blue colors indicate regions of fast seismic velocities, while reds indicate slow seismic velocities. The blue region cutting across the center of the diagram is the downgoing Farallon slab, which has been subducting under North America for 100 million years. (From W.K. Hamblin and E.H. Christiansen: *Earth's Dynamic Systems,* 8/e, 1998. Reprinted by permission of Prentice Hall, Upper Saddle River, N.J.)

| Oceanic ridge | Trenches | ● Shallow-focus earthquake | ● Intermediate-focus earthquake | ● Deep-focus earthquake |

FIGURE 6-12

Distribution of earthquakes of shallow, intermediate, or deep focus. Deep-focus earthquakes occur only at subduction zones. (From W.K. Hamblin and E.H. Christiansen: *Earth's Dynamic Systems,* 8/e, 1998. Reprinted by permission of Prentice Hall, Upper Saddle River, N.J.)

FIGURE 6-24

The age of the ocean floor is shown as bands of different color on the basis of the magnetic striping developed during sea-floor spreading. The youngest ocean floor is near the mid-ocean ridge, while the oldest is furthest away. (From R.W. Christopherson, *Geosystems: An Introduction to Physical Geography,* 3/e, 1997. Reprinted by permission of Prentice Hall, Upper Saddle River, N.J.)

m.y. = millions of years

0–2 m.y.	58–66 m.y.
2–5 m.y.	66–84 m.y.
5–24 m.y.	84–117 m.y.
24–37 m.y.	117–144 m.y.
37–58 m.y.	144–208 m.y.

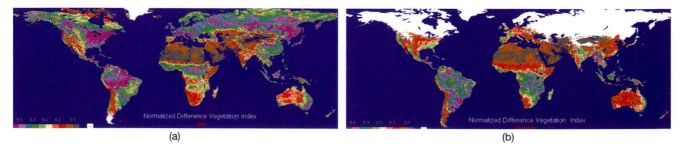

(a) (b)

FIGURE 7-2

False color satellite image of the vegetation coverage of the land surface in Northern Hemisphere in (a) winter and (b) summer.

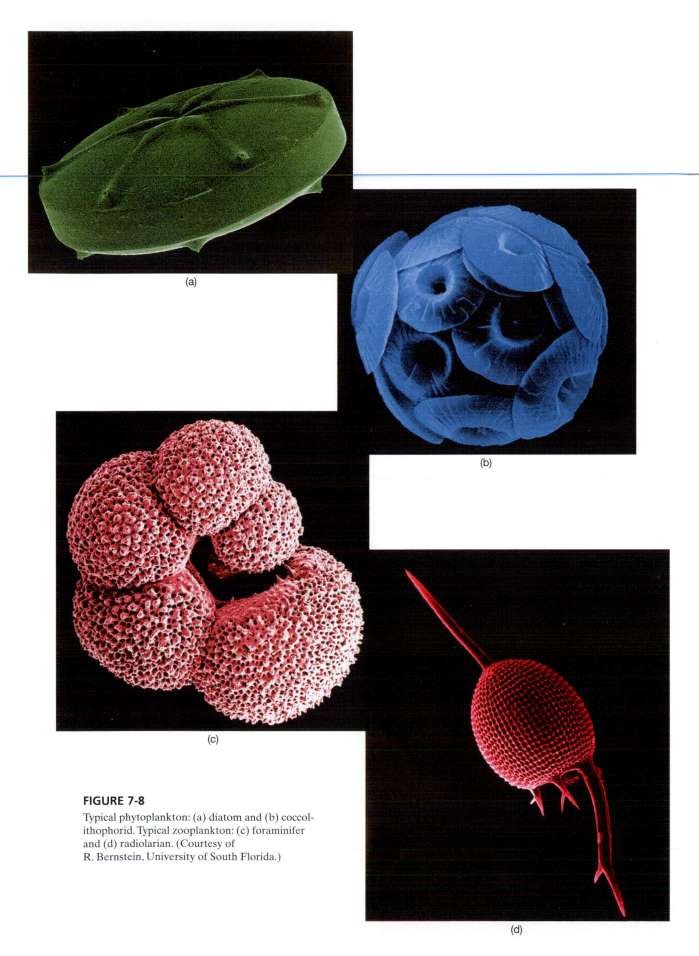

(a)

(b)

(c)

(d)

FIGURE 7-8

Typical phytoplankton: (a) diatom and (b) coccolithophorid. Typical zooplankton: (c) foraminifer
and (d) radiolarian. (Courtesy of
R. Bernstein, University of South Florida.)

FIGURE 7-10

The concentration of photosynthetic pigments as determined by the Coastal Zone Color Scanner (CZCS) on the *Nimbus 7* satellite. Pigment concentrations are indirect indicators of rates of primary production. (Courtesy of NASA.)

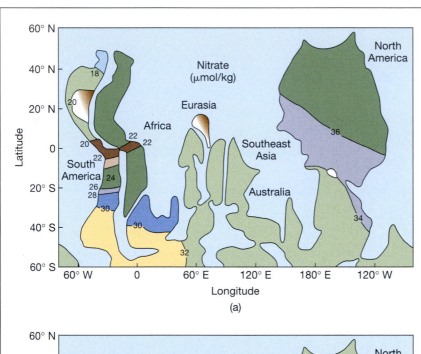

FIGURE 7-11

Concentrations of dissolved (a) nitrate and (b) O$_2$ in the deep ocean, in micromoles per kilogram of seawater. Maps are drawn for a water depth of 4000 m. At this depth the mid-ocean ridges appear as mountain ranges, and the continental outlines are substantially modified. (After W. S. Broecker and T.-H. Peng, 1982. *Tracers in the Sea.* New York: Eldigio Press, Columbia University, p. 31.)

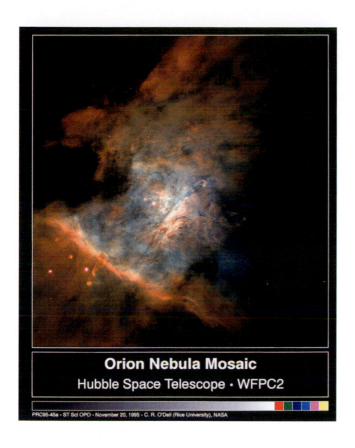

FIGURE 8-3

A star-formation region in the constellation Orion.

FIGURE 9-5

(a) Microfossils and (b) stromatolites from the 3.5-billion-year-old Warrawoona formation in Australia. ((a) J. W. Schopf, *Science* 260:640-646; (b) J. W. Schopf, ed., *Earth's Earliest Biosphere,* Princeton, NJ: Princeton University Press, 1983.)

(a)

(b)

FIGURE 9-9
A banded iron-formation, or BIF.

FIGURE 9-10
Samples of the detrital form of (a) uraninite and (b) pyrite.

(a)

(b)

(a)

FIGURE 10-8
Photomicrograph of (a) shocked quartz and (b) microspherules from K–T boundary clays.

(b)

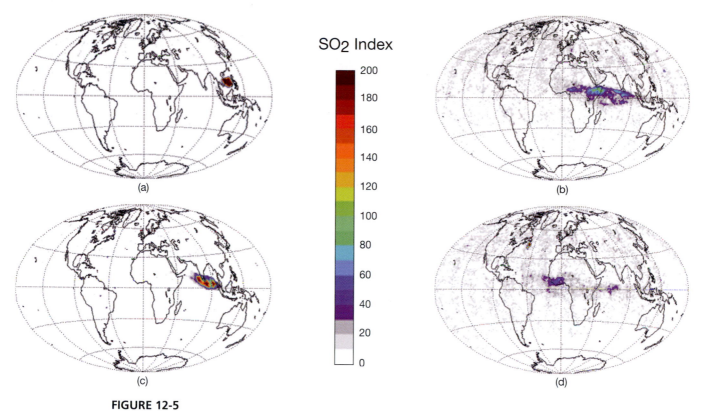

FIGURE 12-5

Satellite observations of the Mount Pinatubo aerosol cloud, 1991. (a) June 16. (b) June 18. (c) June 23. (d) June 30. (After Bluth, G. J. S., et al., 1992, "Global tracking of the SO$_2$ clouds from the June 1991 Mount Pinatubo eruption." *Geophysical Research Letters,* **19:**151-154.)

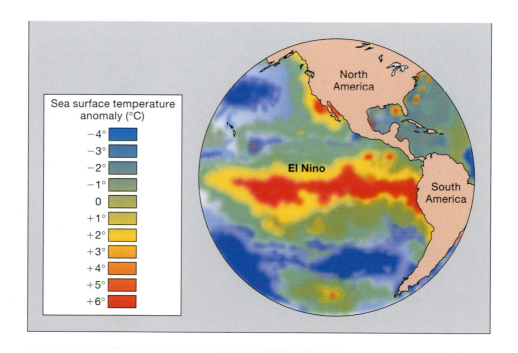

FIGURE 12-13

Sea-surface temperature anomalies in the central and eastern Pacific during the 1997–1998 ENSO event.

a) June – August

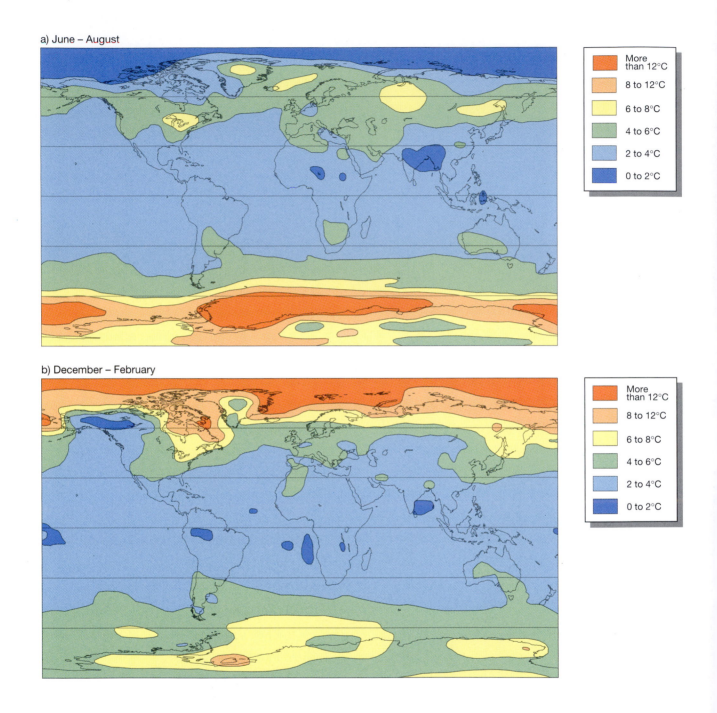

■	More than 12°C
■	8 to 12°C
■	6 to 8°C
■	4 to 6°C
■	2 to 4°C
■	0 to 2°C

b) December – February

■	More than 12°C
■	8 to 12°C
■	6 to 8°C
■	4 to 6°C
■	2 to 4°C
■	0 to 2°C

FIGURE 13-9

Predicted equilibrium changes in surface temperature during Northern Hemisphere (a) summer and (b) winter for doubled CO_2 conditions. (From T. McKnight: *Physical Geography: A Landscape Appreciation*, 6/e, 1999. Reprint by permission of Prentice Hall, Upper Saddle River, N.J.)

a) June – August

■ (dark blue)	More than 2mm day^{-1}
■ (blue)	1 to 2mm day^{-1}
■ (green)	0 to 1mm day^{-1}
■ (light green)	0 to -1mm day^{-1}
■ (yellow)	-1 to -2mm day^{-1}
■ (orange)	Less than -2mm day^{-1}

b) June – August

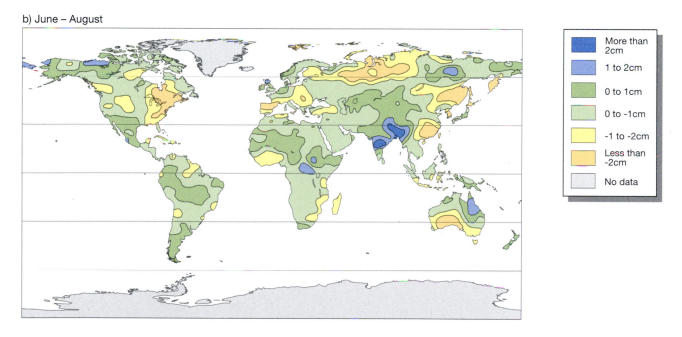

■ (dark blue)	More than 2cm
■ (blue)	1 to 2cm
■ (green)	0 to 1cm
■ (light green)	0 to -1cm
■ (yellow)	-1 to -2cm
■ (orange)	Less than -2cm
■ (grey)	No data

FIGURE 13-10

Predicted equilibrium changes in (a) precipitation and (b) soil moisture during Northern Hemisphere summer for doubled-CO_2 conditions. Precipitation contours are in millimeters/day (areas of decrease are stippled); soil moisture contours are in centimeters of water. (From T. McKnight: *Physical Geography: A Landscape Appreciation*, 6/e, 1999. Reprint by permission of Prentice Hall, Upper Saddle River, N.J.)

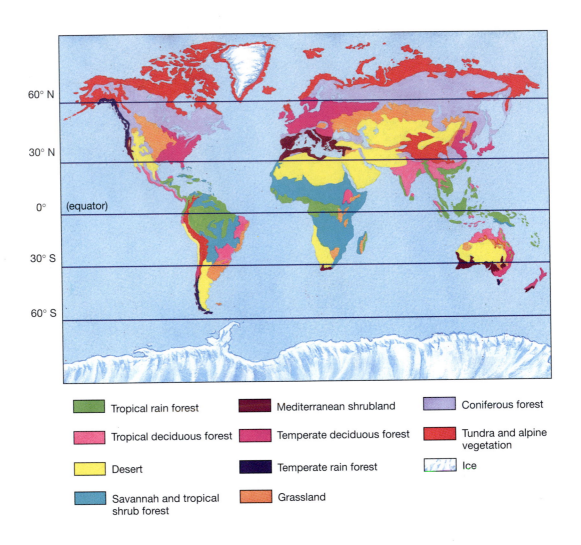

🟩 Tropical rain forest	🟥 Mediterranean shrubland	🟪 Coniferous forest
🟥 Tropical deciduous forest	🟥 Temperate deciduous forest	🟥 Tundra and alpine vegetation
🟨 Desert	⬛ Temperate rain forest	⬜ Ice
🟦 Savannah and tropical shrub forest	🟧 Grassland	

FIGURE 15-5

World distribution of the major terrestrial biomes. (From Audesirk and Audesirk: *Biology: Life on Earth*, 5/e, 1999. Reprinted by permission of Prentice Hall, Upper Saddle River, N.J.)

Low-Latitude Glaciations

Various unusual hypotheses have been offered to explain the evidence for low-latitude glaciation during the Late Proterozoic Period, including what is called the "snowball Earth" model. In this model, the polar caps extended to within a few degrees of the equator. There is a problem with this idea, however: The ice albedo feedback would have been so strong that the oceans should have frozen over entirely. We have just seen that Earth could eventually have escaped this fate by building up a dense, CO_2 atmosphere from volcanic outgassing. However, if the ice caps had extended all the way down to the tropics, the oceans would have frozen faster than volcanic CO_2 could have accumulated. This would probably have caused the extinction of most, or all, surface life. As there is no geologic evidence for such a global calamity at this time, the snowball Earth model is probably wrong.

An alternative explanation for the Late Proterozoic glaciation is that the paleomagnetic data on which Figure 8-12 is based have for some reason been misinterpreted. Recall that the paleolatitude of a given rock sample is determined from the dip angle of the remnant magnetic field relative to the bedding plane of the rock. Suppose, for example, that Earth's magnetic field was temporarily aligned perpendicular to its spin axis (instead of parallel, as it is today). Low magnetic dip angles would then correspond to high geographic latitudes, so what appear to be equatorial deposits might actually have been polar ones. But this explanation is unsatisfying as well because there is no reason to believe that the magnetic field orientation can actually change in this way.

The most intriguing explanation for low-latitude glaciation, originally suggested by the Australian geologist George Williams, is that Earth's obliquity was higher in the past. Recall that Earth's obliquity, currently 23.5°, is what gives rise to our seasons. It also determines what part of the planet is warm and what part is cold. At low obliquities (like today's value), the poles are the coldest part of the planet and the equator is the warmest part. At obliquities above 54°, however, the situation is reversed: The poles become warm and the equator becomes cold (Box Fig. 8-1). If Earth's obliquity were much higher in the past, the tropics could have been glaciated while the poles remained ice-free.

This last hypothesis has its problems as well. Getting Earth up to high obliquity in the first place does not appear to be difficult. The same large impact that formed the Moon could have left Earth tilted on its side. (Indeed, we have an example of such an occurrence in our Solar System. The planet Uranus has a very high obliquity, 89°, which is thought to have been produced by a similar type of large impact.) The real problem is figuring out how Earth could have had its obliquity reduced to its present low value. Recent research suggests that this might be accomplished by a feedback mechanism involving expansion and retraction of continental ice sheets in phase with Earth's obliquity cycle, but the idea remains highly speculative. The problem of low-latitude glaciation is still largely unexplained. It demonstrates that we are far from understanding all aspects of Earth's climate history.

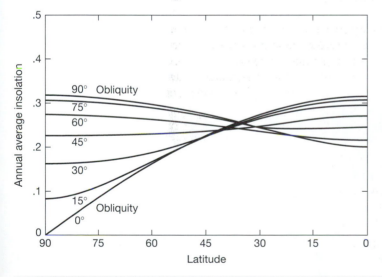

BOX FIGURE 8-1

Mean annual solar insolation as a function of latitude for different values of Earth's obliquity. The current obliquity is 23.5°, which is relatively low. Hence, the poles are cold and the equator is warm. If the obliquity were higher than 54°, though, the equator would become the coldest part of the planet and the poles would be warm. (After R. Ward, *Journal of Geophysical Research*, 79:3375, 1974.)

---

cold—see the box "Low-Latitude Glaciations"). This, after all, is what we see today and during the recent past: the major areas of continental glaciation are Antarctica, which sits directly on the South Pole, Greenland, and the northern parts of Europe and North America, which have been intermittently ice-covered during much of the Pleistocene. So, some of the glaciations observed in the geologic record could simply be a consequence of having large continents near the poles.

This idea, however, cannot explain the climate of the Late Proterozoic Era, when both continents and glaciers appear to have been situated at relatively low latitudes. And it cannot explain the warm Mesozoic Era (discussed below) when high-latitude continents were present but were almost completely ice-free. Both of these appear to represent real, global climate changes, not just apparent ones caused by changes in continental position. To cause such changes, something must have altered either the planetary albedo or the greenhouse effect. Changes in continental position and area can affect the albedo to some extent. However, this is probably not sufficient to explain the major climate shifts that are observed.

Variations in Atmospheric CO₂

＊ What, then, are the most important factors influencing long-term climate? Based on the previous discussion, the two primary influences would appear to be changes in solar luminosity and in atmospheric CO_2. As indicated by Figure 8-9, these two factors are thought to have largely balanced each other out during most of Earth's history: as solar luminosity gradually went up, atmospheric CO_2 levels gradually went down. This does not explain, though, why Earth's climate was alternately warm and cold. If the weathering rate feedback were all that mattered, Earth's surface temperature should have varied smoothly with time. What caused it to go up and down?

The primary forcing factor may indeed have been atmospheric CO_2. But, if so, the reasons why it varied must have been more complicated than we have indicated so far. As discussed in Chapter 7, many different factors, including plate tectonics, volcanism, precipitation patterns, and the biota, are capable of influencing atmospheric CO_2 levels. Even continental positions may make a difference. During the Late Proterozoic Era, for example, the conti-

nents appear to have been clustered near the equator. Equatorial continents should have been warmer and wetter than Earth as a whole (again assuming the planet's obliquity was low) and should therefore have had high rates of silicate weathering. This, in turn, could have led to drawdown of atmospheric CO_2, thereby cooling the entire planet. So, equatorial continents may actually cause global cooling—an idea that runs contrary to conventional geologic wisdom. This is a good example of why we need a systems approach to understand Earth's climate history.

Biological Influences on Atmospheric CO₂

The biota may also have caused atmospheric CO_2 levels to fluctuate. One way in which they could have done so is by altering the rate of silicate weathering. As discussed ＊ in Chapter 7, plants and microorganisms are thought to accelerate weathering by increasing the CO_2 partial pressure in soils and by releasing organic acids that help dissolve rocks. **Vascular plants** have a well-developed stem or trunk, for transporting water and nutrients from the ground up to their leaves, and typically a well-developed root system, which takes up water and nutrients and helps support the plant. The spread of vascular plants with deep root systems during the Devonian Period, about 400 m.y. ago, could in theory have caused atmospheric CO_2 levels to decrease (Fig. 8-13). There appears to have been no

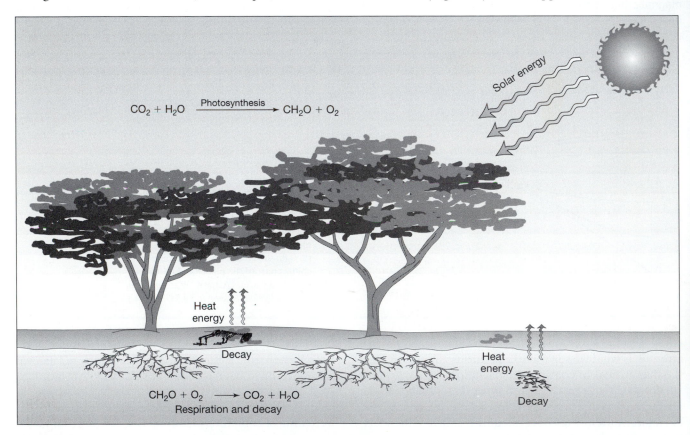

$$CO_2 + H_2O \xrightarrow{\text{Photosynthesis}} CH_2O + O_2$$

Solar energy

Heat energy

Decay

Heat energy

Decay

$$CH_2O + O_2 \longrightarrow CO_2 + H_2O$$
Respiration and decay

FIGURE 8-13

Land plants enhance the partial pressure of CO_2 in soils through root respiration and decay. (From T. McKnight: *Physical Geography: A Landscape Appreciation*, 6/e, 1999. Reprinted by permission of Prentice Hall, Upper Saddle River, N.J.)

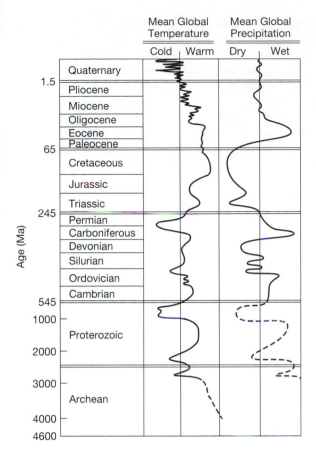

FIGURE 8-14

Estimated change in surface temperature during the Phanerozoic eon. (From K.C. Condie and R.E. Sloan: *Origin and Evolution of Earth: Principles of Historical Geology*, 1998. Reprinted by permission of Prentice Hall, Upper Saddle River, N.J.)

obvious climate change at this time, however, so this effect does not seem to have been very pronounced.

The Permo-Carboniferous glaciations, however, do correlate with a possible CO_2 drawdown brought about by the formation of large coal deposits during the Carboniferous Period, 286 to 360 m.y. ago. This is the time when much of the coal in North America formed. In this case, it is the organic carbon cycle, rather than the carbonate–silicate cycle, that would have been responsible for causing the change in atmospheric CO_2. Burial of organic carbon releases oxygen, of course, so atmospheric O_2 levels may have soared during this time. It has been suggested that the large flying insects of the Carboniferous Period, which included dragonflies with two-foot wingspans, evolved in response to this increase in atmospheric O_2. Modern insects have poorly developed circulatory systems and, hence, would be unable to transport enough oxygen to their cells if they grew too large. High atmospheric O_2 levels might have enabled Carboniferous insects to overcome this limitation.

Proponents of the Gaia hypothesis have been quick to point out that the biota do appear to have influenced

Earth's climate. And they are at least partly right: there is little doubt that climate would have evolved differently if life had not been present. Just how strong the biological influence might be is the subject of ongoing research. We suggest below, however, that the main perturbations to atmospheric CO_2 and climate over the past 200 million years have come from plate tectonics.

Climates of the Past 200 Million Years

The Warm Mesozoic Era

The Mesozoic and Cenozoic Eras span the past 245 m.y. of Earth history. The Mesozoic, the age of the dinosaurs, is thought to have been considerably warmer than today (Fig. 8-14). In contrast to the Precambrian Period, for which very little paleoclimate information is available, the fossil evidence relating to Mesozoic climate is quite extensive. For example, during the mid-Cretaceous Period, around 100 m.y. ago, lush ferns and alligators resided in what is now Siberia. Dinosaur skeletons have been recovered from north of the Arctic Circle in Alaska. These and other pieces of evidence indicate that the mid-Cretaceous climate was on the order of 2 to 6 degrees Celsius (3.6 to 11°F) warmer at the equator and 20 to 60 degrees Celsius (36 to 110°F) warmer at the poles (Fig. 8-15).

A second type of evidence that indicates that the Mesozoic climate was warm comes from measurement of oxygen isotope ratios in carbonate sediments recovered from deep sea cores. Although we will postpone detailed discussion of this topic until Chapter 10, oxy-

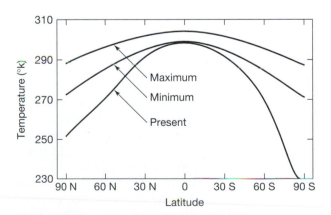

FIGURE 8-15

Estimated limits on longitudinally averaged surface temperatures during the mid-Cretaceous period, 100 m.y. ago, as compared with today. (After E. J. Barron and W. M. Washington, "The Carbon Cycle and Atmospheric CO_2," *Geophysical Monograph* 32, AGU, Washington, D.C., 1985.)

gen isotopes provide information on both the temperature of the water in which they formed and on the amount of water stored in the polar ice caps. The measured isotope ratios tell us that Mesozoic ocean water was much warmer than today. This is especially true of the deep ocean, which currently has an average temperature of only 2°C. Deep-ocean temperatures during the Mesozoic Era were as high as 15°C. Furthermore, the polar ice caps, which today hold enough water to raise sea level by nearly 80 m, appear to have been absent throughout the entire Mesozoic and during the early Cenozoic as well.

What factor or combination of factors was responsible for the extreme warmth of the Mesozoic? In keeping with the discussion of the previous section, suspicion currently centers on higher atmospheric CO_2 levels. Climate models suggest that an increase in CO_2 by a factor of 4 from the present value could explain the warm climate of the mid-Cretaceous Period. This hypothesis is bolstered by paleomagnetic evidence that indicates that the sea floor was spreading faster at that time than it has in the more recent geologic past. Recall that the spreading rate can be estimated by looking at magnetic patterns in the sea floor. Faster spreading rates would have led to faster rates of subduction of carbonate sediments and this, in turn, should have led to increased rates of CO_2 production from carbonate metamorphism. More CO_2 may also have been released by outgassing at the mid-ocean ridges themselves. And higher sea level at that time, itself caused by faster sea-floor spreading as well as the absence of polar ice, would have meant that there was less land area available on which to weather silicate rocks. So, the idea that atmospheric CO_2 levels were higher during the Cretaceous Period is consistent with what is known about the plate tectonic cycle at that time.

Carbon Isotopic Evidence of High Mesozoic CO_2 Levels

Our discussion of paleo-CO_2 levels and climate has thus far been based mostly on theory. It has recently become possible, however, to test the theory with data, at least for the past few hundred million years. The method that is being used is based on carbon isotopes. Recall that carbon has an unstable (radioactive) isotope, ^{14}C, in addition to the normal, stable isotope, ^{12}C. Carbon-14 is not useful for studying events that occurred millions of years in the past because its half-life, 5730 years, is much too short. Carbon, though, also has a third, stable isotope, ^{13}C, that can provide information about the carbon cycle in the distant past.

Carbon-13 is useful for the following reason: Photosynthetic organisms can use any of the three carbon isotopes to produce organic material; however, they can

most easily metabolize the lightest one, ^{12}C. So, if CO_2 is abundant in the organism's environment, the organic matter that is produced is depleted in ^{13}C relative to ^{12}C. If CO_2 is scarce, the organisms use whatever isotope is available and the relative isotopic abundances do not change as dramatically.

Isotopic abundances are measured in parts per thousand (‰), also called 'parts per mil.' The standard way of expressing ^{13}C abundances is to use 'delta (δ) notation'

$$\delta^{13}C(o/oo) \equiv \left[\frac{\left(\frac{^{13}C}{^{12}C}\right)_{sample} - \left(\frac{^{13}C}{^{12}C}\right)_{standard}}{\left(\frac{^{13}C}{^{12}C}\right)_{standard}} \right] \times 1000$$

Negative $\delta^{13}C$ values mean that the sample is depleted in ^{13}C relative to the standard. The standard employed is a particular carbonate deposit, a fossil belemnite (an ancient organism) from the Peedee formation. Hence, carbon isotopes are said to be measured on the 'PDB' scale.

Carbon isotope abundances have now been measured in sediments of various ages by a number of different research groups. Figure 8-16 shows the difference in $\delta^{13}C$ values between carbonates and organic carbon. The story that has emerged is consistent with that told above: organic matter from sediments of Mesozoic age contains less ^{13}C than organic matter from sediments deposited during the last 20 m.y., indicating that atmospheric CO_2 levels were probably higher at that time. Although there is still considerable disagreement concerning how much CO_2 has decreased, the idea that CO_2 is a main driver of climate on these time scales has received additional support.

Other Possible Influences on Mesozoic Climate

Although higher atmospheric CO_2 levels can account for the overall warmth of the Mesozoic climate, this mechanism cannot by itself explain the extremely small latitudinal temperature gradient at that time. The equator-to-pole temperature contrast during the Mid-Cretaceous Period was only 20 to 30°C, as compared with 50 to 60° today. Part of this difference can be explained by the absence of polar ice at that time. Recall that ice cover interacts with climate by way of a strong, positive feedback loop. Removal of the ice caps would cause a large decrease in the albedo of the polar regions which, in turn, should cause them to warm significantly.

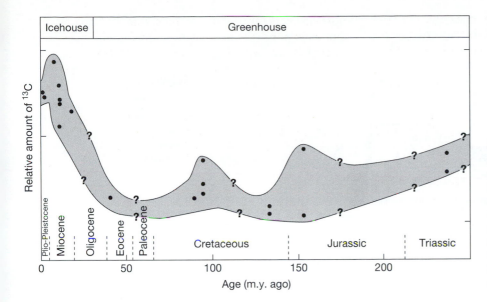

FIGURE 8-16

Carbon isotope data for the past 150 m.y. The vertical scale is a measure of the difference in ^{13}C content between carbonates and organic carbon. Large differences in ^{13}C correspond to lower atmospheric CO_2 values. The data are consistent with a decrease in CO_2 since the mid-Cretaceous Period. (After B. Popp et al., *American Journal of Science* 289: 436, 1989.)

Calculations with climate models suggest, however, that ice albedo feedback alone is not sufficient to explain the extreme warmth of the Mid-Cretaceous poles. It seems likely that the atmosphere–ocean system was for some reason more effective in transporting heat from the equator to the poles than it is today. One possibility is that the thermohaline circulation of the oceans ran backward at that time: warm, but highly saline, deep water formed at low latitudes welled up near the poles, where it then warmed the climate through evaporation. Unfortunately, no one has yet demonstrated that this mechanism could work. A second possibility is that the tropical Hadley circulation extended further poleward than it does today. Because Hadley cells are very efficient at transporting heat, this mechanism could explain the low latitudinal temperature gradient. Moreover, it could explain the apparent absence of subfreezing temperatures in continental interiors (Siberia, for example) during the long polar night. Again, however, no one has demonstrated that this mechanism is dynamically feasible. Hence, the details of Mid-Cretaceous climate remain largely unexplained.

Cooling during the Cenozoic Era

Starting about 80 m.y. ago, Earth's climate began to cool (except for a short-lived warming during the early Eocene Period). The initial decrease may simply have been caused by a decrease in mid-ocean ridge spreading rates, leading to a reduction in atmospheric CO_2. However, the cooling trend accelerated around 30 m.y. ago during the Oligocene epoch in a way that does not correlate with the spreading-rate data. Thus, paleoclimatologists have searched for other explanations for the observed cooling. One intriguing theory is that the car-

bonate–silicate cycle was perturbed by plate tectonics, but by a mechanism that differs from those discussed previously.

During the Mesozoic and early Cenozoic eras, India was a separate continent drifting slowly toward Asia. The two continents collided around 40 m.y. ago, a process that is still continuing today. The collision created a gigantic chain of mountains (the Himalayas) and a huge area of uplifted terrain called the Tibetan Plateau (Fig. 8-17). The Himalayan Mountains provided fresh, readily erodable surfaces on which silicate weathering could proceed rapidly. At the same time, the uplift of the Tibetan plateau created seasonal rainfall (the southeast Asian monsoon), which provided the water needed for weathering to occur on the face of the Himalayan range. The combination of these factors may have accelerated silicate weathering rates over a substantial portion of Earth's surface, thereby helping to bring atmospheric CO_2 concentrations down to the relatively low levels that prevail today.

The point to be drawn from this discussion is that plate tectonics probably does influence climate over long time scales, but not necessarily in the way that geologists have traditionally imagined. Continental-scale glaciers can indeed grow when land masses drift close to the poles but only if climatic conditions are ripe for such a development—specifically, if atmospheric CO_2 concentrations are relatively low. The main influence of plate tectonics on climate appears to be indirect: by changing the way in which the carbonate–silicate cycle operates, plate tectonics helps to modulate atmospheric CO_2 levels. This, in turn, affects climate by way of the greenhouse effect. Such changes, in combination with the long-term increase in solar luminosity, can account for the main features of the long-term climate record.

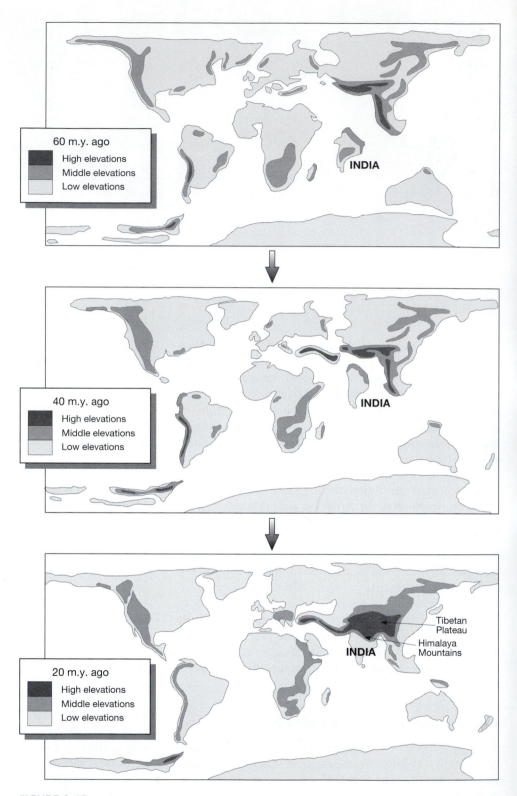

FIGURE 8-17

The collision of India with Asia, shown in a series of paleogeographic maps for the Paleocene (60 m.y. ago), the Eocene (40 m.y. ago), and the Miocene (20 m.y. ago) Eras. (After E. J. Barron, *Paleogeography, Paleoclimatology, Paleoecology* 50:45, 1985.)

Chapter Summary

Earth formed from accretion of solid materials that condensed out of the solar nebula soon after the Sun itself formed. The age of Earth is identical to that of meteorites, 4.55 b.y., as determined by radiometric age dating. Earth's core probably formed as the Earth itself was forming as a result of heating and stirring by large impacts. The Moon is thought to have formed as a consequence of a glancing impact by a Mars-sized planetesimal and, hence, is something of a cosmic accident.

Earth's atmosphere and ocean formed along with the planet from impact degassing of incoming planetesimals and from volcanic outgassing. The resulting atmosphere was probably rich in CO_2 and N_2 but contained little O_2 prior to the origin of life and the evolution of photosynthesis. Atmospheric CO_2 levels may have remained extremely high (more than 1 bar) for hundreds of millions of years if the continents were originally small

because silicate weathering would have been inhibited. Thus, the early climate may have been warm despite substantially lower solar luminosity at that time.

As the Sun grew brighter, atmospheric CO_2 levels should have fallen as a consequence of the negative feedback in the carbonate–silicate cycle. This negative feedback loop is the principal reason why the Earth has managed to maintain liquid water and life on its surface for the last 3.5 b.y. Earth's climate has nonetheless experienced numerous warm and cold cycles during this time, including five major episodes of glaciation. The current epoch, which represents a transient interglacial period within the Pleistocene glaciation, is one of these cold episodes. The primary cause of the major climate shifts that are observed is probably variations in atmospheric CO_2 induced by changes in the carbonate-silicate cycle and in the rate of burial of organic carbon.

Key Terms

accretion	impact degassing	proto-Sun
chondrites	intersteller clouds	shelly fossils
corona	isochron diagram	solar nebula
dropstones	lead-lead ating	solar wind
eon	magma ocean	terrestrial planets
geothermal heat	meteorites	tillite
giant impact hypothesis	moraines	vascular plants
glacial striations	neutrinos	volatile compounds
heavy bombardment period	paleoclimate	zircon
ice-rafting	planetesimals	

Review Questions

1. How old is the Solar System, and how is this age determined?
2. How is the age of Earth determined if no rocks older than 4.1 billion years have been preserved?
3. How do Jupiter and the Moon affect the habitability of Earth?
4. How and when did the atmosphere and ocean form? Which gases are thought to have been present in the early atmosphere?
5. Why does the Sun get brighter with time?
6. How might the carbonate–silicate cycle have helped to solve the faint young Sun problem?
7. What types of geologic evidence are used to infer past glaciations? How many separate episodes of glaciation have occurred during Earth's history?
8. How do the Precambrian glaciations differ from those of the more recent past?
9. How are carbon isotopes used to infer past atmospheric CO_2 concentrations?
10. How are atmospheric CO_2 levels affected by the presence of land plants?
11. What mechanisms might explain the warm climate of the Mesozoic Era? How might the equator-to-pole temperature gradient have been reduced?
12. Why did climate cool during the past 40 million years?

Critical-Thinking Problems

1. a. The variation in the solar flux, S, reaching the Earth at different times can be approximated by the formula

$$\frac{S}{S_0} = \frac{1}{1 + 0.4(1 - t/t_0)},$$

where $S_0 = 1370$ W/m^2 = present solar constant, t_0 = present age of the Sun, 4.6 billion years, and t = time in billions of years, starting from the Sun's formation. Calculate the solar flux at $t = 0$, 1.5 billion, and 3.0 billion years ago as a percentage of its present value.

b. Calculate Earth's effective radiating temperature today and at $t = 0$, 1.5 billion, and 3.0 billion years ago, assuming that the albedo has remained constant at 0.3.
c. Calculate the mean surface temperature of Earth at these same times, assuming that the greenhouse effect is given by

$$\Delta T_g = 22 + 0.5(T_e - 233).$$

This equation presumes that atmospheric CO_2 does not change but that water vapor does. Using graph paper (or a computer), make a plot of T_s versus time. When does the mean surface temperature reach the freezing point of water? Express your answer in terms of billions of years *before present.*

Further Reading

General

1. *The Ages of Gaia: A Biography of our Living Earth,* J. E. Lovelock, W. W. Norton and Co., New York, 1988.

2. *Hothouse Earth: The Greenhouse Effect and Gaia,* J. Gribbin, Grove Weidenfeld, New York, 1990.

Chapter 9
The Chemical Environment: Evolution of the Atmosphere

A banded iron-formation, or BIF. (From J.W. Schopf.) [See color section]

Key Questions

- What was the composition of the atmosphere early in Earth's history?
- When did life originate, and how did it affect atmospheric composition?
- When and why did atmospheric O_2 become abundant?
- When did the ozone layer form, and how did its formation affect Earth's surface environment and the evolution of the biota?
- What controls the atmospheric O_2 concentration today?

Chapter Overview

Earth's present atmosphere is rich in molecular oxygen and has a well-developed ozone layer that shields the planet's surface from harmful solar ultraviolet radiation. This oxygen is produced by photosynthesis and, hence, would not have been present prior to the origin of life. The atmosphere at that time was probably rich in carbon dioxide and nitrogen, with smaller amounts of reduced gases such as hydrogen and methane. Both life and photosynthesis appear to have originated quite early, but atmospheric O_2 levels remained low throughout the first half of Earth's history. The O_2 level rose around 2.0 b.y. ago for reasons that are still poorly understood but that may be related to the gradual oxidation of Earth's mantle. Atmospheric O_2 has been maintained at a relatively constant level for the past few hundred million years by a negative feedback mechanism that appears to involve the oxygenation of the deep oceans and the availability of dissolved phosphorus.

Introduction

In Chapter 8 we dealt with the evolution of Earth's climate over very long time scales. We have a separate discussion of climate evolution there, because it is an inherently interesting topic and because it is a major issue for the future. But climate is not the only aspect of the Earth system that has changed over time. Atmospheric composition has evolved in a fundamental way that is extremely important to all life, including humans: It has become rich in oxygen, O_2. Were it not for the rise of atmospheric oxygen, there would be little discussion of climate, because the most advanced forms of life on Earth would still be microbes.

A second change in atmospheric composition that has affected both microbial and advanced life has been the development of a protective ozone layer. We have already seen that ozone is important because it blocks out harmful solar ultraviolet (UV) radiation. But, if the concentration of atmospheric O_2 was initially very low, there must have been a time when the ozone layer did not exist. How did life cope with the solar UV flux at that time? How did life originate in the first place? As we shall see, this last question is one of the most fascinating, unsolved problems of modern science.

The Prebiotic Atmosphere

Before discussing the question of life's origin, let us briefly review what we have already learned about the composition of Earth's early atmosphere. In Chapter 8 we saw that CO_2 concentrations of at least a tenth of a bar (0.1 times the present atmospheric pressure) are the most likely solution to the faint young Sun problem. Even higher atmospheric CO_2 levels (up to 10 bars) could have been present during the first few hundred million years of Earth's history, prior to the growth of the continents. Once significant land areas appeared, the weathering of silicate rocks would have caused atmospheric CO_2 levels to decrease, but CO_2 should still have remained a significant component of the atmosphere.

The other major constituent of the early atmosphere was probably molecular nitrogen, N_2. As we pointed out earlier, nitrogen is relatively inert. Thus, most of the N_2 released by volcanic outgassing and by impact degassing of infalling planetesimals should have remained in the atmosphere. We shall soon see that the behavior of N_2 is actually more complicated than this: Nitrogen cycles between Earth's atmosphere and its crust, although at a much slower rate than does carbon or oxygen. However, the conclusion that most of Earth's N_2 should have resided in the atmosphere since early in the planet's history still seems justifiable.

The two major atmospheric constituents, N_2 and CO_2, would have been accompanied by a host of minor constituents produced by volcanic outgassing and by the continued infall of planetesimals. The most abundant of these would probably have been molecular hydrogen, H_2, and carbon monoxide, CO. (Carbon monoxide is familiar as a potentially lethal gas that is present in automobile exhaust fumes.) Both of these are reduced gases that are released by volcanos. **Reduced gases** are gases capable of reacting with O_2. In today's O_2-rich atmosphere, H_2 and CO are rapidly oxidized to water and carbon dioxide, respectively, so their concentrations are relatively low (0.5 ppm for H_2, 0.1–0.2 ppm for CO). In the primitive atmosphere, both gases are estimated to have had concentrations on the order of 100 to 1000 ppm.

Prebiotic O_2 Concentrations

In the absence of photosynthesis, atmospheric O_2 concentrations must have been very low. But they would not have been exactly zero. Some O_2 would have been produced within the atmosphere by a variety of different photochemical reactions. A **photochemical reaction** is a chemical reaction induced by light or by other electromagnetic radiation. One specific mechanism for producing O_2 is shown in Figure 9-1. First, H_2O and CO_2 molecules are **photolyzed** (split apart by UV or visible radiation). This photolysis produces hydroxyl (OH) radicals and atomic oxygen (O) atoms. (A **radical** is a molecule that is highly reactive, because it has an unpaired electron in its outer shell.) Then, these products react to form O_2 molecules and H atoms:

FIGURE 9-1

The abiotic production of oxygen in the early atmosphere.

$$OH + O \longrightarrow O_2 + H.$$

This reaction by itself is not a net source for O_2, because the O_2 molecule that is formed can be destroyed by reacting with the leftover hydrogen atom. Hydrogen, however, is an extremely light atom capable of escaping to space. (Hydrogen and helium are the only gases light enough to escape Earth's atmosphere.) When hydrogen escapes, the O_2 that was produced during its formation is left behind in the atmosphere. The net rate of O_2 production is equal to one-fourth the rate at which the H atoms produced by H_2O photolysis escape. This process is limited by the very small amount of water vapor available in Earth's stratosphere and is more than 100 times slower than the rate at which O_2 is produced by photosynthesis followed by organic carbon burial today.

The O_2 produced in this manner would not have accumulated in the atmosphere but instead would have been consumed by various O_2 sinks, or loss processes. The major O_2 sink today is the oxidative weathering of organic carbon and reduced minerals in rocks. In **oxidative weathering,** the minerals being weathered react with O_2 during the process. Oxidative weathering, however, requires relatively high atmospheric O_2 levels. The fastest loss process for O_2 on the early Earth would have been reactions with reduced volcanic gases, especially H_2 and CO. The overall reactions can be written as

$$2H_2 + O_2 \longrightarrow 2H_2O$$
$$2CO + O_2 \longrightarrow 2CO_2.$$

The actual photochemical reactions involved require several steps, but the net result would have been production of H_2O and CO_2, as shown here. Studies of modern volcanos suggest that they emit reduced gases at more than 30 times the rate at which O_2 is produced abiotically. Volcanos on the early Earth would probably have released even more reduced gases, because Earth's interior would have been hotter and because the mantle might have been more reduced. Because the available O_2 sinks would have been larger than the O_2 source, the atmosphere should have been essentially O_2-free.

The type of atmosphere just described is termed a **weakly reducing atmosphere:** It contains small concentrations of reduced gases and virtually no free O_2 (Fig. 9-2 and Table 9-1). Actually, trace amounts of O_2 should have been present, especially at high altitudes, as a result of photochemical reactions like the one shown in Figure 9-1. But the O_2 concentration near the surface would have been far too small to support breathing and too low to affect surface weathering. The O_2 levels would also have been too low to form an ozone layer. Thus, Earth's surface might have been exposed to much higher fluxes of solar UV radiation.

Although many geologists accept the idea of a weakly reducing atmosphere on the primitive Earth, the theo-

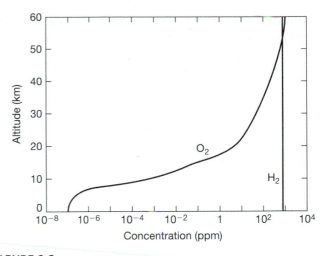

FIGURE 9-2
Vertical profiles of O_2 and H_2 in a weakly reducing primitive atmosphere.

ry has been criticized by some biologists and chemists because it does not fit conventional ideas concerning the origin of life. Those ideas require that the early atmosphere be more strongly reducing. In particular, the **prebiotic synthesis** of complex organic compounds would have been more efficient if the atmosphere had contained large amounts of methane, CH_4, and ammonia, NH_3. Let us take a brief detour from our discussion of the early atmosphere to see why these gases are considered to have been important.

The Origin of Life

The Miller–Urey experiment

The conventional theory of life's origin arose out of a series of laboratory experiments performed in 1953 by a graduate student at the University of Chicago, Stanley Miller, working under the guidance of the famous geochemist Harold Urey. Miller and Urey filled flasks with mixtures of gases that were considered, at that time, to be representative of Earth's primitive atmosphere. (These

TABLE 9-1

Changes in Atmospheric Composition over Time			
	Prebiotic Atmosphere	*Archean Atmosphere*	*Modern Atmosphere*
Surface pressure	1–10 bars	1–2 bars	1 bar
N_2	10–80%	50–80%	78%
O_2	about 0	about 0	21%
CO_2	30–90%	10–20%	0.036%
CH_4	10–100 ppm	1000–10,000 ppm	1.6 ppm

gases had just recently been discovered in Jupiter's atmosphere. Because Jupiter does not lose hydrogen, its atmosphere was considered by Urey to be "unevolved." Urey reasoned, not quite correctly, that Earth's atmosphere would have been similar in composition before hydrogen had had time to escape.) The flasks contained methane and ammonia along with water vapor and molecular hydrogen. The researchers then sparked the flasks with powerful electric discharges, simulating lightning in the prebiotic atmosphere.

After several minutes of electrification, the walls of the flasks became coated with a sticky, brownish material. When analyzed, this material was found to contain an assortment of organic compounds, including amino acids. **Amino acids** are compounds—containing an amino group (NH_2) and a carboxyl group (COOH)—that are important building blocks for proteins (see the box "The Compounds of Life"). **Proteins,** composed of one or more chains of amino acids, are key molecules in organisms. Proteins may be *enzymes* (guiding chemical reactions), structural components, *hormones* (maintaining constant body conditions), or transport molecules.

The Miller–Urey experiment, as it is now called, made headlines around the world. It was indeed a revelation to discover that many of the basic compounds on which life depends can be synthesized by a straightforward process that could have occurred in nature. Many scientists working in this field thought that we might be close to understanding how life itself began.

Today, the Miller–Urey experiment is still held in high regard scientifically, but researchers are less certain that it represents a critical step in the origin of life. One reason is that just mentioned: Current theories of the early atmosphere suggest that it was not as strongly reducing as the gas mixtures in Stanley Miller's flasks. Methane and ammonia, if they were present at all, would probably have been held to relatively low concentrations because they are photolyzed by solar UV radiation. The hydrogen then escapes to space, and the carbon and nitrogen atoms left behind are converted into CO_2 and N_2. Furthermore, modern volcanos do not emit much methane or ammonia. Early volcanic gases might have been more highly reducing; however, even this was probably not enough to produce a Miller–Urey type atmosphere.

The RNA World

A second way in which our ideas about the origin of life have changed is that most biologists now believe that proteins were not among the earliest structural elements of life. Rather, life might have relied exclusively on RNA or some simpler variant of RNA.

The evidence that RNA preceded proteins was discovered in the mid-1980s by Thomas Cech of the University of Colorado and Sydney Altman of Yale University, and it earned them the 1989 Nobel Prize in chemistry. Cech and Altman discovered that one particular type of RNA molecule was capable of cleaving (cutting) itself into smaller pieces. This capability meant that it was theoretically possible for an RNA molecule to replicate (duplicate) itself without help from any other molecule. DNA-based organisms cannot do this. The DNA molecule can replicate only with the aid of complex enzymes made of proteins. An **enzyme** is a biological molecule that speeds up, or *catalyzes,* a particular biochemical process.

Life depends on a complex interaction among DNA, RNA, and proteins. The DNA carries the basic genetic information (the blueprint for the organism); the RNA is used to transfer this information to other parts of the cell, where proteins are made; and the proteins perform many different cell functions, including the replication of DNA and RNA.

A primitive, RNA-based organism could have been much simpler than today's organisms. RNA itself could have been the molecule in which the genetic information was stored. RNA is less stable than DNA, but it carries essentially the same information. And because it can cleave itself, RNA could have reproduced without the aid of enzymes. The elegance and simplicity of this idea has led biologists to suggest that DNA-based life was preceded by an **RNA world,** in which only RNA-based organisms were present.

Even if this idea is correct, it does not solve the problem of life's origin. We still need to make the basic compounds of which RNA is composed and then assemble them into a self-replicating molecule. The required phosphate molecules were probably present in the primitive ocean as a result of weathering of rocks, but ribose and the four nitrogen-containing bases have more-complicated structures that may or may not have been easy to form.

Prebiotic Synthesis of Organic Compounds

In the Miller–Urey experiment, the investigators found that they could synthesize amino acids by starting from gaseous mixtures that contain methane and ammonia. If RNA-based organisms came first, the first requirement for originating life would have been to synthesize ribose and the four bases adenine, guanine, cytosine, and uracil. Alternatively, the earliest organisms could have used simpler compounds that later evolved into RNA, but this scenario presents the difficult problem of how life could have switched from one molecular basis to another. (Switching from RNA to DNA is not considered difficult from an evolutionary standpoint, because the molecules are so similar. Indeed, the first step in synthesizing DNA within a cell is to form the corresponding RNA molecule.)

Could ribose and the bases have formed from compounds present in the primitive atmosphere and oceans? We can get a clue by looking at chemical formulas. The molecular formula for ribose is $C_5H_{10}O_5$. (The corre-

The Compounds of Life

Life depends on a complex array of organic (carbon-containing) compounds that organisms use to perform different tasks. Perhaps the most fundamental of these compounds are amino acids and nucleic acids. Amino acids are the building blocks of proteins, which are essential to many different cell functions. Nucleic acids, which include both **ribonucleic acid (RNA)** and **deoxyribonucleic acid (DNA),** are the carriers of genetic information. DNA stores this information, and RNA transfers the information to different parts of the cell and makes proteins and other compounds.

Chemically, an amino acid is an organic compound that contains an amino group (NH_2) and a carboxyl group (COOH). The two simplest ones are glycine and alanine (Box Fig. 9-1). These compounds are similar except for the *side chain:* The hydrogen atom, H, in glycine is replaced by a

methyl group, CH_3, in alanine. Other amino acids have more-complicated side chains containing oxygen, nitrogen, or sulfur. Twenty different amino acids are found in naturally occurring proteins, but many other amino acids are chemically possible.

RNA and DNA are more-complicated compounds that consist of chains of molecules called **nucleotides.** Each nucleotide consists of three parts. In RNA, these include a phosphate molecule, a ribose (sugar) molecule, and a nitrogen-containing base. The base can be any of four molecules: adenine (A), guanine (G), cytosine (C), or uracil (U) (Box Fig. 9-2). The nucleotides in RNA are linked together in a long, single strand.

A DNA nucleotide is similar to an RNA nucleotide except that the sugar molecule is deoxyribose instead of ribose and that one of the four bases, uracil, is replaced by thymine (T). The DNA nu-

Glycine   Alanine

BOX FIGURE 9-1

Two of the simpler amino acids found in proteins.

cleotides are strung together in two chains that form a **double helix.** This double-stranded, twisting structure was discovered in 1953 by James Watson and Francis Crick. The sequence of bases in DNA carries information in the form of the **genetic code.** Groups of three individual nucleotides code for specific amino acids. For example, the sequence CCA codes for glycine and CGA codes for alanine.

(a) DNA nucleotide   (b) RNA nucleotide

BOX FIGURE 9-2

Structural diagrams of the components of (a) DNA and (b) RNA. (From G. S. Kutter, *The Universe and Life,* Jones and Bartlett Pub., Inc. Boston, 1987.)

sponding structural formula, which shows how the atoms are arranged, is shown in Box Fig. 9-2b.) Simple division shows that ribose can be formed from five molecules of H_2CO—the compound *formaldehyde,* which is commonly used to preserve dead animals. (If you dissected a frog in high school biology class, you probably recall its distinctive smell.) The molecule H_2CO should not be confused with the shorthand notation for organic carbon, CH_2O, that we have used elsewhere in this book. Whereas H_2CO is an actual molecule, CH_2O merely represents complex hydrocarbons that have approximately the same relative ratios of C, H, and O atoms.

Thus, the first step in forming ribose is to synthesize formaldehyde. This step, it turns out, is easily accomplished. Photochemical reactions in weakly reducing, CO_2-rich atmospheres are predicted to produce large quantities of formaldehyde. Because formaldehyde is soluble in water, much of it should have dissolved in rainwater and been transported into the early ocean. Converting formaldehyde into ribose is also not difficult, because formaldehyde spontaneously reacts to form sugars in water solution. The problem for life's origin, which we will not discuss here, is that lots of other sugars form in addition to ribose and that these molecules might have interfered with the synthesis of RNA. What we can say is that the necessary starting material for forming ribose should have been available.

What about the four bases—could they have been formed on the prebiotic Earth? Let us take the same approach and begin with chemical formulas. For simplicity, we consider only the simplest base, adenine. Its molecular formula is $C_5H_5N_5$. Evidently, adenine can be formed from five molecules of HCN, *hydrogen cyanide.* Hydrogen cynanide is an extremely deadly poison. To prebiotic chemists, it is considered an essential building block for life.

Forming hydrogen cyanide in the prebiotic atmosphere is more difficult than forming formaldehyde. In a Miller–Urey–type atmosphere containing methane and ammonia, HCN would have been generated by lightning. In a weakly reducing, CO_2–N_2 atmosphere, lightning would not have sufficed because the carbon and nitrogen atoms produced by the lightning would have combined with oxygen atoms from the CO_2. However, the primitive atmosphere might have contained a few tens of parts per million of methane, CH_4. Methane photolysis in the stratosphere produces molecular fragments that can combine with N atoms that flow down from the ionosphere, forming HCN. The N atoms are produced from the breaking apart of the ion N_2^+ when it recombines with an electron.

The key to this mechanism is to identify a source for atmospheric methane. The methane in today's atmosphere is almost entirely of biological origin. Some abiotically generated methane is released, though, in fluids coming from hydrothermal vents at mid-ocean ridges. These vents, which are described in more detail below, are places where hot, mineral-laden water flows into the deep ocean. The water contains dissolved carbon compounds, including both CO_2 and CH_4, that appear to have originated in the upper mantle. Today, about 1% of the carbon in these fluids is in the form of CH_4; the rest is CO_2. Early Earth, though, might have had a more reduced upper mantle, so the vent fluids might have been much richer in CH_4. If so, the production of HCN may well have been possible.

Other Theories of Life's Origin

Some researchers remain skeptical that life could have formed on Earth's surface or in its oceans. Although the fundamental building blocks of life, H_2CO and HCN, were probably available, concentrating them sufficiently to allow further reactions to occur might have been difficult. And the more complex organic compounds that might have formed in this way would not have lasted long in the surface-ocean environment, because they would have been destroyed by photochemical and thermal (heat-driven) reactions. So, researchers have been considering alternative ways of forming complex organic compounds.

One possibility is that the relevant organic compounds were formed in space and brought to Earth by asteroids or comets or as tiny dust particles. **Interplanetary dust particles (IDPs)** are small particles recovered from the stratosphere that are known to be of extraterrestrial origin. We know that organic compounds, including amino acids, exist in IDPs as well as in some meteorites. Indeed, amino acids and many other complex organic compounds have now been identified in interstellar dust clouds (Fig. 9-3). They are believed to form from reactions between ions and neutral molecules that occur at very low temperatures. Typical temperatures in interstellar dust clouds are on the order of 10 K, not much above absolute zero. It may seem surprising that organic chemistry could occur in this environment, but it is pre-

FIGURE 9-3

The Horsehead Nebula, an interstellar dust cloud, in the constellation Orion. (From Science Photo Library, Photo Researchers, Inc.)

cisely the extremely low temperatures involved that allow complex organic molecules to form and survive.

Some of the molecules formed in the interstellar environment are thought to have survived the collapse of the cloud that formed our own Sun and solar nebula. They would have been incorporated into solid materials that condensed out of the nebula and accreted to form asteroids and comets. Such materials might have been delivered to Earth in great quantities during the heavy bombardment period of Solar System history, between 4.5 b.y. ago and 3.8 b.y. ago.

A third theory of life's origin is that it took place in or around **hydrothermal vents** in the mid-ocean spreading ridges. Recall from Chapter 6 that mid-ocean ridges are places where new sea floor is being created. The ridges are cooled by seawater that flows a kilometer or more down through cracks in the rock, is heated, and then rises rapidly back to the surface. In the process, the water picks up reduced substances such as hydrogen (H_2), hydrogen sulfide (H_2S), and dissolved ferrous iron (Fe^{2+}). *Ferrous iron* is a reduced form of iron that is soluble in seawater. When it hits the cold water, the hot (350°C) vent water produces a dark plume of precipitating material called a **black smoker** (Fig. 9-4). The majority of the dark material is iron sulfide, FeS, produced by the reaction between ferrous iron and hydrogen sulfide.

Are submarine hydrothermal vents a likely place for life to have originated? The vent systems are rich in the types of reduced materials from which organic molecules can be synthesized. They contain liquid–solid interfaces that some researchers think are needed to organize organic molecules into specific patterns. One model suggests that life originated on the surface of *pyrite* (FeS_2) mineral grains, which are abundant in hydrothermal vent systems. However, complex organic molecules are not stable at the high (350°C) temperatures observed in vents located directly on the ridge axes. If life did originate at the mid-ocean ridges, it probably

did so in cooler, off-axis vents. Some researchers argue that even the off-axis vents are too warm and that the best place for life to have originated would be in some near-freezing surface environment. The debate as to whether life originated in a hot or cold environment is likely to continue until we have a better idea of how the process actually occurred.

Living bacteria have, in fact, been found in off-axis vents at water temperatures up to 130°C. Furthermore, studies of the relationships between the RNA of these and other organisms show that sulfur-metabolizing, **thermophilic** (heat-loving) **bacteria** appear to be very ancient. Both high temperatures and sulfur compounds are found within the hydrothermal vent systems. Studies of the bacteria living within the modern vent systems indicate that they have changed very little over the past 3.5 b.y. and, thus, may bear a close resemblance to primitive life forms. This observation is consistent with the idea that life originated in hydrothermal vents, but it does not require that it did so. It may be that life originated at the surface, colonized the vents, and then all surface life was annihilated by the impact of a giant planetesimal. (Recall from Chapter 8 that Earth was regularly bombarded by asteroids or comets until at least 3.8 b.y. ago.) This scenario would be consistent with the RNA evidence of a thermophilic last common ancestor, but it would not require life to have originated at high temperatures.

Effect of Life on the Early Atmosphere

Despite the uncertainties regarding how it actually happened, we know that life had originated by at least 3.5 b.y. ago, because geologists have identified microfossils and stromatolites in rocks of that age. **Microfossils** are the preserved remains of single-celled organisms (Fig. 9-5a). **Stromatolites** are the fossilized remains of bacteria that form layered, sedimentary mats (Fig. 9-5b). Indeed, the presence of organic carbon in 3.8-b.y.-old rocks from Isua, West Greenland, and from nearby Akilia Island suggests that life might have been present 300 million years earlier. Some of this organic carbon is strongly depleted in ^{13}C, suggesting that it was formed by photosynthesis. No actual microfossils of this age have been found; however, the evidence suggests that life colonized Earth's oceans and continental shelves soon after the heavy bombardment period ended, around 3.8 b.y. ago.

How did the presence of life affect the composition of the atmosphere? The Warrawoona microfossils (Fig. 9-5) have relatively large diameters, larger than most types of bacteria but similar to modern cyanobacteria. **Cyanobacteria,** formerly referred to as blue-green algae, are capable of performing both oxygenic and anoxygenic photosyntheses. **Oxygenic photosynthesis** is the "normal"

FIGURE 9-4

A black smoker. (From Dr. Ken MacDonald/Science Photo Library, Photo Researchers, Inc.)

FIGURE 9-5

(a) Microfossils and (b) stromatolites from the 3.5-billion-year-old Warrawoona formation in Australia. ((a) Reprinted with permission from J. W. Schopf, *Science,* 260, 640–646. Copyright 1993 American Association for the Advancement of Science; (b) J. W. Schopf, ed., *Earth's Earliest Biosphere,* Princeton, NJ: Princeton University Press, 1983.) [See color section]

type of photosynthesis in which O_2 is produced. **Anoxygenic photosynthesis** uses H_2S instead of H_2O to reduce CO_2 to organic carbon; this process does not yield O_2. In practice, cyanobacteria switch back and forth between these two metabolic processes in response to changes in their local environment. When hydrogen sulfide, H_2S, is present, cyanobacteria photosynthesize anoxygenically; when H_2S is absent, they generate O_2.

If the Warrawoona organisms were indeed cyanobacteria, it is likely that they were producing oxygen at least some of the time. But this does not necessarily mean that the atmosphere would immediately have become O_2-rich. As discussed below, free oxygen does not seem to have appeared in the atmosphere until about 2.2–2.0 b.y. ago. Evidently, the O_2 produced by photosynthesis was entirely consumed by reactions with reduced gases or with other reduced materials at Earth's surface.

Production of Methane

Although the origin of life did not immediately result in abundant free O_2, organisms could have affected atmospheric composition in other ways. One of the most significant effects would have been the production of methane. We suggested earlier that some methane might have been present prior to life's origin as a consequence of outgassing from mid-ocean ridge hydrothermal vents. The potential biotic source of methane, however, is substantially larger. **Methanogenic bacteria,** or **methanogens** for short, produce energy from chemical reactions that generate methane. The simplest of these is

$$CO_2 + 4H_2 \longrightarrow CH_4 + 2H_2O.$$

RNA sequencing of modern organisms indicates that some methanogens are very ancient, so there is good reason to believe that they existed on the early Earth. We have already seen that the primitive atmosphere should have been rich in CO_2 and contained hundreds to thousands of ppm of H_2 (possibly even more, if hydrogen escape was slower than it is today because the upper atmosphere was colder). Many methanogens can survive at these H_2 concentrations, although most require about 1% H_2 (0.01 bar) to reproduce. Even if they did not grow everywhere on Earth's surface, methanogens should have been present in locally H_2-rich environments such as hydrothermal vents and subsurface rocks. (The reaction of water with reduced iron minerals produces H_2 in basaltic rocks.) Their effect should have been to draw down atmospheric H_2 and to replace it with CH_4. If methanogens on the early Earth produced methane at the same rate that it is produced biologically today, the atmospheric methane concentration could have exceeded 1000 ppm, more than 600 times its present level (1.6 ppm). The higher abundance results from the fact that methane would have been destroyed less rapidly in a low-O_2 atmosphere.

Atmospheric CO_2 levels would not have been directly affected by methanogens, because the methane that they generated would eventually have been re-oxidized to CO_2 and H_2O. However, CO_2 might have been affected indirectly, because methane is a greenhouse gas that would have warmed Earth's climate. Warmer surface temperatures would have accelerated the rate of silicate weathering, thereby drawing down atmospheric CO_2 by the negative feedback mechanism discussed in Chapters 7 and 8. The presence of 1000 ppm of methane would have warmed the early atmosphere by about 15°C, equivalent to the temperature change produced by a change in CO_2 by a factor of 30. Thus, atmospheric CO_2 concentrations might have dropped by a factor of 30 once methanogens evolved.

Although this discussion is highly speculative, the geologic record provides indirect evidence that something like this actually happened. Some of the sedimentary or-

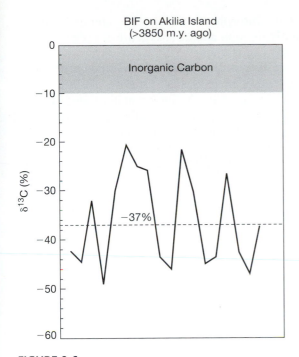

FIGURE 9-6

Carbon isotopes measured in organic carbon from 18 samples of 3.9-b.y.-old rocks on Akilia Island, near Greenland. (After Mojzsis et al., *Nature*, 38,4:55–59, 1996).

ganic carbon from this interval is highly depleted in ^{13}C (Fig. 9-6)—more so than can be explained by photosynthesis alone. This depletion is thought to be an indication of widespread methanogenic activity. The methane produced by methanogens is depleted in ^{13}C relative to ^{12}C. The methane produced in this manner was consumed by other **methanotrophic** (methane-loving) **bacteria.** These bacteria also preferentially consume the lighter carbon isotope, so their remains would have been doubly depleted in ^{13}C. These remains may account for the very low ^{13}C organic carbon observed in the geologic record of that time.

One further observation can be made that ties in with our discussion of climate in Chapter 8. The first well-documented glaciation in Earth history, the Huronian glaciation, occurred about 2.3 b.y. ago—nearly the same time that atmospheric O_2 first rose to appreciable levels. The rise of atmospheric O_2 should have caused a catastrophic decline in methane, because methane is photochemically short-lived in an oxidizing atmosphere. If methane was a significant contributor to the greenhouse effect prior to that time, its abrupt disappearance could have triggered the onset of the Huronian glaciation.

Cycling of Atmospheric Nitrogen

A second way in which organisms might have affected the atmosphere is by cycling nitrogen through the atmosphere–ocean system. Organisms need nitrogen for making proteins and nucleic acids as well as for

other biochemical functions. However, most of them cannot use nitrogen in its normal molecular form, N_2. Instead, they require **fixed nitrogen,** in which nitrogen atoms are bonded to other types of atoms. Ammonia (NH_3) and nitrate ion (NO_3^-) are two examples of fixed-nitrogen compounds.

Modern marine organisms acquire fixed nitrogen in two ways. One of these is from lightning. In the high-temperature region surrounding a lightning discharge, nitrogen and oxygen react to form nitric oxide, NO:

$$N_2 + O_2 \longrightarrow 2\,NO.$$

Nitric oxide is a radical that plays an important role in ozone photochemistry. In today's atmosphere it is eventually oxidized to nitric acid, HNO_3. Nitric acid is soluble in water and, thus, is quickly rained out of the troposphere. In solution, it **dissociates** (comes apart) to form hydrogen ions and nitrate ions:

$$HNO_3 \longleftrightarrow H^+ + NO_3^-.$$

The resulting nitrate ion can be directly used by organisms as a source of fixed nitrogen.

Some marine organisms can make their own fixed nitrogen by a process termed **nitrogen fixation.** Most of the biologically available nitrogen in the oceans is fixed by cyanobacteria, the same type of organisms that appear to have been present at Warrawoona. So, we can surmise that nitrogen fixation was one of the earliest metabolic activities to be developed. Indeed, the ability to fix nitrogen is widespread among **prokaryotes,** which are single-celled organisms that lack *cell nuclei*—structures that house genetic material (Fig. 9-7a). (In prokaryotes, the genetic material is not concentrated within a specific structure.) **Eukaryotes** (organisms with cell nuclei) cannot fix nitrogen (Fig. 9-7b). Prokaryotes are thought to be more primitive than eukaryotes because their cell structure is simpler. They are also more resistant to UV radiation, which is consistent with the idea that they evolved under a low-O_2 atmosphere that lacked a protective ozone layer.

How did the very earliest organisms acquire fixed nitrogen? Lightning is the most likely answer, although the reaction between N_2 and O_2 could not have been the source, because there would have been little O_2 available. However, the analogous reaction

$$N_2 + 2\,CO_2 \longrightarrow 2\,NO + 2\,CO$$

could have provided fixed nitrogen at a modest rate.

If nitrogen fixation were not balanced by some reverse process, N_2 would be completely removed from the atmosphere in only about 20 million years. This does not happen, however, because atmospheric nitrogen is recycled by several different processes. Today, the dominant

FIGURE 9-7

(a) Prokaryotes have no nucleus, and the DNA is dispersed within the cell. (From Eric V. Grave, Photo Researchers, Inc.) (b) Eukaryotes have their DNA enclosed within a cell nucleus. (From A.B. Dowsett/Science Photo Library, Photo Researchers, Inc.)

recycling mechanism is bacterial **denitrification.** Some organisms can derive energy by reacting nitrate with organic matter. In the process, nitrogen is released as either N_2 or as nitrous oxide, N_2O. Most of the nitrous oxide is converted back into N_2 by photolysis, although some of it re-

acts in the stratosphere to form NO. Denitrification is particularly rapid in **anoxic** (oxygen-free) regions of the ocean and in anoxic soils.

Before life was present, nitrogen would have been recycled by other mechanisms. One mechanism that is thought to have been important is that illustrated in Figure 9-8. NO formed by lightning would have been converted to more-soluble forms such as nitrous acid, HNO_2, and then removed from the atmosphere by rainfall. Once in the ocean, HNO_2 could have been reduced to ammonia, NH_3, by reaction with dissolved ferrous iron. The ammonia thus formed would have leaked back into the atmosphere and been converted back into N_2 by photolysis. Because the rate of ammonia photolysis is high and the nitrogen fixation rate (biological or abiotic) is relatively low, it seems likely that most of Earth's nitrogen has resided in the atmosphere as N_2 ever since the planet formed.

The Rise of Oxygen: Geologic Evidence

We suggested earlier in this chapter that oxygen was being produced photosynthetically by 3.8 b.y. ago or even earlier but that its abundance was initially kept low by reactions with reduced volcanic gases. At some time during Earth's history, the production of oxygen overwhelmed the sinks, and O_2 began to accumulate in the atmosphere. Before discussing why this occurred, let us first try to decide *when* it occurred. This topic remains controversial, because the geologic evidence for O_2 accumulation is somewhat ambiguous.

Banded Iron-Formations

One type of geologic evidence that bears on the rise of oxygen is the occurrence of banded iron-formations. **Banded iron-formations (BIFs)** are laminated sedimentary rocks that consist of alternating millimeter-thick lay-

FIGURE 9-8

The nitrogen cycle on the early Earth.

FIGURE 9-9

A banded iron-formation, or BIF. (From J.W. Schopf.) [See color section]

ers of iron-rich minerals and chert (Fig. 9-9). Such minerals include *magnetite* (Fe_3O_4) or *hematite* (Fe_2O_3), and *chert* (SiO_2). They are of enormous economic importance today. Much of the iron that is used in making steel and automobiles come from BIFs in Canada and Australia. But these deposits do not form today; nor have they done so at any time in the recent past. Radiometric age dating shows that almost all BIFs formed prior to 1.9 b.y. ago. The only exceptions are a few BIFs that formed in the Late Proterozoic, about 0.6 to 0.8 b.y. ago. As we mentioned in Chapter 8, the Late Proterozoic is unusual for several reasons, the most intriguing of which is the evidence of low-latitude glaciation. Some researchers have suggested that the two phenomena are linked: Extensive ice cover inhibited oxygen transfer between the atmosphere and ocean, and this led to the reappearance of BIFs.

The reason that BIFs are useful as oxygen indicators is that iron can exist in more than one oxidation state. The **oxidation state** of an atom, molecule, or compound is its degree of oxidation. Substances with a low oxidation state have a large number of available electrons; substances with a high oxidation state do not (see the box "Oxidation States of Iron"). We have already mentioned one oxidation state of iron, **ferrous iron** (Fe^{2+}). A second, more oxidized state

is termed **ferric iron** (Fe^{3+}). Iron ions in these two oxidation states have very different chemical properties: Fe^{2+} is soluble in seawater, whereas Fe^{3+} is not. Because iron switches from Fe^{2+} to Fe^{3+} when oxygen is present, it can provide indirect information about past O_2 levels.

To make use of this information, we need to understand the process by which BIFs formed. Although no consensus has been reached regarding the precise mechanism, there is general agreement on some parts of the story. To explain the voluminous quantities of iron deposited in BIFs, large portions of the deep oceans must have been anoxic. This condition would have allowed iron to be transported as dissolved Fe^{2+}. The iron was probably supplied originally from continental weathering and from mid-ocean ridge hydrothermal vent fluids. Trace element patterns in BIFs, especially those of the **rare earth elements** (atomic numbers 58–71 in the periodic table), show that at least some of the iron must have come from the vents. What happened next is not well understood. We know that BIFs did not form on the floor of the deep ocean. If they had, they would have been destroyed when the sea floor was subducted. Instead, the dissolved iron must have been transported to the margins of the continents, where it was deposited on stable continental shelves. To be precipitated, the iron must first have been oxidized from Fe^{2+} to Fe^{3+}.

Exactly how this occurred is still a subject of debate. One suggestion is that the iron was brought to the surface by wind-induced upwelling of the type that occurs in some modern coastal settings. The fine, millimeter-scale banding observed in the BIFs could have been caused by seasonal changes in winds that produced upwelling at certain times of the year but not at others. Iron-rich sediments would have formed when the upwelling was strong; silica-rich sediments would have formed when upwelling was weaker. Once the dissolved ferrous iron was brought to the surface, it could have been oxidized by dissolved O_2 produced by photosynthetic cyanobacteria. Alternatively, the iron could have been oxidized abiotically. Laboratory experiments have shown that dissolved Fe^{2+} can be oxidized to Fe^{3+} by UV radiation. (In the process, H_2O is reduced to H_2.)

Oxidation States of Iron

Iron occurs in three oxidation states in nature: elemental (or metallic), ferrous, and ferric. The elemental form is located mostly in Earth's core, which is composed largely of iron–nickel alloy. Ferrous and ferric iron occur in the mantle and crust. They exist in dissolved form as Fe^{2+} and Fe^{3+} ions, respectively. The positive charges arise because these ions are missing electrons: Fe^{2+} is missing two electrons, and Fe^{3+} is missing three. We say that these two ions are in the 2+ and 3+ oxidation states. Elemental iron, by comparison, has an oxidation state of zero.

When iron reacts with oxygen, the oxidation state of the iron atom determines the number of oxygen atoms with which it combines. Oxygen, when it reacts, can be thought of as having a valence of 2−.

Thus, to produce a neutral molecule, one Fe^{2+} ion combines with one O^{2-} ion to form the mineral FeO (wüstite). Two Fe^{3+} ions combine with three oxygen ions to form Fe_2O_3 (hematite).

Most of the iron in BIFs consists of magnetite, Fe_3O_4. This is equivalent to one molecule of FeO bonded to one molecule of Fe_2O_3. So, BIFs contain a 50–50 mixture of ferrous iron and ferric iron.

Whether the iron in BIFs was oxidized by UV radiation or by biologically generated O_2 is still unresolved.

Because the oxygen in BIFs could have come from several sources, we cannot use them directly to infer the O_2 content of the atmosphere. We can, however, safely conclude that the deep oceans must have been anoxic prior to 1.9 b.y. ago, during the time that BIFs formed. This, in turn, implies that the atmospheric O_2 concentration was lower than it is today. So, BIFs are a strong indicator that the atmosphere has changed during recorded geologic history.

Detrital Uraninite and Pyrite

More-direct information on atmospheric O_2 levels can be obtained from other types of geologic indicators. Elements other than iron are also capable of changing their oxidation state. Uranium, for example, has two common oxidation states: U^{4+} and U^{6+}. As with iron, these two ions differ in their solubilities. In this case, however, it is the oxidized ($6+$) form that is soluble and the more reduced ($4+$) form that is not.

The U^{4+} ion combines with oxygen to form **uraninite, UO_2**. This mineral occurs in rocks today, but it is normally oxidized to the soluble, $6+$, state during weathering. (Notable exceptions to this rule occur in some of the rivers that drain the Himalayas, where the eroded sediments are redeposited very quickly, before the uranium can be oxidized.) Dissolved uranium is transported to the oceans, where it diffuses into anoxic sediments, is reduced, and precipitates as UO_2.

This modern cycle of uranium weathering and deposition does not seem to have operated early in Earth's history. Sedimentary rocks older than about 2.2 b.y. contain uraninite in *detrital* form (Fig. 9-10a). A **detrital mineral** is a mineral that survived the weathering process and was transported to the site of deposition as a solid particle rather than in solution. Such a mineral can often be identified by its appearance, which closely resembles the texture of the source rock. The pres-

FIGURE 9-10

Samples of the detrital form of (a) uraninite and (b) pyrite. (From J.W. Schopf.) [See color section] ence of the detrital form implies that, at the time when the source rock was weathered, the atmospheric O_2 content was too low to oxidize uraninite. Quantitative analysis suggests that the O_2 concentration during this time was less than 10^{-3} bars, or 0.005 PAL. (*PAL* means "times the present atmospheric level.") Unfortunately, this is only an upper bound. The actual O_2 concentration prior to 2.2 b.y. ago could have been much lower than this. Indeed, it might have been essentially the same as on the prebiotic Earth, where the surface O_2 concentration is thought to have been on the order of 10^{-13} PAL (see Fig. 9-2).

A second mineral that was deposited in detrital form prior to about 2.2 b.y. ago, but not since then, is **pyrite**, FeS_2 (Fig. 9-10b). It tells essentially the same story as uraninite. When pyrite is weathered under today's oxidizing atmospheric conditions, the sulfur is oxidized to sulfate, SO_4^{2-}, and the iron is oxidized to Fe^{3+}. But this oxidation apparently did not happen on the early Earth. Eroded grains of pyrite were transported long distances by streams and rivers, as evidenced by the rounded appearance of some samples, and were deposited in chemically unaltered form. This evidence is another indication that atmospheric O_2 was low during the Archean Eon.

Paleosols and Redbeds

Other geologic indicators that have been used to study the rise of atmospheric O_2 include **paleosols** (ancient soils) and **redbeds** (reddish-colored sandy and silty sediments). Heinrich Holland of Harvard University has conducted chemical analyses of numerous Precambrian paleosols. He has found that most paleosols older than about 2.2 b.y. have lost significant amounts of iron, whereas paleosols younger than 1.9 b.y. have retained it. This finding is consistent with the story outlined above. Prior to 2.2 b.y. ago, atmospheric O_2 was low, so the iron released during weathering remained as soluble Fe^{2+} and was carried away by groundwater. The resultant paleosols are iron-poor. After 1.9 b.y. ago, atmospheric O_2 was relatively abundant, so the iron released by weathering was oxidized to insoluble Fe^{3+} and was retained in the soil. Holland's detailed calculations predict that atmospheric O_2 was less than 0.01 PAL before 2.2 b.y. ago and greater than 0.15 PAL after 1.9 b.y. ago.

Redbeds are sandy sediments that were deposited on land by rivers or as windblown dust (Fig. 9-11). They form today in arid regions, such as the American Southwest. The reddish color of these deposits comes from a thin layer of hematite, Fe_2O_3, that coats the surfaces of the sediment grains. The iron in hematite is of the oxidized, Fe^{3+}, variety, so redbeds indicate oxidizing atmospheric conditions at the time of their formation. The earliest confirmed redbeds are thought to have formed about 2.2 b.y. ago, which is consistent with the other evidence for a rise in atmospheric O_2 at about this time.

FIGURE 9-11

A sequence near Uranvan, Colorado, showing Triassic redbeds overlying Upper Jurassic sandstones. (From P. Cloud, *Oasis in Space*, Norton, New York, 1988.)

The Rise of Ozone

The rise of atmospheric oxygen would have been accompanied by an increase in stratospheric ozone. As we mentioned previously, ozone is critical for life because it shields out harmful solar ultraviolet radiation. The principal wavelength region in which ozone shielding is important is between about 200 and 300 nm (Fig. 9-12). (Recall that 1 nm [nanometer] $= 10^{-9}$ m $= 10^{-3}$ μm.) Ultraviolet radiation with wavelengths shorter than 200 nm is also harmful to organisms, but it is effectively absorbed by both O_2 and CO_2. Thus, it should not have been a problem even in an anoxic atmosphere. Ultraviolet radiation between 300 and 400 nm is much less harmful to organisms and is not considered to be a threat, even though a substantial flux of such radiation reaches Earth's surface today.

As we mentioned in Chapter 1, the amount of UV radiation absorbed by ozone depends on the ozone *column depth*—that is, the total amount of ozone between the surface and the top of the atmosphere. The column depth is usually measured in *Dobson units*. One Dobson unit (DU) is equivalent to a layer of pure ozone 0.01 mm thick at the ground. The average ozone column depth today is about 320 DU, equivalent to a 0.32-cm-thick layer of pure ozone at 1 atm pressure.

How much ozone would have been present if the atmospheric O_2 concentration were lower than it is today? **Photochemical models** can be used to address this question: From a model calculation that is capable of reproducing today's average ozone column depth, O_2 is gradually removed from the model atmosphere. The results of one such calculation are shown in Figure 9-13. The calculation shows that the ozone column depth increases nonlinearly with atmospheric O_2 level: Even a

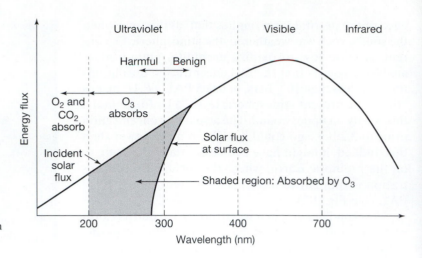

FIGURE 9-12

The wavelength region where absorption by ozone is important.

small amount of O_2 produces a substantial ozone column depth. The reasons have to do with the details of ozone photochemistry, which we shall save for a later chapter. Exactly when a biologically effective UV screen would have been established is not clear, because organisms differ greatly in their tolerances for UV radiation. However, examination of the calculated UV fluxes shows that most of the harmful UV radiation would already have been absorbed once the ozone column depth exceeded about 100 DU, or roughly one-third of today's value. As Figure 9-13 demonstrates, this column depth would have been reached at an O_2 level of only 0.01 PAL. For comparison, the paleosol evidence discussed earlier indicates that atmospheric O_2 was greater than 0.15 PAL by 1.9 b.y.

ago. So, we can infer that a reasonably effective UV screen was already established by this time.

This prediction can be tested indirectly by comparing it with the fossil record. We mentioned earlier that eukaryotes are much less tolerant of UV radiation than are prokaryotes. They also require oxygen in order to live, whereas some prokaryotes do not. The earliest eukaryotes that have been found, corkscrew-shaped organisms called *Grypania* (Fig. 9-14), appeared about 2.1 b.y. ago, 100 million to 200 million years after the last appearance of detrital uraninite and pyrite deposits. Phytoplankton (free-floating photosynthetic organisms) become abundant in the geologic record soon after this time. Phytoplankton are more susceptible to UV radiation than are

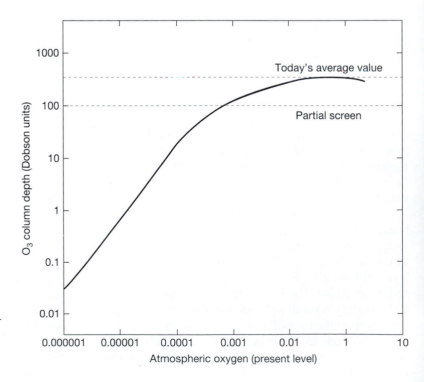

FIGURE 9-13

Ozone column depth at different atmospheric O_2 levels, as calculated by a photochemical model. "PAL" stands for "times the present atmospheric level."

FIGURE 9-14

A fossil eukaryote, *Grypania*, dated at 2.1 b.y. old. (From T. M. Han and B. Runnegar, *Science* 257, 232, 1992.)

benthic (bottom-dwelling) organisms, because the phytoplankton are periodically brought up to the surface of the ocean, where UV fluxes would be highest. Both these observations are consistent with the idea that an effective UV screen had already been established by 2.1 b.y. ago.

Possible Causes of the Rise of Oxygen

Let us return to the question posed earlier in the chapter: If oxygen-producing cyanobacteria were present as early as 3.5 b.y. ago, why did it take an additional billion years or more for atmospheric O_2 levels to increase? Either the source for O_2 must have been smaller in the past,

or the sinks (removal mechanisms) for O_2 must have been larger. The first step in analyzing this problem is to decide which of these possibilities is more likely. To do this, we require additional information.

Constraints Imposed by Carbon Isotopes

Fortunately, additional constraints can be imposed from the study of carbon isotopes. In Chapter 8 we used carbon isotopes to infer past concentrations of atmospheric CO_2. Photosynthetic organisms selectively use ^{12}C instead of ^{13}C when plenty of CO_2 is available in their environment. As a result, sedimentary organic carbon is typically about 25‰ lighter (that is, depleted in ^{13}C) than sedimentary carbonates.

Carbon isotopes have been measured in rocks dating as far back as 4.0 b.y. ago. Figure 9-15 shows $\delta^{13}C$ values that have been obtained from carbonate rocks and from dispersed sedimentary organic carbon (hydrocarbons), or **kerogens,** throughout this time span. The organic carbon data are quite scattered. Part of the scatter is caused by natural variability; part is caused by changes in carbon isotope ratios that occur after sediments are buried. Note, however, that the average $\delta^{13}C$ value of carbonates has remained close to 0‰ for most of the past 3.8 b.y. and that the average $\delta^{13}C$ value of organic carbon appears to have remained near -25‰.

The ^{13}C content of carbonate sediments reflects the isotopic composition of bicarbonate in seawater, because that is where most of the carbonate originates. The ^{13}C content of dissolved bicarbonate depends on the relative burial rates of organic carbon and carbonates. When the burial rate of isotopically light (low-^{13}C) organic carbon is high, the remaining bicarbonate in the oceans becomes isotopically heavy. When the organic carbon burial rate is

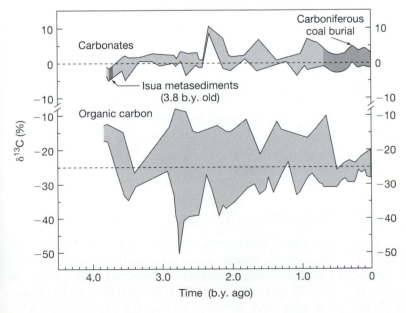

FIGURE 9-15

Graph of $\delta^{13}C$ values of carbonates and of organic carbon over the past 3.8 b.y. (After J. W. Schopf, *Earth's Earliest Biosphere,* Princeton, NJ: Princeton University Press, 1983.)

low, the dissolved bicarbonate becomes isotopically lighter. So, the carbon isotope record tells us something about how fast organic carbon was buried in the past. This, in turn, tells us how fast O_2 was being produced.

Taken at face value, the data in Figure 9-15 imply that the fraction of carbon buried as organic carbon has remained roughly constant with time. Today, that value is about 20%. The other 80% of the carbon is buried as carbonate minerals. A constant burial fraction, in turn, implies that the source of atmospheric O_2 in the past was at least as large as it is today. Indeed, if volcanos were more active in the past, as seems likely, the rate of organic carbon burial must have been higher than today, and the O_2 source would have been correspondingly larger.

The alternative explanation for the delayed rise of atmospheric oxygen is that the sinks for O_2 were larger in the past. As we mentioned earlier, the two long-term sinks for atmospheric O_2 today are oxidation of reduced minerals (ferrous iron, sulfides, and kerogen) during weathering and oxidation of reduced volcanic gases. Oxidative weathering is the dominant loss process today; it balances 70 to 90% of the O_2 produced by organic carbon burial. But this sink would not have operated when atmospheric O_2 levels were low. We have already seen that reduced minerals such as uraninite and pyrite were not oxidized under these conditions.

The volcanic gas sink, however, was probably higher in the past, because more geothermal heat was available to drive volcanism. But this factor, by itself, cannot explain the rise of oxygen, because higher volcanic outgassing rates would have meant that more CO_2 was going into the atmosphere–ocean system. If a constant fraction of this CO_2 was buried as organic carbon, the rate of O_2 production would have risen in parallel with the rate of O_2 loss.

Changes in Mantle Oxidation State

A possible solution to this problem involves a change in the oxidation state of volcanic gases over time. Today, volcanic gases consist largely of H_2O and CO_2, with smaller amounts of H_2, CO, and reduced sulfur compounds. The measured ratio of $H_2:H_2O$ is typically about 1:100; the ratio of $CO:CO_2$ is about 3:100. If, however, the primitive mantle was more reduced than today, these ratios could originally have been much higher, and the corresponding O_2 sink could have been much larger. Are there any reasons to believe that this was the case? The answer seems to be yes. Data from mineral inclusions in 3.3-b.y.-old diamonds support this hypothesis, as do recent measurements of the $Fe^{2+}:Fe^{3+}$ ratio of other old, mantle-derived rocks. Furthermore, new data on the high-pressure solubility of various *siderophile* ("iron-loving") *elements* in molten silicates and molten iron imply that the upper mantle was originally in chemical equilibrium with metallic iron. This is precisely what we would expect in the aftermath of the

impact that is believed to have formed the Moon. The present mantle, in contrast, is considerably more oxidized than this.

How could the mantle have become more oxidized with time? The most likely mechanism is that illustrated in Figure 9-16. When seawater circulates through the mid-ocean ridges, some of it is trapped within the fresh sea floor. The sea floor is then said to be *hydrated*. Typical, hydrated sea floor contains about 3% water by weight. This hydrated sea floor is subducted down into the upper mantle at plate boundaries. As the subducted slab moves downward and heats up, the water is released. Most of it returns to the surface and is outgassed by volcanos. However, some fraction of it is converted to H_2 by reactions of the form

$$2\ FeO + H_2O \longrightarrow Fe_2O_3 + H_2.$$

The H_2 is then outgassed by volcanos, and some of it escapes to space. The rate of hydrogen escape is low at present but should have been much higher prior to 2.2. b.y. ago, when the atmosphere was reduced. The net result is that both the upper mantle and the volcanic gases released from it should have become more oxidized with time.

Modern Controls on Atmospheric O_2

Let us now turn our attention to the modern system. We have offered some suggestions for why atmospheric O_2 concentrations may have increased when they did. A related question, which we might hope would be somewhat easier, is this: What controls the atmospheric O_2 concentration today?

The answer, surprisingly, is that we do not know for sure, although researchers do have a number of ideas. Whatever the oxygen control mechanism is, it appears to be very efficient. The modern atmospheric O_2 level is 21% by volume, or 0.21 bar. It seems unlikely that the O_2 concentration has strayed from this level by more than ±50% since the late Devonian Period, about 360 million years ago. The evidence is that forests have existed since that time and, while they have always been able to burn, they have never disappeared entirely.

Forest Fires and Atmospheric Oxygen

Fires burn more intensely when the oxygen content of the air is increased. A graphic, and tragic, illustration of the phenomenon occurred in the mid-1960s, when an *Apollo* space capsule burned up on its test pad in Houston with three astronauts aboard. At that time, NASA

FIGURE 9-16

The cycling of water and hydrogen between the crust and upper mantle. Loss of hydrogen to space during the first half of Earth's history should have resulted in progressive oxidation of the upper mantle.

was using pure O_2 in its spacecraft to minimize launch weight. After the fire, the space agency returned to using normal air because they realized, belatedly, that pure O_2 was too flammable.

A similar problem could occur globally if the atmospheric O_2 content got too high. Forest fires, ignited by lightning or other mechanisms, might rage out of control and burn everything within reach. It is difficult to determine the exact O_2 level at which this would occur, but laboratory experiments using wet matchsticks and shredded paper suggest that an O_2 concentration of 35% by volume would be enough to destroy most of the global biota (Fig. 9-17). A catastrophe of this magnitude may actually have taken place in the aftermath of the K/T impact event, 65 m.y. ago. But, as we shall discuss in Chapter 10,

this was a very special event in which fires may have been ignited all over the globe by finely dispersed ejecta re-entering Earth's atmosphere. Other than that scenario, there is no evidence that such widespread forest fires have occurred. We can infer that the atmospheric O_2 concentration has probably remained below 35% by volume ever since forests first appeared.

Forest fires can also be used to place a lower bound on atmospheric O_2 levels. Fires will not ignite when the O_2 concentration falls below about 13% by volume. This limit is somewhat firmer than the upper bound on O_2, because it depends on simple physics: At O_2 concentrations below 13%, a flame loses heat by convection more rapidly than it gains heat by combustion. Sedimentary rocks preserve a more-or-less continuous record of charcoal since the late Devonian period. Charcoal is produced from the incomplete combustion of organic matter by fire, so forest fires must have burned on and off throughout this period. Hence, we conclude that the atmospheric O_2 concentration has not fallen below 13% by volume during the last 360 m.y.

What mechanism could have stabilized atmospheric O_2 within the "fire window" during the past few hundred million years? The main loss processes for O_2, surface weathering and the oxidation of reduced volcanic gases, are thought to be independent of O_2 concentration within this range. The control mechanism must therefore act on the O_2 source—namely, photosynthesis followed by burial of organic carbon. Let us consider the factors that affect the organic carbon burial rate.

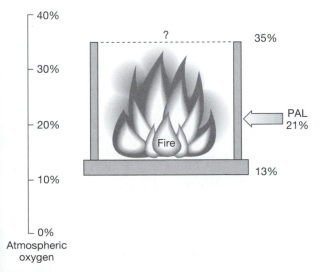

FIGURE 9-17

The oxygen "fire window," showing proposed minimum and maximum atmospheric O_2 levels consistent with the continuous record of charcoal deposition since the Devonian period. (From T. P. Jones and W. G. Chaloner, *Global and Planetary Change* 97, 39, 1991.)

Oxygenation of the Deep Ocean

Most of the organic matter being buried today is deposited in marine sediments rather than on land. The same seems to have been true during most of the past few hundred million years, except during the Carboniferous Period, when the great coal beds were forming. Thus, the control of organic carbon burial appears to lie in the ocean.

FIGURE 9-18

The vertical profile of dissolved O_2 in the (low-latitude) ocean.

For many years, researchers have believed that the feedback mechanism that stabilizes atmospheric O_2 involves the deep oceans. The deep oceans today contain relatively high concentrations of dissolved O_2. Indeed, over much of the globe, the dissolved O_2 content of deep water is higher than that of surface water (Fig. 9-18). The reason is that the water in the deep ocean originates at the surface at high latitudes, where temperatures are very cold. Like most other gases, O_2 increases in solubility as the temperature goes down. Thus, cold, high-latitude surface water contains more dissolved O_2 than does surface water at lower latitudes. This high O_2 content is passed on to the deep ocean when the cold surface water sinks as part of the global thermohaline circulation.

Oxygen concentrations in the deep ocean would drop dramatically, however, if the atmospheric O_2 concentration were lowered. Even today, over much of the ocean a pronounced **oxygen minimum zone** exists at a depth of about 1 km (see Fig. 9-18). The low O_2 concentrations at this depth are caused by the decay of organic matter that falls from the surface ocean. In some regions—the Black Sea, for example—the rain of organic matter is high enough to cause the deep ocean itself to become anoxic. If the atmospheric O_2 concentration were to decrease from its present level, the area covered by such **anoxic basins** would expand. The lower O_2 content of deep water could, in principle, allow more organic carbon to be buried on the ocean floor, because the organic matter would decay less quickly. This, in turn, could provide a negative feedback that might stabilize atmospheric O_2. Unfortunately, however, measurements do not support this idea. There is no direct correlation between the concentration of dissolved O_2 and the organic carbon content of marine sediments. Evidently, organic carbon in sediments can be oxidized quite efficiently by bacteria that utilize either sulfate or nitrate rather than O_2.

Dissolved Oxygen and Sedimentary C:P Ratios

The actual O_2 control mechanism appears to be slightly more complicated than that outlined above. Recall from Chapter 7 that marine productivity is usually limited by the availability of key nutrients, especially nitrogen (N) and phosporus (P). Of these, P is thought to be the most critical, because N can be fixed by organisms such as cyanobacteria. The concentration of P in seawater is controlled by the rate of supply of P from the weathering of rocks on the continents and by the loss of P due to incorporation in sediments. The C:P ratio in sediments is related to the dissolved oxygen concentration in seawater. Sediments overlaid by well-oxygenated water have lower C:P ratios (that is, they contain more phosphorus) than do sediments overlaid by anoxic water. The reasons are complex: Apparently, bacteria in sediments store phosphorus when oxygen is available and use this stored material as an energy source when oxygen levels become too low. Phosphorus also tends to be bound up with iron compounds in sediments when oxygen is present. Both processes tend to create higher C:P ratios in sediments deposited under anoxic conditions.

The existence of such a relationship means that the dissolved O_2 content of deep water can affect the burial rate of organic carbon indirectly by altering the availability of phosphorus. This effect creates a negative feedback loop that may stabilize atmospheric oxygen (Fig. 9-19). If atmospheric O_2 were to decrease for some reason, deep ocean O_2 would decrease, and the C:P ratio in sediments would increase. This change would allow more organic carbon to be buried without removing any additional phosphorus. Increased burial of organic carbon would result in increased O_2 production, which would help restore atmospheric O_2 to its original level.

Although this mechanism may help stabilize atmospheric O_2, it does not imply that O_2 levels have remained constant. Atmospheric O_2 levels can be affected by

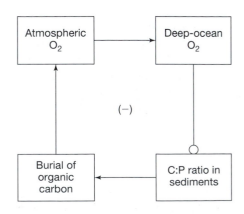

FIGURE 9-19

A likely feedback loop for controlling atmospheric O_2, involving dissolved O_2 concentrations and the C:P ratio in marine sediments.

changes in the rate of organic carbon burial on land, such as those that occurred during the Carboniferous Period. Furthermore, the degree of deep-ocean oxygenation may depend on the details of ocean circulation. Periods such as the Mesozoic, when the poles were much warmer, are unlikely to have had the same type of thermohaline circulation that operates today. Data from oxygen isotopes indicate that during most of this time deep water was 10 to 15°C warmer than it is today. Warmer deep water would have contained less dissolved O_2. Thus, anoxia should have tended to be more prevalent if other factors had remained the same.

If the oxygen control mechanism suggested above is correct, however, the atmospheric O_2 concentration should have responded to this change. Larger areas of anoxia would have promoted the deposition of sediments with high C:P ratios. This deposition, in turn, should have led to increased oxygen production and higher atmospheric O_2 levels. So, the dinosaurs may have breathed a somewhat enriched mixture of air. The possible connection between deep ocean circulation and atmospheric O_2 concentration illustrates yet another interesting linkage in the highly intertwined Earth system.

Forest Fires and C:P Ratios in Sedimentary Organic Matter

Although the control mechanism described above is attractive, it leaves one important question unanswered: If the control of atmospheric O_2 lies in the oceans, why have O_2 levels always remained within the fire window for forests? Either this is just a coincidence, or forests themselves must have something to do with regulating O_2. But forests are not part of the control mechanism outlined above.

The answer to this question may also involve sedimentary C:P ratios. The C:P ratios of marine sediments and terrestrial sediments are very different. The C:P ratio of typical marine organisms is 105:1, whereas organic matter derived from trees has a characteristic C:P ratio of about 1000:1. Thus, the burial of terrestrial organic carbon removes much less phosphorus than does the burial of an equivalent amount of marine organic matter. And, although the burial rate of organic matter on land is relatively low, some of the organic matter produced there is carried away by rivers and buried in river deltas.

Could this mechanism create a feedback loop that keeps atmospheric O_2 within the fire window? Suppose that atmospheric O_2 levels were to become so high that the forests all burned down. The phosphorus that was being utilized by terrestrial vegetation would go directly into rivers and, thence, to the oceans, where it would be utilized to make low C:P marine organic matter. The burial of this marine organic matter would generate less oxygen than would the burial of the same amount of terrestrial organic matter. Oxygen production would go down, causing atmospheric O_2 levels to decrease. The net result would be a negative feedback cycle, which could conceivably help to keep atmospheric O_2 within the limits acceptable to terrestrial vegetation (Fig. 9-20).

Unfortunately, although this idea is conceptually sound, it is not supported by data. For reasons that are not entirely understood, the C:P ratio of river delta sediments does not appear to be much higher than that of normal marine sediments. Either other plants with low C:P ratios provide most of the terrestrial organic material, or else sufficient reprocessing of material occurs during sediment deposition to smooth out the C:P differences between terrestrial and marine organic matter. So, the question of why the forests have not burned down remains an unsolved mystery.

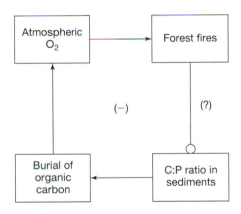

FIGURE 9-20

A possible feedback loop by which forests might help control atmospheric O_2. The link between forest fires and the C:P ratios in near-shore marine sediments is not well established. Some such control mechanism may be required to keep atmospheric O_2 levels within the "fire window."

Chapter Summary

1. Prior to the origin of life, Earth's atmosphere was probably composed mostly of CO_2 and N_2, with smaller amounts of reduced gases such as H_2 and CO. Some CH_4 may have been available from outgassing at mid-ocean ridge hydrothermal vents. Photochemical reactions in the atmosphere could have generated H_2CO and HCN, the two essential building blocks for life. Alternatively, researchers have suggested that the organic compounds from which life formed might have been either imported from space or synthesized at submarine hydrothermal vents. The vents had the additional advantage of being shielded from all but the very largest impacts and could have provided a safe haven for the earliest organisms.

Early colonization of the vent systems is consistent with the RNA evidence for a thermophilic last common ancestor.

2. Once life originated, it changed the composition of the atmosphere by converting hydrogen into methane. The biota should also have increased the rate at which nitrogen was cycled between the atmosphere and ocean.

3. Photosynthesis, which produces O_2, is thought to have originated very early in the Earth's history, as evidenced by 3.8-b.y.-old microfossils from Warrawoona that resemble modern cyanobacteria. Nevertheless, the geologic record indicates that atmospheric O_2 did not become abundant until 2.2 to 1.9 b.y. ago. Prior to that time, O_2 levels were suppressed by reaction with reduced volcanic gases. Although we do not know for sure why atmospheric O_2 rose when it did, one possibility is that the change was caused by the progressive oxidation of Earth's mantle. This decreased the percentage of H_2 and CO in volcanic gases and allowed the production rate of O_2 from photosynthesis and organic carbon burial to overwhelm the volcanic sink.

4. Accompanying the rise in atmospheric O_2 was a corresponding rise in the abundance of ozone. A reasonably efficient UV shield should have developed once the O_2 concentration exceeded about 1% of its present level. Based on the geologic evidence, this threshold should have been passed by 1.9 b.y. ago, or perhaps a few hundred million years earlier.

5. Atmospheric O_2 concentrations appear to have remained relatively constant for the past 360 million years, as evidenced by the continuity of the fossil charcoal record. The most plausible control mechanism involves the degree of oxygenation of deep water and the C:P ratio of marine sediments. Less-oxygenated deep water promotes a higher sedimentary C:P ratio that, in turn, allows more organic carbon to be buried for a given burial rate of phosphorus. The burial of terrestrial organic matter in river deltas may also influence the C:P ratio of sediments, allowing forests to exert some control over ambient O_2 levels.

Key Terms

amino acids
anoxic
anoxic basin
anoxygenic photosynthesis
banded-iron formation
black smoker
cyanobacteria
denitrification
deoxyribonucleic acid (DNA)
detrital mineral
dissociate
double helix
enzymes
eukaryotes
ferric iron
ferrous iron
fixed nitrogen

genetic code
hematite
hydrothermal vents
interplanetary dust particles (IDPs)
kerogen
methanogenic bacteria
methanotrophic bacteria
methanogens
microfossils
nitrogen fixation
nucleotides
oxidation state
oxidative weathering
oxygen minimum zone
oxygenic photosynthesis
paleosols
photochemical models

photochemical reactions
photolyze
prebiotic synthesis
prokaryotes
proteins
pyrite
RNA world
radical
rare earth elements
redbeds
reduced gases
ribonucleic acid (RNA)
stromatolites
thermophilic bacteria
uraninite
weakly reducing atmosphere

Review Questions

1. What atmospheric conditions were assumed in the Miller–Urey experiment?
2. Why have biologists suggested the possibility of an RNA world?
3. How can you explain the evidence for a thermophilic last common ancestor of all extant life? Give two possible reasons.
4. What is the overall chemical reaction by which methanogens metabolize?
5. What is the difference between prokaryotes and eukaryotes? Which are seen first in the fossil record?

6. What types of geologic evidence are used to study the rise of atmospheric O_2?
7. When did the ozone layer become thick enough to provide an effective UV screen?
8. What do carbon isotopes tell us about the rise of atmospheric O_2?
9. What does the fossil charcoal record tell us about atmospheric O_2 levels?
10. How is the atmospheric O_2 content maintained at its current level?

Critical-Thinking Problems

1. a. The atmosphere consists of 78% N_2, 21% O_2, 1% ^{40}Ar, and about 350 ppm CO_2. What is the mean molecular weight of air? Round your answer to three significant figures. (*Note:* The atomic weights of N and O are 14 and 16, respectively.)

b. The total mass of the atmosphere is about 5×10^{18} kg. How many moles of air does it contain? How many moles of O_2 and CO_2 are present? (*Note:* Calculate the latter two answers from the first one, *not* by computing the mass of O_2 and CO_2. The concentrations listed for the various gases are percentages by volume, not by mass, so you need to work in moles.)

c. Forests and soils contain roughly 2160 Gton of carbon in the form of wood, leaves, and humus. (The actual value is not known this accurately, but this choice makes the numbers work out well.) Suppose that all of the world's forests were to burn down instantaneously and that all of the soil carbon was oxidized as well. By how much would atmospheric CO_2 increase? By how much would O_2 decrease? Express your answers in percentages. Assume that the burning equation is the same as that for respiration and decay:

$$CH_2O + O_2 \longrightarrow CO_2 + H_2O.$$

2. The combined burial rate of organic carbon in marine sediments and in coal is approximately 0.3 Gton(C) per year. This burial is the net source of atmospheric O_2. In steady state, this source of oxygen is balanced by the weathering of reduced materials in rocks (kerogen, sulfides, and iron.) If the weathering rate were to remain constant following the disaster in Problem 1c, and if all photosynthesis were shut off (in the oceans as well as on land), how long would it take for atmospheric O_2 to disappear?

Further Reading

Both General

P. Cloud, *Oasis in Space: Earth History From the Beginning,* W. W. Norton and Co., New York, 1988.

J. C. G. Walker, *Earth History: The Several Ages of the Earth,* Jones and Bartlett Publishers, Inc., Boston, 1986.

Chapter 10
Biodiversity through Earth History

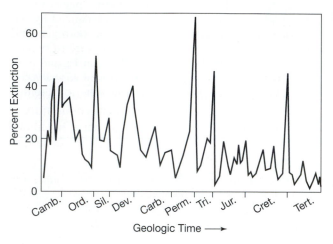

The fossil record of extinction rate, shown as the percentage of existing genera that went extinct in a particular interval (*stage*) of geologic time. (Compiled by J. Sepkoski, University of Chicago.)

Key Questions

- What is biodiversity, and how is it measured?
- How has the diversity of life changed over the history of Earth?
- How can the fossil record be used to elucidate this history?
- What caused the massive losses of global diversity that have occurred several times over the past half-billion years of Earth history?

Chapter Overview

This chapter introduces the concept of *biodiversity* and discusses changes in biodiversity through geologic time. *Paleontologists* (scientists who study ancient life) use many methods to reconstruct this history. Shelly marine organisms (those with the most robust fossil record) appeared abruptly about a half-billion years ago. Their diversity rose rapidly (geologically speaking), reaching a level that persisted for a few hundred million years. Then, about 250 million years ago, Earth experienced a dramatic loss of species diversity: Perhaps 95% of all species living at that time went extinct. A subsequent event, occurring some 65 million years ago, led to the extinction of the dinosaurs. A rich body of evidence supports the theory that this most recent mass extinction was a direct result of the collision of a 10-km-diameter meteorite with Earth. The environmental consequences of this event were severe and clearly capable of annihilating most life on Earth at the time. Speculative theories link the pacemaker for the regular occurrence of impacts and extinctions over Earth history to extraterrestrial phenomena, including unrecognized planets and companion stars to our Sun, and to movements of our solar system through the galaxy.

Biodiversity

How do we measure the "health" of the Earth system? By analogy to living systems, a healthy planet should actively transport nutrients from where they are not needed to where they are and should eliminate wastes (as demonstrated in Chapter 7). Its important environmental variables (temperature and atmospheric and oceanic compositions) should not fluctuate wildly (see Chapters 8 and 9). And it should be capable of responding to disturbances, such as volcanic eruptions and meteorite impacts, in such a way as to minimize their consequences.

These characteristics of a stable Earth system are ones that we normally associate with living organisms. Indeed, it is these very characteristics that are most suggestive of an important role for the biota in the regulation of the Earth system. We have seen that the biota have affected Earth's long-term climate evolution and are actively involved in the processes that regulate the oxygen content of the atmosphere. In these cases, the perturbations to which the biota have responded have been gradually imposed, over tens of millions to hundreds of millions of years. But how has the Earth system responded to *rapid* environmental change? Is the resilience revealed over the long term also a characteristic in the short term? The answer is yes. Earth has been subjected to insults the magnitude of which we are unable to imagine: Meteorites broader than the ocean is deep have struck Earth many times during its 3.5 billion years of inhabitation by organisms. The robustness of the planetary system is revealed by a fossil record displaced but not interrupted and by a geological and geochemical record that suggests that the long-term environmental consequences of these sudden disturbances were small.

How is the diversity of life forms on the planet—its **biodiversity,** the number of species in an area—related to the stability of the Earth system? We might propose that Earth's health can be measured by the number of species it supports. This assumption is implicit in the concern over the loss of biodiversity Earth is currently experiencing as a result of deforestation and loss of habitat. But do the abilities we associate with a healthy planet depend on its biodiversity? Is global biodiversity an indicator of the functional status of the Earth system?

Measures of Biodiversity

Biodiversity can be determined in a number of different ways. Perhaps the simplest measure of biodiversity is the number of species present in a *community* (a group of organisms that live in a given place). A community with 5 species of insects is much less diverse than one with 100 species.

There are some problems with this simple definition, however. One problem has to do with heterogeneity. Suppose there are two communities, both with two species of organisms, as shown in Table 10.1. According to the simple definition, the two communities are equally diverse. Community I, however, has 99 individuals of species A and 1 individual of species B, whereas community II has 50 individuals of species A and 50 individuals of species

TABLE 10-1

Diversity of Two Simple Communities		
	Number of Individuals, Species A	*Number of Individuals, Species B*
Community I	99	1
Community II	50	50

B (Fig. 10-1). The chance of encountering species B in community I is quite remote, only 1 in 100; this community is quite homogeneous. In contrast, community II seems more diverse, because there is an equal likelihood of encountering an individual of species A and of species B. Community II is thus more heterogeneous.

To capture the importance of heterogeneity, measures of diversity other than simply the number of species have been proposed. The *Simpson's diversity index* measures the likelihood that two individuals drawn from the same community will be of different species. This likelihood is expressed quantitatively as follows:

$$\text{Simpson's diversity} = 1 - [(\text{proportion of species A})^2$$
$$+ (\text{proportion of species B})^2 + \ldots].$$

The proportion of each species in the community is identical to the probability that an individual chosen at ran-

Community I

Community II

FIGURE 10-1

Two communities comprising two species (A and B) each. Although equal numbers of species are represented, Community I appears more diverse than Community II.

Diversity of Interactions

Even the more elaborate measures of diversity, such as the Simpson's diversity index, fail to account for a characteristic of ecosystems that is important to our understanding of the feedbacks between the biota and the physical world: the diversity of interactions. A community consisting of 500 species of ants, with relatively uniform populations of each, along with a few species of plants and predators, is highly diverse according to this index. However, in terms of the diversity of roles played by these organisms, the community is extremely homogeneous.

Here the infancy of Earth system science is clearly a limitation: No diversity index has been proposed that captures the degree of interaction between biological and physical components of the Earth system. Such a diversity index should increase as the number of couplings among the biota and with the physical world increases. It should also incorporate the attribute of *redundancy;* the Earth system is more resilient if there are alternative ways of performing important functions, such as photosynthesis or decomposition. If one of these pathways is lost (for example, through extinction), the others can compensate. An ecosystem with 10 interactions and only 20 species is then not as diverse as one with 40 species interacting in these 10 ways. The final attribute to incorporate into a systems diversity index is potential diversity. Species in small abundance today may come to dominate after a disturbance. In doing so, they will ensure that some vital function of the Earth system continues with little interruption or modification.

dom will be of that species. The probability of choosing two individuals of that species in a row is in the square of the proportion, just as the probability of throwing two "heads" in a row during a coin toss is $(0.5)^2 = 0.25$, or 1/4. The value of this index for our two simple communities I and II are $1 - (0.99^2 + 0.01^2) = 0.02$ for the homogeneous Community I and $1 - (0.5^2 + 0.5^2) = 0.5$ for the heterogeneous Community II, as shown in Table 10-2. When the number of species is large and the composition heterogeneous, the maximum Simpson's diversity approaches 1.0.

The Simpson's diversity index is clearly superior to a simple species count in expressing biodiversity. Nevertheless, in most discussions of global biodiversity of the past, present, and future (to some extent, this book included), only the species count is used.

Diversity and Stability

A long-standing debate exists among ecologists about the relationship between diversity and stability. For most communities, diversity increases from the poles to the tropics: Most of the highly diverse communities exist within 10 to 20 degrees of the equator. Why are the tropics so diverse? Tropical climates tend to be stable over a range of time scales. In the short term, the lack of large

seasonal variations in solar insolation leads to only small monthly contrasts in temperature and rainfall. Moreover, investigation of the geological record reveals that on long time scales, tropical temperatures have changed relatively little. Even during the Ice Ages, tropical temperatures fell only slightly while temperate to polar climates cooled substantially. The high diversity and climatic monotony of the tropics have been taken to indicate that environmental stability leads to high diversity—a premise called the *time stability hypothesis.* The persistence of uniform environmental conditions in the tropics presumably allows evolution to proceed without disruption, leading to higher diversity.

In contrast, the *intermediate disturbance hypothesis* states that the high diversity of tropical ecosystems is the result of disturbances that occur with intermediate frequency and intensity. This hypothesis is in direct contradiction with the time stability hypothesis. Highly diverse tropical rainforests tend to have some species with few or no young trees (indicating that they are dying out) and other species with a very high proportion of young trees (indicating that they are increasing in abundance). This turnover is presumed to be the result of a fairly recent natural disturbance. Regions of rainforest that are known to have been relatively undisturbed over historical time tend to have lower diversities. Also, high-

TABLE 10-2

Diversity of Two Simple Communities, and Simpson's Diversity Index			
	Number of Individuals, Species A	*Number of Individuals, Species B*	*Simpson's Diversity Index*
Community I	99	1	0.02
Community II	50	50	0.50

ly diverse coral reef ecosystems tend to occur at the outer edge of barrier reefs, where these ecosystems are periodically confronted with the damaging effects of waves and storms.

Both hypotheses link the diversity of life on Earth to the stability (or instability) of the environment. As ecologists continue to debate this issue, Earth systems scientists begin to wonder whether environmental stability might, in turn, depend on biodiversity. In Chapters 8 and 9, we studied Earth's environmental evolution. Let us now explore Earth's biological evolution and look for evidence that the biota and environment have evolved together in a way that has enhanced the stability of the environment and the diversity of life.

The Fossil Record of Biodiversity

How might we study the relationship between diversity and environmental stability in the Earth system? The best evidence for the coupled response of the environment and life to perturbation comes from the fossil record. Paleontologists have catalogued the first and last appearances of thousands of species of fossilizable organisms (primarily those with hard parts: skeletons, shells, or teeth). These observations reveal that the diversity of life on the planet has generally increased over the past several hundred million years but that there have been significant fluctuations.

The biological diversity of Earth today has been acquired at least partly as a result of evolution. **Evolution** is the descent, with modification (for example, genetic mutation) of preexisting life forms. The dominant mechanism of evolution is **natural selection,** the unequal survival and reproduction of organisms, owing to environmental pressures that result in the preservation of favorable characteristics. Characteristics that enhance an organism's survival and/or reproductive success are called **adaptations.** We can safely presume that, through evolution by natural selection, the total number of species has increased over the 3.5-billion-year history of life on Earth. Whether this increase has been linear, exponential, or represented by some more complicated pattern is a fundamental question of paleontology.

Diversity, expressed most simply as the total number of species, has increased as new species originated through genetic mutations that created a favorable change—in body plan, in body function, or in behavior—and diminished as other species went extinct. **Extinction** is the loss of all individuals within a species. Thus, diversity is best thought of as a dynamic characteristic of the biota:

$$\begin{matrix} \text{rate of change in number} \\ \text{of species on Earth} \end{matrix} = \begin{matrix} \text{origination} \\ \text{rate} \end{matrix} - \begin{matrix} \text{extinction} \\ \text{rate} \end{matrix}$$

Researchers estimate that 10 to 25 species originate every year, and 10 to 25 go extinct. These are considered to be the average rates of origination and extinction as a result of natural selection, which was first described by Charles Darwin. However, certain intervals of Earth history exhibit much greater rates of species origination and extinction. In fact, there appears to be a continuum of origination and extinction rates from the slow to the extremely rapid. Thus, origination and extinction, like other natural phenomena (floods, earthquakes), are likely to be the result of a variety of causes occurring on many different time scales and over a range of magnitudes.

There is a close analogy between the above equation and the equation that describes population growth. Such relationships generally predict that population increases rapidly (exponentially) at first, when the number of individuals is small. If, however, as the population increases the birth rate decreases or the death rate increases (due to shortages of food, competition, and so on), the population ultimately reaches a steady-state value known as the *carrying capacity,* at which the birth rate equals the death rate. Such behavior is called **logistic growth** (Fig. 10-2). We might expect to find that the diversity record is logistic in shape, if the rate of origination decreases or the rate of extinction increases as diversity increases.

Biologists define species on the basis of the ability of individuals to interbreed. This criterion obviously cannot be applied to fossils. Thus, paleontologists have used similarities in *morphology* (body shape) to group fossils into species. More than a hundred years of paleontolog-

FIGURE 10-2

The logistic growth curve. When a population is small, the birth rate outpaces the death rate and population growth is rapid. As the population increases, the birth rate decreases or the death rate increases, and the population reaches a steady state called the carrying capacity.

(a) Apparent Species Diversity

(b) Geologic Map Area

(c) Estimated Volume of Sediment

FIGURE 10-3

(a) The fossil record of species diversity of marine organisms, compared with the (b) outcrop area and (c) the volume of sedimentary rocks of the same age. All values are normalized to a million-year interval of Earth history. (After Signor P.W., *Ann. Rev. Ecol. Syst.* 21:509–539, 1990.)

ical research has been compiled in an effort to determine how the diversity of life has changed over Earth history.

The result of this effort is shown in Figure 10-3a, which presents the group with the best fossil record, marine *invertebrates* (organisms such as clams, starfish, and worms, which have no vertebrae). Note that the number of species has been normalized to what we would find if rocks representing exactly 1 million years of Earth history were investigated. This normalization has been done because the geological periods are of uneven duration. Longer periods would otherwise appear to be overly diverse.

These data seem to indicate that the species diversity of marine organisms has increased in an exponential fashion over the past 600 million years. Although consistent with our realization that diversity must have increased over some interval of the past 3.5 billion years, Figure 10-3a gives no indication of an approach to steady state predicted from

the logistic growth model. Our faith in these data is further weakened by the observation that the outcrop area (the area of rocks exposed at the Earth surface; Fig. 10-3b) and the volume of sedimentary rocks (Fig. 10-3c) that contain the fossils of various ages display essentially the same pattern. Particularly troubling is the anomalously high abundance of Devonian and Tertiary species (see the geologic time scale in Chapter 8), outcrops, and sediment volume. In addition, we realize that the number of species discovered by paleontologists depends on the availability of rocks to study. Thus, we are forced to conclude that the fossil record of species diversity is not a true reflection of the trend through Earth history. The higher species diversity of more-recent times could be more apparent than real.

A strategy paleontologists have adopted to minimize this sampling bias is to group their fossil data into increasingly higher levels of taxonomy. **Taxonomy** is the systematic organization of living or fossil organisms into a hierarchy (see the box "Taxonomy"). The occurrence of a single specimen of a species confirms the existence of its higher taxonomic levels—*genus, family, order, class, phylum* (for animals and some protists) or *division* (for bacteria, fungi, plants, and some protists), *kingdom,* and *domain.* No other species in any of these other taxonomic levels need be found. Thus, it is unlikely that a higher taxonomic level will be unrepresented in the fossil record of a time when it did indeed exist. Paleontologists generally agree that the diversity record at the family level and above, for marine organisms with well-preserved skeletons and shells, is reasonably robust.

The most outstanding feature of the fossil record at higher taxonomic levels is the explosion in diversity during the first few million to tens of millions of years of the Phanerozoic Eon (beginning about 545 million years ago), in the Cambrian and Ordovician Periods (see the geologic time scale in Chapter 8). During this short interval, every phylum of fossil-producing marine invertebrates, and many of the classes and orders, appeared in the fossil record for the first time (Fig. 10-4). Within 100 million years, the number of orders of marine organisms appears to have reached a rather constant value (Fig. 10-4c) that has persisted for the subsequent 500 million years.

A speculative hypothesis for this explosion of life forms in the earliest Phanerozoic is that Earth's mantle and crust suddenly shifted by nearly 90 degrees, placing previously equatorial continents in polar orientations. The climatic changes would have severely stressed the biota, opening up many possibilities for newly favorable adaptations.

The shape of the time history of order-level diversity (Fig. 10-5a) is reminiscent of the logistic growth curve described earlier in this chapter (Fig. 10-2). By analogy, we can imagine that early in the Cambrian Period, as organisms evolved fossilizable hard parts, new species evolved rapidly as they developed distinctive ways to utilize their environment (see Fig. 10-4). Common morphologies (related to form or function) arose that allowed them to be grouped into these higher taxonomic levels. As these *niches* were

Taxonomy

Given the millions of species that exist on Earth today and the thousands of distinct fossils recognized in the geologic record, a system of organization is clearly needed. Hence the field of taxonomy, (from the Greek *taxis,* meaning arrangement or order) was born. Two approaches are possible: one based on evolutionary, genetic relationships, and the other based on discernible physical similarities. The former is theoretically more satisfying, but the latter is usually more practical, especially for extinct organisms observable only as fossils. Species of *extant* (currently living) organisms are differentiated on the basis of their inability to interbreed. Higher levels of groupings (Box Fig. 10-1) into genera (the plural of genus), families, orders, classes, phyla (the plural of phylum) or divisions, kingdoms, and domains are based on the degree of dissimilarity in all detectable characteristics between *taxa* (plural of **taxon,** or taxonomic group). For example, all families within an order should share some similarity that is distinct from all other orders.

Plantae	Kingdom
Tracheophyta	Phylum (Division)
Angiospermae	Class
Ranales	Order
Ranunculaceae	Family
Ranunculus	Genus
acris	Species

BOX FIGURE 10-1

A taxonomic tree for the common buttercup, up to the kingdom level. The domain level is omitted for simplicity. (From T. McKnight: *Physical Geography: A Landscape Appreciation,* 6/e, 1999. Reprint by permission of Prentice Hall, Upper Saddle River, N.J.)

filled and the opportunities for new lifestyles diminished, the rate of appearance of novel body shapes for marine organisms (as reflected in the number of new orders) diminished. The rest of the Phanerozoic had slower rates of origination of new orders, and this slow down was very nearly balanced by the loss of orders (steady state was achieved).

The record of family-level diversity is somewhat richer (Fig. 10-5b), with an early history comparable to that for orders, but with substantial perturbations later in the Phanerozoic. Of particular interest are the significant losses of families at the ends of the Ordovician (438 million years ago), Devonian (367 million years ago), Permian (245 million years ago), Triassic (208 million years ago), and Cretaceous (65 million years ago) Periods and the persistent increases in family diversity in the Cambrian–Ordovician boundary and the Mesozoic–Cenozoic transition. Extraordinary extinction events in which more than 25% of all extant families are lost are called **mass extinctions.** In Chapter 1 we learned that the end of the Cretaceous Period (the end of the Mesozoic Era) was marked by the extinction of the dinosaurs, along with many other species of

organisms. This mass extinction pales somewhat, however, in comparison with that at the end of the Permian Period (the end of the Paleozoic Era), when approximately 95% of all species on Earth went extinct. The resiliency of higher taxa is demonstrated by their smaller losses: Only about 15% of the orders (Fig. 10-5a) disappeared, as opposed to 50% of families (Fig. 10-5b) and up to 85% of marine genera (Fig. 10-5c). These mass extinctions seem to have been sudden, especially when compared with the recoveries, which required several million to tens of millions of years.

The general trends over Phanerozoic time indicate that although marine diversity at the family and order levels achieved steady state early on, the Permian mass extinction initiated an increase in family-level and genus-level diversity that has continued to the present. The persistent increase in biodiversity through the Mesozoic and Cenozoic Eras suggests that marine organisms have continued to find new ways of exploiting the environment. Some have developed new abilities as predators or as scavengers. Others have adapted to previously unfilled niches. In all cases they have evolved new biological characteristics that distinguish

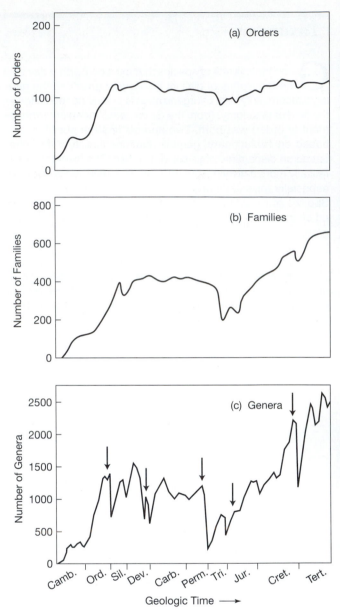

FIGURE 10-4

The number of new (a) phyla, (b) classes, and (c) orders of marine invertebrates that first appear in sedimentary rocks of a given age. The Cambrian and Ordovician were clearly a remarkable time of origination of a variety of new body shapes (which allow paleontologists to recognize the appearance of the various taxonomic groups). (After Erwin, D.H., Valentine, J.W., and Sepkoski, J.J. Jr., *Evolution* 41: 1177–1186, 1987.)

them from other groups of organisms, allowing their classification into new taxonomic groups.

These trends are remarkable in terms of the number of large losses of diversity the biota have suffered over the past 600 million years. The lack of fluctuation at high taxonomic levels and the rapid recovery from mass extinctions in the lower taxonomic levels exhibited by the fossil record point toward an amazing resiliency of the biota to perturbation. Mass extinctions cause substantial losses in diversity, but they also tend to stimulate origination rates as new organisms evolve and fill the ecological "job vacancies" resulting from the extinctions.

Although the response is geologically rapid, it takes up to tens of millions of years. The fossil record should not be used to argue for a quick recovery of the modern loss of species diversity.

Mass Extinctions

The five major disruptions to the global biological diversity over the past 500 million years—the mass extinctions shown in Figure 10-5—share many similarities. All appear to be rather sudden, at least within the resolution of the fossil record. In all cases, at least 40% of genera previously in existence became extinct. The effect at the species level was much larger. In the cases that have been well studied,

FIGURE 10-5

The apparent diversity of fossilized marine invertebrates through geological time, grouped at the (a) order, (b) family, and (c) genus levels. Arrows show mass extinctions. (After Signor, P.W., *Ann. Rev. Ecol. Syst.* 21:509–539, 1990.)

there is an amazing lack of selectivity in the types of organisms that went extinct. In particular, older species (that is, species that had been in existence for millions of years) seemed to suffer the same likelihood of extinction as did "young" taxa. In each case, the extinction event was followed by an interval of recovery in diversity.

Although the largest of all known mass extinctions occurred at the end of the Permian Period, 245 million years ago, the details of that extinction are just now becoming clear. In contrast, the Cretaceous–Tertiary (K–T) mass extinction (65 million years ago) has been the focus of a tremendous amount of research in the past few decades. It has also captured the attention of the general public, because it is the event at which the dinosaurs went extinct. For these reasons, our discussion will center on the K–T event.

The Cretaceous–Tertiary Mass Extinction

It has been estimated that 75% of all species went extinct in a very short interval of geologic time at the end of the Cretaceous Period. The species affected included both marine and land-based organisms. The most well known, of course, were the dinosaurs.

Two important questions arise concerning the K–T mass extinction: What caused it? and How did the Earth system respond to and recover from it? The first of these has been one of the most thoroughly studied questions in Earth science and has perhaps engendered more hypotheses than the origin of life itself. The second question is of particular importance to us, because it returns us to our discussion of the resiliency of the Earth system to perturbations.

Possible Causes of the K–T Mass Extinction

As many as 20 hypotheses on the cause of the mass extinction at the end of the Cretaceous Period have appeared in the scientific literature. Add to that a similar number of wild speculations appearing in the tabloids (including invasions by extraterrestrial life forms), and you have the sort of scientific enigma that is certain to stimulate discussions in all forums for some time to come. Of these, only four seem to have survived the scrutiny of scientists working on the problem: (1) sudden sea-level changes, (2) sharp temperature fluctuations, (3) volcanic eruptions, and (4) meteorite impacts. (We use the term *meteorite* liberally to mean a comet or asteroid that impacts Earth. Formal usage, however, would require us to refer to such objects as *meteoroids* until they strike Earth, at which point the materials that remain are referred to as meteorites.)

A central tenet of the scientific method is that hypotheses cannot be proved, only disproved. A hypothesis that continues to be consistent with observations may be elevated to the status of theory. Eventually, however, even well-established theories may be disproved as new observations are made that are inconsistent with theory. The debate about the causes of mass extinctions, especially that at the K–T boundary, provides a good example of how theories rise and fall from general acceptance within the scientific community.

Sea-Level Change and Climate Change. Until quite recently, changes in sea level or climate were the generally accepted theories for explaining mass extinctions. Major drops in sea level are known to have contributed to extinctions of marine life by exposing vast regions of the continental shelves to the atmosphere, leading to the loss of habitat for shallow-marine organisms (which tend to dominate the fossil record). Climate excursions, especially glaciations or periods of extreme warmth, have also severely stressed marine communities, especially those in the tropics, which are especially temperature-sensitive, causing substantial losses of biodiversity.

However, some of the largest known sea-level drops were not associated with mass extinctions. It is also unclear how sea-level changes could have exterminated land-dwelling organisms, such as the dinosaurs. As for climate change, the best studied of the glaciations, those of the Pleistocene Epoch (see Chapter 11), were associated with only a modest level of extinction, despite rapid climate swings and sea-level fluctuations.

Meteorite Impact and Volcanic Eruptions. Astronomical explanations for mass extinctions have long existed but have suffered from lack of substantiation. One such hypothesis called for a nearby supernova, which would have destroyed Earth's ozone layer, leading to high levels of exposure to ultraviolet radiation. On similar, purely theoretical footing was the idea that a large meteorite impact caused the K–T extinction.

Anomalously high concentrations of **iridium**—an element that is rare at Earth's surface but is concentrated in Earth's interior and in extraterrestrial materials—were found in 1979 in unusual, fine-grained sediments deposited at the end of the Cretaceous (the K–T boundary clays of Gubbio, Italy). This iridium anomaly was the piece of evidence needed to establish meteorite impact as the currently accepted theory for the cause of the K–T extinction (see Chapter 1). However, a minority of scientists are still convinced that widespread volcanism was to blame. These scientists argue that volcanism would lead to many of the same environmental changes as a meteorite impact would have. We will examine this controversy as an example of how theories compete for acceptance by the scientific community.

Proponents of the meteorite-impact theory made a number of predictions, any of which, if invalidated, would disprove the theory. Luis Alvarez, one of the discoverers of the iridium enrichment layer, has enumerated predictions to test the theory and the ways in which previously made and new observations are consistent with the theory:

1. *An iridium enrichment should be found in K–T boundary sediments worldwide.* The K–T iridium anomaly has been documented in more than 75 localities around the globe in both marine and land-based sediments. A 10-km-diameter meteorite is required to account for the amount of iridium deposited in all these regions.

2. *This enrichment should always be found within the same interval of geologic time.* All iridium layers found near the K–T occur in an interval defined by magnetic polarity reversals (see Chapter 6) known as *polarity chron C29R,* which has been precisely dated to span the interval from 65.6 to 64.9 million years ago. The K–T boundary is placed at 65.0 million years ago by these researchers.

3. *Large meteorites strike Earth sufficiently frequently to explain the extinction record.* The late Eugene Shoemaker of the U.S. Geological Survey (a discoverer of the Shoemaker–Levy comet that collided with Jupiter in 1994) demonstrated that a strong relationship exists between the frequency and size of impacts (Fig. 10-6). The data he used span observations of the rate of Space Shuttle pitting (one dust-sized impact every 30 μsec on average) to the frequency of large cratering events (every 10,000 years for 100-m-wide impactors). According to this relationship, a meteorite 10 km in diameter should strike Earth on average every 100 million years. This is a rare event, but many such events have occurred since the origin of life.

4. *On shorter time scales, such events should be rare.* In polarity chron C29R, there are no other iridium enrichments. In a study of a 34-million-year interval surrounding the K–T event, it is the only known anomaly. In fact, no iridium anomalies of comparable size have been detected anywhere else in the geologic record. According to prediction 3, several iridium layers should be lurking in rocks that have not been analyzed, unless most meteorites do not have large iridium abundances (for example, gaseous comets).

5. *Plants as well as animals should have suffered as a result of the meteorite impact.* There is now clear evidence of significant turnover in the types of vegetation inhabiting the land surface, as recorded particularly well by spores and pollen (Fig. 10-7).

6. *The gross chemical composition of the boundary clays should be identical worldwide, given that they all originated from the same excavated material.* The K–T boundary clays from Denmark and from the central Pacific are reported to be so similar in composition as to preclude diverse, local origins. The boundary clay is thought to contain debris from the impactor itself, along with a larger amount of material from Earth's crust and upper mantle that was excavated by the impact.

7. *The boundary clays should differ in composition from more-typical clays deposited above and below the boundary clays at individual sites.* Compositional differences between the iridium-rich boundary clays and the surrounding sediments have been confirmed at a number of sites.

8. *Any chemical or isotopic signature in the boundary clay will have a significant extraterrestrial component.* The iridium anomaly is the best studied of these anomalies, although other geochemical and isotopic signatures are arguably extraterrestrial.

9. *The boundary clays should bear some evidence of the high temperatures generated during impact.* Small

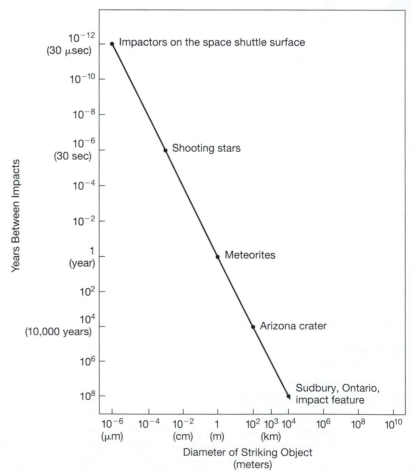

FIGURE 10-6

The frequency of collisions of extraterrestrial material with Earth, as a function of the size of the material. (After Alvarez, Luis W., *Phys. Today*, July, 1987, 24–33.)

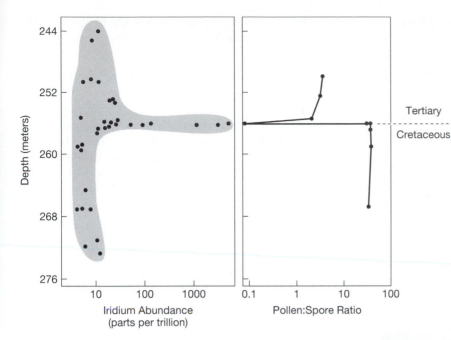

FIGURE 10-7

The Cretaceous–Tertiary iridium anomaly is coincident with a major change in the types of plants inhabiting the Raton Basin of New Mexico, as indicated by the fossil pollen-to-fern spore ratio. Samples were drawn from a drill core at the depths indicated on the graph. (After Alvarez, Luis W., *Phys. Today*, July, 1987, 24–33.)

spherules of silicate minerals (Fig. 10-8; Color Plate X) have been found in the boundary clays. These spherules are thought to have formed when molten rock droplets, ejected into the atmosphere during the impact, cooled and solidified into glassy spheres before being redeposited at the Earth's surface.

10. *The boundary clays should bear some evidence of the high pressures generated during impact.* Small, fractured grains of *shocked quartz* are commonly found in boundary clays (see Fig. 10-8). These grains form only when quartz is subjected to very high pressures— the sort we would expect from a meteorite impact. Under these conditions, the crystal structure of quartz is deformed in a characteristic fashion, so geologists can easily differentiate shocked quartz from quartz that is gradually subjected to the high pressures deep within Earth. Shocked quartz has also been found near known meteorite craters and in rocks surrounding the sites of underground nuclear explosions.

11. *The K–T event should have generated wildfires that might have left a sedimentary record of charred material.* Charcoal is indeed commonly found in K–T boundary clays. Recent work by NASA scientists provides a natural explanation for the apparent worldwide occurrence of wildfires during the K–T event. The impact excavated a huge amount of material in forming the crater, and this material was blasted in all directions away from the point of impact. Some of it was injected high into the atmosphere. The heating associated with reentry of ejected material through the atmosphere, estimated to be 50 to 100 times the solar flux, created unbearably hot conditions in cloud-free regions all over the world. Thus, a mechanism for rapid drying and ignition of vegetation on all continents would have existed. In cloudy areas, or beneath

(a)

(b)

FIGURE 10-8

(a) Shocked quartz and (b) microspherules from K–T boundary clays. (From Dr. David Kring/Science Photo Library, Photo Researchers, Inc.) [See color section] the sea or lakes, this energy would instead have gone into evaporating liquid water in cloud droplets.

12. *The iridium-rich layer should be just above the last dinosaur fossil.* At Gubbio, Italy, the iridium layer is less than a millimeter above the last occurrence of Cretaceous *foraminifera* (microscopic plankton; see Chapter 7). However, foraminifera are not dinosaurs. Dinosaur fossils are rare, and we are unlikely to find the last dinosaur fossil directly below the iridium layer (although, theoretically, if we keep looking we might find a charred dinosaur fossil buried in iridium-rich clay). Paleontologists have found articulated dinosaur fossils (fossils with joints intact) no closer than 2 m below the K–T boundary, as defined by the iridium layer. Although disarticulated dinosaur fossils have been found in sedimentary rocks well above the K–T boundary, generating some confusion and controversy, these fossils likely have been eroded from older rocks and redeposited in younger sediments.

13. *The pattern of extinction should show no evidence of preferential survival of species that were well adapted to the Cretaceous environment.* As mentioned before, mass extinctions in general, and the K–T one in particular, have been notably nonspecific. However, it does appear that large land animals were particularly hard hit by the K–T mass extinction; virtually all species with average weights greater than 25 kg (55 lb) went extinct. In contrast, mammals, which at the time were lightweights, fared much better. Perhaps smaller animals fared better simply because their population sizes were larger: Statistically speaking, some would survive the event. Or perhaps they were better able to avoid the immediate effects of the meteorite impact. After the extinction, mammals evolved, filling many of the niches left vacant by the demise of the dinosaurs.

The confirmation of the Alvarez predictions supports the meteorite-impact theory for the K–T extinction, but it does not prove it. The volcanic theory of mass extinction shares many features of the meteorite-impact theory. Large increases in volcanic activity would, in the short term, increase the amount of sulfuric acid aerosol in the atmosphere, cooling the climate. The settling ash could create a worldwide clay layer similar to the K–T boundary clay and might also produce spherules and shocked quartz. In the long term, carbon dioxide released by volcanism would lead to global warming. These climatic fluctuations would present environmental stresses to which many groups of organisms could not adapt. However, according to Alvarez, the volcanic theory is inconsistent with three observations:

1. Sand-sized spherules, even if ejected by volcanoes, would not reach ballistic orbits (as do impact ejecta) or be distributed globally, as the K–T spherules are observed to be.

2. Shocked quartz of the sort found at the K–T bound-

ary has never been found in deposits of volcanic origin but is common in deposits associated with known impact craters.

3. Volcanic ejecta tend to have very low iridium concentrations.

Who is right? The arguments are complicated, and reputable scientists populate both sides of the debate. Perhaps both are right, in the sense that the energy released by a large impact might have stimulated considerable volcanism. Current observations, however, seem most consistent with the meteorite-impact hypothesis, and most scientists favor this explanation. Perhaps a new observation will soon be made that will invalidate the meteorite-impact theory and require a new one to be developed. In the meantime, the meteorite-impact theory continues to serve as a useful framework for further inquiry into the causes and consequences of mass extinctions.

The Smoking Gun: Chicxulub Crater

Another prediction from the meteorite-impact theory is that somewhere on Earth a very large crater should exist, representing the site of impact of the 10-km meteorite. Until very recently, scientists had explained their inability to locate such a crater by pointing out that it was probably excavated in oceanic crust, given that three-quarters of Earth's surface is ocean-covered, and thus might have already been subducted into the mantle by plate tectonics (see Chapter 6). Even if the impact had occurred in continental crust, they argued, the crater is likely to have been eroded. Nevertheless, the search continued. Several putative K–T craters were discovered but found to be of a different age or of insufficient size to explain the observed iridium anomaly.

Then, in the early 1990s, geologists discovered convincing evidence that a subsurface structure, located near the town of Chicxulub (sheek'-soo-loob), Mexico (Fig. 10-9), was indeed of K–T age. Previous investigators had suspected, on the basis of remotely sensed geophysical anomalies, that this structure was an impact crater. Then, exploratory wells drilled into the Chicxulub structure recovered samples containing shocked quartz and glass microspherules that closely resembled samples from known K–T intervals elsewhere. Further analysis of the drilled core materials revealed very strong enrichments of iridium. Isotopic age dating and paleomagnetic evidence constrained the age of the materials to polarity chron C29R, the interval containing all K–T iridium anomalies. Current estimates of at least a 200-km diameter for the crater indicate that it is the largest on Earth and one of the largest in the solar system.

Environmental Consequences of the K–T Meteorite Impact

What was Earth like during the late Cretaceous Period when the meteorite hit? How did the global environment change after the impact? If we could transport ourselves back 65 million years to the day of the impact, we would

FIGURE 10-9

(a) The location of the Chicxulub impact crater. (b) The shape of the crater as inferred from gravity anomalies. (Courtesy of U.L. Sharpton, Lunar and Planetary Institute.)

The impact would have been unexpected and sudden. The passage of the meteorite through the atmosphere would have converted nitrogen gas (N_2) to nitric oxide (NO), as does lightning on a smaller scale today (see Chapter 9). Produced in large quantities, the NO would have destroyed the stratospheric ozone layer by catalyzing reactions that consume ozone, allowing ultraviolet radiation to reach Earth's surface unimpeded. Organisms unfortunate enough to be on the Yucatan Peninsula would have instantly been annihilated as an explosion of unimaginable intensity sent shock waves across the landscape. A huge crater would have been carved out immediately. The excavated material would have been sent on ballistic trajectories, enveloping Earth in a blanket of dust and debris. As the impact ejecta passed back through the atmosphere, the material would have been heated to temperatures sufficient to ignite wildfires in the affected areas. Soot would have been introduced into the atmosphere, further reducing the amount of sunlight reaching Earth. Because the bedrock of the Chicxulub area (mostly limestone) contained a thick layer of the calcium sulfate mineral *anhydrite,* a huge quantity of sulfuric acid aerosol would have been injected into the stratosphere. The scattering of sunlight from this aerosol layer would have reduced the amount of incoming solar radiation even further. Presuming a coastal impact, huge *tsunamis* (tidal waves) would have been initiated, wreaking havoc on Caribbean coastal environments. Evidence of tsunamis has been found on the island of Haiti and elsewhere in the Caribbean area.

In the next weeks, an unrelenting series of environmental insults would have ensued that few organisms could survive. The amount of solar radiation reaching Earth's surface would have been drastically reduced because of the presence of dense debris clouds and aerosol layers. The resultant dramatic cooling at all latitudes and cessation of photosynthesis would have removed the source of food for most of the life on the planet, both on land and in the sea. Where precipitation fell, it would have been in the form of rain and snow of sulfuric and nitric acids, produced from the rainout of sulfate aerosols and oxidation of NO.

Over the next several months, these tremendously detrimental effects would have lessened. The dust and soot would have rained out of the atmosphere, and the acidity of the precipitation would have been reduced. However, very little life would have remained. The organisms that had survived the direct effects of the impact would have been subjected to a scarcity of food and suitable habitat. Cool climates would have continued for more than a year because of the persistence of sulfate aerosol in the stratosphere.

Environmental recovery would have continued over the next few years, as the ozone layer was restored and the sulfate aerosol was removed from the atmosphere. However, global warming caused by thousands of gigatons of carbon dioxide released from the vaporization of limestone at the impact site would have persisted for tens of thousands to hundreds of thousands of years. Given that many organisms are intolerant of rapid temperature excursions,

find ourselves in a world quite different from that of today. Equator-to-pole gradients in temperature were less, and the poles were essentially ice-free, most likely the result of high atmospheric CO_2 levels. Dinosaurs inhabited tropical to near-polar latitudes—not in the abundances of their peak some tens of millions of years before, but still in sufficient numbers to dominate higher levels of the food chain.

The K–T Strangelove Ocean

Scientists studying the isotopic composition of carbon in skeletons of foraminifera deposited in sediments that span the K–T impact have found an intriguing perturbation of the marine ecosystem that persisted for hundreds of thousands of years after the K–T event. Normally, the operation of the oceanic biological pump (see Chapter 7) causes an enrichment in the carbon isotope ^{13}C in surface waters and an enrichment in ^{12}C in deep waters. The ^{12}C is preferentially incorporated into algal tissue during photosynthesis, it is then added to deep waters as the organic carbon in that tissue settles to the deep ocean and is decomposed by aerobic processes to dissolved CO_2. As a result, the $^{13}C:^{12}C$ ratio is enhanced in surface waters and diminished in deep waters. Foraminifera and other calcareous organisms (those producing $CaCO_3$ skeletons) record the ratio of ^{13}C to ^{12}C in their skeletons. Planktonic foraminifera, which live near the surface, record surface-water ratios, whereas benthic foraminifera, which live on the sea floor, record deep-sea ratios. In the modern ocean and in sediments recovered from layers deposited prior to the K–T event, the measured difference in the $^{13}C:^{12}C$ ratio between planktonic and benthic foraminifera is about 2 per mil (parts per thousand). However, as shown in Box Figure 10-2, the K–T event is marked by a collapse of this gradient. The implication is that the biological pump ceased to exist and that the ocean was essentially lifeless. Scientists have dubbed this the *Strangelove Ocean,* an allusion to the classic 1964 movie about nuclear war, *Dr. Strangelove.*

Perhaps even more intriguing than the collapse of the biological pump itself is its persistence. According to the isotopic record, the export of organic carbon from the surface ocean persisted at very low levels for hundreds of thousands of years, despite rapid rates of origination of new species and reestablishment of ecosystems in other settings. Although it is conceivable that slow rates of evolution retarded the reestablishment of the biological pump, other explanations seem to be required. One possibility is that toxic levels of trace metals such as copper, cadmium, and zinc resulted from the dissolution of the impact ejecta. Only after hundreds of thousands of years were these metals reduced to sufficiently low concentrations that metal-intolerant organisms could become reestablished.

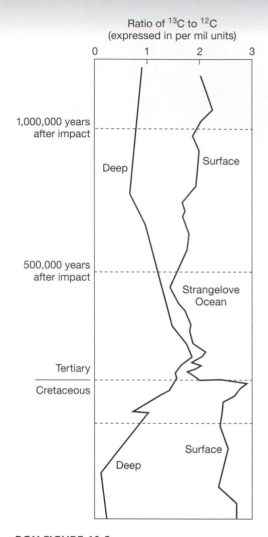

BOX FIGURE 10-2

Changes in the ratio of carbon isotopic composition between the surface ocean and deep ocean as a result of the Cretaceous–Tertiary mass extinction. The ocean's isotopic value is recorded in the skeletons of planktonic foraminifera, which represent surface waters, and benthic foraminifera, which represent deep waters. (After Zachos, J.C., Arthur, M.A., and Dean, W.E., *Nature* 337:61–64, 1989.)

this warming (following the initial cooling) might have assisted in the mass extinction and been at least one of the factors delaying the reestablishment of an active oceanic *biological pump* (see Chapter 7 and the box "The K–T Strangelove Ocean"), for hundreds of thousands of years.

Extraterrestrial Influences and Extinction

There are two varieties of impacts: those caused by comets and those caused by asteroids. **Comets** are essentially large, dirty snowballs. They are composed primarily of solidified compounds that exist on Earth as gases: water, ammonia, methane, and carbon dioxide. They also include smaller amounts of rocky and metallic debris. Most comets are in stable orbits around the Sun, beyond Pluto, in a region known as the **Oort cloud** (Fig. 10-10). However, some comets are on perturbed orbits that pass through the solar system. These comets include many that have been observed over historic time, such as Halley's comet and comet Hale–Bopp.

Asteroids, in contrast, are composed of minerals and metallic elements characteristic of Earth and the other inner-solar-system planets (Mercury, Venus, and Mars). Most asteroids are in orbit around the Sun in a region

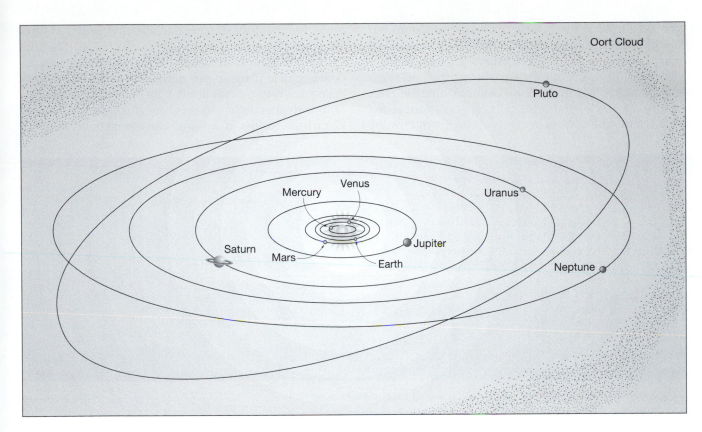

FIGURE 10-10

The solar system, showing the location of the Oort cloud (not drawn to scale) and the asteroid belt. (From T. McKnight: *Physical Geography: A Landscape Appreciation,* 6/e, 1999. Reprint by permission of Prentice Hall, Upper Saddle River, N.J.)

known as the **asteroid belt,** between Mars and Jupiter (see Fig. 10-10). The asteroid belt represents the remains of an inner planet that failed to form or was destroyed by a large meteorite impact early in our solar system's history. Some asteroids are as large as 1000 km in diameter, but collisions within the asteroid belt create small, mostly dust-sized fragments.

Iridium is associated with the mineral and metallic material in extraterrestrial materials. We can infer, then, that rocky asteroids contain a significantly higher concentration of iridium than do gaseous comets. The large iridium spike in the K–T boundary clay thus indicates that the impactor was most likely an asteroid, although we cannot rule out a large comet.

Asteroids in the asteroid belt and comets in the Oort cloud pose no immediate threat of impact. Their orbits can be disturbed, however, sending them on paths through the inner solar system. For example, the gravitational force of Jupiter periodically disturbs the orbits of asteroids and comets, resulting in some being sent on Earth-crossing orbits. This same gravitational attraction, however, tends to deflect the orbits of Earth-bound comets and protect us from cometary impacts. Sometimes, of course, Jupiter itself suffers the effect of its gravitational attraction of comets and asteroids, as we witnessed in July 1994 when the comet Shoemaker–Levy collided with the planet.

Examination of the Phanerozoic history of global biodiversity reveals that mass extinctions have occurred about once every 100 million years (see Fig. 10-5). In fact, a more detailed analysis of the fossil record of the past 250 million years has revealed a 26-million-year **periodicity**—a time interval of regular recurrence—in the global rate of extinction (Fig. 10-11). This regularity tends to

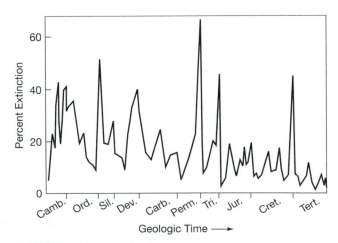

FIGURE 10-11

The fossil record of extinction rate, shown as the percentage of existing genera that went extinct in a particular interval (*stage*) of geologic time. (After Sepkoski, J.J. Jr., *Geotimes,* March, 1994, 15–17.)

favor an extraterrestrial cause for mass extinctions but requires another theory to explain why large asteroid or cometary impacts should occur every 26 million years.

At least three hypotheses have been proposed to explain the periodicity in the fossil data of extinction rates (Fig. 10-12). Each hypothesis calls on gravitational disturbances to send comets or asteroids on Earth-crossing orbits, but the three differ in the causative factors. (1) According to the *galactic plane hypothesis* (Fig. 10-12a), the oscillation of our solar system back and forth through the dense central plane of the spiral arm of the Milky Way galaxy leads to a periodically high likelihood of interactions with interstellar clouds of gas and dust, which could disturb the orbits of asteroids and comets in the Oort

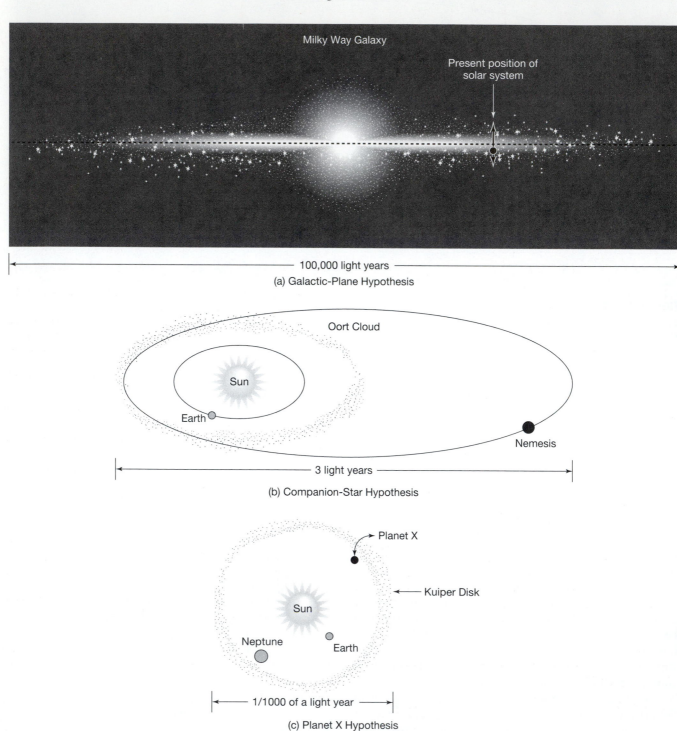

FIGURE 10-12

Three astronomical hypotheses explaining the 26-million-year periodicity in the fossil record of extinction. Figures not drawn to scale. (After Raup, D.M., 1986. See Further Reading.)

cloud, sending some on Earth-crossing orbits. (2) The *companion-star hypothesis* (Fig. 10-12b) proposes that the Sun has a companion star equal to about 10% of its mass, dubbed Nemesis by the popular media. Nemesis is presumed to be on a highly elliptical orbit that passes through the Oort cloud every 26 million years. (3) The *Planet X hypothesis* (Fig. 10-12c) calls for a disturbance of another cluster of comets (called the *Kuiper disk*), thought to exist outside the orbit of Neptune, by a hypothetical tenth planet that orbits beyond Pluto and is known as Planet X.

None of these hypotheses has been disproved, but the evidence in support of any of them is scant. The existence of companion stars or a tenth planet must first be proven before we can argue that their presence is likely to lead to meteorite impacts. Thus, hypotheses concerning the cause of periodicity in the fossil record of extinction are likely to remain speculative for some time.

What are the odds that a large impactor will hit Earth in our lifetime? They are higher than you might think. The direct correlation between impactor size and diameter (see Fig. 10-6) allows us to make this determination with some confidence. Let us assume that humans live about 100 years (admittedly, a bit optimistic). Because 100-m-diameter impactors (the size of the one that created Meteor Crater in Arizona) strike Earth about every 10,000 years, the likelihood of one striking Earth during your lifetime is about (100 years)/(10,000 years), or 1 in 100. The environmental consequences of such an impact are not tremendous, however, except close to the impact site. The impact frequency decreases by a factor of 100 for every factor of 10 increase in impactor diameter. Thus, the probability of a 1-km meteorite impact during your lifetime is about 1 in 10,000, and the probability of a K–T-size impact during that time is 1 in 1 million. By comparison, your chances of being struck by lightning are about 1 in 3000. Your chances of being killed by a lightning strike are quite high compared with your chances of being killed by the direct effects of a 1-km meteorite. But the likelihood that civilization will be destroyed by a meteorite impact is much larger than that of destruction by lightning strikes.

Chapter Summary

1. The diversity of life on Earth is a function not only of the number of species, but also of the degree to which the populations of those species are uniformly distributed (heterogeneous).

2. Environmental stability seems to lead to high biodiversity in some instances; however, modest disturbance enhances diversity in others.

3. The diversity of life on Earth has oscillated in response to imbalances between the origination of new species and the extinction of existing species.

4. The fossil record of species-level diversity is biased by preservational artifacts associated with the likelihood of discovering fossils from a given interval of time.

5. The fossil record of family-level diversity is less biased by these artifacts and can thus be interpreted in terms of true trends in biodiversity through Earth history.

   a. Biodiversity of marine invertebrates increased dramatically about 545 million years ago, reaching a high level that persisted for 300 million years.

   b. The largest extinction known occurred 245 million years ago, when approximately 95% of all species went extinct.

   c. Biodiversity recovered gradually after this event, and the general trend continues to the present. A significant disruption to this trend occurred 65 million years ago, at the end of the Cretaceous Period.

6. The Cretaceous–Tertiary mass extinction is the best studied of all major mass extinctions. Current scientific opinion favors the impact of a 10-km-diameter meteorite as the primary cause of this extinction.

   a. All available geological data are consistent with this hypothesis.

   b. A large crater dated as 65 million years old has been found in Chicxulub, Mexico.

   c. The meteorite impact would have destroyed the ozone layer, blocked out sunlight for days to weeks, ignited wildfires worldwide, baked organisms exposed to red-hot reentering ejecta, bathed the land surface in acid rain, and caused substantial global warming from elevated atmospheric CO_2.

   d. Large-meteorite impacts are predicted to occur on average every 100 million years or so. However, regular periodicity in the fossil record of extinction at a 26-million-year period might be caused by extraterrestrial forces, including close approaches of an undiscovered tenth planet, a companion star to the Oort cloud of comets, or crossings of our solar system through the galactic plane.

Key Terms

adaptation	extinction	Oort cloud
asteroids	iridium	periodicity
asteroid belt	logistic growth	taxon
biodiversity	mass extinction	taxonomy
comets	meteorite	
evolution	natural selection	

Review Questions

1. What two processes cause the diversity of life on Earth to change through time?
2. How do paleontologists deal with the incomplete nature of the fossil record to establish a geologic history of biodiversity changes?
3. Why do diversity trends differ among taxonomic levels?
4. What explanations have been proposed for the mass ex-

tinction at the K–T boundary? Why is a meteorite impact the favored theory today?
5. What are the environmental consequences of the impact of a 10-km-diameter meteorite with Earth?
6. What are some hypotheses for the apparent 26-million-year regularity of the fossil record of mass extinctions?

Critical-Thinking Problems

1. Here you will calculate the change in a population of organisms (N) as a result of births and deaths. A mathematical expression can be used to calculate next year's population (N_{t+1}) on the basis of this year's population (N_t), where t = years

$$N_{t+1} = N_t + rN_t\left(1 - \frac{N_t}{K}\right)$$

Let us dissect this equation. The potential growth rate—that is, the birth rate minus the death rate—is r, and the carrying capacity—the population size that can be supported (given constraints of food or other resource availability, competition, and so on)—is K. If the population is small relative to K, then the term in parentheses is essentially 1, and the potential growth rate is achieved. In other words, next year's population would simply be some multiple of this year's population (exponential growth). As N approaches K, the population will tend to slow its growth, finally reaching the carrying capacity. This behavior is called logistic growth. You will also witness something bizarre: behavior that has been labeled *chaos*. PART 1 (Logistic growth): Fill in the following table, and then graph N_t versus t; assume that $r = 1.0$ and $K = 1000$.

Describe the growth curve, and explain why it has the logistic growth shape (on the basis of the numbers you calculate). PART 2 (Chaos): Now repeat your calculations (do 15 years worth) for the following values of r:

$$r = 2.0$$
$$r = 2.8$$

Graph your results, either on separate graphs or using different symbols or colors on the same graph (be sure the graph(s) is(are) legible). If the population growth rate goes negative, call it quits on that series of calculations; the population has gone extinct.
Describe how the behavior changes as the growth rate increases from 1.0 to 2.0 to 2.8. When $r = 2.8$, the system is described as being chaotic. A scientist who observed this population might conclude that purely random factors are controlling the size of this population. What is wrong with this conclusion?

2. Perform the calculation that Luis Alvarez used to establish the size of the K–T impactor. Use the following information:
   a. Assume that the iridium layer was uniformly distributed around Earth by the impact.
   b. On average, the layer had a concentration of iridium of 10 parts per billion (10 ppb) by weight.
   c. On average, the layer was 4 cm thick.
   d. The density of the layer was 2.5 g/cm^3.
   e. Assume the meteor was spherical, with a density of 6.0 g/cm^3, and an iridium content of 0.5 parts per million by weight (0.5 ppm).
   f. The radius of Earth is 6378 km.
      What is the diameter of the meteorite?
3. Determine the probability that an asteroid of the following diameters will hit Earth during your (optimistic) 100-year-lifetime: 1 m, 100 m, 10 km.

TABLE 1

Time (years)	N_t	$1 - \frac{N_t}{K}$	$rN_t \cdot \left(1 - \frac{N_t}{K}\right)$	N_{t+1} (use as N_t next time)
1	20.0	0.98	19.6	39.6
2	39.6	0.96	38.0	77.6
3	77.6	0.92	71.6	149.2
4				
5				
6				
7				
8				
9				
10				

Further Reading

General

D. H. Erwin, 1993. The Great Paleozoic Crisis, Life and Death in the Permian. New York, Columbia University Press, 327 p.
W. Glen, 1994. The Mass-Extinction Debates: How Science Works in a Crisis. Stanford University Press, 370 p.
D. M. Raup, 1986. The Nemesis Affair, a Story of the Death of Dinosaurs and the Ways of Science. New York, Norton, 220 p.

Advanced

L.W. Alvarez, 1987. Mass extinctions caused by large bolide impacts. *Physics Today,* July, 1987, pp. 24–33.
A. Hallam and P. B. Wignall, 1997. Mass Extinctions and their Aftermath. Oxford, Oxford University Press, 320 p.
G. Ryder, D. Fastovsky, and S. Gartner, 1996. The Cretaceous–Tertiary Event and Other Catastrophes in Earth History. Boulder, Colo., Geological Society of America, 569 p.

Chapter 11
Pleistocene Glaciations

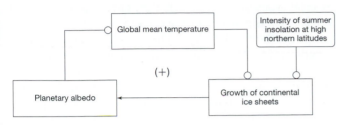

Feedback diagram showing the effect of changes in glacial growth on global temperature.

Key Questions

- What has caused Earth to oscillate in and out of glacial states over the past 2 million years?
- What geologic evidence substantiates theories about the causes of glaciation?
- Have components of the Earth system amplified the glacial climate response?

Chapter Overview

When viewed on million-year time scales, Earth is presently in a glacial interval. Thick continental ice sheets cover Antarctica and Greenland, and only 20,000 years ago vast portions of northern North America and Scandinavia and other parts northern Europe and Asia were covered with many-kilometers-thick accumulations of ice. Although polar ice has existed on Antarctica for tens of millions of years, only during the past 2.5 million years have ice sheets extended from the Arctic into the northern midlatitudes. Thus, there is something peculiar about the operation of Earth's climate system during the past 2.5 million years that distinguishes it from the rest of the past 300 million years of Earth history.

In addition, there is now convincing evidence that this glacial interval, known as the **Pleistocene Epoch** (1.8 m.y. ago to 10 k.y. ago), was characterized by regular cycles of growth and decay of Northern Hemisphere continental ice sheets. The oscillations are likely to continue. The warmer interval we are enjoying now, known as the **Holocene Epoch,** is likely to come to an end several thousand years from now as the Northern Hemisphere ice sheets return. (That is, unless the rise in atmospheric CO_2 from the burning of fossil fuels disturbs the natural cycle). What causes this cyclicity of glaciation?

As we shall see, both changes in the distribution of sunlight across Earth's surface (insolation) and in atmospheric CO_2 seem to be involved. These changes in insolation are a predictable feature of Earth's orbit about the Sun—during warm as well as cold intervals of

Earth history. They were not discussed in Chapter 8 because the magnitude of the forcing is very small compared with that of the changes brought about by solar evolution on 100-m.y. time scales. Yet as we narrow our focus to shorter time scales of climate change, these more subtle forcings rise in importance. During the Mesozoic, orbital changes caused cyclical variations in sedimentation that likely reflect climate change. But the Pleistocene climate system seems to have been especially attuned to these forcings, as indicated by the oscillations of continental ice sheets.

Geologic Evidence of Pleistocene Glaciation

The scenic beauty of Canada and the northern portions of the United States, Asia, and Europe is to a great extent the result of the action of tremendously large sheets of ice that covered these regions during the Pleistocene (Fig. 11-1). Buried under more than a kilometer of ice, the land surface was plucked, ground up, and excavated. With the melting of the ice, the depressions formed in this way were filled with water and became the thousands of lakes that pepper the North country. In this section we will see that the modification of the landscape by glaciers, together with their effect on the isotopic composition of seawater,

provide convincing evidence that continental ice sheets expanded and decayed several times during the Pleistocene with almost clockwork regularity.

Glacial Deposits Document Major Glaciations

Several types of geological features are characteristic of glaciation. As glaciers advance, rocks frozen to their base gouge the bedrock below. These gauges, called **glacial striations** (Fig. 11-2a), indicate the direction of past glacial movement. The advance of continental ice sheets modifies the landscape in other ways as well. Ridges of sediment, known as **moraines** (Fig. 11-2b), are deposited at the front and sides of the ice sheets. When the ice melts, these ridges are left behind, marking the farthest advance of the ice sheet. The indiscriminate nature of glacial erosion and deposition is reflected in the sediment contained in moraines and other glacial deposits. This sediment, called **till** (Fig. 11-2c), contains a mixture of material of various sizes, from mud to boulder-sized rocks, and various composition, reflecting rocks eroded from a variety of places and transported to the point of deposition by the advancing ice sheet. Some of the material produced by glacial abrasion is silt-sized (about hundredths of a millimeter in diameter). When deposited in the typically arid region that surrounds continental ice sheets, this material is picked up by the wind and carried great distances. Such windblown deposits, known as **loess** (Fig. 11-2d), form the rich soils of the midwestern U.S. grain belt.

FIGURE 11-1

The Pleistocene ice sheet at maximum extent. (From W.K. Hamblin and E.H. Christiansen: *Earth's Dynamic Systems,* 8/e, 1998. Reprinted by permission of Prentice Hall, Upper Saddle River, N.J.)

FIGURE 11-2

Various geological features characteristic of glaciation. (a) Glacial striations are formed by the gouging out of bedrock by pebbles frozen onto the base of an advancing glacier. The lineations indicate the direction of glacial movement. (b) Moraines form by the bulldozing action of advancing glaciers. (c) Till is the mix of sediments of various types and sizes resulting from the indiscriminate transportation by glaciers. (d) Loess is fine-grained sediment produced by glacial abrasion and transported to the site of deposition by wind. (Part b from T. McKnight: *Physical Geography: A Landscape Appreciation,* 6/e, 1999. Reprinted by permission of Prentice Hall, Upper Saddle River, N.J.) (Part a, c, d from T. McKnight: *Physical Geography: A Landscape Appreciation,* 5/e, 1996. Reprinted by permission of Prentice Hall, Upper Saddle River, N.J.)

Geologists of the nineteenth century used glacial deposits, including the geographical distribution of moraines, to define four main intervals of glaciation in Europe. In the early part of the 20th century, four glaciations were also identified in North America. These glaciations on separate continents are now known to have been coincident with each other, and they represent the largest of the glacial episodes. Ice sheets in the Northern Hemisphere were advancing and retreating in unison. But the number four greatly underestimates how often this process has occurred. The problem is that each successive advance obliterates much of the geological record of earlier advances.

Detailed analysis of layers upon layers of windblown loess, preserved throughout Europe, proved that there had been at least 17 glacial intervals prior to the last. And evidence from marine sediments, described next, showed that such cycles have occurred with clocklike regularity for more than 2.5 million years.

The Oxygen Isotope Record of Glacial–Interglacial Oscillations

While ice sheets were destroying the sedimentary record of glaciation on land, a continuous record of climate change was being deposited on the sea floor. This record, however, was chemical rather than physical; it was contained in the isotopic composition of the skeletons of marine organisms. The ratio of the stable isotopes ^{18}O and ^{16}O in the calcium carbonate skeletons of marine plankton depends on the temperature of that water. The colder the water, the greater the tendency for minerals to incorporate ^{18}O, and thus the larger the ratio of ^{18}O to ^{16}O in $CaCO_3$. As with carbon isotopes (Chapter 9), oxygen isotope ratios are typically reported in "delta" notation, where $\delta^{18}O$ is the ratio of ^{18}O to ^{16}O normalized to a standard (either seawater or a standard limestone) and multiplied by 1000 to accentuate the small differences observed.

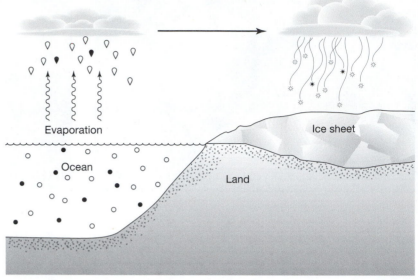

FIGURE 11-3

Changes in the oxygen isotopic composition of seawater during the growth of continental ice sheets. Evaporation preferentially removes ^{16}O (as H_2O) from the ocean; the ^{18}O then falls as snow on the ice sheets. The ocean thus becomes enriched in ^{18}O as the ice sheet grows and enriched in ^{16}O as the ice sheet melts and the water returns to the ocean.

○ H_2O containing ^{16}O

● H_2O containing ^{18}O

Theoretically, then, the isotopic analysis of skeletal material recovered from cores of deep-sea sediments should reveal the history of temperature fluctuations in the overlying surface waters. However, an additional isotopic effect is equally important. As continental ice sheets grow, significant quantities of water are removed from the oceans (Fig. 11-3). Aside from causing considerable drops in sea level (the last glaciation dropped sea level by about 130 m), significant changes in the oxygen isotopic composition of seawater occur as well. Water molecules containing ^{18}O tend to evaporate more slowly than do those containing ^{16}O, the lighter isotope. Thus, after evaporation from the oceans and transport onto the ice sheet, snow falling onto the ice caps tends to be enriched in ^{16}O, leaving the oceans enriched in the heavier ^{18}O. This effect adds to the direct temperature effect: Calci-

um carbonate precipitated from a glacial-age ocean has a larger $\delta^{18}O$, both because the water is cold and because the seawater is enriched in the heavier isotope.

In the 1950s the first deep-sea sediment cores were recovered, and isotopic analyses were performed on them. Instead of the four glaciations originally indicated by the continental record, the marine record indicated that dozens of climate swings have occurred over the course of the Pleistocene (Fig. 11-4). The major intervals of Northern Hemisphere glaciation—**glacials**—of the past 700,000 years appear to have occurred every 100,000 years or so. During glacials, globally averaged surface temperatures were about 10°C and atmospheric CO_2 concentrations were about 200 ppm. These cold intervals are separated by warmer, shorter intervals known as **interglacials.** During interglacials, continental glaciation is limited to

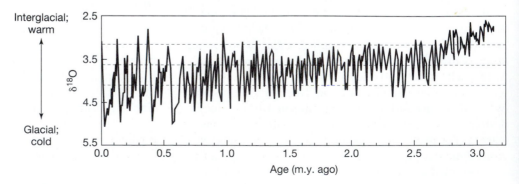

FIGURE 11-4

Deep-sea record of the $\delta^{18}O$ of seawater during the Pleistocene. The analyses were performed on two genera of bottom-dwelling foraminifera deposited in the sediments of the midlatitude North Atlantic. Interglacials appear as peaks, with smaller values of $\delta^{18}O$; glaciations appear as valleys. (After Raymo, M.E., *Ann. Rev. Earth Plan. Sci.*, 22:353–383, 1994.)

Greenland and Antarctica; globally averaged surface temperatures are about 15°C and atmospheric CO_2 concentrations are about 280 ppm. The Holocene Epoch (the past 10,000 years) represents one such interglacial.

The regularity of glacial–interglacial variation is remarkable. The **period** of a cyclical phenomenon is the time it takes to complete one cycle. Curiously, the dominant periodicity (cyclical nature) of glacial–interglacial variation seems to have been different prior to 700 k.y. ago. Before that time, glacial–interglacial swings were smaller and occurred on approximately a 40,000-year time scale. Something fundamental to the climate system changed 700,000 years ago, and scientists are actively working to resolve what that fundamental change was. The pre-Pleistocene cooling and the onset of significant continental glaciation in the Northern Hemisphere (from 3.0 m.y. to 2.5 m.y. ago) are also apparent in the oxygen isotope record (see Fig. 11-4).

Why has Earth's climate system been oscillating between these two states—glacial and interglacial—with apparent periods of 100,000 and 40,000 years? The answer seems to involve small changes in the way Earth orbits the Sun, changes that repeat in a predictable fashion over tens of thousands of years. Three sorts of changes are involved (Fig. 11-5):

1. changes in the degree to which Earth's orbit around the Sun is oblong (elliptical);

2. changes in the tilt of Earth's spin axis with respect to the plane of its orbit around the Sun; and

3. changes in the spin axis about the vertical (wobble).

The pacemaker that determines the variability of climate requires amplification, however, because only small changes in insolation result from the orbital variations. We will first explore these orbital variations and will compare their predictions to the isotopic record from deep-sea sediment cores. We will then discuss feedback mechanisms (including those affecting atmospheric CO_2) that may have provided the amplification necessary to create the large climate swings of the Pleistocene.

Milankovitch Cycles

What causes these remarkably regular shifts in Earth's climate? Long before the oxygen isotope evidence was obtained, scientists of the 19th century had suspected that the Pleistocene glacial–interglacial cycles were caused by variations in Earth's orbit around the Sun. In the early part of the 20th century, this hypothesis was put on a quantitative footing by the Serbian mathematician Milutin Milankovitch. He not only elaborated the mathematical theory of how orbital variations affect climate, but also calculated the changes in orbital parameters over the past several thousand years and demonstrated the connection between this theory and the rather scant geological record that existed during his time. The regular variations in Earth's orbit are often referred to as *Milankovitch cycles* in honor of this achievement. Milankovitch suggested that the critical factor for Northern Hemisphere continental glaciation was the amount of

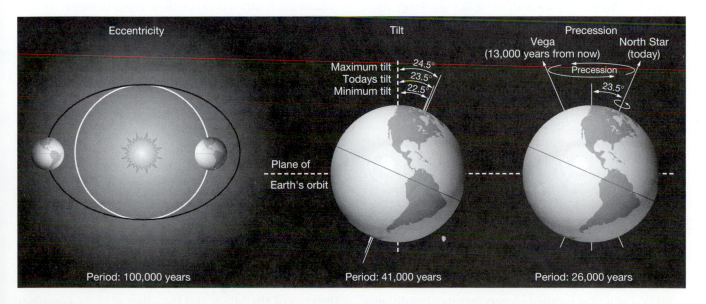

FIGURE 11-5

Aspects of Earth's orbit around the Sun that have implications for climate change. (a) The elliptical nature of the orbit (eccentricity) changes on 100,000-year time scales. (b) The tilt of the spin axis with respect to the plane of Earth's orbit around the Sun (obliquity) changes on a 41,000-year time scale. (c) The orientation of the spin axis in space wobbles (precesses) with periodicities of 19,000 and 23,000 years. (From J.P. Davidson, W.E. Reed, and P.M. Davis: *Exploring Earth: An Introduction to Physical Geology*, 1997. Reprinted by permission of Prentice Hall, Upper Saddle River, N.J.)

Kepler's Laws

First law: The orbit of each planet is an ellipse with the Sun at one focus.

Half of the major axis of an ellipse is called the *semimajor axis.* The semimajor axis is usually represented by the letter *a.* The distance from the center of the ellipse to one focus is equal to the length of the semimajor axis (*a*) multiplied by the eccentricity (*e*) (see Fig. 11-6). In other words, the eccentricity of the ellipse can be determined by dividing *ae* by *a.* An ellipse with zero eccentricity is a circle (the two foci coincide with the center of the circle). Earth's current eccentricity is 0.017; Earth's orbit is nearly circular, but not quite.

Second law: A line joining a planet to the Sun sweeps out equal areas in equal times.

Figure 11-6 shows that to sweep out an equal area (each shaded region) in the same amount of time (Δt), Earth must travel farther around the perimeter of the ellipse when it is close to the Sun than when it is far from it. From a practical standpoint, then, this law means that the planet moves faster when it is closer to the Sun and slower when it is farther away.

Third law: The square of a planet's orbital period is proportional to the cube of its semimajor axis.

A planet's orbital *period* is the time that it takes the planet to go around the Sun. If we express the period *P* in Earth years and the planet's semimajor axis *a* in astronomical units (AU, the average Earth–Sun distance), we can replace the word *proportional* with *equal* and write Kepler's third law as $P^2 = a^3$.

---

summertime insolation at high northern latitudes. High insolation leads to warmer summers, and the winter snowpack melts (as we see today). However, under low insolation the snowpack would survive over the summer, allowing snow and ice to accumulate and an ice sheet to form. Subsequent summers would allow further growth of the ice sheet toward lower latitudes.

Despite the elegance of Milankovitch's "astronomical theory of the Ice Age," it was strongly criticized by the scientific community in the 1920s and 1930s. His response to these criticisms is perhaps best reflected in this excerpt from his 1941 book: "I do not consider it my duty to give an elementary education to the ignorant, and I have also never tried to force others to apply my theory, with which no one could find fault." [M. Milankovitch, *Canon of Insolation and the Ice Age Problem,* 1941]

Orbital Theory

Orbital theory itself predates Milankovitch; it was developed in the 17th century by Johannes Kepler and Isaac Newton. The results are summarized by three rules that are known as Kepler's laws (see the box). The most important of these laws for our purposes is the first: The planets travel around the Sun in *elliptical* orbits with the Sun at one *focus.*

An ellipse is defined mathematically as the collection of points whose combined distance to two fixed points (the *foci*) is equal to a constant. The constant is the length of the long, or *major,* axis of the ellipse (Fig. 11-6). The degree to which the orbit of a planet (or other rotating object) is elliptical is called its **eccentricity.** The distance from the center of the elliptical orbit to either focus is only 1.7% of the distance from the center to the edge of the ellipse along the major axis. In other words, the foci and the center are nearly indistinguishable, and Earth's orbit is very nearly circular. The eccentricity (often designated as *e*) is expressed numerically as this percentage in decimal form (*e* = 0.017).

Because planetary orbits are eccentric, Earth is closer to the Sun at some times of the year than at others. The point of closest approach is called **perihelion,** and the point of maximum Earth–Sun distance is called **aphelion** (see Fig. 11-6). The amount of sunlight hitting Earth is slightly greater at perihelion than at aphelion (as we shall see in Critical-Thinking Problem 2).

Another factor that affects Earth's climate is the planet's **obliquity,** the fact that its spin axis is tilted 23.5 degrees from the perpendicular to the plane of its orbit. Obliquity creates the contrast between the seasons (Chapter 4); with no obliquity, the annual variation in the amount of solar insolation (resulting from the eccentricity of the orbit) would be very small. Winters and summers would be very similar.

The combination of an eccentric orbit and a tilted spin axis means that the **seasonal temperature contrast**— that is, the range of temperature extremes between summer and winter—is different in the Southern and Northern Hemispheres. Less total energy is reaching Earth at aphelion, 13 days after the Northern Hemisphere summer solstice, so the Northern Hemisphere summer tends to be milder than Southern Hemisphere summers. Northern Hemisphere winters are also milder, because they occur near perihelion. Southern Hemisphere winters occur at aphelion and are relatively harsh.

Changes in Earth's Orbit through Time

Milankovitch's theory predicts that the gravitational influences of the Moon and the other planets, combined with Earth's slightly nonspherical shape, induce small but important variations in Earth's orbital parameters. These variations affect the amount of summertime insolation at high northern latitudes, triggering the onset and end of glacial intervals.

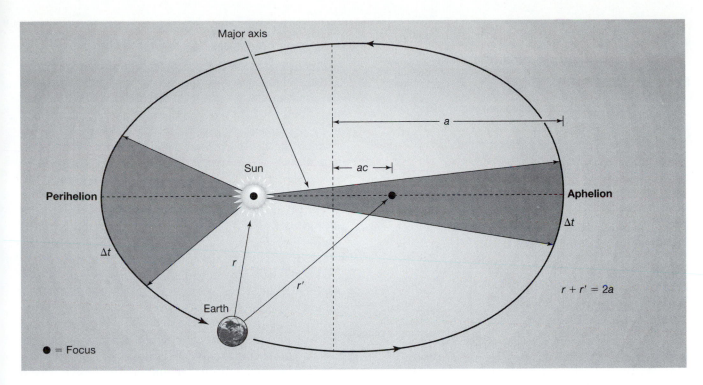

FIGURE 11-6

Earth's elliptical orbit, showing the Sun at one focus. (From T. McKnight: *Physical Geography: A Landscape Appreciation*, 6/e, 1999. Reprinted by permission of Prentice Hall, Upper Saddle River, N.J.)

Precession of the Spin Axis

The most noticeable change in Earth's orbit has to do with the direction of its spin axis. Currently, the spin axis is oriented such that the North Pole points almost directly at the bright star *Polaris*, otherwise known as the North Star.

The spin axis has not always pointed in that direction, however. Egyptian pyramids built in 3000 B.C. were designed to observe the north star of the time, *Alpha Draconis*, not *Polaris*. Over time, the North Pole describes a circle in space as Earth wobbles like a spinning top that is tilted from the vertical. As it does, the spin axis points to different stars in the sky. The period of **precession** (that is, the time it takes for the spin axis to wobble around or *precess* one complete circle) is about 25,700 years. However, the direction of the major axis of Earth's elliptical orbit is also precessing, but in the opposite direction, a phenomenon referred to as the *precession of perihelion*. Because perihelion is precessing in the opposite direction from the spin axis precession, the amount of time required to go through a complete cycle is shorter than 25,700 years. Two main periods result, at 23,000 and 19,000 years.

Precession modifies the relationship between the seasons and the distance from the Sun shown in Figure 11-6. Every half precession cycle, the hemisphere with the greatest degree of seasonal contrast switches between the north and the south. When the Southern Hemisphere has mild summers and winters, the Northern Hemisphere has hot summers and cold winters, and vice versa. Northern Hemi-

sphere glaciation is promoted by a precessional state, as today, with northern summer at aphelion and thus low seasonal contrast. The maximum interglacial condition was achieved 9000 years ago, with hot summers in the north.

Obliquity Variations

The same phenomenon that causes Earth's spin axis to precess also causes the obliquity to vary from 22 to 24.5 degrees, with a dominant cycle length of about 41,000 years. If you observe a spinning top carefully, you will notice that it, too, undergoes periodic changes in its tilt as its spin axis precesses. This effect becomes more pronounced as the spin rate of the top decreases: The top begins to wobble. Like precession of the spin axis, a change in obliquity does not alter the total amount of sunlight striking Earth. Rather, it determines the extent of seasonal contrasts: The warmth of summers and the coldness of winters is increased by higher obliquities (Fig. 11-7).

Eccentricity Variations

Earth's eccentricity also undergoes oscillations that can affect climate. Variations in eccentricity are caused by a somewhat different phenomenon than are the precessional and obliquity variations: the gravitational pull of the other planets. The combined effect of all the planets causes Earth's eccentricity to vary periodically between 0 and 0.06. (Recall that the current value is 0.017.) As was

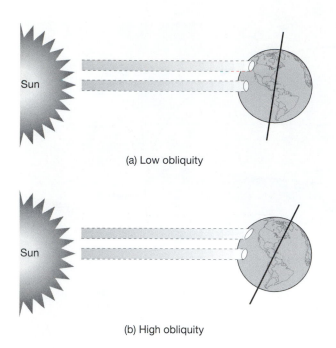

FIGURE 11-7

(a) At low obliquity, Earth has less contrast in insolation between the seasons. (b) At high obliquity, the seasonal contrast is greater.

true of the precessional cycle, two main periods are predicted. They are much longer in this case, however: About 100,000 years and about 400,000 years.

The eccentricity variations differ from the precessional and obliquity variations in one other significant respect: Eccentricity variations cause changes in the annually averaged amount of sunlight hitting Earth, whereas precessional and obliquity variations do not. (The direction and steepness of tilt of a planet's spin axis have no direct effect on the total amount of sunlight the planet receives.) We can show mathematically that Earth receives about 0.2% more sunlight at maximum eccentricity than at minimum eccentricity. This difference is thought to be too small to cause major climate shifts by itself, but it might have some effect if it is amplified by a feedback mechanism (discussed later in this chapter).

Probably more important is the fact that eccentricity influences the climatic effect of the precession cycle. When Earth's eccentricity is zero, there is no difference between the perihelion distance and the aphelion distance from the Sun, so it does not matter when summer or winter occur. When the eccentricity is large, Northern Hemisphere glaciation is especially favored when precession causes Northern Hemisphere summer to occur at aphelion.

In summary, the optimal conditions for glaciation are low obliquity (low seasonal contrast) and high eccentricity, which, with a precession angle that places the Northern Hemisphere summers at aphelion, minimizes the amount of summertime insolation at high northern latitudes.

Comparing Orbital Forcing and Climatic Response by Means of Oxygen Isotopes

The combination of these various orbital forcings causes Earth's climate system to oscillate between two states. The situation can be displayed in a diagram similar to that developed for Daisyworld (Chapter 2). The glacial and interglacial states are represented as valleys separated by a ridge (Fig. 11-8). Presumably the glacial state is situated in a deeper valley than the interglacial state, because a greater fraction of Pleistocene time was spent in glaciation. Orbital forcings continually rock the system back and forth. Since 700 k.y. ago, the amplitude of this rocking has exhibited a strong a 100,000-year periodicity. When these variations exceed a threshold, the system moves over the ridge into the other state.

We can now piece together the various parts of the astronomical theory of the Ice Ages described above and can test the theory against the oxygen isotopic record of temperature and sea level changes during the Pleistocene. The precession, obliquity, and eccentricity variations can all be described mathematically, and the resulting equations can be solved for the amount of insolation received on a monthly or annual basis for any particular latitude. This is nothing new, of course; Milankovitch made these calculations many decades ago. Some improvements have been made on the original calculations by Milankovitch, however. The particular result shown at the top of Figure 11-9 is the average monthly insolation (Q) for June, at 65° N latitude.

Shown at the bottom of Figure 11-9 is the past 400,000 years of the oxygen isotope record from Figure 11-4. Is the observed climate response ($\delta^{18}O$) the expected response to the climate forcing (Q)? Such a comparison is difficult; indeed, we might conclude from a visual inspection that the two curves are unrelated. However, the wiggles of these curves can be considered the combination of a number of waves of different frequency (the mathematical inverse of period) and amplitudes, much as a mu-

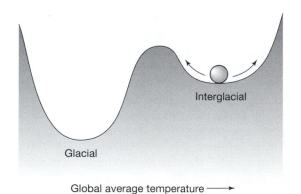

FIGURE 11-8

Stability of the glacial and interglacial states, relative to the changes in high-latitude Northern Hemisphere insolation that rock the state of the system back and forth.

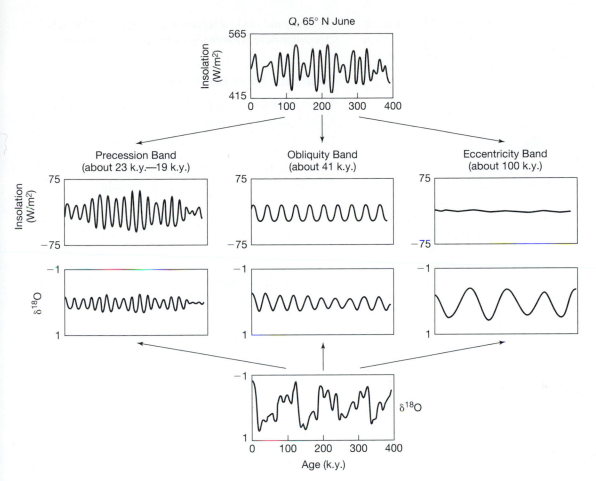

FIGURE 11-9

Northern Hemisphere June insolation (Q, the climate forcing) and marine oxygen isotopic composition ($\delta^{18}O$, the climate response) and their dominant periodic components. (After Imbrie et al., *Paleoceanography* 7:701–738, 1992.)

sical chord is the combination of a number of notes, each of a different frequency (pitch). These curves can then be separated into their component periodic waves (or *bands*); the most important ones are shown in Figure 11-9. For Q these are the precession, obliquity, and eccentricity bands. As predicted by Milankovitch's astronomical theory, the dominant periodicities of the $\delta^{18}O$ record occur at these same bands (19 k.y. and 23 k.y., 41 k.y., and 100 k.y., respectively). Thus, there appears to be good evidence that the pacemaker for the ice ages is the "Milankovitch" signal.

Moreover, the amplitude of the response to the precession and obliquity changes seems to be roughly proportional to the amplitude of the forcing. These observations suggest that a simple link exists between fluctuations in the amount of radiation received at high latitudes and the extent of glaciation.

A closer inspection of Figure 11-9, however, reveals an important departure from a straightforward link between climate forcing and climate response: The direct forcing in the eccentricity band is very small (some 10% of that in the other bands), yet the climate response is the largest of the three bands. Are there positive feedback loops in the climate system that amplify the weak eccentricity forcing into the major climatic response to orbital fluctuations? The remainder of this chapter will explore this possibility.

Glacial Climate Feedbacks

The timekeepers for the glacial–interglacial climate fluctuations during the Pleistocene were subtle, periodic changes in Earth's orbital parameters. However, these changes have been small. Moreover, the dominant periodicity of glacial–interglacial fluctuations has been 100,000 years; if this phenomenon is the result of eccentricity changes, an amplifier is needed. The important climate variables that we need to consider are albedo and the greenhouse effect. Clearly, the growth of continental ice sheets influences the albedo of the planet, so this effect must be incorporated into any model that attempts to explain Pleistocene climates. Clouds exert a major control on planetary albedo, and we may wonder whether the cloud albedo varies in concert with the Milankovitch cycles. Finally, we have seen how changes in the greenhouse effect of atmospheric CO_2 have affected climate on long time scales; perhaps the Pleistocene climate system has responded to more-rapid fluctuations in atmospheric CO_2.

Ice–Albedo Feedbacks

Any change in the seasonal distribution of solar luminosity that affects the growth of ice during the winter or the meltback during the summer has the potential to affect the planetary albedo (see Chapter 3). As ice sheets begin to grow, they convert a region that previously had reduced albedo during the summer, as snow melted, to one that maintains high albedo throughout the year. The annual average albedo thus increases, which will lead to both a regional and a global cooling. This cooling will accelerate the growth of the continental ice and will allow it to spread to lower latitudes.

The positive ice–albedo feedback, involving global temperature, ice-sheet growth, and albedo (Fig. 11-10), was introduced in Chapter 3. In Figure 11-10 the forcing is also indicated; note that a small change in the intensity of summer insolation at high northern latitudes could potentially lead to large changes in ice-sheet coverage and global temperature. Researchers have shown that the growth and destruction of the Northern Hemisphere ice sheet has a characteristic response time of about 100,000 years. Thus, the dynamics of glaciation are especially tuned to a frequency of one cycle per 100,000 years and should respond quite sensitively to eccentricity-induced changes. Numerical models show that instabilities develop as an ice sheet becomes very large, such that fairly subtle changes in high-latitude insolation can lead to its catastrophic destruction.

The ice–albedo feedback has the potential to explain the entire amplification of the Milankovitch forcing needed for the observed climate change. However, studies of polar ice cores indicate that carbon dioxide levels have also varied substantially. The greenhouse effect associated with these changes is not negligible. Ice cores also reveal that the amount of cloud condensation nuclei in the atmosphere has changed with the Milankovitch cyclicity. We will now explore some of the proposed mechanisms

for large and rapid changes in atmospheric CO_2 levels and cloud condensation nuclei on glacial timescales.

Evidence from the Vostok Ice Core

During the Southern Hemisphere summer of 1982–1983, French and Soviet scientists met in Vostok, on the high plateau of Antarctica, to take samples from the longest, most continuous ice core ever recovered from the Antarctic Ice Sheet. With each length of ice removed from this 2-km-deep hole, our knowledge of Earth's climate history was extended millennia into the past. By the time the last, deepest section was recovered and sampled, 200,000 years of ice accumulation was revealed. The samples were taken to France, where the oxygen isotopic composition of the ice was analyzed. This analysis revealed a history that matched well what we would expect from the marine record of isotopic changes during the Pleistocene.

However, one of the most important discoveries made by the scientists studying the Vostok ice core was the realization that the ice contained air bubbles frozen into the glacier as it grew. Using extreme care, the scientists were able to measure the concentration of carbon dioxide in the bubbles.

The results of their analyses provide a firm link between global climate change and variations in the quantity of greenhouse gases in the atmosphere. What they found is that the CO_2 concentration falls and rises in concert with variations in local temperatures (recorded in the hydrogen isotopic composition of the ice; Fig. 11-11), and both records are well correlated with global changes in temperature and ice-sheet size determined from $\delta^{18}O$ variations.

The rapidity of some of the changes is truly remarkable. The increase from glacial-stage levels (about 190 ppm) to nearly contemporary CO_2 levels (240 ppm) occurred over only 4000 years, between 16 k.y. and 12 k.y. ago. Analyses of bubbles from ice created during the next-to-last deglaciation, approximately 145 k.y. ago, display a similarly rapid rise in CO_2. The drop in CO_2 levels from the previous interglacial (about 130 k.y. ago) to the height of the last glacial (about 20 k.y. ago) was more subdued—about 1 ppm per millennium. The sawtooth nature of these changes is nearly identical to the oxygen isotope

FIGURE 11-10

Feedback diagram showing the effect of changes in glacial growth on global temperature.

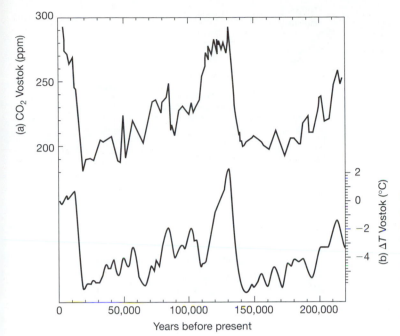

FIGURE 11-11

(a) CO_2 concentrations (in parts per million by volume). (b) Antarctic temperature (compared with the present-day temperature in °C), plotted against age, from the Vostok ice core. (From R.W. Christopherson, *Geosystems: An Introduction to Physical Geography*, 3/e, 1997. Reprinted by permission of Prentice Hall, Upper Saddle River, N.J.)

record, suggesting a close link between CO_2 changes, ice volume, and global temperature.

Feedbacks Affecting Atmospheric CO_2 on Glacial Time Scales

In Chapters 7 and 8 we found that on long time scales (millions of years) the carbonate–silicate geochemical cycle, together with the weathering and deposition of organic-carbon rich sedimentary rocks, determines the steady-state atmospheric CO_2 level. When studying glacial–interglacial fluctuations, however, other processes must be included, because the assumption of steady state with respect to these processes may not be valid on shorter time scales (thousands to hundreds of thousands of years). The partitioning of carbon between the atmosphere and terrestrial biomass, and between the atmosphere and ocean, as affected by the oceanic biological pump (Chapter 7), are potentially of great importance to the CO_2 balance during glacial cycles. So too are the processes of limestone weathering and limestone deposition. On glacial–interglacial time scales, limestone weathering need not be balanced by limestone deposition, and any imbalance will affect atmospheric and oceanic concentrations of CO_2. Let us explore these feedbacks in greater detail.

Role of the Biological Pump.

The photosynthetic conversion of dissolved carbon dioxide to organic matter in the surface ocean, the settling of this material through the water column, and its decomposition at depth (that is, the *biological pump*, described in Chapter 7) dominates the distribution of carbon throughout the world's ocean. Be-

cause the atmosphere equilibrates with the surface ocean, that photosynthetic conversion dominates the atmospheric CO_2 content as well. An atmospheric CO_2 pressure of 280 ppm (the preindustrial level) represents a biological pump that operates at intermediate efficiency, because regions of the ocean exist today (and presumably in preindustrial times as well) where nutrient concentrations are not completely depleted by biological uptake (Chapter 7). If nutrients were completely utilized—that is, if the biological pump were 100% efficient in removing nutrients and CO_2 from surface waters—the atmospheric CO_2 pressure would be reduced to about 165 ppm. At the other extreme, if the biological pump ceased completely, the atmospheric CO_2 level would rise to about 720 ppm as the CO_2-charged deep waters mixed with the surface waters and homogenized the chemical composition of the ocean. Thus, the low CO_2 concentrations of glacial intervals might be the result of a more efficient biological pump.

Why might the glacial ocean support greater biological productivity? Most of the answers proposed in the scientific literature involve increased nutrient supply through upwelling or riverine delivery. The hypotheses described next are intended to explain why nutrients might have been more available to the oceanic biota during glacial times.

Shelf Nutrient Hypothesis.

The CO_2 concentration at the height of the last glaciation (20 k.y. ago) was about 190 ppm. Might the biological pump have been more effective during the glaciation than it is today, perhaps as a result of a higher concentration of nutrients in the ocean as a whole? The nutrient concentration of the ocean represents a balance between supply by rivers (and, for nitrogen, by bacterial processes that convert nitrogen gas from

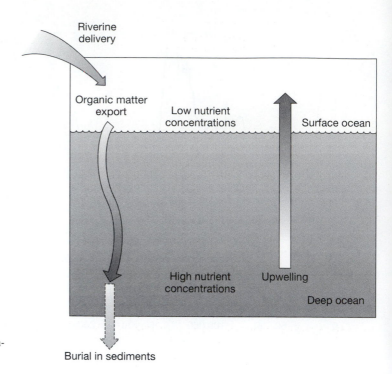

FIGURE 11-12

Simplified view of the nutrient through-put of the oceans.

the atmosphere to nutrient nitrate) and removal, primarily by sedimentation of organic matter (Fig. 11-12). Thus, if the biological pump was intensified by higher nutrient concentrations during glacial intervals, either riverine fluxes were greater or sedimentation rates were lower.

As the glaciers grew, sea level fell, exposing the vast, low-relief margins of the continents (the **continental shelves**). Sediments of the continental shelves are rich in organic matter and nutrients, as the result of the highly productive nature of the overlying waters. When these sediments became exposed, weathering reactions released the nutrients (especially phosphate) to the rivers draining the shelves (Fig. 11-13). This nutrient release enhanced the global delivery of phosphate to the oceans, causing an increase in oceanic phosphate concentrations, marine productivity, and carbon export from the surface

ocean and a drop in atmospheric CO_2. The resulting effect on global temperature and ice volume created a positive feedback loop, as shown in Figure 11-14. An attractive aspect of this hypothesis is that the residence (response) time of phosphate in the ocean is about 100,000 years, which essentially matches the major periodicity of glacial–interglacial cycles. The oceanic phosphate cycle is thus "tuned" to a major Milankovitch frequency and should be able to provide at least part of the required amplification of the Milankovitch forcing.

However, there are problems with this *shelf nutrient hypothesis*. The distribution of the trace element cadmium throughout the ocean follows that of phosphate very closely. Cadmium is incorporated into $CaCO_3$ skeletons of bottom-dwelling organisms in proportion to its oceanic concentration, whereas phosphate is not. Thus, pale-

FIGURE 11-13

The exposure of nutrient-rich shelf sediments as a result of a drop in sea level—part of the shelf nutrient hypothesis for the cause of changes in biological productivity on glacial time scales.

FIGURE 11-14

Systems diagram of the shelf nutrient hypothesis for the reduction of atmospheric CO₂ during glaciation.

oceanographers use the cadmium content of the fossils of bottom-dwelling foraminifera as a proxy indicator of changes in the phosphate content of the oceans in the past. What these researchers find is that the cadmium content of these fossils does not indicate higher phosphate concentrations during glacial intervals. This inconsistency has led scientists to shift the focus of their search for productivity changes to other nutrients, particularly iron.

The Iron Fertilization Hypothesis. Iron plays an important role in limiting primary productivity in certain regions of the ocean (Fig. 11-15). In these regions the major nutrients are not depleted as they are in the rest of the surface ocean (Chapter 7), and productivity appears to be limited by trace nutrients such as iron. Those regions might have received additional supplies of iron during glaciations. The most likely source was windblown dust particles, which typically have a coating of iron that dissolves in seawater. These particles were produced in regions that became more arid during glaciation and transported from these regions by prevailing winds that might have been intensified in response to a greater equa-

tor-to-pole temperature gradient. Recall from Chapter 4 that the zonal winds (for example, the midlatitude westerlies) are fundamentally a consequence of this latitudinal temperature gradient. The record of windblown dust accumulation in marine sediments supports this *iron fertilization hypothesis;* rates increase severalfold during glacial intervals.

The Coral Reef Hypothesis. The continental shelves between 30° N and 30° S latitude provide a habitat that is ideal for the growth of corals and other calcium carbonate-secreting organisms (Fig. 11-16). As these organisms grow, they add incrementally to the rock framework of the reef, building laterally as well as vertically, until the water surface is reached.

As we saw in Chapter 7, the production of CaCO₃ can be written as follows:

$$Ca^{2+} + 2HCO_3^- \longrightarrow CaCO_3 + CO_2 + H_2O.$$

Thus, the growth of coral reefs serves as an additional source of carbon dioxide to the atmosphere. (The effect

FIGURE 11-15

The iron fertilization hypothesis for the intensification of the biological pump during glaciations.

FIGURE 11-16

Coral reefs may be responsible for the changes in atmospheric carbon dioxide concentrations that occur between glacials and interglacials. (From T. McKnight: *Physical Geography: A Landscape Appreciation*, 5/e, 1996. Reprinted by permission of Prentice Hall, Upper Saddle River, N.J.)

is only temporary, however; after tens of thousands of years, this excess carbon dioxide becomes converted to bicarbonate, as the result of mineral weathering, and is redeposited as $CaCO_3$.) In contrast, when ancient reefs are exposed by a drop in sea level, chemical weathering leads to their dissolution. This process is the reverse of reef growth: Atmospheric CO_2 is converted into bicarbonate, which is carried by rivers to the ocean. Thus, the growth and destruction of coral reefs can affect atmospheric CO_2 on glacial–interglacial time scales.

The link to glacial–interglacial CO_2 fluctuations is once again through sea-level changes. As the glacial ends and the ice sheets begin to melt, sea level rises, flooding the continental shelves. In the tropics, reef growth resumes and CO_2 is released to the atmosphere. The increase in atmospheric CO_2 causes an increased greenhouse effect, thereby amplifying the original climate warming (Fig. 11-17). Conversely, as sea level begins to fall at the end of the interglacial, reefs become exposed to the atmosphere, and rain, soil, and groundwaters begin the process of reef dissolution. Again, the feedback loop is positive.

The importance of this feedback depends on the rates of these processes and on how responsive the rates are to sea-level changes. Studies of rates of calcium carbonate formation by reef-building organisms indicate that the reef ecosystem can easily keep pace with sea-level rise. The rate of limestone dissolution, however, is relatively slow. Thus, it is possible that, in this *coral reef hypothesis*, there is an unbalanced response to sea-level rise and fall in terms of reef growth and dissolution.

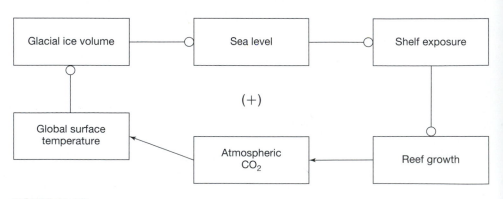

FIGURE 11-17

Systems diagram of the coral reef hypothesis.

a) Reconstructed vegetation cover, 18 k.y. ago

b) Present-day "potential" vegetation cover.

FIGURE 11-18

The difference in South American vegetation between (a) the last glacial maximum and (b) today. Note the large increase in forest cover at the expense of grassland and savannah. The present-day map shows the "potential" vegetation cover; deforestation and other human activities have reduced the forest cover from its potential coverage shown here. (Courtesy J. Adams, Oak Ridge National Laboratory.)

Changes in Terrestrial Biomass: A Negative Feedback

In living tissue, the terrestrial biomass today contains about the same amount of carbon (600 Gton (C)) as does the atmosphere; about twice that much is contained in dead and decaying organic material in soils. On the basis of studies of plant fossils and other climatic indicators, it appears that the amount of forest coverage, and thus terrestrial biomass, was drastically reduced during the last glacial interval (Fig. 11-18). Much of the northern forests were covered with ice, and tropical regions experienced greater aridity and thus the replacement of tropical rainforests with grasslands. Estimates have been made of the total amount of carbon that was transferred from the oceans to the terrestrial biomass at the end of the last glaciation. These numbers have large uncertainties, but it is clear that the change (around 700 Gton (C)) was many times larger than the net change in the amount of CO_2 in the atmosphere (around 160 Gton (C)).

The growth of the terrestrial biomass during deglaciation and its destruction during the initiation of a glacial interval represent the only negative feedbacks that we have been able to identify to changes in atmospheric CO_2 on glacial timescales (Fig. 11-19). That the CO_2 level rose and fell in concert with global temperature during the past 220,000 years indicates that positive feedback mechanisms have predominated.

Cloud–Albedo Feedbacks

Two compounds are particularly important in the formation of cloud condensation nuclei: **methane sulfonic acid (MSA)** and sulfur dioxide. The record of variation in the abundance of these cloud seeds can be read in the sulfur content of glacial ice in Greenland and Antarctica. Analysis of the composition of the Vostok ice revealed that the amount of MSA in the atmosphere of the Southern Hemisphere has varied with temperature over the past 150,000 years (Fig. 11-20). During glaciations, the MSA content of the atmosphere over the Southern Ocean apparently was

FIGURE 11-19

Systems diagram showing negative feedback between the size of the terrestrial biomass and climate change on glacial–interglacial time scales.

substantially greater than during interglacials. Might this have had a climatic consequence?

Both MSA and sulfur dioxide have an important biological source on the unpolluted Earth. Marine algae produce the gas *dimethyl sulfide* (DMS) as a byproduct of the regulation of the salt content of their cells. DMS escapes to the atmosphere, where it undergoes chemical transformation to either MSA or sulfur dioxide. *Aerosols* (suspended atmospheric particles—in this case, tiny droplets) are formed that can serve as nuclei for the condensation of water vapor in the formation of clouds. It is likely that the rates of production of DMS, MSA, and sulfur dioxide increase as marine algal productivity in gen-

eral increases. This increase in aerosol production would increase the aerosol concentration in the atmosphere. As a result, there should be an increase in the number of cloud water droplets and thus in the density of clouds. In turn, cloud albedo would increase and thereby reduce Earth's average surface temperature.

The high content of MSA in ice formed during glacial intervals indicates that the productivity of the glacial ocean was greater than that of the interglacial ocean. Why? Should not marine algae be more productive when water temperatures are warmer? In Chapter 6 we saw that some of the most productive waters of the world are at high latitudes. The detrimental effects of the lack of

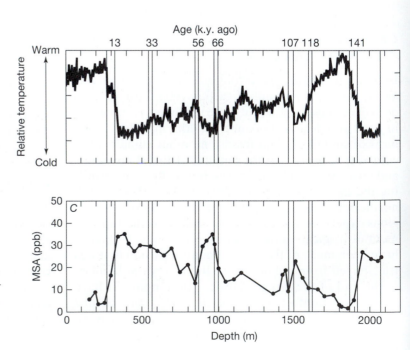

FIGURE 11-20

MSA content of Antarctic ice, compared with the relative local temperature (as indicated by the hydrogen isotopic composition of the ice), over the past 150,000 years. (After Legrand et al., *Nature* 350:144, 1991.)

sunlight and of cold water temperatures on marine algae seems to be more than compensated for in these regions by a high supply of nutrients. Today, a strong thermocline at low latitudes stabilizes the water column and tends to prevent upwelling. In contrast, high-latitude surface waters lack a thermocline. Wind mixing penetrates to great depths and mixes nutrient-enriched deeper waters to the surface. Thus, cooling of higher-latitude temperate waters during glacial intervals should have reduced the thermocline, fostered water-column overturn, and brought a greater nutrient supply to surface waters, supporting higher rates of primary production.

If marine algal productivity tended to increase during glaciations, the feedback loop would be positive: This set of processes would tend to amplify the climate systems response to the Milankovitch forcing (Fig. 11-21). The fact that the MSA content of Antarctic glacial ice rises as global temperature falls is suggestive of an important role for this feedback loop in Earth's climate system of at least the recent geologic past.

As our discussion turns to the issues of short-term climate change, both natural and anthropogenic, we must not forget that we are living in a time of unparalleled sensitivity of the climate system. The very rapid climate shifts that Earth has experienced in its most recent geologic past

FIGURE 11-21
Cloud-albedo climate feedbacks involving the algal production of cloud condensation nuclei.

caution us that tampering with the climate system might result in unexpectedly large climate responses. Furthermore, a prediction of future climate based solely on the Milankovitch forcing indicate that Earth should soon slip into the next glaciation. But, as a result of fossil fuel burning and its effect on atmospheric CO_2 levels, it is not at all clear that this will be the case. We will draw on our knowledge of paleoclimates in subsequent chapters to make a more informed prediction of the future climates of Earth.

Chapter Summary

1. Earth's climate has varied on a number of time scales, from decadal to millennial changes we can directly observe to the billion-year evolution drive by solar evolution.
2. The best documented of these changes are those that occurred on intermediate time scales during the glacial cycles of the Pleistocene.
3. Substantial, periodic swings in climate over the past 2 million years are indicated by the geologic record of glacial deposits and by the oxygen isotopic record of global temperature and of continental ice-sheet volume.
4. Orbital (Milankovitch) theory provides the answer to the question of causation of glacial–interglacial cycles.
   a. Small changes in the configuration of Earth's orbit around the Sun, varying in a predictable fashion, have provided the slight changes in seasonal insolation necessary to initiate these climate swings. These changes include changes in orbital eccentricity, obliquity, and the precession of the spin axis.
   b. The climate response matches the relative magnitude of the forcing at the frequencies of the precessional and obliquity changes.
   c. However, the dominant periodicity of glacial–interglacial fluctuation is 100,000 years. The forcings from eccentricity changes at this periodicity are too small.
5. The climate system seems to have amplified the 100,000-year signal preferentially, through feedbacks involving the dynamics of continental ice sheets, the albedo of ice and clouds, and the controls on marine productivity and the growth of coral reefs and thus also on the cloud albedo and greenhouse effect.

Key Terms

aphelion
continental shelves
eccentricity
erratic
glacial
glacial striations

Holocene Epoch
interglacial
loess
methane sulfonic acid (MSA)
moraine
obliquity

perihelion
period
Pleistocene Epoch
precession
seasonal temperature contrast
till

Review Questions

1. What types of geologic evidence are diagnostic of glaciation?
2. What causes changes in the oxygen isotopic composition of seawater?
3. Which three characteristics of Earth's orbit around the Sun vary on the timescale of Pleistocene glaciations? How does each of these affect the amount of energy received from the Sun?

4. What orbital configuration favors glaciation? Why?
5. How is the oxygen isotopic record of marine limestones used to test Milankovitch's theory of the ice ages?
6. What role do biologically produced sulfur gases play in glacial climate fluctuations?
7. What factors might have caused atmospheric CO_2 variations that kept pace with glacial climate fluctuations?

Critical-Thinking Problems

1. Return to our Daisyworld analogy from Chapter 2. Construct a Daisyworld-like model of the MSA-climate feedback loop shown in Figure 11.21. First sketch a graph of how DMS production by algae would affect global temperature. Then sketch another graph of your view of how changes in global temperature might affect algal DMS production. Defend both graphs in writing. Then combine these graphs, and discuss the stability of the equilibrium states indicated.

2. An ellipse is defined as the locus of all points such that the sum of the distances to two fixed points, called the foci, is a constant. We can easily show that this constant is equal to $2a$, where a is the semimajor axis of the ellipse. (See Figure 11-6 and accompanying discussion.) The eccentricity e of the ellipse is defined such that the distance from one focus to the midpoint of the figure is ae. An ellipse with $e = 0$ is a circle.

   a. Kepler's first law states that the planets move around the Sun in elliptical orbits with the Sun at one focus. Earth's present orbit has a semimajor axis of 1 AU and an eccentricity of 0.017. The point of closest approach to the Sun is the perihelion; the point farthest away is the aphelion. How much closer is Earth to the Sun at perihelion than at aphelion? Express your answer in astronomical units.

   b. The Milankovitch theory of the ice ages holds that the most important forcing factor is the difference in solar heating at high latitudes when Northern Hemisphere summer occurs at perihelion as opposed to aphelion. Using the inverse square law (Chapter 3), find the solar flux at perihelion and at aphelion. Recall that the solar flux at 1 AU is 1370 Watts/m². How much higher is the solar flux at perihelion than at aphelion today? Express your answer as a percentage. How much warmer is the effective radiating temperature of Earth (Chapter 3)?

   c. The eccentricity of Earth's orbit varies with time as a consequence of gravitational perturbations caused by the other planets. Repeat Question 2b for e at its maximum value of 0.06.

   d. Kepler's third law states that the square of a planet's period P is proportional to the cube of its semimajor axis a. When P is expressed in Earth years and a is in AU, the relationship is simply $P^2 = a^3$. Venus and Mars have semimajor axes of 0.72 and 1.52 AU, respectively. How many Earth years does it take for them to go around the Sun?

Further Reading

General

"Chaotic Climate," by W. S. Broecker (1995). *Scientific American,* November 1995, pp. 62–68.
"What Drives Glacial Cycles?" by W. S. Broecker and G. H. Denton (1990). *Scientific American,* January 1990, pp. 48–56.

Advanced

"On the Structure and Origin of Major Glaciation Cycles, Parts 1 and 2," by John Imbrie and others (1992, 1993). *Paleoceanography,* Volume 7, pp. 701–738, and Volume 8, pp. 699–735.

"Modeling the response of the climate system to astronomical forcing," by Andre Berger (1995). In *Future Climates of the World,* A. Henderson–Sellers (ed.), World Survey of Climatology, Elsevier Science Publishers B.V., Amsterdam.
"A new view on the driving mechanism of Milankovitch glaciation cycles," by H.-S. Liu (1995). *Earth and Planetary Science Letters,* Volume 131, pp. 17–26.
"The geophysiology of climate," by Lee Kump and James Lovelock (1995). In *Future Climates of the World,* A. Henderson-Sellers (ed.), World Survey of Climatology, Elsevier Science Publishers B.V., Amsterdam.

Chapter 12
Short-Term Climate Variability

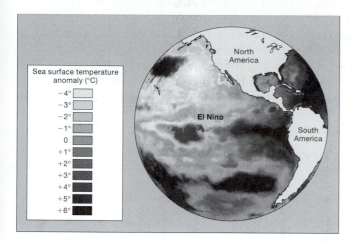

Sea surface temperature anomaly (°C)
- −4°
- −3°
- −2°
- −1°
- 0
- +1°
- +2°
- +3°
- +4°
- +5°
- +6°

North America

El Nino

South America

Sea-surface temperature anomalies in the central and eastern Pacific during the 1997–1998 ENSO event.

Key Questions

- Can we document climate changes that occur over shorter time periods than those we have discussed so far?
- What causes climate change on short time scales?
- Where in the climate system do we look for processes that might cause climate variability on the time scale of years to decades?

Chapter Overview

We have seen in previous chapters how the Earth system has undergone long-term changes in climate. In this chapter, we shall turn to shorter time scales and show how climate has changed during the past 12,000 years. In this case, we will be looking at changes that last for several hundred to a thousand years. For the most part, we will be considering changes caused by processes that are internal to the Earth system, rather than external forces such as the long term variability of the Sun or changes in Earth's orbit about the Sun. We then move to even shorter time scales and discuss the year-to-year variability that we find in the global climate. Here we will focus on two processes, both of which involve the oceans: one in the tropics (El Niño–Southern Oscillation events) and one at high latitudes involving sea ice-ocean–atmosphere feedbacks.

By the end of this chapter, we will have observed that there have been large changes in climate (from a human perspective) over the past 12,000 years, but none of them were as large as the changes that are likely to happen in the future as a result of anthropogenic greenhouse warming.

Introduction

Time Scales of Change

In the past few chapters we have taken a rapid tour through 4.6 billion years of Earth's history, focusing on the processes that have shaped that history and examining the degree to which the physical environment has changed since the early days of Earth's formation. The underlying message of each chapter is the same: Earth acts as a single integrated system over any time scale we choose. This system has remained remarkably stable (staying within certain limits or boundaries) throughout its history, but within those limits Earth has experienced a considerable degree of change and variability. How we define these limits depends on the specific components of the system under discussion (global temperature, gas composition of the atmosphere, and so on). In general terms, we can say that for the past 3.5 billion years Earth has maintained an environment capable of supporting life—but life can exist under a wide range of conditions.

Indeed, when we look at Earth's history we find intervals that were much warmer than the present and intervals that were cooler (when huge masses of ice covered large areas of the land surface); we find that the atmospheric composition has changed considerably through time; and we find intervals when the distribution of land, ocean, and mountain ranges was very different from what we see today. From our earlier discussions we are also left with the impression that these large changes (such as changes in solar luminosity) take place over long time scales, but they occur very slowly—slowly enough that when we move to shorter time scales, these changes are not noticeable. On these shorter time scales, however, other processes are observed that produce smaller and more-frequent variations (such as the orbital changes that trigger ice ages). Finally, we also see from our earlier discussions that some of the forcing for these changes originates outside the Earth system itself. However, as we move to much shorter time scales, the situation is very different.

In focusing on short-term climate change and variability, we consider changes on a hundred- to a thousand-year time scale as well as variability on interannual to decadal scales. Our objective is twofold. On the one hand we present these processes as further illustrations of how system components interact; on the other hand, this discussion provides background material for our examination of global warming. We have seen the role long-term changes in the level of atmospheric greenhouse gases have played in Earth history, and we will discuss the possible impacts anthropogenic greenhouse gases may have in the near future. Before we look at these possible future changes, however, let us set them in the context of the variability in the climate system that occurs naturally over these shorter time frames.

The last major continental glaciation reached its maximum extent about 18,000 years ago. The temperature difference between the last glacial maximum and today is on the same order of magnitude as the temperature change we expect to see in response to greenhouse warming by the later part of the 21st century. The interval from the present to the last glacial retreat (about 12,000 years ago) is referred to as the **Holocene.** When we look at this epoch

from the perspective of Chapter 11, the temperature record of the Holocene looks relatively uninteresting (Fig. 12-1a). Once Earth warmed up from the last glacial maximum, nothing much else seems to happen. If we zoom in a little closer and look at the past 10,000 years (Fig. 12.1b) or 1000 years (Fig. 12.1c), however, we see that a great deal of smaller-scale climate change has taken place.

It is not only the climate that has changed. Although our discussion of the Holocene focuses on climate change and variability, there have been large ecological changes over this same interval. The species we find today existed prior to the last glaciation, but the characteristic groupings, or assemblages, of species that we find in specific environments has changed considerably. For example, the assemblage of species that we recognize as members of

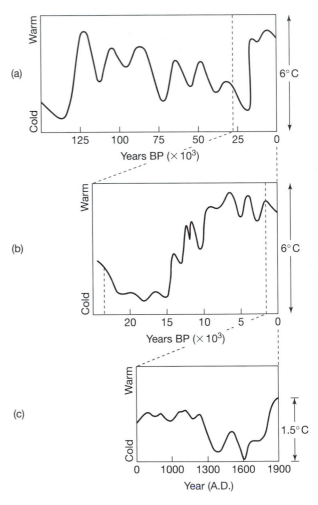

FIGURE 12-1

Mean global temperature change since the last glacial maximum. (a) Generalized oxygen isotope curve from deep-sea sediments. (b) General estimates from pollen data and alpine glaciers (emphasis on midlatitudes from eastern North America and Europe). (c) General estimates from historical documents (emphasis on the North Atlantic region). (d) Global mean temperature change between 1860 and 1990, based on land stations and ship observations, representing differences from the 1950–1980 mean. (Adapted from U.S. Committee for GARP, 1975, *Understanding Climate Change,* National Academy of Sciences, Washington, D.C.)

FIGURE 12-1 *(Continued)*

the boreal forest community in the high midlatitudes came into being only during the Holocene, although the individual species existed prior to this time. What makes this interval even more interesting is that, for the first time, we also have to consider the impact that human societies have on the Earth system. Furthermore, many of the changes that appear to have been very significant in some regions during the Holocene actually represent smaller changes than those we project might occur in the future.

Humans have been adding greenhouse gases to the atmosphere in increasing quantities for the past 200 years. We might think that this should have had some sort of climatic impact by now, and one question that we asked in Chapter 1 was whether there is any evidence that warming is actually occurring. We pointed out that it is very hard to distinguish the climate change signal from the natural variability of the system. Later in this chapter, we will begin to see why this is so by moving to even shorter time scales and discussing processes that cause variability in the system on an annual to decadal basis.

Climate Change and Variability

Climate is an assemblage of weather conditions experienced at a location over some reference time frame. The World Meteorological Organization suggests that a 30-year averaging interval be used to describe climate. *Climate variability* refers to the fluctuations that take place within that interval. If there is a significant difference in the average conditions or in the pattern of variability between two time spans, we refer to this difference as a climate change. In evaluating climate change, we can compare one 30-year interval to another or to the average over the past 100 years or 1000 years, or we can look at any month, year, decade, and so on within our 30-year period and compare that with the 30-year average. However, one thing that quickly becomes apparent when we carry out this sort of exercise is that there is no such thing as "normal" when it comes to climate. Climate varies continuously.

If we examine the climate of any 30-year (or longer) time span, we notice that conditions during that interval are seldom average: At any time the climate is almost always warmer or cooler, wetter or drier, than the average would suggest. Therefore, another favorite occupation of scientists studying these climate records is to try to determine whether the climate is warming or cooling, or getting wetter or drier. In other words, the researchers look for *trends* in the data. Figure 12.1d is a plot of global mean temperatures based on historical records that show a general warming trend since the late 19th century. This warming is in broad agreement with what we might expect from the increase in anthropogenic greenhouse gases. However, to extend the temperature series backward, we can use **proxy data**—data that cannot be obtained by direct measurement but can be inferred from other evidence. When we do, we can interpret the warming trend in the 20th century as part of the recovery from an earlier cool period (Fig. 12.1c). Or, if we go further back, we can see it as a small fluctuation in the *cooling* trend following an even earlier warm period appoximately 6000 years ago (Fig. 12.1b). Hence, any analysis of trends is dependent on the time scale chosen: What looks like a trend at one time scale may simply be a short-term variation at another. Consequently, as we discuss short-term climate (climate over human time scales), we are much more concerned with the variations in climate than with the average conditions.

Proxy Climate Data

We have a vast quantity of data with which to describe and analyze the present-day climate system. An array of measuring stations covers much of Earth's land surface, and ships and floating buoys provide measurements for much of the world's oceans. Where there are gaps in the coverage, we can typically fill them in with satellite data. In fact, in terms of quantity, satellites now provide the bulk of the observational climate data set. Consistent and reliable satellite data, however, have been available only since the early 1970s. The data prior to the satellite era have huge gaps over the oceans, and over the land the climate record is highly variable—some regions have several hundred years of observations, others less than 30. With such a short record of observational data, how do we determine climate variability and climate change of the past?

We can make some estimates of the state of the climate system at different times in the past through the use of proxy data. We have already seen some examples in earlier chapters. In Chapter 11, for instance, we discussed the geological evidence of the advance and retreat of continental glaciers, such as glacial till deposits and moraines, as well as atmospheric and climate reconstructions derived from ice cores.

The chemical composition, physical structure, and sedimentation pattern of materials that make up sedimentary rocks provide some indication of the environmental conditions that were in existence when the sediments were deposited. We can obtain these data both by examining sedimentary rocks on land and by recovering cores drilled in the sea floor that provide a record of the sedimentary history of the ocean basins. Fossils in these sediments can tell us much about the physical environment at the time the organisms lived. If we assume that fossil plants, animals, and protists lived in the same environment as those having similar physical characteristics that exist today, then by analogy we can make some assessment of what the environment was like at that location in the past. The principle that processes operating today also operated in the past, and that past geologic events can be explained in terms of processes active today, is known as **uniformitarianism.** This principle governs much of the work done in *paleoclimatology* (the study of past climates).

As we move to time scales covering the past 10,000 years, we make use of other types of evidence. One of the most useful data sources is ice cores, particularly for the earlier part of this interval. Numerous other techniques are also employed for climate reconstruction. Two of the most common are based on *palynology* and *dendrochronology.*

Palynology is the study of pollen and organic microfossils. Pollen grains are preserved in many different environments (for example, lake sediments and peat bogs). If we drill a core into sediments from a lake or from a peat bog, divide the core into segments going back through time, and then extract the pollen from each layer, we can reconstruct the plant assemblages that lived in the area of that core at each time interval in the past. We then use the present-day distribution of those assemblages to place constraints on what the environment was like in the past. Pollen data from peat bogs have been used to reconstruct climate over the past 30,000 to 35,000 years in the British Isles.

Dendrochronology is a method of dating trees by counting their annual growth rings. This method uses the growth characteristcs of certain tree species. The cross section of a tree trunk consists of a series of rings, each ring representing growth over one year. By counting the annual rings, we know the age of the tree. The width of each ring indicates the amount of growth that occurred during the growing season. In certain circumstances, that amount can be related to the temperature during the growing season or to water availability, both of which can tell us something about the climate while the tree was alive. For example, a climate record that extends back almost 5500 years has been reconstructed from tree rings of bristlecone pines in the White Mountains of California.

The Holocene

Assembling proxy climate evidence from around the world, we see that the Holocene displays a considerable range of climate change and variability. It is difficult to generalize, but there appears to be a dominance of temperature changes in the mid- and high-latitudes, whereas the tropics and subtropics appear to have experienced greater changes in moisture availability. These changes result partly from orbital effects that enhance seasonality and continentality (which directly affect the temperature regime) and partly from the resulting circulation changes (such as changes in the monsoon circulation) that affect precipitation patterns. Other factors may also have played a role in regional climate changes, as we will see later in this chapter.

Although there is evidence of widespread and frequent changes in climate throughout the Holocene, that evidence is not continuous in time or space. It is also difficult to assign accurate dates to some of the changes that we observe. Consequently, when we attempt to explain the climate record, it is not always easy to establish which changes are regional and which are tied to some larger-scale global forcing. Although we focus on only the broad patterns of Holocene climate change that are apparent in the record, it is still difficult to establish the degree to which these changes are global. Evidence of synchronous climate changes can be found in numerous locations around the world, but they are not found everywhere. Even where the evidence suggests that a change occurred in several different parts of the world at the same time, the evidence also indicates that the magnitude of the event was not the same at all locations. We will return to this fact in our discussion of global warming in Chapter 13. Although we tend to talk in terms of mean global changes, remember that there may be very large differences in the regional expression of these changes.

Another point to note is that relatively small changes in the mean global temperature are associated with relatively large changes in the physical environment. Mean global temperatures at the height of the last glaciation were probably only 5 to 7°C lower than they are at present. Eight hundred years ago, when the Vikings were able to colonize parts of Greenland and grow sufficient crops to maintain a continuous settlement, mean global temperatures were only 0.5°C or so warmer than they are today. The significance of these numbers is highlighted by the fact that global temperature changes predicted for a doubling of atmospheric CO_2 are on the order of 1.5 to 4°C. In other words, the changes we predict for potential global warming in the 21st century are much greater than anything that has occurred in the past 10,000 years, and those changes are of comparable magnitude to the warming that took place between the last glacial maximum and the present day.

The Younger Dryas

In very broad terms, Earth began to warm between 10,000 and 15,000 years ago. The ice sheets that covered large areas of North America and northwest Europe began to retreat. Global sea levels rose as the glaciers melted, and the erosion from large volumes of meltwater reshaped the landscape around the edges of the former ice margins. Vegetation began to colonize the previously glaciated regions, and new vegetation patterns developed as soils formed and temperature and rainfall patterns changed. This spread of milder, more benign conditions came to an abrupt end 10,500 years ago in a climatic reversal known as the **Younger Dryas** event. The Dryas flower was widespread at that time—giving its name to the climatic event—but is currently found only in arctic and alpine tundra.

Some of the best evidence of climate and vegetation changes during the Younger Dryas comes from pollen and geologic analyses in northern Europe. As the climate warmed after the glacial retreat, there was a general increase in the density of vegetation, particularly grasses and sedges. This increase was followed by an increase in shrubs such as juniper and in willow; in some areas, the shrubs were later replaced by birch woodland. Pollen analysis shows a similar sequence of events in the British Isles, Ireland, and Scandinavia. Geologic evidence indicates that most of Scotland was probably deglaciated by 11 k.y. ago. At this point a major climatic reversal occurred. By about 10.8 k.y. ago, there was a new ice sheet several hundred meters thick over western Scotland, and there was renewed advance of valley glaciers in the upland regions of northern Europe. (*Valley glaciers* are individual glaciers that form at the head of a valley in mountainous regions and flow down the valley). The pollen evidence shows a synchronous change in vegetation. The northern woodland diminished in area and was restricted to a few sites. The vegetation became more open, and the pollen data show a predominance of cold-tolerant vegetation types.

These changes represent a significant climate shift in northern Europe. However, the global impacts of this shift are not so obvious. There is evidence of a similar climate shift in New England and along the east coast of Canada. There is little other evidence from North America except in the Gulf of Mexico and the Gulf of California, and there is only limited evidence from the Mediterranean. However, climatic reversals also appear to have occurred in the Andes and in Africa. The strongest evidence from Africa comes from lake levels, which increased after the northern deglaciation. During the early Holocene, what we know today as the Sahara Desert was primarily grassland (savannah), and the ecology of the region was very different from what we find today. However, lake levels declined again after 11 k.y. ago; the evidence indicates that while the Younger Dryas event was taking place in northern Europe, much of tropical and subtropical Africa was experiencing increased aridity. Furthermore, data obtained from ocean cores taken in the western tropical Pacific Ocean and off the coast of Japan show some indication of a climate change at that time. Glaciers in the Southern Alps of New Zealand may have also readvanced during this interval.

The Younger Dryas thus appears to be centered primarily on the North Atlantic region, but synchronous climate changes might have occurred in other parts of the globe. What process might explain a shift in climate that has a strong regional, rather than global, focus yet is able to influence widely scattered regions across Earth's surface?

North Atlantic Deep-Water Formation. A prime candidate for explaining the strong regional focus of the

Younger Dryas climate change is the ocean circulation of the North Atlantic. The relatively high sea-surface temperatures in the northeast North Atlantic, which bring mild conditions to northern Europe today, result from the northward movement of the warm surface waters of the Gulf Stream and North Atlantic Drift. We saw in Chapter 5 that this movement was controlled in part by the atmospheric circulation and in part by the thermohaline circulation. Recall that the North Atlantic thermohaline circulation is driven by deep-water formation in the Norwegian and Greenland Seas: As the cold and highly saline water subsides and moves southward, it is replaced by warm, northward-moving water at the surface. The geochemist Wallace Broecker has suggested that some of the climate changes that accompanied deglaciation resulted from events that cut off or reduced this deepwater formation. Meltwater from the North American ice sheet, which normally flowed southward to the Gulf of Mexico, might have been blocked by a retreating lobe of ice as the ice cap melted. The meltwater then would have flowed eastward through the Gulf of St. Lawrence. The result would have been a large infusion of cold freshwater to the northern North Atlantic. Because freshwater is less dense than saltwater, this infusion would have pro-

duced a stable surface layer that would freeze very easily, pushing the sea ice margin southward and cutting off the formation of the North Atlantic deep water. Both the change in the thermohaline circulation and the southward expansion of the sea ice would have cut off the flow of warm surface water in the North Atlantic Drift, which would have resulted in a significant climate change in the region. Such a process could account for the climate reversal experienced during the Younger Dryas and would also explain the apparent focus on the North Atlantic region. This explanation is still open for debate, however, as it is difficult to account for a synchronous change in New Zealand by this mechanism.

An On–Off Switch in the North Atlantic. Although we have known about the Younger Dryas event for years and we have known that the event took place fairly rapidly (in geological terms), an ice core obtained from the Greenland ice cap in the early 1990s revealed the startling information that these changes might have taken place in less than a decade. The snow accumulation record (Fig. 12-2) shows increased accumulation in the warmer intervals; it also shows that the switch from cold to warm intervals occurred over a very short time span. Atmos-

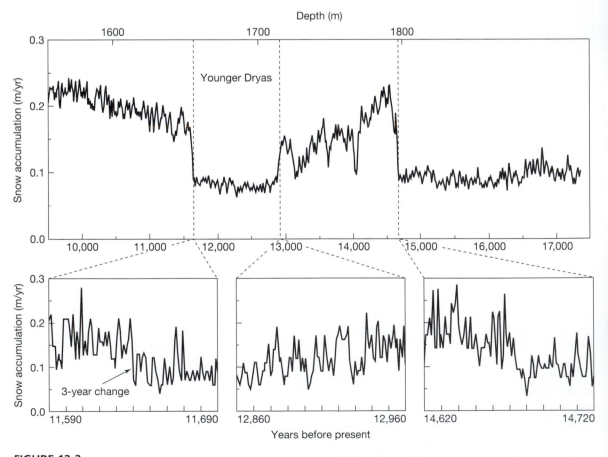

FIGURE 12-2

The snow accumulation variation record from a Greenland ice core. (After Alley, R., et al., "Abrupt increase in Greenland snow accumulation at the end of the Younger Dryas event," *Nature*, **362**:527–529, 1993.)

pheric dust deposited on the ice and recorded in the ice cores reveals similar rapid changes in deposition rates. More dust is deposited during glacials than during interglacials, because the increased north–south temperature gradient results in a stronger atmospheric circulation; that stronger circulation carries more dust. Both the snow accumulation and the dusty deposition, therefore, indicate a shift in the atmospheric circulation. The rapidity with which these changes occur suggests that the system switches almost instantly between two modes of circulation.

These data are of interest for two very different reasons. One obvious question concerns the processes or events that toggle the switch. The most likely cause is the one we have described here: a change in the oceanic circulation. Mechanisms that switch the thermohaline circulation on or off, or that cause sudden large movements in the latitude of the maximum sea-surface (and atmospheric) temperature gradient, could produce the sudden shifts in the atmospheric circulation that the ice core data suggest.

The exact mechanism for explaining the shift in circulation is the subject of much speculation and discussion. One suggestion is that attributing the cause to any single process may be a mistake. An alternative approach is to view the system as chaotic. *Chaos theory* represents a rapidly emerging branch of science dealing with dynamic systems. Chaotic systems are iterative: The state of the system at one point in time is dependent on the state at the previous point. However, a characteristic of these systems is that very slight changes in the starting point are amplified through positive feedbacks, so the possible results diverge rapidly after only a short interval. Almost identical starting points can result in very different outcomes, and different starting points can produce outcomes that are very similar. The consequence is that, after a certain interval, the system becomes essentially unpredictable.

In 1960, Edward Lorenz, a meteorologist at the Massachussetts Institute of Technology, was the first to recognize that the atmosphere is a chaotic system. Since then we have known why accurate long-range weather forecasting is impossible and why accurate daily weather forecasts can be effective only on time scales of days to a couple of weeks. One important attribute of chaotic systems, however, is that they exist in quasi-equilibrium states; in the case of the atmosphere, for example, we know in general what it will be like next year, even though we cannot predict exactly what it will be like on any given day. In the terminology of chaos theory, these quasi-equilibrium states are called *strange attractors:* The system is never precisely at that point, but it is always somewhere close to it. Another characteristic of chaotic systems is that they may switch rapidly between two or more of these quasi-equilibrium states. The point at which this switch occurs is called a *bifurcation point.*

We can view the climate system as having two (or more) stable steady states: glacial and interglacial, with

the transition between the two representing the bifurcation point. When the climate system is near the bifurcation point (for example, at the end of the last glaciation), the system is unstable and any number of small perturbations could be amplified through positive feedbacks to push the system rapidly toward one stable state or the other. This possibility leads us to the second reason why the Greenland ice core data have generated considerable interest. If these data do indicate that climate can switch rapidly between two very different stable states after a relatively small perturbation, a new wrinkle is added to the greenhouse warming question: If increased greenhouse gases should lead to a rapid shift in the climate system (possibly to a third, much warmer, quasi-equilibrium state), then our expectations of global warming (discussed in Chapter 13) may significantly underestimate the strength and the speed of the climate response to greenhouse forcing.

The Holocene Climatic Optimum

The short cold interval of the Younger Dryas event ended with a rapid shift to warmer conditions, followed by a constant climate or relatively slow warming over the next several thousand years—the **Holocene Climatic Optimum.** Evidence suggests that summer temperatures were slightly higher in the mid-Holocene (5000 to 6000 years ago) than they are at present. This interval was long thought of as a time of persistent mild conditions with very little climate change. This view originates from early studies in Europe, where the pollen record shows little evidence of major climatic shifts. Elsewhere in the world the picture is less straightforward.

Ancient lake levels in East Africa and the Sahara Desert show evidence of much wetter conditions than exist today. These high lake levels occurred when temperatures were higher—presumably, evaporation would have been greater then. This means that the amount of rainfall in the region must also have been greater than it is today. The evidence suggests a northward shift of the intertropical convergence zone in the Northern Hemisphere summer, bringing higher rainfall to the Sahara and East Africa and an enhanced monsoonal circulation (increased summer rain) in Arabia and northwest India. There is also evidence that the Mediterranean Sea experienced increased summer rainfall during this same interval. Turkey and western Iran, however, appear to have been more arid than at present.

That the mid-Holocene climate was very different from today's climate is also indicated by archeological evidence. The Tarim Basin, the site of the ancient silk route from China to Europe, is currently a desert, but between 5000 and 6000 years ago it was forested and populated with numerous settlements. Nomads grazed cattle in the central Sahara, and the Harappan culture flourished in the Indus River valley. This agricultural society, located in what is now the Rajasthan Desert, may have been the first to cultivate cotton. At first glance, changes in climate would appear to be a likely explanation for the decline of these earlier civilizations. Undoubtedly, the Sahara, for example, is now much drier than it was then. In several cases, however, it is possible that human land-use practices, rather than a change in climate, led to land degradation. In fact, it is likely that the collapse of some of these cultures resulted from the interaction of both factors—where land degradation due to human land use was amplified by changing climate conditions.

The Medieval Warm Period

Temperatures fell after the Holocene Climatic Optimum, reaching a minimum about 3000 years ago, but rose to a new maximum during the **European Medieval Warm Period.** Temperatures in Greenland appear to have increased after A.D. 600 to 650, with the temperatures in the North Atlantic (Greenland and Iceland) reaching a maximum around A.D. 1100. At this time the Vikings established a self-supporting colony on the southwest coast of Greenland. The colony lasted almost 400 years and, at its largest extent, had 280 farms and a population of 3000. By the end of the 12th century, however, because of the decreasing temperatures, the sea ice east of Greenland grew more extensive. By the middle of the 14th century, ships had to take a more southerly route to avoid the ice. By A.D. 1410, communication with the Greenland colonies was lost completely.

Farther south, in northern, western, and central Europe, the Medieval Warm Period reached a maximum between A.D. 1150 and A.D. 1300. During this period, wheat was grown in Norway at about 64° N, oats and barley were grown in Iceland, vineyards were cultivated in England, and farm settlements spread to higher elevations in Norway, northern England, and Scotland—all evidence of milder climates than those areas have today. The average temperature of central England during this interval is estimated to have been 0.5 to 0.8°C above the mean for the first half of the 20th century.

From the beginning of the 14th century, the climate became more variable. Wetter (and probably colder) summers in Europe from A.D. 1313 to A.D. 1317 led to a succession of failed harvests and widespread famine, and the expansion of farms into the upland regions of northern Europe and Scandinavia came to an abrupt end. The interval from A.D. 1250 to A.D. 1350 was one of numerous large storms and floods. It has been suggested that the storminess was caused by a cooling at high latitudes that caused the sea ice to expand southward, resulting in an increased temperature gradient in the North Atlantic. Flooding along the North Sea coasts of Denmark and Germany was extensive, and 100,000 to 400,000 people were reported to have drowned in the floods that occurred at that time. The impact of these large environmental changes on human societies was then compounded by the arrival in Europe of the bubonic plague in 1346. The plague lasted until 1361, killing an estimated 25 million people (one-quarter of Europe's population).

The Little Ice Age

The extreme climatic fluctuations that marked the end of the Medieval Warm Period led to an interval of cooling until the early 1500s, after which the climate stayed relatively stable or even began to recover in some areas. The North Atlantic climate, however, entered a renewed episode of rapid cooling in the late 1500s, a period now known as the **Little Ice Age.** Although the cooling was originally thought of as a regional climate fluctuation centered on western Europe and the North Atlantic, a growing body of evidence from the European Alps, Asia and the Himalayas, South America, New Zealand, and Antarctica suggests that changes might have occurred over much of the globe. However, not all parts of the world show evidence of a climate change at this time. Where they do, it is not clear that the changes were synchronous or that they lasted the same length of time everywhere. The Little Ice Age continued through the middle of the 19th century, but the reduced temperatures were not continuous. The Little Ice Age is characterized by considerable variability, with episodic cold spells that varied in timing and duration from place to place.

The evidence of the Little Ice Age takes numerous forms, among them the readvance of mountain glaciers, the lowering of tree lines, increased erosion and flooding, sea-ice expansion, and the freezing of canals and rivers. The canals of Holland have long been used for transportation, and there are reliable records of freeze-up since 1633. The canals seldom freeze over in today's climate, but in the 15th and 17th centuries it was common for them to be frozen for 3 months at a time. There is also documented evidence of glaciers in the Swiss Alps advancing and covering houses on the outskirts of several villages (Fig. 12-3). A climate change is also suggested by indicators of population decline: The abandonment of settlements and decreasing agricultural productivity. However, the societal evidence is not straightforward; population changes and declining productivity were almost certainly influenced by societal and political factors and by disease.

As with the Younger Dryas event described earlier, the Little Ice Age appears to have had a strong regional focus, yet evidence is mounting that many other parts of the world also experienced a similar cooling. In this case, however, there was no retreating continental ice sheet to help force large changes in the oceanic circulation. Are there any other likely explanations for this particular shift in the climate system?

Volcanoes and Climate. From earlier chapters we know that atmospheric composition can change and that these

(a)   (b)

FIGURE 12–3

The Argentiere glacier in the French Alps. (a) An etching made about 1850, showing the extent of the glacier during the waning phase of the Little Ice Age. (b) The same view, photographed in 1966. (Reprinted with permission from Understanding Climate Change, A Program for Action, National Academy of Sciences, Washington, D.C.)

changes can have a significant impact on climate. Are there any mechanisms for changing the atmospheric composition on the time scales discussed in this chapter? Yes. Two primary mechanisms are worth discussing: (1) a change in gas composition due to anthropogenic activity and (2) volcanic activity. Anthropogenic activity, the subject of a later chapter, cannot explain changes prior to the 20th century. Here we will focus on the impacts of volcanoes on climate.

There has been an ongoing debate over the possible effects of volcanic eruptions on the global climate. We have already seen that volcanism played an important role in the origin and early evolution of Earth's atmosphere. It has also been suggested that volcanic eruptions have a direct impact on today's climate, and they might also account for the severity of the climate during the Little Ice Age. As early as 1784, Benjamin Franklin hypothesized that the abnormally cold winter of 1783–1784 might have been due to the 1783 eruption of Hekla, a volcano on Iceland. Actually, any climate anomaly in that year was more likely due to the 1783 eruption of Laki, another Icelandic volcano. Laki is, in fact, the largest effusive eruption in the historic record, producing about 12 km³ of basalt lava flows.

Scientists once believed that the volcanic ash injected into the atmosphere during an eruption would result in an increase in global albedo, which would reduce the solar input to the surface and result in a surface cooling. The volcanic ash, however, aggregates very quickly and falls out of the atmosphere within tens to hundreds of kilometers of the eruption. More important to climate is the gas (primarily sulfur dioxide--SO_2) injected to high altitudes in the stratosphere. Sulfur dioxide gas oxidizes, forming sulfuric acid droplets, an atmospheric aerosol. The aerosol scatters and reflects solar radiation, reducing the solar radiation that reaches the surface (hence cooling the sur-

face). The aerosol also absorbs upwelling infrared radiation, resulting in a warming of the stratosphere. These effects, although more important in a global sense than are those due to volcanic ash, are still relatively short-lived. The **residence time** is the length of time an aerosol is present in the stratosphere after a volcanic eruption. The aerosol concentration in the atmosphere decreases at a rate that is analogous to the rate of radioactive decay. You can think of the residence time as being similar to the half-life of the aerosol in the atmosphere. The residence time of a volcanic aerosol in the stratosphere is on the order of about a year. Consequently, any climatic effects from a single eruption are also very short-lived, lasting only 1 to 2 years after the event.

The degree to which a volcanic eruption affects climate depends in part on the location of the eruption and on the way the atmospheric circulation distributes the stratospheric aerosols globally. Mount Pinatubo in the Philippines experienced a major eruption in June 1991, when the equivalent of 3 to 5 km³ of dense rock was ejected into the atmosphere (Fig. 12-4). The ejecta also included approximately 20 Mton of SO_2 gas, which was subsequently converted to sulfuric acid aerosol particles, with the largest concentration occurring in the lower stratosphere at an altitude between 15 and 20 km. The aerosol cloud was distributed very rapidly around the globe by the stratospheric circulation. After about 10 days, the cloud had been distributed fairly uniformly over an area of 15 million km², in a 10,000-km-wide band from Indonesia to central Africa (Fig. 12-5; Plate 13). The cloud circled the globe in 22 days.

It is clear from the observations that a major volcanic eruption in the tropics can produce an aerosol cloud that is distributed rapidly around the globe and has an observable effect on the radiation budget of the stratosphere. But, can such an eruption also have a significant effect on the tropospheric climate?

FIGURE 12-4

The eruption of Mount Pinatubo in the Philippines, June 1991. (From Robert M. Carey, NOAA/Mark Marten, Photo Researchers, Inc.)

One way to address this question is to measure or estimate the effect the aerosol cloud has on the radiation balance. A perturbation in the radiation budget caused by the presence of volcanic aerosols is referred to as the **volcanic aerosol forcing.** To measure this forcing, we need long-term measurements of the global radiation budget. These long-term measurements then provide the background against which we can see the anomalies caused by the eruption. Such long-term measurements are now available from satellite observations, which, in this case, showed a dramatic increase in albedo over the tropics immediately after the Mount Pinatubo eruption. The satellites were able to measure strong anomalies in the radiation budget over much of the globe in the months that followed. The effects reached a maximum in late 1992 with a reduction in the mean global surface temperature on the order of 0.5°C.

Interestingly, although temperatures tended to be lower in the year after the eruption, North America and parts of Europe and Asia had an extremely warm winter in 1991–1992. Such regional differences have been observed after several major eruptions during the past century. It has been suggested that the direct radiative impact of the aerosol produces cooling at the surface over the tropics but that the absorption of longwave radiation by the aerosol in the tropical stratosphere enhances the east–west circulation at midlatitudes. This enhanced circulation draws

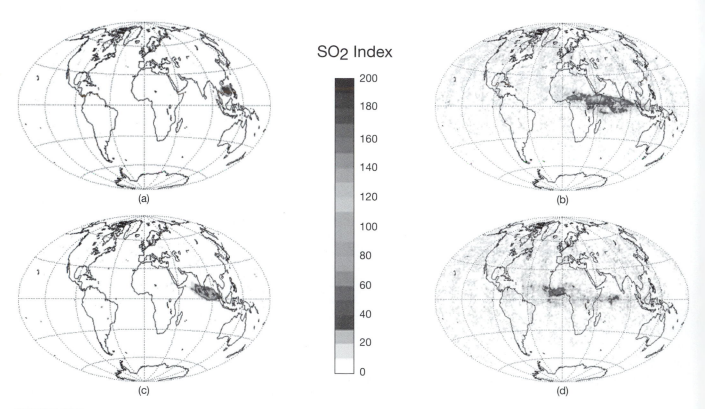

FIGURE 12-5

Satellite observations of the Mount Pinatubo aerosol cloud, 1991. (a) June 16. (b) June 18. (c) June 23. (d) June 30. (After Bluth, G. J. S., et al., 1992, "Global tracking of the SO² clouds from the June 1991 Mount Pinatubo eruption." *Geophysical Research Letters,* **19:**151-154.) [See color section] more warm air over the continents, producing the warmer winters experienced by Europe and North America.

The historic record holds many accounts of volcanic eruptions. We can also obtain evidence of earlier eruptions from ice cores. Some of the acidic material ejected into the atmosphere during a volcanic eruption may be deposited on a glacier's surface during snowfall events. When the snow is converted to ice, that ice layer has an anomalously high acidity. By counting the annual ice layers, we can obtain an accurate date for the eruption as well. For more-recent eruptions, we can typically match the ice-core evidence with historic accounts. For earlier volcanic events, for which neither historic records nor ice-core data are available, we can still obtain evidence of eruptions from the geologic record in the form of old lava flows and deposits of volcanic material, such as ash. These different data sources can be combined to present a record of the frequency and magnitude of past volcanic events. The climatic effects of these events are normally expressed in terms of temperature changes, with the temperature data being derived from a similar variety of sources (for example, historic records, ice cores, and tree-ring reconstructions).

The relationships between volcanos and temperature have been examined in numerous ways: by direct comparison of the two records, by making a composite of the temperature record from several eruptions (using several months or years before and after each eruption), and by using numerical models to predict the effects of any one eruption. When we consider several eruptions together, the results suggest a global cooling of 0.2 to 0.3°C for 1 to 3 years after an eruption (Fig. 12-6). Individual eruptions producing a significant aerosol cloud can also result in a 1-to-3 year cooling of 0.3 to 0.7°C. Such eruptions include Tambora (Indonesia) in 1815 and Krakatau (Indonesia) in 1883. These effects are relatively small, but it is possible that the effects of larger or more-frequent volcanic events in the past were much greater than those of more-recent events. The eruption of Tambora in April 1815 was probably the largest in recent history, with about 50 km³ of erupted magma. The Toba (Indonesia) event of about 75 k.y. ago, which was probably a series of eruptions, dispersed an estimated 2800 km³ of magma. The Laki (Iceland) eruption in 1783 produced 12 km³ of basalt, whereas the Roza flow (northwest United States) in the Columbia River Group (dated at 14 m.y. ago) released 700 km³ of basaltic lava in a single event.

The 1815 Tambora eruption is probably the most renowned eruption in terms of its effect on climate. Large acidity peaks for that time in ice cores from both Greenland and Antarctica show a global distribution of the aerosol cloud, which is estimated to have been about 10^{11} kg. The following year, 1816, has been referred to as the "year without a summer" in Europe and northeast North America. Hudson Bay remained ice-covered that summer, and estimated average daily temperatures in this region were 5 to 6°C below the long-term average. Although these conditions were maintained only for 2 years, the cold weather from the winter of 1815–1816 through the summer of 1817 produced crop failures in China and bad harvests in India. The failure of the Indian harvest resulted in famine, which was followed by an outbreak of cholera that, over the next two decades, spread through Asia and Europe. However, the cold weather in the eastern Hudson Bay began with a series of anomalously cold years from the winter of 1811–1812, suggesting that not all the cold weather in 1816 was due to the eruption. Nevertheless, several other large eruptions occurred between 1811 and 1814, including the eruption of Vesuvius in 1813. The resulting cold temperatures could have been the cumulative result of the combined volcanic activity.

From the analysis of Greenland ice cores, we know that the interval from A.D. 1250 to 1500 and A.D. 1550 to 1700 were times of high volcanic activity, whereas A.D. 1100 to 1250 had far fewer volcanic events. The first two intervals coincide with the Little Ice Age; A.D. 1100 to 1250 lies in the Medieval Warm Period. Again, the evidence suggests a link between volcanic activity and climate, but that link is still not conclusive.

One problem with establishing the climatic impact of a volcanic eruption in historic times is that the perturbation due to the volcano is on the same order of magnitude as the natural (random) variability of the climate system. Despite the size of the Tambora eruption, it is

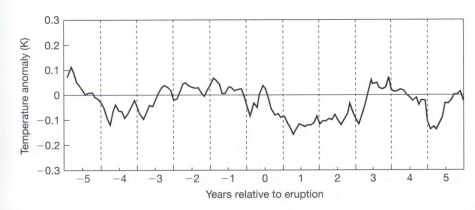

FIGURE 12-6

Cooling due to volcanic eruption: The global mean temperature changes for 5 years preceding and following a large volcanic eruption (at year zero). The temperatures are the average changes noted for six major eruptions: Krakatau, August 1883; Santa Maria, October 1902; Katmai, June 1919; Agung, March 1963; and El Chichón, April 1982. The effects of ENSO (discussed later in this chapter) on temperatures have been removed. (After Robock A., and J. Mao, 1995. "The volcanic signal in surface temperature observations," *Journal of Climate*, **8**:1086–1103.)

still difficult to separate the volcanic signal from other sources of climate variability in the temperature record. As a further example, although the eruption of Mt. Agung (Indonesia) in 1963 was about two orders of magnitude less than that of Tambora, it resulted in a 6°C stratospheric warming that lasted 3 years and an estimated surface cooling of about 0.3°C in the Northern Hemisphere. Conversely, the eruption of El Chichón (Mexico) in 1982 produced about the same amount of magma and the same amount of aerosol as that of Agung (although there was a greater concentration of sulfuric acid), resulting in a 6°C stratospheric warming, but the El Chichón event had very little effect at the surface. Estimates ranging from 0.1 to 0.5°C were made for the expected surface cooling from El Chichón, but the signal, if present, was entirely swamped by a particularly large climatic event known as *El Niño–Southern Oscillation* (ENSO), which will be explained later in this chapter.

Other Causes of Holocene Climate Change

We can invoke volcanic eruptions and changes in ocean circulation to explain some of the more extreme changes in climate that followed the end of the last glacial period, but they do not explain all of the observed changes—particularly the changes to warmer conditions. What else do we need to account for these other changes? Although a large number of theories have been put forward, the actual causes of Holocene climate change are not very well understood. One suggestion is that the Milankovitch orbital forcing, which we discussed in Chapter 11, also plays a role here, with the precession of Earth's spin axis having the greatest influence.

Orbital Changes and Greenhouse Variations. The seasonal and latitudinal distributions of insolation 18 k.y. ago were very similar to those of today (recall from Chapter 11 that the main precessional forcings are at 19 k.y. and 23 k.y.). Later in the Holocene, the precession changed in such a way that seasonality (the difference between summer insolation and winter insolation) increased in the Northern Hemisphere and decreased in the Southern Hemisphere. The atmospheric concentration of CO_2 was also at a minimum (200 ppm) 18 k.y. ago, increasing to about 265 ppm by the mid-Holocene.

These changes have been incorporated into a General Circulation Model that simulated global climate at 3-k.y. intervals throughout the Holocene. The model simulated many of the climate patterns that are indicated in the proxy record. At 18 k.y. ago, the model produced colder temperatures over the continental ice sheet, together with steep temperature gradients at the ice margin, but little change in the tropics. At 6 k.y. ago, temperatures in middle and high latitudes were generally warmer than present. There is no clear difference between the precipitation at 18 k.y. ago and that today, except for a large reduction over Southeast Asia. At 6 k.y. ago, how-

ever, summer rainfall was greater over the Sahara, the Middle East, and in the Southeast Asian monsoon region.

The broad pattern of change from the last glacial maximum through the Holocene Climatic Optimum to the present appears to be explained, therefore, by the same forcings that we outlined in Chapter 11. Explaining the Younger Dryas and the shorter-term changes within the Holocene (such as the Medieval Warm Period and the Little Ice Age), however, is not so simple. Volcanic activity and changes in the circulation of the North Atlantic provide part of the explanation, but there is also the possibility that some of the climate change was triggered by variations in our basic energy supply: the output of radiation from the sun.

Solar Variability. We have already seen that the output of the Sun has changed considerably over the long period of Earth's history. At the short time scales we are considering here, however, we tend to regard the solar output as constant. In fact, the output from the Sun probably varies slightly at *all* time scales—but is the solar variability sufficient to explain the observed climate changes?

The greatest changes in solar output at the decade-to-century time scale accompany variations in the number of sunspots. **Sunspots** are dark areas of lower-than-normal temperatures on the surface of the Sun. Imagine that magnetic field lines wind around the Sun like elastic bands. Where these bands are twisted together, the "knot" at the surface inhibits convection from below, resulting in an area of lower temperatures (sunspots). Telescopes were first used by Galileo to observe sunspots in about 1610, but reliable sunspot data exist only for about the past 150 years. The historic record shows an 11-year cycle in sunspot activity (11 years between the peaks in sunspot abundance) superimposed on a 22-year cycle of magnetic reversals. Just like the magnetic field on Earth, the Sun's magnetic field also experiences periodic reversals, but the Sun's magnetic reversals are more frequent and regular.

The pattern of sunspot activity is illustrated by the *butterfly diagram* in Figure 12-7, which plots the location of sunspots on the Sun's surface through time. In late 1933, there were very few sunspots, but the number increased into 1937. Sunspots appeared first at higher solar latitudes, but, as they increased in abundance, the concentration shifted toward the solar equator. Sunspots then decreased in number until the next minimum in 1944, when, again, the activity switched to the higher latitudes. Plotted in this fashion, the cloud of points in Figure 12-7 resembles a butterfly wing. Note that the period of cycle is not always exactly 11 years. Even when the period is exactly 11 years from minimum to minimum, the curve is not symmetric: Less time passes between a minimum and a maximum than between the maximum and the next minimum.

An increase in the abundance of sunspots, which are darker (cooler) areas on the Sun's surface, corresponds

FIGURE 12-7

The butterfly diagram, showing the distribution of sunspots over the Sun's surface. (From E. Chaisson and S. McMillan: *Astronomy: A Beginner's Guide to the Universe,* 2/e, 1998. Reprinted by permission of Prentice Hall, Upper Saddle River, N.J.)

with an *increase* in the amount of solar radiation Earth receives. The Stefan–Boltzmann law (Chapter 3) suggests that the opposite should occur: An increase in sunspot abundance should result in decreased luminosity and a drop in solar radiation. The reason for this behavior is that the sunspots are surrounded by bright areas of higher-than-normal temperature called **plages,** and the area of plages is larger than the area of spots. The net result is that the Sun is actually slightly brighter during periods when sunspots increase; hence, the amount of solar radiation emitted also increases. In reality, the dark areas of the sunspots and the bright areas of the plages almost cancel each other out.

How does this variation in sunspot activity relate to Earth's climate? Studies have discovered an 11-year or a 22-year cycle in numerous climate records (records of regional floods, droughts, temperatures, and so on). However, the change in solar output over the 11-year cycle is very small. Satellite measurements for the 1980s show that the incoming solar radiation decreased by only about 0.1% from the maximum to the minimum in sunspot activity. The direct climatic consequences of such a change are so small that they would be undetectable. At present, there are no commonly accepted mechanisms for explaining how such a small change in insolation could have a measurable impact on climate.

However, if we plot the mean global temperature changes over the past 100 years and superimpose on that

the *length* of the sunspot cycle, we find a near-perfect match (Fig. 12-8). This correlation suggests a direct relationship between the sunspot cycle and fluctuations in Earth's temperature. But is the correlation real or only circumstantial? It has been suggested that the length of the sunspot cycle is related in some fashion to the efficiency of convection in the solar interior. More energy is transported upward when the cycle length shortens, thus raising the surface radiating temperature of the Sun and increasing solar luminosity. However, there is no evidence of this transport, and no physical theory or model predicts that the sun should act in this way. Note also that the choice of scale on the vertical axes in Figure 12-8 is arbitrary, making the apparent match more obvious. However, this match may be purely coincidental, and there may be no physical relationship between these two parameters.

Systematic records of sunspot activity were initiated in 1848 by Rudolf Wolf, director of the Bern Observatory, after an amateur astronomer, Heinrich Schwabe, published sunspot observations from 1826 to 1843. Wolf also compiled all of the pre-1826 data that were available, extending the record of sunspot abundance back to 1700. He also published the years of sunspot minima and maxima from 1700 back to the invention of the telescope by Galileo in 1610. The pre-1826 data are far from reliable, but they do indicate an interval between 1645 and 1715

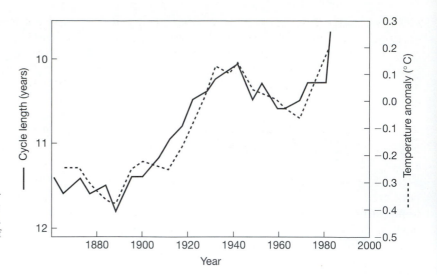

FIGURE 12-8

Variations of the sunspot cycle length (solid) and Northern Hemisphere temperature anomalies from 1861 to 1989. (After Friis-Christensen, E., and K. Lassen, 1991. "Length of the solar cycle: An indicator of solar activity closely associated with climate," *Science,* **254:**698–700.)

when very few sunspots were recorded. The low sunspot abundance during this interval was noted by astronomers at the time and was discussed in detail by two solar astronomers in the late 19th century, Gustav Spörer and E. W. Maunder. This period of low sunspot activity is now referred to as the **Maunder Minimum.**

Large groups of sunspots, which occur during periods of maximum solar activity, can be seen by the naked eye. Most of the earliest recorded sunspot observations, dating back to 28 B.C., were made in Japan, Korea, and China. More-reliable data come from another form of proxy data: the ^{14}C content of tree rings. The variations in solar output over the sunspot cycle cause changes in the magnetic properties of the solar wind (the output of energetic particles from the Sun), which is thought to affect the production of ^{14}C in Earth's upper atmosphere. High sunspot activity increases the magnetic field strength of the solar wind, causing a greater deflection of galactic cosmic rays away from Earth. The reduction in number of cosmic rays entering Earth's atmosphere results in low ^{14}C production. Conversely, low sunspot numbers result in high ^{14}C production. The ^{14}C content recorded in annual tree rings is thought to be related to the amount of ^{14}C produced in the atmosphere.

The naked-eye observations and the proxy data indicate two other intervals when few sunspots occurred: The **Spörer Minimum** (1450–1534) and the **Wolf Minimum** (1282–1342). The fact that the Maunder and Spörer minima coincide with one of the major climate changes in the Holocene, the Little Ice Age, has led some researchers to speculate that the two may be related. Furthermore, the 12th and 13th centuries (the Medieval Warm Period) was the interval with the greatest sunspot activity in the record. Detailed studies of ^{14}C tree-ring records have also shown that mid-latitude glaciers advanced and temperatures fell when sunspot numbers were low and that temperatures increased and glaciers retreated when sunspot numbers were high. Unfortunately, not all of the changes in the ^{14}C record match suspected global climate changes.

Might the sunspot cycle have had a greater effect on solar luminosity in the past than it does today? At first glance, the answer is yes. Observations show that the Sun is significantly less variable in total light output than are other stars of similar magnetic activity, suggesting that our Sun could have been more variable in the past than it is now. However, the level of magnetic activity necessary to achieve this greater variability has not been observed in our Sun since the beginning of reliable sunspot records, so there is little chance that solar variability can explain climate changes over the past 150 years. In addition, the proxy data suggest that the present-day level of solar variability is as high as any achieved in the past several millennia. The changes in solar output estimated for the Holocene are still less than the amount estimated to be necessary for the observed climate changes.

In summary, we have considerable evidence that solar activity varies over a wide range of time scales, with the most-regular variation being an approximate 11-year cycle in sunspot activity. Recent observations indicate that the solar output over this cycle varies only slightly, and this difference is not sufficient to explain observed variations in climate. Yet, if we look at variations in the length of the sunspot cycle, we see a close match with global temperature changes over the past century. There is also clear evidence that the major climate changes that have taken place over the past 1000 years coincide with extremes in solar activity (as indicated by the number of sunspots observed). Although we have no direct evidence that these changes in solar activity result in variations in solar luminosity large enough to explain the climate changes, the possibility exists.

We can thus present a picture of climate change over the past 20,000 years in which the broad trends from the last glacial maximum (at 18 k.y. ago) to the present are explained by Milankovitch orbital variations, possibly modified on the decade-to-century time scale by changes in solar activity. On shorter time scales, further variablity may be introduced by changes in ocean circulation and volcanic activity. The nature of the climate changes that occurred at any point in time may be a result of some combination of various factors. The Little Ice Age, for example, was a time of low solar activity, with extreme cold spells possibly caused by periods of enhanced volcanic activity.

Present-Day Climate Variability

Figure 12-1d presents a compilation of mean global temperatures since 1860. The record shows a steady rise in temperature through the first part of the 20th century, peaking in the 1940s (encompassing the Dust Bowl period in the United States). Temperatures then dropped in the 1960s and 1970s before rising again in the 1980s. In Figure 12-8, we showed that the temperature record closely matches the length of the sunspot cycle. Yet, the temperature trend is also consistent with the suggestion of CO_2 induced global warming. As we discussed in Chapter 1, we can explain the 1940–1970 temperature decrease by increases in sulfate aerosols as a result of the burning of coal. Nevertheless, the causes of 20th-century climate trends are still open to debate; thus, for the remainder of this chapter, we are going to turn to the year-to-year variability in the record.

Figure 12-9 is a schematic illustration of the major components of the climate system. We have seen that as we change time scales, different parts of this system take on different levels of importance. Over very long time scales,

the input of energy to the system is variable, and the climate system is regulated by silicate weathering, plate tectonics, and changes in atmospheric composition. The distribution of oceans and continents changes on time scales of tens of millions of years, and the discussions in Chapters 4 and 5 indicate that this distribution would have had some effect on climate. We saw in Chapter 11 that on time scales of thousands to hundreds of thousands of years, climate changes are associated with atmospheric composition, Earth's orbit about the Sun, and changes in the extent of large ice sheets. In the first part of Chapter 12 we considered several processes that affect climate on the order of decades to centuries, or thousands of years. When we turn to time scales of years to tens of years, however, we see that most of the features that are significantly variable at long time scales are not significant at short time scales. Similarly, although the land surface can change (through cultivation, deforestation, and so on), these changes are not likely to have a major global impact on a year-to-year basis. Given this, all we have left to consider are the atmospheric state itself (temperature, cloud cover, and so on), and the condition of the oceans.

In practice, the state of the atmosphere is not particularly important in this context. Recall from Chapter 4 that the atmosphere transports heat very rapidly, and any anomalies are quickly dissipated. The atmosphere has very little "memory" of change. In contrast, the oceans absorb and store large amounts of heat, and they release this heat very slowly. Consequently, the ocean's memory of any change is much longer than the atmosphere's. Transient anomalies that develop in sea-surface temperatures can

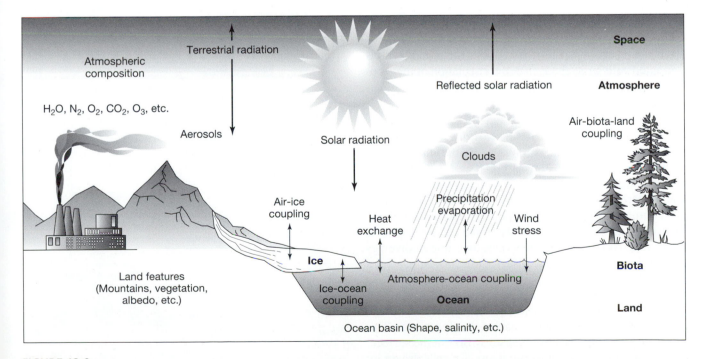

FIGURE 12-9

Schematic representation of the climate system. (After Schneider, S., and R. Londer, 1984. *The Coevolution of Climate and Life*. San Francisco: Sierra Club Books.)

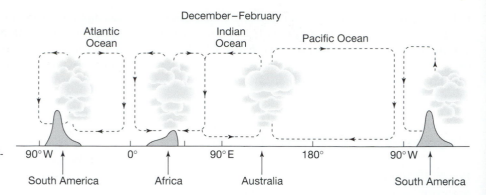

FIGURE 12-10

The east–west circulation in the equatorial troposphere. The shaded areas represent heavy precipitation.

be expected to have a lingering impact on climate for some time afterward. The oceans, therefore, are the most likely place to look for processes that might cause climate anomalies on the interannual or decadal time scales.

So, where do we look in the oceans? One obvious place to start is in the tropics, where we find the large convective towers of the Hadley cells that drive the atmospheric circulation over a large portion of the planet. We will also look at the opposite extreme, and examine the role of the polar oceans in short-term climate variability.

El Niño–Southern Oscillation (ENSO) Events

The name **El Niño** was originally given to a warm ocean current that appeared off the coasts of Peru and Ecuador. The current flowed for only a few weeks, and, because it usually occurred near Christmastime, local fishermen named it El Niño after the Christ child. The name has taken on a different meaning more recently; it is now used by researchers to describe a major shift in the oceanic circulation that occurs in this region every 2 to 10 years. This broad oceanic shift is associated with large changes in the circulation of the tropical atmosphere, which give rise to significant climate anomalies over much of the tropics and midlatitudes.

The Equatorial Atmospheric Circulation. Superimposed on the north–south Hadley circulation (described in Chapter 4) is a significant east–west circulation in the troposphere that is most prevalent over the equatorial Pacific. The western equatorial Pacific has the highest sea-surface temperatures on the globe. This region, which encompasses Australia and Indonesia, is a site of intense atmospheric convection. As is true of Hadley cells, the rising air diverges at high altitudes, but, in this case, we are concerned with a component of the flow that moves eastward and westward along the equator rather than northward and southward (Fig. 12-10). The eastward-moving air crosses the Pacific, where it subsides off the west coast of South America. The circulation is completed by an easterly flow at the surface. This circulation is linked to other smaller cells driven by convection over South America and Africa. Figure 12-10 shows the normal pattern of the equatorial east–west cir-

culation. The figure also indicates the normal pattern of precipitation, with heavy precipitation in the convective regions and drier conditions in the areas of subsidence. The circulation cells produce an oscillation in the sea-level pressure distribution between the western and the central/eastern portions of the tropical Pacific Ocean. When pressures are low in the west, they tend to be higher in the east, and vice versa. This oscillation in sea-level pressures is referred to as the **Southern Oscillation** (SO).

The Ocean Circulation. The persistent easterly wind at the surface in the Pacific Ocean produces a westward-flowing ocean current, which results in the water piling up in the western part of the ocean. This causes very warm water to accumulate in the western Pacific (Fig. 12-11). Figure 12-11

(a)

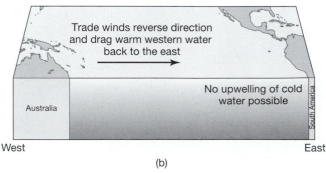

(b)

FIGURE 12-11

The ocean surface layer in the tropical Pacific Ocean.

FIGURE 12-12

Schematic diagram of the interaction between the atmosphere and ocean in the tropical Pacific Ocean.

shows that water piles up in the west, which causes the ocean surface to slope downward from west to east. The slope is exaggerated in the diagram; the difference in surface elevation from west to east is only on the order of a couple of meters. This east-to-west movement of water thickens the warm surface layer in the west and thins it in the east. The thinner surface layer in the east allows the upwelling of colder, nutrient-rich water from below, which promotes high levels of biological productivity and large fish populations.

We can regard this pattern of atmospheric and oceanic circulation as the norm, but in any year we see substantial differences. In some years this pattern intensifies: The sea-surface temperatures in the central and eastern Pacific are colder than normal, and convection over Indonesia is enhanced. These conditions are referred to as **La Niña** conditions—the pattern is similar to the "normal" conditions, but the circulation is enhanced. More-drastic

changes occur when the pattern breaks down in what is referred to as an **El Niño—Southern Oscillation** (ENSO) event. The Southern Oscillation may also be referred to as being in a "cold" phase during La Niña (or *anti-ENSO*) events and in a "warm" phase during ENSO events.

Before we describe what happens when an ENSO event occurs, let us look at this circulation again. The convection in the western Pacific occurs as a result of the high sea-surface temperatures, but the high sea-surface temperatures are a result of the atmospheric circulation, which is driven by the convection. In other words, it is a classic chicken-and-egg situation. It is not a case of the ocean forcing the atmosphere or vice versa. Instead, it is a single integrated system (Fig. 12-12). If we perturb the system at any point, we should expect to see changes throughout all of its components.

ENSO Events. Scientists are still debating what causes ENSO events to occur. For the purposes of this discussion we will simply break into the cycle in Figure 12-12 and ask what happens if, for some reason, there is a decline in the strength of the easterly winds.

If these winds weaken or reverse direction, which happens in some ENSO events, there is nothing to restrain the pile up of warm water that has accumulated in the western Pacific. This water then comes sloshing back across the ocean in what is known as a *Kelvin wave.* It takes about 60 days for this wave to travel back across the Pacific. When it does, it has two major consequences. First, it shifts the pool of high sea-surface temperatures from the western to the central Pacific (Fig. 12-13; Plate

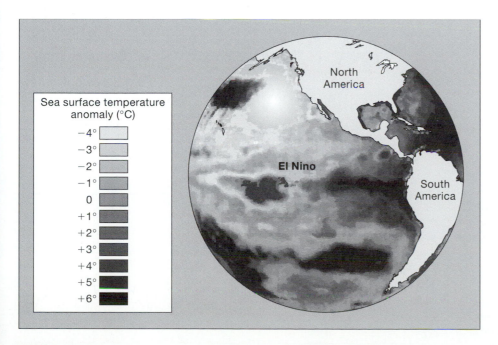

FIGURE 12-13

Sea-surface temperature anomalies in the central and eastern Pacific during the 1997–1998 ENSO event. [See color section]

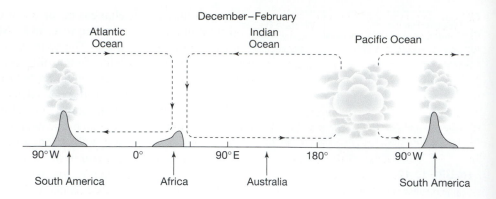

FIGURE 12-14

Atmospheric circulation during an ENSO event.

13), which then completely changes the atmospheric circulation. Second, it shuts off the upwelling in the eastern Pacific, which has drastic consequences for biological productivity. The loss of the nutrient-rich water leads to a massive die-back of marine organisms and the bird life that feeds on them.

The changes in the atmospheric circulation are shown in Figure 12-14. The greatest area of convective activity during ENSO events lies over the central Pacific. The rising air diverges to the east and west, meeting and subsiding over Africa, although there is also localized uplift on the western side of the Andes. In a non-ENSO year, there is low pressure (rising air) at the surface over Australia/Indonesia and high pressure (subsiding air) at the surface in the central

and eastern Pacific (Fig. 12-10). In an ENSO year, this pattern reverses: Pressure increases over Australia and decreases in the central Pacific. We can calculate the pressure difference between these two locations (Fig. 12-15a) and plot this difference through time to produce the **Southern Oscillation Index** (SOI) (Fig. 12-15b). This index is a measure of the pressure difference between the western and eastern parts of the tropical Pacific Ocean. Strong positive values indicate La Niña (non-ENSO) conditions; strong negative SOI values indicate ENSO conditions, which have occurred throughout the interval for which recorded data are available. We also find evidence from ice cores that ENSO events of various magnitudes have been a feature of the climate system for at least the past 500 years.

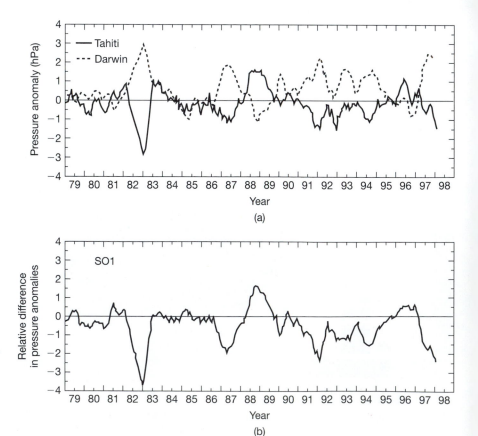

FIGURE 12-15

(a) The sea-level pressures at Tahiti and Darwin, Australia, and (b) the Southern Oscillation Index (SOI). The SOI is computed from the sea-level pressure differences. Negative values of the SOI indicate warm (El Niño) events. Note the strengths of the 1982–1983 and 1997–1998 events. (After the NOAA Climate Prediction Center.)

Climatic Impacts of ENSO. The most dramatic impacts of ENSO events are seen in their effects on rainfall patterns. In non-ENSO years, summertime convection and rainfall occur over Australia and Indonesia. There is also convective activity and rainfall over equatorial Africa and the Amazon Basin. In contrast, there is subsiding air (dry conditions) west of the Andes. However, in ENSO years the dominant convective region shifts toward the central Pacific, and convection and rainfall over Australia and Indonesia diminish. Figure 12-14 shows that there is also subsidence over Africa and some localized convection over the western Andes. The result is drought in central America, Brazil, Australia, Indonesia, and southeast Africa and anomalously high rainfall amounts in the central Pacific and on the western slopes of the Andes in Ecuador and Peru. These high rainfall amounts typically result in floods and landslides, with their accompanying high levels of soil erosion. ENSO events also appear to have some effect on the monsoon circulation over India, resulting in increased rain over southern India and reduced rainfall over northern India and the Himalayas.

The impact of ENSO events is not confined to the tropics. The changes in atmospheric circulation can also influence the midlatitudes. We saw in Chapter 4 that the location and strength of midlatitude weather systems are controlled in part by the subtropical highs and in part by the sea-surface temperature gradients. Because both of these factors change during ENSO events, it is not surprising that they should have some effect on midlatitude climate. The general pattern of temperature and rainfall anomalies associated with an ENSO event are shown in Figure 12-16 for the Northern Hemisphere winter. In practice, each ENSO event tends to be somewhat different from others in the record. For example, the strength and location of the sea-surface temperature anomalies are not always the same, and ENSO events may last 1 year or may extend over several years. As a result, the pattern of the climate anomaly in the tropics tends to be fairly consistent, although it may vary in magnitude, but the midlatitude effects are highly variable. The western United States, for example, was very dry in the 1976 event but very wet in 1982.

It is interesting to note that the Southern Oscillation has been in a warm phase (low SOI values) almost continuously since 1976 and that there were five ENSO events between 1976–1977 and 1997–1998. There was only one La Niña event during this interval. Furthermore, the system was in an almost constant El Niño mode from 1990 through the winter of 1997–1998 (except for a short break in 1995–1996). Climate model simulations that use the observed sea-surface temperature distribution in the tropical Pacific recreated much of the temperature variability of the 1970s and 1980s, including the global shift to warmer temperatures in the 1980s. Again, although the temperature record is consistent with the possibility of enhanced greenhouse warming, we can still explain the recent pattern in terms of the natural variability of the system. Nevertheless, it is also possible that greenhouse-warming-induced changes in the atmosphere or ocean are causing the extended El Niño pattern.

Sea Ice and Climate

In the last part of this chapter we turn to the polar oceans to examine climate variability induced by sea-ice modifications to atmosphere–ocean interactions at high latitudes. In earlier chapters we hinted at the importance of the ice distribution when we discussed the ice–albedo feedback

December–February

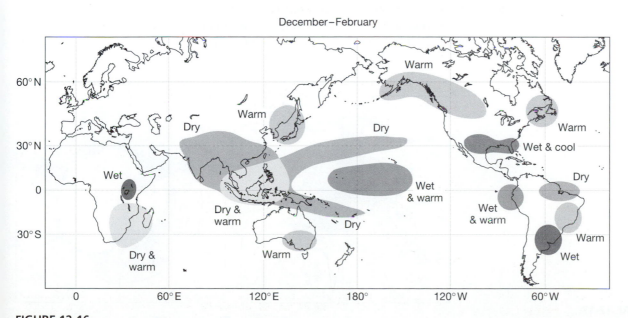

FIGURE 12-16

Rainfall and temperature anomaly patterns associated with an ENSO event. (After the NOAA Climate Prediction Center.)

The 1982–1983 and 1997–1998 ENSO Events

The 1982–1983 ENSO event was one of the most severe of the 20th century. It developed unexpectedly. The normal early warning sign of changing tropical sea-surface temperatures was missed because of the preceding eruption of a Mexican volcano (El Chichón). The volcanic aerosols reduced the outgoing infrared radiation, resulting in lower satellite-derived sea-surface temperature measurements. The resulting pattern of climate anomalies was also slightly different from a "normal" ENSO: They were much more intense than earlier ENSO events. The impact of this particular event was considerable, with major droughts and floods occurring throughout the tropics. It is estimated that climate-related catastrophes resulting from the 1982–1983 ENSO event left more than 1000 people dead and caused almost $9 billion worth of damage:

- Ecuador and northern Peru experienced floods and landslides that left 600 dead and resulted in crop and property losses totalling approximately $400 million. Guayaquil, Ecuador, had 20 times its normal rainfall in May 1983.

- In Indonesia, there were crop failures and starvation.
- In Botswana, the ENSO-induced drought followed two previous years of drought and eventually led to the loss of thousands of livestock.
- Eastern Australia suffered the worst drought of the century. Animal feed supplies were so diminished that thousands of sheep, dying of starvation, had to be shot. The dry conditions resulted in huge dust storms, one of which deposited 11,000 tons of topsoil on the city of Melbourne.
- Tahiti and French Polynesia had last experienced a typhoon at the beginning of the 20th century. The warm-water pool that formed in the central Pacific during the 1982–1983 ENSO event generated several large storms, and the islands were hit by six typhoons in 5 months.
- In the United States, increased rain in the Midwest resulted in the flooding of the Mississippi River. An increase in storms on the west coast resulted in severe flooding and landslides in California, where 10,000 houses were lost or

damaged and farm losses totalled half a billion dollars. There was a record snowfall in the Rockies, which, when it melted, resulted in flooding in Salt Lake City and along the lower Colorado River.

It is likely that the 1997–1998 event will have been even more severe. The effects of this event are still being evaluated, but estimates so far indicate extensive impacts in the equatorial Pacific, North and South America, and East Africa:

- Flooding, mud slides, and disease killed more than 80 people in Peru, and flooding in Ecuador resulted in 90 deaths and the evacuation of 22,000 people.
- The central Pacific had eight tropical cyclones, but only two the year before.
- Drought conditions produced hunger in Indonesia and Papua New Guinea. The extreme drought in Indonesia resulted in forest fires that burned over 1 million acres of forest.
- In the United States, this event produced heavy rains, flooding, and mudslides along the California coast but a very mild winter in the northeastern region.

and bottom water formation. In this section we will discuss the seasonal distribution of sea ice and some of the factors that control this distribution, and we will examine some of the ways in which changes in the ice cover affect the polar ocean climate. These effects are particularly interesting because, as we will see in Chapter 13, they tend

to dominate the high-latitude response to increasing atmospheric CO_2 levels in global climate models.

Seasonal Distribution of Sea Ice. The seasonal distribution of the sea ice of polar oceans is shown in Figure 12-17. The seasonal range of the ice cover is extreme: The

Northern hemisphere sea ice   Southern hemisphere sea ice

FIGURE 12-17

The seasonal distribution of sea ice in the Northern and Southern Hemispheres.

Absolute Average   Average Absolute
Maximum   Minimum

Northern Hemisphere ice cover almost doubles in size from approximately 8.5×10^6 to 15×10^6 km² between summer and winter; the Southern Ocean ice cover grows from 4×10^6 to 20×10^6 km². Sea ice forms when the temperature of the ocean surface drops below the freezing point ($-1.8°C$ for typical ocean salinities). Sea ice grows in thickness by new ice forming on the bottom of the ice pack, typically reaching thicknesses of about 1 m during a single winter. The ice cover melts and freezes seasonally at the equatorward margins of the pack but lasts throughout the year at higher latitudes. This permanent *pack ice* in the Arctic Ocean reaches equilibrium thicknesses of about 5 m; the Southern Ocean ice is much thinner, because very little survives through the summer melt.

Several features of this ice distribution are important to our discussion. The large seasonal range is a result of thermodynamic controls on the ice cover (that is, the ice cover is responding to the seasonal changes in temperature). The asymmetric nature of the ice margin in the Arctic Ocean, however, reflects the influence of the continental configurations and the ocean circulation. We find the ice margin extending well to the south in both the western North Atlantic and western North Pacific. At the eastern margins of both oceans, open water extends much farther north off Norway and Alaska. Note that the shape of the ice margin reflects the pattern of ocean currents you saw in Chapter 5. The warm water of the North Atlantic Drift and the Kuroshio Current prevent the ice margins from extending farther south in the eastern oceans, whereas the East Greenland Current, the Labrador Current, and the Oyoshio Current bring colder water and carry increased ice cover southward in the west. The ice distribution helps illustrate one of the important facts about the sea-ice cover: Sea ice moves. The ice does not simply form and melt in place; it is in a constant state of motion. The typical pattern of sea-ice drift in the Arctic Ocean is shown in Figure 12-18. The map reveals two primary features of the ice circulation: a clockwise circulation, or gyre (the Beaufort Sea Gyre), in the west and the Transpolar Drift Stream, which flows across the pole from the Siberian coast to the East Greenland Sea. This ice flows into the East Greenland Current, where it is transported southward out of the Arctic Ocean. On average it takes about 5 years for ice that forms off the Siberian coast to be transported across the Arctic Basin. Ice that forms in the Beaufort Gyre, however, may last much longer, possibly circling the western Arctic several times before melting in the summer or becoming caught in the Transpolar Drift.

Moved around by winds and ocean currents, the ice is broken into individual pieces *(ice floes)* that continuously join and break up. Where floes collide, large mounds of ice are thrust up into pressure ridges; where floes move apart, they produce areas of open water called

FIGURE 12-18

The typical pattern of sea-ice drift in the Arctic Ocean.

leads or *polynyii* (singular, *polynya*). Leads are linear open-water features, and polynyii are irregular areas of open water. In winter, these open-water areas freeze very rapidly, but because the ice pack is so dynamic, there is always a small amount of open water present, even in midwinter. This is important for two reasons: One, the presence of open water allows for the constant production of new ice thus releasing salt to the upper ocean and increasing the density of the surface layer. Two, the open water has a considerable impact on the Arctic energy budget: The ocean loses heat to the atmosphere about 100 times faster from the open water than it does through the insulating ice cover.

Ice–Climate Interactions. We have seen in Chapters 4 and 5 that the sea-ice cover forms as a response to ocean temperatures and that the distribution is then modified by winds and ocean currents. However, the ice cover is not simply a response to the climate: The ice cover modifies the atmospheric and oceanic circulations. We have seen that ice production in the North Atlantic contributes to the formation of North Atlantic Deep Water, which is a major factor in driving the oceanic thermohaline circulation. We have also discussed the ice–albedo feedback, whereby a change in temperature causes a change in the ice cover and the surface albedo, which further modifies the temperature. Recall that this is a positive feedback: Increasing temperature → decreasing ice cover → decreasing albedo → increasing temperature. This feedback can operate year-round near the sunlit ice margins but can play a role only in summer at higher latitudes. (Recall that there is no solar radiation at high latitudes in winter.) Another sea-ice feedback, however, is important year-round. The ocean surface is a source of thermal en-

FIGURE 12-19

The sea ice–ocean heat flux feedback loop.

ergy and latent heat for the polar atmosphere, and the most rapid heat transfer from the ocean to the atmosphere occurs through the areas of thin ice or open water. Thus, we can picture a second positive feedback (Fig. 12-19): Increasing temperature leads to decreasing ice cover, which leads to increasing ocean heat flux to the atmosphere, which leads to increasing temperature. Both the ice–albedo feedback and the ocean heat-flux feedback are related to the ice extent and concentration (that is, the area of ice-covered ocean). The heat-flux feedback is also controlled to some degree by ice thickness: The heat loss from the ocean to the atmosphere increases as ice thickness decreases. The extent of the ice cover, the ice concentration, and the ice thickness are all determined by the energy budget and the ice dynamics. If we want to generate numerical models of these processes, we need to be able to account for both thermodynamic and dynamic interactions among the ice, ocean, and atmosphere. Several models with this degree of sophistication exist, but, as we will see in later chapters, few have been incorporated in atmospheric or coupled atmosphere-ocean global climate models.

Chapter Summary

1. Earth's climate changes over all time scales from decades and centuries to millions of years.
   a. Since the last glacial maximum, Earth has undergone several changes in climate that appear to be global in extent, even though the magnitude of the changes varies from region to region.
   b. Although the evidence for these changes is conclusive, our explanations for why they occurred are ambiguous.
2. As Earth warmed up from the effects of the last glaciation, a sudden and very rapid climate reversal occurred (the Younger Dryas event), resulting in the readvance of ice sheets and glaciers over much of northern Europe and other parts of the globe.
   a. Several hypotheses have been proposed to explain the Younger Dryas, the most likely of which invloves the switching on and off of the thermohaline circulation in the North Atlantic Ocean.

   b. Regardless of the exact cause of this shift in climate, the Younger Dryas is of particular interest because recent ice-core data show that the change occurred very rapidly, over years to decades; the apparent ability of the climate system to shift rapidly between two very different states poses some interesting questions about how the climate may respond to future greenhouse warming.
3. Two major warm periods occurred during the Holocene: one in the mid-Holocene and the other, the European Medieval Warm Period. The forcing for these events is not clear, but they can probably be attributed to the Milankovitch orbital forcing discussed in Chapter 11, with the precession of the poles having the greatest influence.
4. A return to colder conditions, the Little Ice Age, took place from the late 1500s through the mid-19th century. One of the

contributing causes of the colder temperatures during the Little Ice Age may be the effect of a higher frequency of volcanic eruptions.

   a. The sulfur dioxide injected into the stratosphere by an eruption hydrolyzes, forming sulfuric acid droplets, which both reflect solar radiation and absorb some of the long-wave radiation emitted from the troposphere.

   b. The combined effect of an increased albedo and increased long-wave absorption is to heat the stratosphere and cool the troposphere.

   c. The result is lower-than-normal surface air temperatures for 1 to 2 years after a major eruption.

5. Further factors contributing to a change in climate during the Little Ice Age may include changes in the abundance of sunspots and the length of the sunspot cycle, which may affect the solar output and Earth's climate.

   a. Although there is no direct evidence of a climate forcing by sunspots, two of the longest time periods for which no, or few, sunspots were observed occurred during the Little Ice Age.

6. On century-long time scales, Earth's climate experiences large variations over intervals of years to decades. Most of the forcing for these variations originates in the oceans.

   a. ENSO events and sea ice–atmosphere–ocean interactions are two examples of how the ocean may have an impact on climate on these time scales.

   b. ENSO events are large-scale changes that occur in the atmospheric and oceanic circulations of the tropical Pacific Ocean and have a significant impact on tropical precipitation patterns. They also seem to produce anomalous climate conditions in different parts of the midlatitudes.

   c. At higher latitudes, the interactions taking place between the ocean and the atmosphere are modified by the sea-ice cover through several positive feedback processes.

   d. The global consequences of these feedbacks are not as great as those of the ENSO events, but they still have a significant impact on polar and midlatitude climates.

Key Terms

dendrochronology	Little Ice Age	Southern Oscillation Index
El Niño	Maunder Minimum	Spörer Minimum
El Niño–Southern Oscillation	palynology	sunspots
European Medieval Warm Period	plages	uniformitarianism
Holocene	proxy data	volcanic aerosol forcing
Holocene Climatic Optimum	residence time	Wolf Minimum
La Niña	Southern Oscillation	Younger Dryas

Review Questions

1. What is the Holocene?
2. Why are climatologists usually not very interested in "average" climate conditions?
3. What are proxy data? Describe several examples of proxy climate data.
4. Briefly describe the Younger Dryas, the Holocene Climatic Optimum, the European Medieval Warm Period, and the Litte Ice Age.
5. Explain why the formation of North Atlantic Deep Water might have played a role in causing the Younger Dryas event.
6. How do volcanoes affect climate?
7. What are sunspots? Why are they thought to have a possible effect on climate?
8. Use a diagram to help describe the "normal" east–west atmospheric circulation in the tropical Pacific Ocean.
9. Explain what happens to the atmospheric and oceanic circulations in the tropical Pacific during an ENSO event.
10. How do ENSO events affect precipitation patterns in the tropics? Explain.
11. Describe the major feedback processes between sea ice and climate.

Critical-Thinking Problems

1. In this chapter we suggested that the most likely place to look for processes that would force short-term climate variability is the ocean, because the ocean has a longer "memory" of its previous states than does the atmosphere. As an example we pointed out that temperature anomalies persisted for much longer in the ocean. Atmospheric temperature anomalies are dispersed very rapidly, so they affect the climate for only a very short time. As a further example of the rapid transfers that take place in the atmosphere, use the information from the hydrologic cycle from Figure 4-23 to calculate the residence time of water in the atmosphere. How many times a year does the water in the atmosphere go through the cycle of evaporation and precipitation?

2. In the section "Sea Ice and Climate," we described a positive feedback among ice concentration, ocean heat flux, and temperature. However, we can also assume that there will be considerable evaporation from the open water surface, which, because of the cold air temperatures over polar oceans, should result in condensation and the formation of low-level clouds. Clouds have a high albedo, which tends to

lower temperatures, but they are also very efficient at trapping infrared radiation (that is, they have a strong greenhouse effect). Given this, modify Figure 12-19 so that it includes the effect of clouds. You may need to construct different diagrams for summer and winter.

Further Reading

General

Wigley, T. M. L., M. J. Ingram, and G. Farmer, 1981. *Climate and History*. Cambridge, England: Cambridge University Press.

Knox, P. N., 1992. "A Current Catastrophe: El Niño." *Earth,* **1**(5): 31–37.

Advanced

Rind, D., and J. Overpeck, 1993. "Hypothesized causes of decade-to-century-scale climate variability: Climate model results." *Quaternary Science Review,* **12:**357–374.

Chapter 13
Global Warming

A mountain glacier. (From Gilbert S. Grant, Photo Researchers, Inc.)

Key Questions

- How does the amount of CO_2 produced by the burning of fossil fuels compare with natural CO_2 sources?
- By which processes will anthropogenic CO_2 eventually be removed from the atmosphere?
- How much are atmospheric CO_2 and other greenhouse gases expected to rise over the next few decades to centuries, and how will this rise affect Earth's climate?
- How is sea level projected to change over the next century?
- How will forests and other ecosystems respond to atmospheric CO_2 increases?
- What policies might be adopted to reduce future CO_2 emissions?

Chapter Overview

Atmospheric CO_2 concentrations are increasing as a consequence of the combustion of fossil fuels and the cutting down of tropical forests. If no actions are taken to reduce emissions, CO_2 levels are expected to more than double over the next century and could increase by a factor of 7 to 10 over the next few centuries. Other greenhouse gases, including CH_4, N_2O, and various CFCs, are also increasing in concentration in the atmosphere. Temperature increases within the next century are predicted to be only a few degrees Celsius, but long-term climate changes could be substantially larger than this. Even a few degrees of warming could have substantial effects on water availability, on speciation within terrestrial ecosystems, and on sea level. Whether to take action to limit greenhouse gas emissions is an issue that is currently being hotly debated.

Introduction

We have seen in earlier chapters that Earth's climate has varied on a number of different time scales. Over the past few billion years, the Sun has brightened considerably and Earth's climate has gone through a series of warm and cold periods. But the system has always remained within a range conducive to the existence of liquid water and to the continued presence of life. Over the past few million years, the climate has oscillated between glacial and interglacial intervals triggered by variations in Earth's orbit and amplified by internal feedbacks within the climate system. Over the past hundred years, the climate has warmed by about 0.5°C, but this change has been superimposed on other, shorter-term warming and cooling trends that make interpretation of the recent climate record difficult.

In this chapter, we use the information and techniques described earlier to try to predict how climate may change over the next few decades to millennia. The possibility of future global warming is, after all, why both scientists and policymakers are so interested in climate and why we have devoted so much attention to the subject. In trying to make such a prediction, we will show why an interdisciplinary approach is essential and how a knowledge of climate history can help compensate for the deficiencies in our physical models of the climate system.

Carbon Reservoirs and Fluxes

We saw in Chapter 1 (Fig. 1-2) that atmospheric CO_2 concentrations have increased by about 25 percent since the beginning of the 19th century. This change is larger than any natural fluctuation that has occurred since the retreat of the glaciers 11,000 years ago and is almost certainly attributable to human-induced causes, principally the burning of fossil fuels and deforestation. To understand why the human influence is so noticeable, it is useful to compare the sizes of the various carbon reservoirs and the rates at which carbon cycles in both the natural and perturbed systems.

Natural Reservoirs and Fluxes

The carbon cycle is diagrammed in Figure 13-1. This diagram combines the essential features of the organic and inorganic carbon cycles described in Chapter 7. It also includes a box labeled "fossil fuels," which includes several types of compounds—most importantly coal, oil, and natural gas. Fossil fuels are, in many respects, a natural part of the carbon cycle. (It is the rate at which they are being

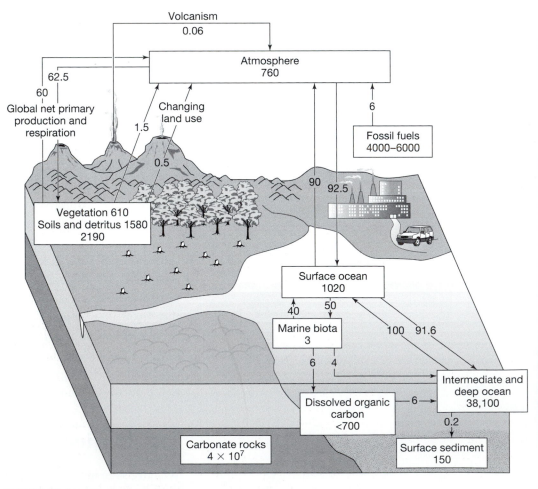

FIGURE 13-1

The major reservoirs and fluxes in the global carbon cycle. Units are Gton (C) for reservoir sizes and Gton(C)/yr for the fluxes. (After Climate Change, 1994: Radiative Forcing of Climate Change, Cambridge Univ. Press, Cambridge, p. 21.)

TABLE 13-1

Fossil Fuel Reservoir Sizes and Burning Rates		
Reservoir	*Size, Gton(C)*	*Burning rate, Gton(C)yr*
Coal	4000	2.5
Oil	500	2.5
Natural gas	500	1.0
Total	5000	6.0

oxidized that is unnatural.) They were created over many millions of years by the accumulation of dead organic matter in soils and sediments. Table 13-1 lists the relative amounts of different types of fossil fuel. Most of it is coal, which is stored in vast quantities in many parts of the world.

The total amount of fossil fuels is uncertain, because new reservoirs are continually being discovered and because whether a given reservoir is counted depends on the price of fuel. Estimates of the amount of fossil fuel that is economically recoverable at 1998 market prices range from 4000 to 6000 Gton(C), 7 to 10 times the amount of carbon that was present in the preindustrial atmosphere as CO_2, 590 Gton(C). This gives us our first indication of why the burning of fossil fuel is a potential problem: Simply, a lot of it is available. In Critical-Thinking Problem 1, we shall compute what would happen to atmospheric CO_2 concentrations if all the fossil fuel were to be burned instantaneously.

Of equal importance to the carbon reservoirs themselves are the fluxes that connect them. These fluxes are in many cases wholly disproportionate to the size of the reservoirs involved. For example, the huge carbonate rock reservoir is connected to the rest of the system by only a tiny flux, about 0.06 Gton(C)/yr. Shown in Figure 13-1, this volcanic CO_2 flux must, on average, equal the rate at which carbonate sediments are preserved. Despite its large size, this reservoir exerts little influence on the rest of the carbon cycle on any time scale shorter than hundreds of thousands of years. So, processes such as silicate weathering and volcanism, which were extremely important on geologic time scales, should have little effect on atmospheric CO_2 levels over the next few centuries.

The largest fluxes in the natural carbon cycle are the exchange of carbon between the atmosphere and the terrestrial biosphere, about 60 Gton(C)/yr, and between the atmosphere and surface ocean, about 90 Gton(C)/yr. The terrestrial biosphere, largely comprised of terrestrial vegetation and soils and detritus, exchanges CO_2 with the atmosphere by way of the short-term organic carbon cycle. In that cycle, the photosynthetic uptake of CO_2 by forests is balanced by the respiration and decay of plant material. This cycle is responsible for the seasonal fluctuations in atmospheric CO_2 (the Keeling curve) discussed in Chapters 1 and 7. The ocean exchanges CO_2 with the atmosphere by diffusion through the surface interface. Once in the surface ocean, CO_2 is taken up by marine photosynthesizers.

The net rate of photosynthesis in the oceans, about 50 Gton(C)/yr, is comparable to that on land, but it does not affect the atmosphere in the same way, because the amount of living biomass is very small: The marine biota contain only about 3 Gton of carbon, as compared with approximately 610 Gton of carbon in forests. Free-floating marine organisms and seaweed do not need the massive amounts of structural carbon required by trees. As a result, the marine organic carbon cycle is more closely balanced than the terrestrial cycle and does not contribute appreciably to the seasonal fluctuations of CO_2 in the atmosphere.

Rates of Fossil-Fuel Burning and Deforestation

The currently observed increase in atmospheric CO_2 concentrations is attributable, at least in part, to the combustion of fossil fuels. For natural gas (which is mostly methane), the combustion reaction can be written as

$$CH_4 + 2\,O_2 \longrightarrow CO_2 + 2\,H_2O.$$

The combustion reactions of coal and oil are more complicated, as these fuels consist of mixtures of more-complex hydrocarbons, but the main reaction products are the same, CO_2 and H_2O. The rate of fossil-fuel consumption is known fairly accurately (to within 10 percent or better), because records are kept by companies that produce fossil fuels and by the countries in which these companies operate. The 1998 value is about 6 Gton(C)/yr (Table 13-1). World population is currently about 6 billion, so this consumption rate amounts to 1 metric ton of carbon per person per year. The rate of fossil-fuel burning is about 10 times less than the rate of CO_2 exchange with the terrestrial biosphere but 100 times larger than the rate of CO_2 release by volcanos. Its effect on the atmosphere is disproportionately large, however, because this part of the global carbon cycle is not in balance.

Fossil fuel consumption is distributed unequally among the various nations of the world. As shown in Figure 13-2, most of the fossil fuel is burned in the Northern Hemisphere, with North America leading the way at about 1.5 Gton(C)/yr, or 25% of global emissions. The United States, with only about 5% of the world's population, accounts for some 22% of global CO_2 emissions. The large amount of CO_2 emitted reflects the high standard of living and energy-intensive lifestyle. Other countries are catching up, however. China, in particular, has a rapidly growing economy and an increasing appetite for fossil fuels to go with it. China recently surpassed Russia to become the second biggest coal-consuming nation in the world, after the United States. With its large population and extensive coal reserves, China could become the world's largest CO_2 emitter within the next two decades.

Of the 6 Gton(C)/yr produced by the combustion of fossil fuels, about 3 Gton(C)/yr is currently accumulating in the atmosphere. This number can be calculated directly from the observed 1.5 ppm/yr rate of atmospheric CO_2 increase. (We shall do so in Critical-Thinking Problem 2.)

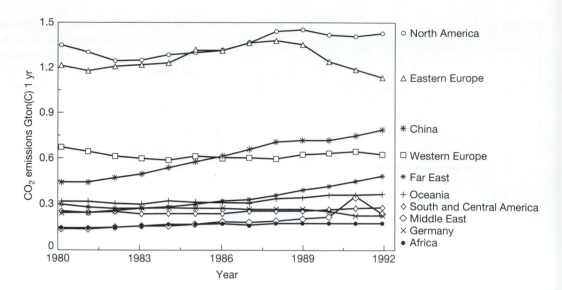

FIGURE 13-2

Estimated CO_2 emissions from the burning of fossil fuels in different regions. (After Carbon Dioxide Information Analysis Center, Oak Ridge, *Communications,* Fall 1995, p. 5.)

About 2.5 Gton(C)/yr is apparently being absorbed by the oceans, as the imbalance in fluxes between the atmosphere and the surface ocean (Fig. 13-1) indicates. The remainder is thought to be accumulating in forests and soils.

To further complicate matters, fossil fuel burning is not the only anthropogenic source of CO_2. Significant amounts of CO_2, equivalent to 1.5 ± 1.0 Gton(C)/yr, are also being produced by deforestation, most of which is currently occurring in the tropics. The clearing of forests and the utilization of land for agricultural purposes generally results in a substantial release of carbon into the atmosphere, both from the trees themselves and from the soil beneath them. If the actual emissions are at the high end of the estimated range, tropical deforestation may be contributing as much to atmospheric CO_2 increase as is the burning of either coal or oil. Surprisingly, however, the biosphere as a whole is currently acting as a net sink for carbon. The reasons are discussed below.

CO_2 Removal Processes and Time Scales

How fast the CO_2 released from fossil fuel burning will disappear and where it will ultimately go are perhaps the most confusing aspects of the whole global warming issue. The confusion arises because the carbon cycle is complex and involves processes that operate on a variety of time scales. Scientists have developed computer models of the carbon cycle that are capable of simulating many of these processes and that can be used to make projections of future atmospheric CO_2 concentrations. Some of these model predictions will be described in the following section. To understand these predictions, however, and to gain some idea of how reliable they might be, we must consider the physical processes involved. Next, we de-

scribe the major CO_2-uptake processes in decreasing order of the speed at which they are expected to occur.

Northern Hemisphere Reforestation

The fastest uptake process for anthropogenic CO_2 is photosynthesis by the terrestrial biosphere. As discussed in Chapter 7, we can represent photosynthesis by the simplified reaction

$$CO_2 + H_2O \longrightarrow CH_2O + O_2,$$

where CH_2O represents more-complex forms of organic matter. This organic matter can accumulate either in living biomass or in soils. When it does so, carbon is removed from the atmosphere, at least temporarily. This carbon is eventually returned to the atmosphere when trees are cut down and burned or when soil organic matter decays.

Whether photosynthesis will act as a sink for anthropogenic CO_2 over the next few centuries depends largely on whether forests expand or shrink in size. Deforestation causes them to shrink, so the terrestrial biosphere acts as a CO_2 source. Reforestation causes them to expand, so the biosphere acts as a CO_2 sink. Studies suggest that a significant amount of reforestation, about 0.5 Gton(C)/yr, is occurring in temperate parts of the Northern Hemisphere. Recall from Chapter 1 that deforestation of North America during the 19th century, the *pioneer effect,* was responsible for most of the rise in atmospheric CO_2 between 1800 and 1850. Much of the deforested land was converted to farms and remains farmland today, so that land cannot be where the carbon is going. But the mountain ridges in Pennsylvania were stripped of trees to provide fuel for making steel and for powering trains. The demand for wood is now lower, and many areas are protected from logging by the state and federal governments. Consequently, the forests are regrowing and are probably contributing to the Northern Hemisphere CO_2 sink. Most of these forests will not reach maturity for another century or more, so this sink should remain active for some time into the future.

Northern Hemisphere forest regrowth, however, cannot be the only place where CO_2 is disappearing. The total anthropogenic CO_2 source is about 7.5 Gton(C)/yr, and we have accounted for only about 5.5 Gton(C)/yr of carbon sinks: 3 Gton(C)/yr in the atmosphere, 2 Gton(C)/yr in the oceans, and 0.5 Gton(C)/yr in Northern Hemisphere forests. An additional sink of approximately 2.0 Gton(C)/yr is required to balance the global carbon budget. What could this "missing sink" be?

CO₂ and Nitrogen Fertilization

A substantial part of the missing carbon may be going into existing forests as a consequence of higher atmospheric CO_2 concentrations. Most plants raised under greenhouse conditions (where plenty of water and other nutrients are available) grow faster when exposed to higher CO_2 levels. This process is termed **CO_2 fertilization.** The increased growth rate is caused, in part, by the fact that CO_2 is a limiting nutrient under these conditions. But there is an indirect stimulation effect as well. Plants tend to use water more efficiently under elevated CO_2 conditions. Plants have small openings, called *stomata,* on the undersides of leaves that allow air to pass in and out of them; these stomata need not open as wide as normal when CO_2 levels are high. As CO_2 from the atmosphere enters the leaf, water vapor from inside can escape. Not opening their stomata as wide allows plants to survive under drier conditions, allowing the plants to grow faster under high CO_2 levels.

Whether forests and other natural ecosystems should respond similarly to CO_2 fertilization is a question that has aroused considerable debate. Ecologists have noted that many natural ecosystems are limited by other factors, such as nutrient availability and competition for sunlight. In such cases, higher atmospheric CO_2 concentrations should have little effect on plant growth. Even if CO_2 fertilization is occurring, other factors might come into play in the future as the climate warms. One concern is that soil carbon might decay more rapidly under such conditions. Tropical soils, for example, are deficient in carbon as a result of rapid rates of decomposition. Because most of the carbon in forests resides in the soil rather than in the trees themselves, the total amount of carbon stored in forest ecosystems could actually decrease in the future even if the trees themselves grow faster.

A related phenomenon that might be encouraging CO_2 uptake by the terrestrial biosphere is **nitrogen fertilization.** Nitrogen is an essential nutrient for all organisms, and it is often in short supply because it is difficult to convert atmospheric N_2 into fixed nitrogen that organisms can use. Humans have been helping out in this regard by adding nitrogen fertilizer to agricultural fields and by emitting large amounts of nitrogen oxides from combustion. Agricultural activity does not normally lead to net uptake of CO_2, because the crops that are grown are harvested and eaten and the remaining organic matter is burned or decays. But the nitrogen oxides released into the atmosphere are oxidized to nitric acid, HNO_3, and are removed by precipitation. If the concentration of nitric acid in rainwater becomes too high, the resulting acid rain is harmful to plants and to other organisms, especially fish (although nitric acid generally contributes less to acid rain than does the sulfuric acid generated from SO_2 released by coal burning). At more modest concentrations, the nitric acid becomes a source of fixed nitrogen, which can stimulate plant growth. So, some of the increased forest growth currently taking place may be a response to anthropogenic nitrogen emissions.

Dissolution in the Oceans

The next-fastest mechanism for removing anthropogenic CO_2 from the atmosphere is dissolution in the oceans. The chemical process by which the oceans dissolve CO_2 is described in the box "The Chemistry of CO_2 Uptake." According to Figure 13-1, the flux of CO_2 between the atmosphere and oceans is on the order of 90 Gton(C)/yr, so the residence time for atmospheric CO_2 with respect to this process, t_{OA}, is on the order of 8 years.

This time scale, however, is somewhat misleading. Recall that the ocean can be thought of as consisting of two layers: a well-mixed surface layer approximately 75 m thick and a poorly mixed deep-ocean layer nearly 4 km thick. These layers can be represented schematically by separate boxes, as shown in Box Figure 13-1. Only the shallow surface layer is capable of rapid CO_2 exchange with the atmosphere. Because of its small volume, its CO_2 uptake capacity is relatively low. The uptake capacity is determined largely by the abundance of carbonate ion (see the box "The Chemistry of CO_2 Uptake"). The deep ocean has a much larger volume and, hence, a much larger CO_2 uptake capacity, but its turnover time, t_{SD}, is on the order of 1000 years. So, CO_2 exchange between the atmosphere and the oceans occurs on a variety of time scales, ranging from 8 years to more than 1000 years.

Because the uptake of CO_2 by the oceans is a chemical process, the lifetime of anthropogenic CO_2 depends on how much of it we produce. The first puffs of CO_2 released at the dawn of the industrial age were taken up almost immediately by the surface ocean. But the more CO_2 we release, the deeper it must penetrate into the ocean in order to be buffered by reaction with carbonate ion. The lifetime of CO_2 released today is estimated to be only about 60 years, but the lifetime of CO_2 released in the future is predicted to be much longer. Computer models that take this chemistry into account are required to calculate how fast the ocean will actually take up anthropogenic CO_2.

Dissolution of Sea-floor Carbonates

Carbon dioxide can also be taken up by the dissolution of carbonate sediments on the sea floor. This may at first seem surprising, as we learned previously that the precipitation of carbonate sediments, in conjunction with silicate weathering, removes CO_2 from the atmosphere–ocean system. Carbon-

The Chemistry of CO₂ Uptake

The rate at which CO_2 can be taken up by the oceans depends on ocean chemistry as well as ocean mixing. The reason is that CO_2 does not simply dissolve in seawater as would a gas, such as N_2 or O_2. As we discussed in Chapter 7, when CO_2 dissolves in water, it forms carbonic acid, H_2CO_3, which then dissociates into bicarbonate ions, HCO_3^-, and carbonate ions, $CO_3^=$. Long before humans began perturbing the Earth system, the ocean contained substantial quantities of carbonate and bicarbonate ions as part of the natural inorganic carbon cycle. The presence of these ions in solution makes it possible for seawater to absorb more anthropogenic CO_2 than would otherwise be possible. The reason is that these ions, carbonate ion in particular, moderate the change in the ocean's acidity as CO_2 is added. A chemist would say that they serve as a pH **buffer,** a dissolved substance that helps maintain a stable pH. At pH values that are typical of the surface ocean (pH of about 8), the chemical reaction that occurs can be written as

$$CO_2 + CO_3^= + H_2O \longrightarrow 2\,HCO_3^-.$$

Each anthropogenic CO_2 molecule that enters the ocean combines with one carbonate ion and one water molecule, yielding two bicarbonate ions. A similar reaction converts borate ion ($H_2BO_3^-$) into boric acid (H_3BO_3). The presence of borate increases the ocean's buffering capacity by an additional 25%.

The fact that such chemical reactions occur implies that the ocean's capacity to absorb CO_2 is limited. If the amount of anthropogenic CO_2 added to the ocean exceeds the amounts of carbonate ion and borate ion that were initially present, the ocean's buffering capacity will be exhausted and its ability to absorb CO_2 will be greatly diminished. We can estimate the CO_2 uptake capacity of the ocean by measuring the dissolved carbonate ion concentration in the surface and deep ocean and multiplying by the volumes of the respective reservoirs. (We shall do so in Critical-Thinking Problem 3). The results are shown in Box Figure 13-1. The surface ocean has only a small buffering capacity

compared with the amount of CO_2 that could be produced from the burning of fossil fuels. The deep ocean contains enough carbonate and borate ion to react with approximately 30% of the fossil-fuel reservoir.

The oceans can also absorb anthropogenic CO_2 by dissolving carbonate sediments on the sea floor. The chemical reaction involved is

$$CO_2 + CaCO_3 + H_2O \longrightarrow$$
$$Ca^{++} + 2\,HCO_3^-.$$

This reaction is similar to that by which seawater itself absorbs CO_2, except that the required carbonate ion is initially attached to a calcium ion.

Neither of these two processes is a permanent sink for CO_2 because, even after the reactions have occurred, the CO_2 is still present in the oceans as bicarbonate. This bicarbonate will eventually be removed when enough calcium ions have been provided by silicate weathering to reprecipitate it as carbonate sediments. Only then will the anthropogenic CO_2 truly be gone.

CO₂ uptake reaction:
$$CO_2 + CO_3^= + H_2O \longrightarrow 2\,HCO_3^-$$

BOX FIGURE 13-1

Two-box ocean model illustrating the capacity of the ocean for CO_2 uptake. The numbers in the ocean boxes represent the amount of carbonate ion that can react with CO_2.

ate sediments are the long-term sink for CO_2. On shorter time scales, however, just the opposite occurs: Atmospheric CO_2 is taken up when carbonate sediments dissolve, because both compounds are converted to bicarbonate.

Carbonate sediments dissolve when CO_2-enriched seawater comes in contact with the sediments. For this to occur, the CO_2-rich water needs to be carried down into the deep ocean. As discussed above, this is a slow process, requiring

many hundreds of years. Furthermore, the sediments on the sea floor need to be stirred to expose fresh surface area for the reaction. This stirring, or **bioturbation,** is accomplished by burrowing organisms, such as worms, that make their homes in the sediments. Bioturbation occurs primarily in the uppermost 10 cm of sediments, but repeated episodes of carbonate dissolution followed by renewed burrowing can eventually cause the uppermost 40 to 50 cm of marine sediments to dissolve. As Wallace Broecker of Lamont-Doherty Earth Observatory has observed, "even the worms will do their part in [taking up fossil fuel CO$_2$]!" Thus, sea-floor carbonate dissolution may eventually play an important role in CO$_2$ removal, but it is not likely to prevent atmospheric CO$_2$ from rising over the next few decades to centuries.

Weathering of Continental Rocks

The slowest, but most permanent, sink for anthropogenic CO$_2$ involves the weathering of silicate rocks on the continents, followed by the precipitation of carbonate sediments on the sea floor. As we discussed in Chapter 7, the combination of these two processes can be represented by the chemical reaction

$$CaSiO_3 + CO_2 \longrightarrow CaCO_3 + SiO_2,$$

where CaSiO$_3$ represents a variety of more-complicated silicate minerals. This process is much slower than the carbonate dissolution process discussed above, because it requires that Ca^{2+} ions produced by weathering accumulate in the ocean. The time required to precipitate anthropogenic CO$_2$ as carbonate can be estimated by dividing the total amount of carbon in the combined atmosphere–ocean system, about 38,000 Gton, by the rate at which CO$_2$ is consumed by silicate weathering, 0.06 Gton(C)/yr. This time scale, which is in excess of half a million years, is the characteristic response time of the carbonate–silicate cycle. Our human-induced perturbation to the natural carbon cycle is likely to last at least that long.

Carbonate rocks on the continents can also be weathered and dissolved. This process is analogous to sea-floor carbonate dissolution and provides another "temporary" sink for CO$_2$. The CO$_2$ is not really gone, because it is stored in the oceans as bicarbonate. Carbonate rocks dissolve more rapidly than do silicate rocks, so this loss process could become important on time scales of only a few thousands of years.

Projections of Future Atmospheric CO$_2$ Concentrations

The next step in predicting future atmospheric CO$_2$ concentrations is to incorporate these CO$_2$ sources and sinks into a computer model of the carbon cycle. This is a difficult task, because the real world is considerably more complicated than a simple diagram such as Figure 13-1 would suggest. One of the thorniest complications involves ocean circulation. Computer models are only beginning to be able to simulate ocean circulation realistically. Thus, most studies of CO$_2$ uptake by the oceans have involved much simpler models in which the ocean is subdivided into a small number of boxes. The rates of water exchange between these boxes are then estimated by using the observed distributions of various chemical tracers, such as carbon -14. A **tracer** is an atom, ion, or chemical compound that can be measured easily and whose sources and sinks are relatively well understood. Tracers can be used to determine oceanic (or atmospheric) circulation.

Box Models of Ocean Circulation

We mentioned in the previous section that, to a first approximation, the ocean can be thought of as consisting of two boxes: a well-mixed surface-ocean box and a poorly mixed deep-ocean box. This two-box ocean model is useful for many purposes, but it ignores several features of ocean circulation that are important for CO$_2$ uptake. For example, the two-box model misses the fact that most deep-water formation occurs at high latitudes. This thermohaline circulation can be better approximated by adding a third box to the ocean model (Fig. 13-3a), to distinguish between warm and cold surface waters. In the figure, the heavy, one-way arrows represent the thermohaline circulation. The thin, two-way arrows represent eddy mixing that occurs as a result of small-scale currents within the ocean.

Three-box ocean model

(a)

Box-diffusion model

(b)

FIGURE 13-3

Simplifed models of ocean circulation.

The three-box ocean model does better than the two-box model in representing the thermohaline circulation, but it ignores the fact that the water in the thermocline region interacts with the surface faster than does water at greater depths. An alternative representation of ocean circulation that takes this factor into account, but ignores the thermohaline circulation, is the *box-diffusion* model (Fig. 13.3b). This model divides the ocean into numerous horizontal layers, each of which exchanges water with the layers above and beneath it. This model is better than the three-box model for simulating CO_2 uptake on short time scales, but it may not be very realistic on the time scale of deep-ocean circulation. More-sophisticated box models try to incorporate aspects of both of these approaches.

Atmospheric CO_2 Levels for Different Emission Scenarios

To use computer models to predict future atmospheric CO_2 concentrations, we must estimate how rapidly fossil fuels will be consumed in the future and how much de-forestation (or reforestation) will take place. These factors, in turn, depend on the type of energy and land-use policies different countries will adopt. These factors are difficult to predict, so carbon-cycle modelers typically carry out calculations for a number of different scenarios. Four such scenarios are illustrated in Figure 13-4. Figure 13-4a shows estimated CO_2 production rates; Figure 13-4b shows atmospheric CO_2 concentrations calculated by a cross between the three-box model and the box-diffusion model. All these scenarios assume an anthropogenic CO_2 production rate of 6 Gton(C)/yr in 1990. Curve *a* assumes that the CO_2 production rate after 1990 drops immediately to zero; curve *b* assumes that it decreases by 2% per year; curve *c* assumes that it remains constant; and curve *d* assumes that it increases by 2% per year. Curve *d* (the "business-as-usual" scenario) is simply a continuation of the observed increase in anthropogenic CO_2 production over the past few decades. Two percent per year is also roughly equal to the present rate of global population increase, so this scenario is what we might expect if the amount of CO_2 emitted per person remains constant.

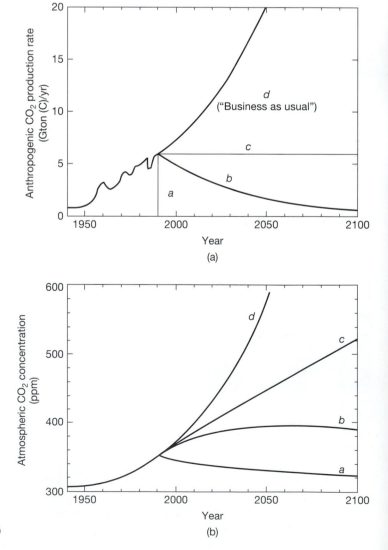

FIGURE 13-4

Different scenarios of (a) CO_2 emission rates and (b) resulting atmospheric CO_2 concentrations. (From Climate Change: The IPCC Scientific Assessment, Cambridge University Press, Cambridge, 1990.)

Long-Term CO$_2$ Projections

What will happen to atmospheric CO$_2$ levels in the distant future if we continue to burn fossil fuels? This question can be studied by using specially designed computer models that are able to take large time steps. For illustrative purposes, let us assume that most of the fossil fuel reserve listed in Table 13-1, 4200 Gton(C), is consumed during the next 400 years. Let us further assume that current deforestation trends continue until only 30% of the world's forests remain.

This particular scenario has been investigated with a computer model that includes a six-box ocean, along with separate boxes to represent the atmosphere, forests, and carbonate sediments on the sea floor. The results of the simulation are shown in Box Figure 13-2. Box Figure 13-2a shows calculated atmospheric CO$_2$ concentrations over the next 3000 years. The computer model predicts that atmospheric CO$_2$ levels will increase to a peak of about 2100 ppm by the year A.D. 2300. At that time, the fossil fuel reserves will be exhausted and atmospheric CO$_2$ will begin to decline.

If we extend this calculation further into the future, the atmospheric CO$_2$ concentration must eventually return to its preindustrial value of 280 ppm, because the model assumes that the carbon cycle was balanced at that time. The amount of time required to return to that steady-state level is shown in Box Figure 13-2b. There, the date is displayed on a logarithmic scale extending millions of years into the future. The processes that are responsible for CO$_2$ uptake on different time scales are indicated on the figure. Most of the CO$_2$ is removed over the next few thousand years by dissolution in the deep ocean and by the dissolution of carbonate sediments and rocks. The last vestiges of the CO$_2$ pulse are removed by silicate weathering over a period of more than 1 million years.

Although the calculations shown here are speculative, they suggest an intriguing possibility. If the climate record of the past few million years were extended into the future, we would expect that Earth should experience at least 10 major glacial–interglacial cycles over the next 1 million years. But atmospheric CO$_2$ levels were relatively low, 200 to 280 ppm, during the previous glacial cycle, whereas they are predicted to exceed 350 ppm during most of the next 1 million years. Could the addition of this much anthropogenic CO$_2$ break the glacial–interglacial cycle? If it did, the results would be at least partly beneficial; after all, no one looks forward to the beginning of the next Ice Age. Yet, the accompanying increase in sea level over this time

(a)

(b)

BOX FIGURE 13-2

Long-term projections for atmospheric CO$_2$ for (a) the next 3000 years and (b) the next several million years. The total amount of fossil fuel consumed is equivalent to 4200 Gton (C). (After Walker and Kasting, *Global and Planetary Change*, **97**:151, Elsevier, NY: 1992.)

could cause shorelines to move substantially inland and force massive relocations of people. The possible long-term effects of the burning of fossil fuels on the Earth system are evidently quite large and will eventually need to be considered.

The calculation depicted in Figure 13-4 predicts that, if anthropogenic CO$_2$ production continues to increase as it has in the recent past (curve *d*), atmospheric CO$_2$ concentrations will reach twice the preindustrial value, or 560 ppm, by the middle of the 21st century. Even if we hold CO$_2$ production constant (curve *c*), atmospheric CO$_2$ levels will continue to increase rapidly and could exceed 500 ppm by the year 2100. This important point is sometimes overlooked: Stabilizing CO$_2$ emissions is *not* equivalent to stabilizing atmospheric CO$_2$. The reason is that CO$_2$ has a relatively long lifetime, so the emissions are cumulative. In contrast, the greenhouse gas

methane has a short atmospheric lifetime (about 12 yr), so its concentration is directly proportional to its emission rate.

Another way of doing such calculations, which might make more sense from a policy standpoint, is to ask what level of CO$_2$ emissions would be required to stabilize atmospheric CO$_2$ at different concentrations. For example, suppose we wanted to stabilize atmospheric CO$_2$ at 450 ppm. According to Figure 13-5, this stabilization would require a decrease in net CO$_2$ production from 6 Gton(C)/yr to just over 1 Gton(C)/yr by the year 2300. Cutting global fossil fuel usage by this amount would be a daunting task. The task

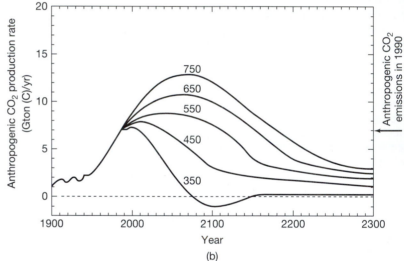

FIGURE 13-5

(a) Various prescribed levels of atmospheric CO_2. Stabilization at these levels requires (b) these net CO_2 emission rates. (After Climate Change 1994: Radiative Forcing of Climate Change, Cambridge University Press, Cambridge.)

might be made somewhat easier if the terrestrial biosphere were still absorbing CO_2 at that time, but such absorption may or may not occur. Thus, we can conclude that large reductions in fossil fuel consumption would be necessary to stabilize atmospheric CO_2 anywhere near its present value.

Projections of Future Climate Change

Suppose that atmospheric CO_2 concentrations do indeed continue to increase over the next few decades, as the carbon-cycle models predict and as seems nearly inevitable. By how much will Earth's climate warm, and when will this warming be detectable?

We can derive a preliminary answer to these questions from the one-dimensional, radiative–convective climate models discussed in Chapter 3. Doubling the atmospheric CO_2 concentration in such a model produces about a 2.5°C increase in global mean surface temperature when the water

vapor feedback is included. In the "business-as-usual" scenario (curve *d* of Fig. 13-4), the first CO_2 doubling occurs by about A.D. 2050. By this reasoning, then, Earth's climate should warm by approximately 2.5°C over the next half century—roughly 5 times the temperature increase that has occurred during the 20th century. We might expect to detect this warming within the next one to two decades.

A prediction of this nature, however, is not good enough for making policy decisions, because it overlooks several factors that should affect Earth's climate over the next few decades. These factors include radiative forcing by trace gases other than CO_2, the oversimplification of one-dimensional models, and the thermal properties of the ocean.

Radiative Forcing by Other Trace Gases

Carbon dioxide is not the only atmospheric greenhouse gas that is currently increasing in concentration. Methane, nitrous oxide, and various chlorofluorocarbon (CFC) compounds have also been increasing in concentration over the past several years (Fig. 13-6). Methane has strong

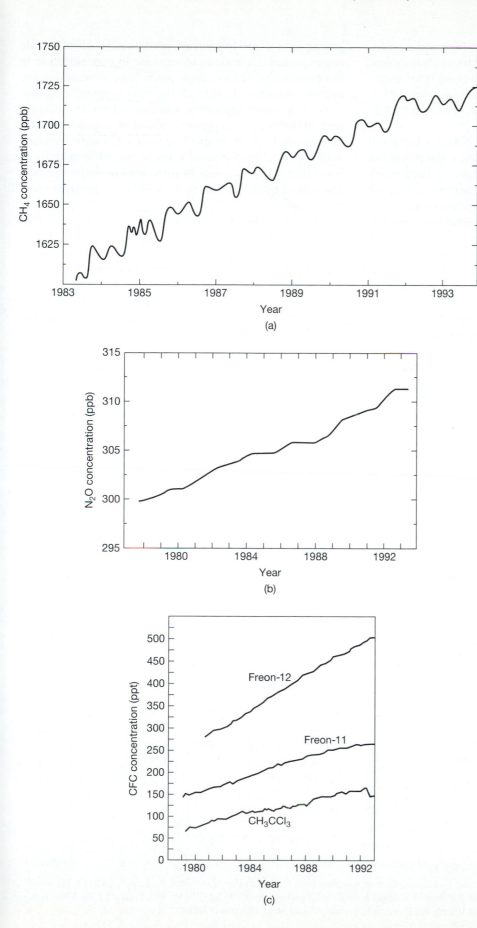

FIGURE 13-6

Trends of atmospheric (a) CH_4, (b) N_2O, and (c) CFC compounds over time. (After Climate Change 1994: Radiative Forcing of Climate Change, Cambridge University Press, Cambridge.)

anthropogenic sources, mostly cattle raising and rice cultivation, that account for 60 to 80% of its total emissions. About 30% of nitrous oxide emissions come from bacterial denitrification in fertilized soils; the remaining 70% is natural. In contrast, CFCs, which include freons, are 100% anthropogenic.

A group led by James Hansen at NASA's Goddard Institute for Space Studies has compiled estimates of the amount of greenhouse warming that has been caused by the addition of CO_2 and other trace gases over the past century and a half (Fig. 13-7). The vertical axis of Figure 13-7 shows the amount of additional direct greenhouse gas forcing per decade caused by these gases. The term *direct greenhouse forcing* refers to the expected temperature change in the absence of feedbacks. Evidently, since the 1960s, the other trace gases have produced about as much additional greenhouse forcing as has CO_2. If this trend were to continue, the climate would warm twice as fast as if CO_2 were the sole contributor to global warming.

Fortunately, the concentrations of at least some freon gases may be stabilized within the next few years as a result of policies that have been adopted to try to halt the depletion of stratospheric ozone. Whether additional radiative forcing will occur will depend on which compounds are used to replace freons and whether these replacement compounds contribute significantly to the

greenhouse effect. In contrast, methane and nitrous oxide are expected to continue to increase in concentration in the near future. Both of these gases have sources that are associated with agriculture. If the world's population continues to increase, agricultural output will need to increase correspondingly, and the production of methane and nitrous oxide is likely go up along with it. Climatic warming may also lead to increased rates of microbial activity in soils; that increase could, in turn, lead to increased production of methane and nitrous oxide (a positive feedback loop).

The good news, from a climatic standpoint, is that world population must eventually level off, probably at no more than double the present population. When that happens, methane and nitrous oxide concentrations should level off as well. So, in the long term, the amount of warming from these gases should be much more limited than the warming caused by the buildup of CO_2.

General Circulation Models (GCMs)

A second reason for distrusting our simple climate prediction is that one-dimensional models fail to capture the intricacies of the real world. As we discussed in Chapter 3, one-dimensional climate models are incapable of simulating many of the processes, such as ice

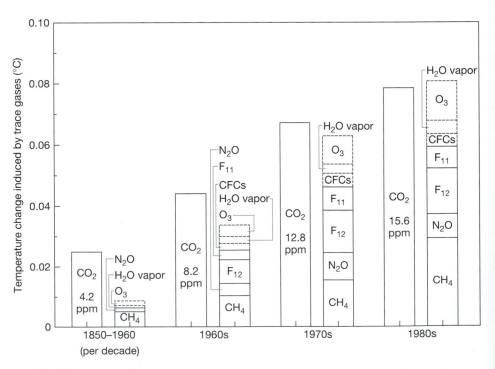

FIGURE 13-7

Direct greenhouse forcing, resulting from CO_2 and other trace gases, by decade. (After J. Hansen et al., in *Effects of Changes in Stratospheric Ozone and Global Climate*, Vol. 1, ed. by J. G. Titus, U.S. Environmental Protection Agency, Washington, DC: pp. 199–218.)

and cloud albedo feedbacks, that are thought to be important to the climate system. Furthermore, these simple models fail to resolve latitudinal temperature gradients as well as differences in climate over land and over the oceans. If Earth were to warm by 2.5°C, would the temperature change be the same in New York as in Calcutta? Probably not. To answer such a question and to account for the more complex feedbacks in the climate system, we need to develop more-complicated models.

The computer models that are now used to study global warming are termed **general circulation models (GCMs).** Many different climate modeling groups around the world, including the National Center for Atmospheric Research (NCAR) in Boulder, Colorado, and the Geophysical Fluid Dynamics Laboratory (GFDL) in Princeton, New Jersey, have developed GCMs. The basic idea behind a GCM is fairly straightforward. Earth's atmosphere (and, in some models, the oceans as well) is divided into a latitude–longitude grid and a number of vertical levels (Fig. 13-8). A typical GCM might have a resolution of 5° × 5° in the horizontal dimension and perhaps 20 vertical layers extending from the surface up to the lower stratosphere. This resolution yields 36 points in latitude (−90° to +90°) and twice that many in longitude (0° to 360°). The total number of grid points, then, is 36 × 72 × 20, or 51,840. At each of these points, the model must keep track of a number of different variables, including temperature, pressure, humidity, and horizontal wind speed. A GCM also needs to account for the absorption and reflection of solar radiation by gases and by clouds and for the transfer of heat and water vapor between the atmosphere and the surface.

The net result of all this complexity is that GCMs require large amounts of computer time, even on the world's fastest computers. A typical, multiyear climate simulation can require hundreds of hours of computation on a supercomputer. Even then, a physically correct solution is not guaranteed. Many of the processes that are important to climate occur on spatial scales that are below the resolution of a GCM. Clouds are a good example. A typical cumulus cloud might have horizontal dimensions of a few kilometers at most, whereas a 5° × 5° grid square at the equator is more than 500 km on a side. Thus, clouds have to be **parameterized** in a GCM—that is, their effects must be simulated on the basis of experiments or observations; their effects cannot be calculated directly. Because clouds play an important role in determining the planetary energy balance, the uncertainties introduced by these parameterizations can lead to substantial uncertainties in the results.

Equilibrium Calculations. As a result of their increased complexity, GCMs exhibit a broader range of sensitivity

FIGURE 13-8
Latitude-longitude grid for a typical atmospheric general circulation model (GCM).

to CO_2 doubling than do one-dimensional models. Most of the models considered in the 1990 report of the Intergovernmental Panel on Climate Change (IPCC) predict an average equilibrium surface warming of between 1.5 and 4.5°C. The **equilibrium surface warming** is the warming that will occur when Earth's surface achieves thermal balance—that is, when the climate system has reached steady state.

Currently, many of the more elaborate, high-resolution GCMs tend to cluster around a predicted equilibrium warming of about 4.0°C for doubled CO_2 conditions. The larger temperature response in these models, as compared with one-dimensional climate models, is mostly caused by an increase in high-altitude cirrus clouds. Such clouds tend to warm the surface, because their contribution to the greenhouse effect exceeds their contribution to the planetary albedo. They increase in abundance when CO_2 doubles, because the surface ocean warms and more water evaporates from it. The added latent heat of this water vapor triggers more-vigorous convection in the troposphere, which leads to cloud formation at higher altitudes. Whether this positive cloud feedback will occur in the real climate system remains to be seen.

General circulation models calculations also predict that the warming caused by increased CO_2 should be unevenly distributed in both space and time. Although the various GCM calculations exhibit numerous individual differences, they agree on several points, such as the following:

- For equilibrium, doubled-CO_2 conditions, the greatest warming should occur in high latitudes

a) June – August

b) December – February

FIGURE 13-9

Predicted equilibrium changes in surface temperature during Northern Hemisphere (a) summer and (b) winter for doubled CO_2 conditions. (From T. McKnight: *Physical Geography: A Landscape Appreciation*, 6/e, 1999. Reprinted by permission of Prentice Hall, Upper Saddle River, N.J.) [See color section]

during late autumn and winter. As shown in Figure 13-9 (Plate 14), the temperature increase in these regions can reach 18 to 20°C, whereas the average warming in this particular simulation is 4°C. The most obvious explanation is that sea ice forms later in the fall as the climate warms and that the snow begins to melt earlier in the spring on land. Both of these factors amplify the climat-

ic warming by way of their positive feedback on Earth's albedo.

• The summertime warming over sea ice in the Arctic and around Antarctica is smaller than the global average. Melting of sea ice during the summer uses up much of the energy of the incident sunlight, and even when the ice melts completely, the cold polar ocean keeps temperatures from rising much above freezing.

- The warming in the tropics is smaller than the global average and varies little with season. Because the saturation vapor pressure of water (Chapter 4) increases nonlinearly with temperature, more of the increased radiative heating from doubled CO_2 goes into evaporating seawater and less goes into raising the surface temperature. The tropical warming is typically about half the global average value.

- Precipitation increases during all seasons in high latitudes and in the tropics and during the winter at midlatitudes. Figure 13-10a (Plate 15) shows this predicted increase for Northern Hemisphere summer. The reason for the increase in precipita-

a) June – August

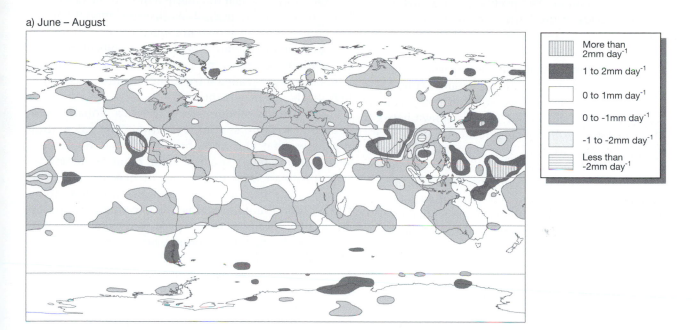

Legend
More than 2mm day^{-1}
1 to 2mm day^{-1}
0 to 1mm day^{-1}
0 to -1mm day^{-1}
-1 to -2mm day^{-1}
Less than -2mm day^{-1}

b) June – August

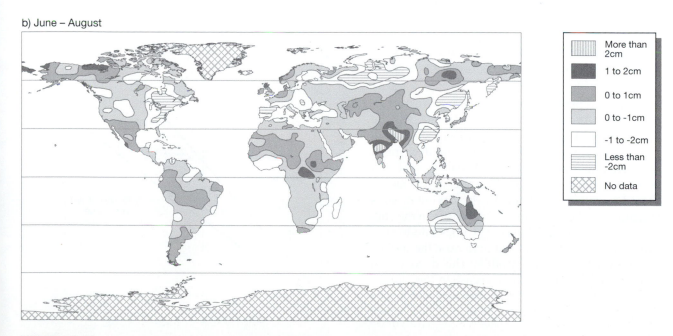

Legend
More than 2cm
1 to 2cm
0 to 1cm
0 to -1cm
-1 to -2cm
Less than -2cm
No data

FIGURE 13-10

Predicted equilibrium changes in (a) precipitation and (b) soil moisture during Northern Hemisphere summer for doubled-CO_2 conditions. Precipitation contours are in millimeters/day (areas of decrease are stippled); soil moisture contours are in centimeters of water. (From T. McKnight: *Physical Geography: A Landscape Appreciation,* 6/e, 1999. Reprinted by permission of Prentice Hall, Upper Saddle River, N.J.) [See color section]

tion is quite simple: Warmer temperatures cause increased rates of evaporation of water from the ocean surface. For atmospheric water vapor to remain in steady state, the increase in evaporation must be balanced by a corresponding increase in precipitation.

In addition to these points on which essentially all GCMs agree, there are several points on which most models are in agreement. The points that are most important to countries in the Northern Hemisphere are the following:

- Most models predict that soil moisture will decrease at northern midlatitudes during summer (Fig. 13-10b; Plate 15). In three high-resolution models studied by the IPCC (including the one shown here), the reduction in soil moisture between latitudes 35 to 55° N ranged from 17% to 23%. The decrease occurs because the models calculate that evaporation will increase faster than precipitation at these latitudes. This prediction is important because it might affect agriculture. If true, it implies that more irrigation would be needed to grow crops in the central United States, for example.

  Note that not all GCMs get this result. Among those that do, questions remain as to whether their spatial resolution is high enough, or their parameterizations of soil processes accurate enough, to make this a reliable prediction. Furthermore, the effect of decreased soil moisture might be at least partly offset by the expected increase in water-use efficiency for plants grown in elevated CO_2 levels. Despite these uncertainties, many climatologists believe that this prediction is cause for concern.

- Most models predict that the warming over northern, midlatitude continents in summer will be greater than the global average. For example, in Figure 13-9a, the calculated warming over much of North America, Europe, and Asia is between 4 and 6°C for doubled CO_2. The reason for this regional warming is related to the drying of soils (mentioned above). With less moisture in the soil, less of the incident solar energy goes into evaporation, so more of it is available to raise the surface temperature. In some models, the decrease in evaporation also causes a decrease in low clouds, which further enhances the surface warming by decreasing Earth's albedo.

Time-Dependent Calculations. A third factor that complicates predictions of future climate is that Earth requires a finite amount of time to warm up as atmospheric CO_2 and other greenhouse gases increase. Even

if atmospheric CO_2 were to double instantaneously, it would be decades, or even centuries, before Earth's surface would reach equilibrium with the additional greenhouse forcing. The reason that Earth is slow to warm, or to cool, is related to the large heat capacity of the ocean. Substantial amounts of thermal energy must be added to the ocean in order to raise its temperature. Hence, the ocean is said to exert a **thermal drag** on the climate system.

One way in which the thermal drag of the ocean can be estimated is by using box-diffusion models of the ocean, like the one shown in Figure 13-3b, in combination with simple climate models. The results of one such calculation are shown in Figure 13-11. Such a time-dependent effect is called a **transient response,** a nonequilibrium change in a quantity.

The climate model used here was a globally averaged one with no spatial resolution. The calculation presumes that the greenhouse effect will increase smoothly into the future in a manner that would be consistent with a CO_2 doubling time of 70 years. The equilibrium response of surface temperature to doubled CO_2 in this climate model is a surface temperature increase of 4°C, similar to that predicted by many GCMs. The time-dependent response, however, is considerably smaller: The calculated surface temperature increase after 70 years, when atmospheric CO_2 has doubled, is only about 2.2°C. The surface warming does not reach 4°C until an additional 50 years have gone by. The difference between the equilibrium and non-

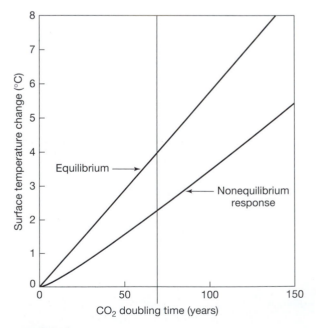

FIGURE 13-11

Equilibrium and nonequilibrium responses of global mean surface temperature to greenhouse gas forcing, as calculated with a simplified model of the atmosphere–ocean system. (After Climate Change: The IPCC Scientific Assessment, Cambridge University Press, Cambridge, 1990.)

equilibrium calculations is caused by the thermal drag of the ocean.

The ocean's thermal drag can also affect the spatial and seasonal distributions of greenhouse warming. Predicting exactly what will happen requires the use of coupled atmosphere–ocean GCMs, which are still in a relatively early stage of development. Nonetheless, several such calculations have already been attempted. As we would expect, the results indicate that the ocean surface warms more slowly than do the continents. The experiments also indicate that the wintertime warming in the polar regions is delayed by the coupling of the high-latitude surface ocean with the deep ocean. The thermocline is absent at high latitudes because the surface water is cold, so mixing between the surface and deep oceans is relatively rapid. Upwelling of cold, deep water keeps the surface ocean colder than is predicted in the equilibrium calculations.

This effect of thermal drag is important because some skeptics of global warming have attempted to discredit the theory by pointing out the lack of observational evidence of recent, high-latitude surface warming. The Alaskan Arctic does appear to have warmed by 2 to 4°C over the past century, but most of the warming apparently occurred before 1930, prior to most of the increase in atmospheric CO_2. The lack of agreement between theory and observations may demonstrate that current climate calculations are oversimplified, but it does not prove that the global warming theory as a whole is wrong.

Long-Term Climate Warming

The climate calculations discussed so far have involved a doubling of atmospheric CO_2 concentrations, which is likely to occur within the next 50 to 100 years unless steps are taken to prevent it. But we saw in the box "Long-Term CO_2 Projections" that atmospheric CO_2 could continue to rise for several centuries beyond that time and that its concentration might eventually reach 2100 ppm—almost 8 times its preindustrial value. What effect would this extended rise have on Earth's climate?

The greenhouse effect of CO_2 is roughly logarithmic, which means that each factor-of-2 increase in CO_2 produces roughly the same amount of warming. An eightfold increase in CO_2 should therefore cause about three times as much warming as would a twofold increase (because $8 = 2^3$.) Thus, a GCM that predicted 3°C of warming for doubled-CO_2 conditions should produce about 9°C of warming for an eightfold CO_2 increase.

The actual range of GCM responses to doubled atmospheric CO_2 is between 1.5 and 4.5°C of warming. The corresponding range for an eightfold CO_2 increase

is 4.5 to 13.5°C of warming. It is instructive to compare these numbers to estimated surface temperature changes during the past 100 million years of Earth history. The warmest part of the Mesozoic is thought to have been about 6 to 10°C warmer than today on a global average. The coldest part of the Pleistocene was probably about 10°C cooler than today. Thus, regardless of whether we believe the low model estimates or the high ones, the warming from an eightfold CO_2 increase could make Earth warmer than it has been for tens of millions of years.

Sea-Level Change

How concerned should we be about the prospect of future global warming? The answer depends not only on the magnitude and rate of the climate change itself, but also on the effects of that change on other parts of the Earth system.

Sea-Level Change during the 20th Century

One factor that we must consider in any discussion of long-term global warming is sea level. Researchers have estimated that sea level has increased by approximately 12 cm since 1880 (Fig. 13-12). The data come from tide gauges on various shorelines around the world. Interpreting such measurements is complicated, because the solid Earth itself is moving up or down in some locations. Parts of Canada and Europe, for instance, are moving upward at a rate of a few centimeters per century, because they are still rebounding from the weight of the great ice sheets that covered these regions as re-

FIGURE 13-12

Global sea-level rise since 1880 (relative to the interval 1951–1970).

cently as 11,000 years ago. In contrast, the Nile River delta is subsiding, because silt from the Nile has increased the loading in this region. These localized trends in land surface elevation must be subtracted from the tide gauge data before we can draw any inferences about sea-level change.

The increase in global sea level during the 20th century parallels the rise in global mean surface temperature (see Fig. 1-4). Indeed, approximately half of the rise in sea level can be attributed to simple **thermal expansion** of surface ocean water. Water, like most other materials, normally expands as it warms. (More correctly, pure water expands when it warms except when its temperature is between 0 and 4°C, when warming causes it to contract.) Because the lateral boundaries of the ocean are largely fixed, any expansion of ocean volume must result in a rise in sea level. The 0.5°C atmospheric temperature rise during the 20th century is expected to have warmed the surface ocean by this same amount, which should have increased sea level by about 5 cm. An increase of 0.5°C in the temperature of the deep ocean would cause a much larger increase in sea level, but such a change would take many centuries to occur, because the thermohaline circulation is very slow. Changes in deep-ocean temperatures caused by Milankovitch cycles may be responsible for periodic sea-level fluctuations of about 5 m that are recorded in carbonate platforms formed during the Mesozoic and Paleozoic Eras.

Most of the remaining increase in sea level during the 20th century is thought to have been caused by melting of **mountain glaciers,** ice fields formed on the cold, upper reaches of mountains (Fig. 13-13). Several glaciers in the Alps are known to have retreated during the past two centuries as Earth emerged from the Little Ice Age. Glaciers in the high Andes Mountains in Peru are also known to be receding rapidly at present. If this trend were to con-

FIGURE 13-13
A mountain glacier. (From Gilbert S. Grant, Photo Researchers, Inc.)

tinue, sea level might eventually rise by another 40 cm from this source alone.

Sea-Level Rise in the Future

A serious concern for the future is that the polar ice caps will begin to melt. Only the ice that is now on land is important in this respect. The melting of sea ice does not increase sea level, because floating ice displaces an amount of seawater that is precisely equal to its mass. (You can prove this by filling a glass with water and ice cubes and allowing the ice to melt. If the ice cubes were initially floating, the water level in the glass will not change.)

Today, large continental ice sheets are found on Greenland and Antarctica. The Greenland ice sheet contains enough water to raise sea level by approximately 7 m, were it to melt entirely. The Antarctic ice sheet contains much more water—some 60 to 70 m of equivalent sea-level rise. However, these two ice sheets are expected to behave quite differently as the climate warms. The island of Greenland extends to lower latitudes than does the continent of Antarctica, so the climate in southern Greenland is considerably warmer than the Antarctic climate. The Greenland ice sheet is therefore much more likely to experience increased melting as the climate warms than is the Antarctic ice sheet. Over most of Antarctica increased snowfall (resulting from warmer ocean temperatures and increased evaporation rates) is expected to cause the ice sheet to thicken over the next 50 to 100 years. This phenomenon could, paradoxically, cause global sea level to decrease as atmospheric CO_2 levels increase.

The West Antarctic Ice Sheet

The actual situation in Antarctica is even more complicated than we have indicated. The Antarctic ice sheet can be divided geographically into an eastern and a western part. The East Antarctic ice sheet contains most of the water and is the part that might thicken as the climate warms. The West Antarctic ice sheet contains less water (about 5 to 6 m of equivalent sea level), but its response to greenhouse warming could be quite different. This ice sheet flows primarily into the Ross and Weddell Seas, where it forms **ice shelves,** large expanses of floating sea ice formed at the margins of continents. That in the Ross Sea is called the Ross Ice Shelf, and that in the Weddell Sea is called the Filchner-Ronne. Both ice shelves are grounded at several points on offshore islands.

Glaciologists have speculated that an increase in water temperature of just a few degrees in the Ross and Weddell Seas could melt enough ice off the bottom of these ice shelves to cause them to become completely free-floating. As a result of this melting, those parts of the West Antarctic ice sheet that feed these shelves might flow much more rapidly, because the contact with the offshore islands currently inhibits the glaciers' flow. Such a sudden, rapid increase in a glacier's flow rate is called a **glacial surge.** Glacial surges are occasionally observed in mountain glaciers, and there is indirect evidence (from ice-rafted debris in North Atlantic sediments) that they occur in continental-scale glaciers as well. Once started, a glacial surge tends to perpetuate itself, because the increased flow rate causes frictional heating at the base of the glacier. This heating, in turn, produces a thin layer of water that allows the glacier to slide more smoothly over the surface. If such a positive feedback process were to be triggered by global warming, the West Antarctic ice sheet might thin relatively rapidly and could contribute significantly to sea-level rise over the next few centuries.

Projections of Future Sea-Level Rise

Given all these possible effects, what can we say about sea-level change in the near future? Will the observed upward trend of the 20th century continue?

Projections of possible sea-level increase during the 21st century are shown in Figure 13-14. Three estimates are shown, reflecting the large uncertainties in the calculation. All three assume a "business-as-usual" scenario for greenhouse warming—that is, they assume that fossil fuels will continue to be burned in proportion to the rising population and that the atmospheric concentrations of other trace greenhouse gases will increase as well. By A.D. 2100, sea level is estimated to increase by between 31 and 110 cm, with a "best guess" of 66 cm. This rate of increase is several times faster than the rate of increase that occurred in the 20th century. Sea-level increases of this magnitude could pose problems for low-lying areas such as the Gulf coast of North America, Bangladesh, and numerous islands in the South Pacific and Indian Oceans.

A potentially more serious problem is the change that might occur in the more distant future. The amount of water now tied up in polar ice is so large, and the projected time scale for global warming is so long, that sea level could ultimately increase by many meters. An increase of this magnitude would wreak havoc with Earth's present geography. A 6-m rise in sea level, for example, would submerge the southernmost one-third of Florida. In the very long term (thousands of years in the future), it is conceivable that the polar caps could melt entirely if we were to consume all of the available fossil fuels. This melting would increase sea level by 70 to 80 m and would submerge roughly 20% of the present continents. So, our decisions about fossil fuel usage over the next few centuries could have major repercussions for our descendants.

Effects on Ecosystems

Also of great interest is the effect that atmospheric CO_2 increases and global warming might have on terrestrial and marine ecosystems. We have already mentioned several such effects. Higher atmospheric CO_2 concentrations are expected to cause increased rates of plant growth. Plants are also expected to use water more efficiently at high CO_2 levels, because they do not need to open their stomata as wide to obtain CO_2 for photosynthesis. This increased efficiency might be offset in some

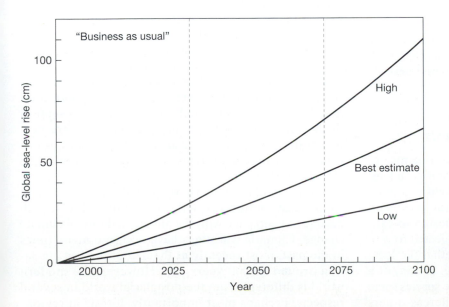

FIGURE 13-14

Three projections for future sea-level rise. All three estimates assume a "business-as-usual" projection for greenhouse gas emissions and climate. (After Climate Change: The IPCC Scientific Assessment, Cambridge University Press, Cambridge, 1990.)

regions by decreased soil moisture during the summertime growing season.

If we examine ecosystems in more detail, the possible changes induced by higher CO_2 levels and higher temperatures become more and more complex. This should not come as too much of a surprise, as biological systems are incredibly complicated by comparison with most physical systems. Here, we mention only a few of the many changes that are expected to take place.

C_3 and C_4 Plants

Different types of plants have different mechanisms of **carbon fixation,** the biochemical process that occurs during photosynthesis by which atmospheric CO_2 is converted to organic carbon. Most photosynthetic organisms alive today (about 95% of all terrestrial plants) fix carbon by a biochemical pathway called the **Calvin cycle.** The first step of this cycle involves the production of a stable, intermediate compound (3-phosphoglyceric acid) that contains three carbon atoms. Hence, this process is called C_3 *photosynthesis,* and plants that metabolize in this way are termed **C_3 plants.**

Some plants, however, including corn, sugarcane, and many tropical grasses, begin the photosynthetic process by producing a four-carbon compound. (Actually, corn and sugarcane are themselves species of grasses, although we do not usually think of them as such.) Plants of this type are termed **C_4 plants.** As a consequence of biochemical modifications, C_4 plants are able to photosynthesize at much lower CO_2 concentrations than are C_3 plants. Indeed, C_4 plants became widespread only about 7 or 8 million years ago, possibly in response to decreased atmospheric CO_2 levels caused by the uplift and weathering of the Himalayan Mountains (see Chapter 8). Also, C_4 plants are much less responsive to CO_2 increases than are C_3 plants. Thus, whereas CO_2 fertilization might have an important effect on the growth rate of C_3 plants, it is not expected to have a large effect on C_4 plants. Thus, certain agricultural crops, such as corn, could be at a disadvantage compared with C_3 weeds in a high-CO_2 world.

Changes in Speciation Within Forests

Even within the C_3 world, different types of plants are expected to exhibit different responses to changes in temperature and moisture availability that might accompany CO_2 increases. Figure 13-15 shows the predicted response of various tree species in Minnesota to estimated climate changes since 1750 and projected to A.D. 2250. The results depend strongly on whether the local soil has a high water-holding capacity (Fig. 13-15a) or a low one (Fig. 13-15b). In the first case, some species (pine and birch) thrive at high CO_2 levels, whereas others

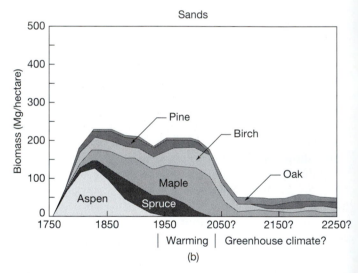

FIGURE 13-15

Predictions of species composition, in terms of (aboveground) biomass in Minnesota forests under doubled-CO_2 conditions, (a) for a clay soil, with a high water-holding capacity, and (b) for a sandy soil, with a low water-holding capacity. (After Climate Change: The IPCC Scientific Assessment, Cambridge University Press, Cambridge, 1990.)

(aspen and spruce) decline when the temperature becomes too warm. In the second, all species do poorly after A.D. 2050, because the soil becomes too dry to support them. A general conclusion that we can reach, irrespective of the details of the calculation, is that species distributions within ecosystems are likely to change as the global climate warms.

A particular concern related to this change in species distribution is whether midlatitude forests will be able to keep pace with the anticipated rate of climate change. Climatic warmings have occurred many times in the past, most recently at the end of the last glaciation around 10,000 years ago. However, the modern world is different from the postglacial world in several respects. Perhaps most importantly, forests in regions such as North America and Europe have been dissected by farms and highways, which tend to inhibit the migration of species. Thus, the poleward spread of species adapted to warmer climates might not occur fast enough to keep pace with the rate at which the climate warms. The extinction of many species of trees and animals could result unless humans intervene actively to facilitate the migration process.

Other Concerns

Accompanying the poleward migration of temperate and subtropical vegetation would be a migration of animal and insect species that live in warm climates. Insects, in particular, pose a significant problem for agriculture. If midlatitude winters become less severe, insect pests that are confined to the tropics today could expand their ranges to midlatitudes. For example, in the United States the potato leafhopper (Fig. 13-16), a serious pest in soybean and other crops, is currently confined to a narrow band bordering the Gulf of Mexico. Warmer winters might allow this pest to double or triple its range. The expansion of insect ranges might well be one of the most economically damaging consequences of climate change.

Various human diseases, such as malaria, that are currently confined to the tropics might become a problem at midlatitudes as well. Malaria is spread by the *Anopheles* mosquito, whose life cycle does not allow it to survive cold winters. If wintertime temperatures remained above freezing in the continental United States, as might happen 100 to 200 years from now if atmospheric CO_2 levels continue to increase, malaria-bearing mosquitos might be able to live there, presenting a whole new set of potential health problems.

Increased atmospheric CO_2 levels could also affect marine ecosystems. Aquatic photosynthesizers, which include both phytoplankton and larger, multicellular organisms such as kelp, are not expected to have increased growth rates from CO_2 fertilization, because their growth rates are generally limited by other nutrients, especially phosphorus and nitrogen. Changes in water temperature could affect some marine species, such as corals, that are adapted to a relatively narrow range of temperature. But the most significant influence of global warming on marine ecosystems is the possible change in ocean currents and upwelling zones that might accompany climate change. Upwelling of nutrient-rich deep water is the key factor that accounts for regions of high productivity in the modern oceans, such as the Grand Banks off the coast of Labrador or the fisheries off the coast of Peru. If the intensity or location of upwelling were to change as the climate warms, it could have major effects on both the high-productivity ecosystems themselves and on the fishing industry that takes advantage of them.

Policies to Slow Global Warming

Should we take action now to slow or halt global warming? Whether immediate action is warranted is one of the most hotly debated scientific issues of our time. The theoretical argument that warming will eventually occur is solid. Whether CO_2-induced global warming has already occurred is more debatable, although we argued in Chapter 1 that it probably has. Because of the high cost of addressing the problem, there is a great reluctance in most societies to act hastily. The attitude of many citizens and governments is to wait and see how bad the problem is before we commit any substantial resources to addressing it.

Although prudence is often justified, many climatologists feel that it will be a matter of only a decade or two before the evidence of global warming becomes impossible to refute. So, it is not too early to consider which measures might be undertaken if this indeed proves to be the case. One step on which virtually everyone involved in the debate agrees is to encourage energy conservation. Every kilowatt-hour of electricity or gallon of gasoline saved is one less that contributes to CO_2 buildup in the atmosphere. Indeed, many conservation measures have already been implemented. Homes built in the United States today are typically much better insulated than were homes built a generation ago. New cars get significantly better fuel mileage than did cars built in the 1960s, although the recent trend toward larger cars and minivans has rolled back some of the gains in fuel efficiency that had occurred during the 1970s and early 1980s.

Energy-conservation measures are to be lauded, but those who have thought seriously about the global

FIGURE 13-16

A potato leafhopper. This insect, along with other agricultural pests, could expand its range poleward if the climate warms. (From Donald Specker, Animals Animals/Earth Scenes.)

warming problem realize that conservation by itself is not likely to solve it. As we discussed earlier in this chapter, it now appears that an 85% cutback in CO_2 emissions would be required to stabilize atmospheric CO_2 at 450 ppm. Stabilization at the 1990 level (350 ppm) is virtually impossible, because it would require negative net CO_2 emissions at some time in the future. With world population increasing, and with developing countries eager to raise their standard of living to levels comparable to that in the West, the demand for energy is likely to increase despite our best efforts to conserve. If we wish to fulfill this demand and still reduce CO_2 emissions, we will have to develop nonfossil energy sources.

Alternative Energy Sources

What alternative, nonfossil energy sources might we turn to? For the production of electricity, nuclear energy is one option that is currently available. Conventional nuclear power plants (Fig. 13-17) produce energy from the **nuclear fission** of uranium atoms, the splitting of an atomic nucleus into two fragments, accompanied by the release of energy. This process, which is the basis for the atomic bomb, produces no CO_2 (although some CO_2 is produced in mining the uranium fuel by means of conventional, gasoline-powered machinery). Many environmentalists, however, feel that nuclear power poses its own threat to the global environment. A large part of their concern stems from the problem of disposal of long-lived radioactive wastes. Although several methods have been proposed for handling wastes, none of them is totally without risk. Nuclear waste disposal sites are vigorously opposed by citizens in areas where the potential sites are to be located. Local opposition has so far prevented the United States from opening its planned nuclear waste storage facility at Yucca Mountain in Nevada. Although many Nevadans see such a delay as positive, it is not without negative consequences. From a practical standpoint, most of the waste produced by existing nuclear reactors is still stored on-site in large underground "swimming pools" rather than in a long-term facility designed for that purpose.

Acceptance of nuclear power varies widely from one country to another. The United States produces about 20% of its energy from nuclear power, but no new nuclear plants have been ordered by utilities for more than 20 years, partly as a result of pressure from vocal antinuclear lobbying groups. Germany and Sweden have also almost abandoned this energy option. Conversely, both France and Japan have active nuclear programs, and both produce the majority of their electricity from nuclear power. One reason for their different outlook may lie in the fact that neither country has appreciable domestic reserves of coal or oil.

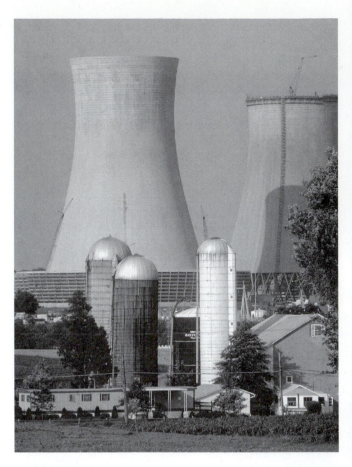

FIGURE 13-17

A nuclear power plant. (From Jerry Irwin, Photo Researchers, Inc.)

A second type of nuclear power that shows some hope for the future is **nuclear fusion,** the combining of lightweight atomic nuclei into a heavier nucleus, with an accompanying release of energy. The fusion of hydrogen atoms into helium is the energy source that powers the Sun. Prototype fusion reactors on Earth use the hydrogen isotopes deuterium (^{2}H) and tritium (^{3}H) as fuel, because these atoms can fuse at lower temperatures than can the normal (^{1}H) isotope. Deuterium is fairly abundant on Earth; 16 of every 100,000 atoms of hydrogen in seawater consist of this isotope. Unfortunately, designing a successful fusion reactor is a technologically daunting task. If ongoing efforts to do so succeed, this process could eventually be an important source of energy.

Other non-CO_2-producing energy sources include wind power, tidal power, geothermal power, and biomass-based fuels. **Wind power** generates electricity by means of windmills, which utilize Earth's solar-energy-driven winds. **Tidal power** produces electricity by using long, floating booms to harness the energy of ocean tides. **Geothermal power** utilizes temperature gradients within the solid Earth as an energy source for generating electricity. **Biomass-based fuels** are liquid fuels, such as methanol (CH_3OH) or ethanol (C_2H_5OH), that are produced from fast-growing plants. These fuels release CO_2 when they are burned, but the plants from which they are made absorb CO_2 while they are growing, so the net effect on the atmospheric CO_2 budget is zero. Although each of these clean alternative energy sources may be useful in certain locales, none appears to be capable of supplanting fossil fuels on a global basis.

Perhaps the most promising energy source in the long term is solar energy. Sunlight is a clean and virtually inexhaustible energy source that is on the verge of becoming economically competitive. In some favorably situated areas, such as southern California, solar power is already competitive and contributes a substantial amount of electrical energy to local power grids. In sunny areas, **solar thermal power** is an efficient method for producing electricity (Fig. 13-18). In this method, sunlight is used to heat a fluid that, in turn, drives a turbine. In more northern regions, the production of electricity from **photovoltaic cells,** specially designed panels that can directly convert sunlight into electricity, may be the most practical solution. Photovoltaic cells, similar to those flown on satellites, are currently more expensive than conventional energy sources, however. Ultimately, huge orbiting **solar-power satellites** may collect the Sun's energy in space and beam it to Earth's surface as laser or microwave radiation. Such satellites could be built by the more industrially developed nations to provide power across the entire globe. Much work would need to be done to determine whether solar-power satellites could be designed to be both economical and safe. They appear to be exorbitantly expensive today, but they might be more cost effective in the future as space technology advances.

Specific Policies That Might Be Adopted

How could a shift to nonfossil energy sources be promoted if such action is deemed necessary? One mechanism

FIGURE 13-18

A solar thermal power plant. (From Hank Morgan/Science Source, Photo Researchers, Inc.)

would be to impose a tax on any energy source that produces CO_2. Such a **CO_2 tax** would make other forms of energy more economically competitive and, hence, more likely to be exploited. It would also encourage people to drive cars that are fuel-efficient and to commute shorter distances to work. The CO_2 tax could be augmented by a "gas-guzzler" tax on fuel-inefficient automobiles. Such taxes need not represent an additional burden on the average citizen, because all or part of the revenue that they generate could be returned to the public in the form of income tax breaks. It would be costly to some industries and localities, however, particularly those involved in mining coal or in producing oil. In the United States, the energy-rich western states, along with some eastern coal-producing states, including Pennsylvania and West Virginia, might lose jobs if such a policy were implemented. Thus, measures such as this are not likely to be accepted unless the population at large becomes alarmed about global warming.

Carbon dioxide emissions could also be reduced by the imposition of direct governmental regulations. An example that is already in effect in the United States is the CAFE (Combined Automobile Fleet Emissions) standard that governs cars sold. Each manufacturer's fleet must meet an average fuel economy rating specified by the federal government. Although this standard undoubtedly increases the number of small, fuel-efficient cars that are sold, it does little to encourage people to drive less or to live closer to their workplace. It also irritates manufacturers, because it limits their flexibility to design cars that will sell. Thus, taxes, as distasteful as they appear to most of us, might be a more effective way of decreasing CO_2 emissions.

One measure that would arouse little or no public opposition would be to encourage the widespread planting of trees. As we discussed earlier in the chapter, existing forests and soils contain 3 to 4 times as much carbon as is currently in the atmosphere in the form of CO_2. Expanding the amount of forested area could, in principle, absorb a substantial amount of fossil fuel CO_2. We say "in principle" because, in practice, the amount of forested land is decreasing as tropical forests are cleared for agriculture. Halting tropical deforestation and encouraging reforestation around the globe would be beneficial for its own sake and might be at least a start toward reducing global warming.

Although all these measures might help reduce the rate at which CO_2 is accumulating in the atmosphere, none is likely to stop the buildup entirely. In the long run, stabilizing climate may well require changes in our lifestyles, including where we live and how we work. In the United States, urban growth over approximately the past century has relied on the availability of automobiles and the relatively low cost of gasoline. Many people prefer to live in distant suburbs, far from their workplaces. For example, in Los Angeles, the average commuter spends an hour and a quarter *each way* driving to and from work alone every day in a private car. Switching to highly fuel-efficient cars can do only so much to reduce CO_2 emissions in such a situation. Cutting back seriously on emissions would require either that the fuel used to power automobiles be derived from renewable (nonfossil) resources, such as biomass-based fuels or that efficient mass-transit systems be developed (preferably ones powered by nonfossil energy sources). Mass-transit systems tend to be efficient, however, only if the population density is high, so the current practice of living in uncrowded suburbs might have to be abandoned. Alternatively, with better electronic communications capabilities becoming available over the Internet, more people might work at home instead of commuting to the office each day. Such lifestyle changes are likely to be adopted only if we become very worried indeed about the problem of global warming.

Chapter Summary

1. The observed rise in atmospheric CO_2 concentrations is caused chiefly by the burning of fossil fuels and, to a lesser extent, by tropical deforestation. The anthropogenic CO_2 source is much smaller than the rate at which CO_2 is released by respiration and decay but much larger than the rate at which CO_2 is emitted by volcanos. It therefore represents a substantial perturbation to the global carbon cycle.

2. The CO_2 generated by human activities can be removed from the atmosphere by several mechanisms.
   a. The fastest of these mechanisms is photosynthesis, but this sink will be effective only if forests are replanted or if CO_2 fertilization of plant growth continues to cause additional carbon to be stored in forests and soils.
   b. Much of the fossil fuel CO_2 will dissolve in the oceans. The rate of CO_2 uptake is limited by the mixing rate of the deep ocean and by the chemical buffering capacity of seawater. Only about 30 to 40% of the available carbon in fossil fuels can be absorbed in this manner.
   c. Additional CO_2 can be removed by the dissolution of carbonate sediments on the sea floor and of carbonate rocks on land, but these processes occur over hundreds to thousands of years.
   d. The fossil fuel CO_2 pulse would be completely removed by silicate weathering on a time scale of about 1 million years.

3. Computer models of the global carbon cycle predict that atmospheric CO_2 levels will double within the next 50 to 100 years and that CO_2 concentrations could exceed 2000 ppm within a few centuries if nothing is done to limit emissions. Radiative forcing of climate could be accelerated by increases in methane and other trace greenhouse gases. Earth's climate could warm by several degrees Celsius over the next century and by as much as 10 to 15°C in the long term. The warming will be unevenly distributed, with the polar regions warming the most and the tropics warming the least. Potentially damaging consequences of such warming include the drying out of continental interiors, the spread of insect pests and tropical diseases, and substantial increases in sea level.

4. Slowing future climate change will require drastic reductions in greenhouse gas emissions, particularly those of CO_2. Emissions reductions of 85% or more would be required to stabilize atmospheric CO_2 anywhere near its current level. Such a reduction could be encouraged by taxation measures that promote energy conservation and the development of alternative energy sources. Whether such measures are warranted is the subject of ongoing debate.

Key Terms

biomass-based fuels
bioturbation
buffer
C_3 plants
C_4 plants
Calvin cycle
carbon fixation
CO_2 fertilization
CO_2 tax
equilibrium surface warming

general circulation models (GCMs)
geothermal power
glacial surge
ice shelves
mountain glaciers
nitrogen fertilization
nuclear fission
nuclear fusion
parameterized

photovoltaic cells
solar-power satellites
solar thermal power
thermal drag
thermal expansion
tidal power
tracers
transient response
wind power

Review Questions

1. How does the amount of CO_2 produced by fossil fuel consumption compare to the natural flux of CO_2 in the carbon cycle?

2. What are the major processes that can remove CO_2 from the atmosphere? What are the approximate time scales for these processes to be effective?

3. What is the size of the fossil fuel reservoir compared with the atmospheric CO_2 reservoir?
4. Why does the ocean have a limited capacity for CO_2 uptake?
5. By how much would we need to reduce CO_2 emissions to stabilize atmospheric CO_2 over the next century?
6. Where is the greatest increase in temperature expected to occur for equilibrium, doubled-CO_2 conditions?
7. Why do some GCM calculations predict that continental interiors might dry out under elevated CO_2 levels?
8. Why is predicting future sea-level change such a tricky task?
9. How does the expected effect of elevated atmospheric CO_2 levels differ for C_3 and C_4 plants?
10. What measures might be taken to reduce future CO_2 emissions?

Critical-Thinking Problems

1. a. The present atmosphere contains approximately 700 Gton(C) in the form of CO_2. Earth's total recoverable fossil fuel reserves contain at least 4200 Gton(C), mostly in the form of coal. (We shall use the value 4200 Gton(C) to be specific.) At present, about half the CO_2 produced by the burning of fossil fuels stays in the atmosphere. The other half dissolves in the oceans or is taken up by the terrestrial biosphere. If this ratio remained constant and we burned up all of our fossil fuels instantaneously, by how much would atmospheric CO_2 concentrations rise? (Express your answer in terms of the new CO_2 level divided by the old one.)
   b. Climate models predict that each doubling of the atmospheric CO_2 concentration will cause the mean global temperature to increase by 2 to 5°C. (The range is due largely to uncertainties about how clouds will respond.) By how much would the mean temperature increase for the scenario described in part a? Express your answer as a temperature range in degrees Celsius and in degrees Fahrenheit.
   c. The actual problem of global warming could be more severe than we have just calculated. Forests and soils together contain an additional 2100 Gton of carbon that might go into the atmosphere if deforestation is not pre-

vented. The ocean becomes more acidic as it absorbs CO_2, so it might not be able to continue taking up as much CO_2 as it has been until now. If we burned up all our fossil fuels and deforested one-third of the globe without losing any CO_2 to the ocean (or to CO_2 fertilization), by how much would atmospheric CO_2 and temperature increase?

2. The atmospheric CO_2 concentration is currently increasing by about 1.5 ppm/yr. How many gigatons of carbon are being added to the atmosphere each year? (*Hint:* The total mass of the atmosphere is 5×10^{18} kg, and its mean molecular weight is about 29. You will need to do the calculation in moles and then convert back to mass units.)

3. The surface ocean contains about 2.6×10^{16} liters of water with a carbonate ion content of about 2×10^{-4} mol/L. The deep ocean contains about 1.4×10^{21} L of water with a carbonate ion content of roughly 9×10^{-5} mol/L. If each mole of carbonate reacts with 1 mol of CO_2 according to the reaction

$$CO_2 + CO_3^{=} + H_2O \longrightarrow 2\,HCO_3^{-}.$$

what percentage of the fossil fuel reservoir, 4200 Gton(C), can be neutralized by the surface ocean? By the deep ocean?

Further Readings

General

S. H. Schneider, *Global Warming: Are We Entering the Greenhouse Century?* Sierra Club Books, San Francisco, 1989.

S. G. Philander, *Is the Temperature Rising: The Uncertain Science of Global Warming,* Princeton University Press, Princeton, New Jersey, 1998.

Advanced

J. T. Houghton et al., *Climate Change 1995: The Science of Climate Change,* Cambridge University Press, New York, 1996.

B. G. Levi, D. Hafemeister, and R. Scribner, *Global Warming: Physics and Facts,* American Institute of Physics, Washington, D.C., 1992.

Chapter 14
Ozone Depletion

Schematic diagram of the Antarctic circumpolar vortex. (From Tom Van Sant/Geosphere Project, Santa Monica/Science Photo Library/Photo Researchers, Inc.)

Key Questions

- How is ultraviolet radiation categorized, and what are its biological effects?
- How is the thickness of the ozone layer measured, and how does this layer vary from place to place?
- How do trace chemicals catalyze the destruction of ozone?
- What is the cause of the Antarctic ozone hole?
- What is being done to prevent ozone depletion in the future?

Chapter Overview

Solar ultraviolet radiation between 200 and 320 nm poses significant health hazards if not effectively blocked by stratospheric ozone. Ozone is formed by reactions initiated by the splitting of O_2 and can be destroyed by various catalytic cycles involving the elements nitrogen, chlorine, and bromine. The latter two elements have large anthropogenic sources and therefore have generated concern. Chlorine, in particular, has been directly implicated in the formation of the Antarctic ozone hole and may be responsible for a slow, long-term decrease in midlatitude ozone levels. International agreements that are already in place are expected to halt the depletion of ozone and to restore the stratosphere to its natural state.

Introduction

We have discussed in detail the evolution of Earth's climate on different time scales. Our reason for dwelling on climate is twofold: First, the prospect of global warming over the next few decades to centuries is probably the most intractable environmental problem that we currently face. Second, the many different aspects of climate serve as an excellent case study for illustrating the interactions among the various elements of the Earth system, along with the system's capacity for self-regulation.

No book on global change, however, would be complete without a discussion of Earth's ozone layer and the possibility that it could be depleted. We learned in Chapter 9 that the development of a protective ozone screen to shield out harmful solar ultraviolet radiation was an important step in the evolution of advanced forms of life. The ozone layer arose naturally as a byproduct of the evolution of photosynthesis and the rise of atmospheric oxygen. It is a relatively fragile feature, however, that is now threatened by chemicals released into the atmosphere by industrial activities. Unlike global warming, ozone depletion is widely recognized as a serious problem, and significant steps have already been taken to reduce its impact. It is important to understand why the ozone layer is so essential and why we must remain committed to protecting it.

Ultraviolet Radiation and Its Biological Effects

We learned in Chapter 9 (see Fig. 9-12) that ozone absorbs ultraviolet (UV) radiation in the 200- to 400-nm region, where few other atmospheric gases absorb. Here, we shall look at this spectral region in greater detail.

UVA and UVC Radiation

Ultraviolet radiation between 200 and 400 nm is usually subdivided into three distinct spectral ranges, as shown in Table 14-1. The longer wavelengths are termed *UVA*, the middle wavelengths are termed *UVB*, and the shortest wavelengths are termed *UVC*. The solar flux increases with increasing wavelength throughout this spectral region, as shown in Figure 14-1a. Thus, more UVA photons are available at the top of the atmosphere than are UVB or UVC photons. The ozone *absorption coefficient* is also low in the UVA region. The absorption coefficient is a measure of how strongly a molecule absorbs electromagnetic radiation of a given wavelength. Because the ozone absorption coefficient is low at UVA wavelengths, most of the incident photons make it down to the ground.

Fortunately, UVA radiation appears to be relatively harmless to humans and other forms of life. Many tanning parlors use UVA radiation to tan their patrons "safely."

TABLE 14-1

Classification of UV Radiation		
Wavelength Range (nm)	*Name*	*Biological Effect*
320–400	UVA	Relatively harmless; causes tanning but not burning
290–320	UVB	Harmful: causes sunburn, skin cancer, and other disorders
200–290	UVC	Extremely harmful but almost completely absorbed by ozone

However, whether UVA radiation is really safe for humans is not fully understood. Overexposure to UVA radiation may lead to premature aging of the skin, and there is some evidence that it can damage the immune system. We are certain, though, that UVA radiation is much less dangerous than shorter-wavelength UV radiation.

At shorter, UVC wavelengths, the solar flux is lower and the ozone absorption coefficient is high. Thus, relatively few UVC photons hit the top of the atmosphere, and even fewer make it to Earth's surface. This is a good thing for humans and other organisms, because UVC radiation is extremely dangerous to most forms of life. Indeed, the absorption peak for DNA, the molecule that contains the genetic information for all organisms, is right in the middle of the UVC region. Some single-celled, prokaryotic microorganisms have developed highly efficient DNA-repair mechanisms and can tolerate substantial doses of UVC radiation. However, more-advanced, eukaryotic organisms, including all multicellular life forms, are extremely sensitive to radiation at these wavelengths.

UVB Radiation and Its Biological Effects

The wavelengths between about 290 and 320 nm, UVB radiation, are the ones of current concern to us. At these wavelengths, the solar flux is relatively high and the ozone absorption coefficient is relatively low. Thus, a substantial radiation flux reaches Earth's surface. The biological effect of this radiation is determined by the **dose rate**, which is the number of UV photons per unit time that lead to a specific biological response, such as sunburn or skin cancer. The dose rate is the product of the surface UV flux and the action spectrum for the particular response being studied. The **action spectrum** of a biological response is the relative efficiency with which UV photons at different wavelengths contribute to that response in a specific organism. For example, the **erythemal action spectrum** shown in Figure 14-1b describes the appearance of sunburn in humans. The dose rate (in relative units) leading to sunburn is indicated by the shaded area in the figure. Clearly, most of the UV rays that cause sunburn are in the UVB spectral region.

Although UVB radiation is not quite as harmful to organisms as is UVC radiation, it is still capable of causing substantial damage. In addition to sunburn, overexposure to UVB radiation can lead to skin cancer in humans. This radiation is also harmful to the eye, where it can cause cataracts and damage the retina. Many animals other than humans, such as hippopotami, are susceptible to sunburn, and at least half of the terrestrial plants that have been studied exhibit slower growth and smaller leaves when exposed to enhanced UVB fluxes. Increases in UVB radiation could be detrimental to aquatic life, including phytoplankton, zooplankton, larval crabs and shrimp, juvenile fish, and corals. Thus, almost all inhabitants of the Earth system have an interest

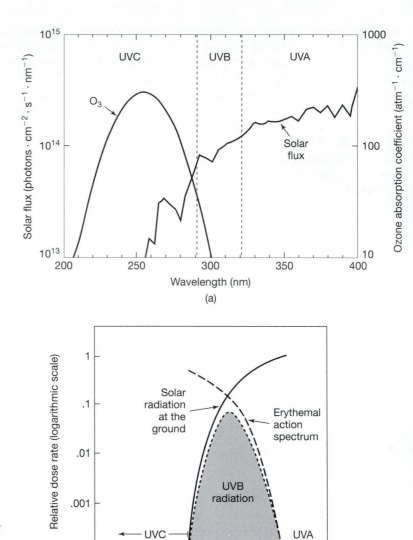

FIGURE 14-1

(a) Graph illustrating the significance of UVB radiation. (b) The shaded area represents the dose rate of the UV radiation that causes sunburn. (After R. Turco, *Air Pollution,* New York: Oxford University Press, 1997.)

in ensuring that the UVB flux does not increase above its present value.

Relationship between UVB Flux and Stratospheric Ozone

How do we know that decreasing stratospheric ozone increases the UVB flux at the ground? In general, this inference is derived theoretically. (We shall do so in Critical-Thinking Problem 1.) In a few cases, however, scientists have been able to measure both quantities directly and to show that these quantities behave as expected. Figure 14-2 shows simultaneous measurements of atmospheric ozone content and ground-level UVB flux at Melbourne, Australia, both before and after the intrusion of ozone-poor air from Antarctica in mid-December, 1987. (The units in which ozone is measured—Dobson units—were introduced in Chapter 1 and are explained further

below.) By January 1, ozone levels had dropped by about 16 percent and the UVB flux had increased by about 25 percent. A similar inverse relationship between ozone and UVB radiation has been observed more recently in Antarctica. The strong negative correlation between the two quantities gives atmospheric scientists great confidence that UVB fluxes will go up if ozone levels go down.

Ozone Vertical Distribution and Column Depth

As we discussed in Chapter 3, most of Earth's ozone (O_3) is confined to the stratosphere. Indeed, the reason the stratosphere exists is because the absorption of solar UV radiation by ozone heats the air and causes the temperature to increase with altitude. The ozone layer is not as uniform,

FIGURE 14-2

Comparison of solar UVB radiation flux at ground level and ozone column depth observed over Melbourne, Australia, from December 1987 to January 1988, during an intrusion of ozone-poor air. (After *Scientific Assessment of Ozone Depletion*, World Meteorological Organization, 1991.)

though, as the discussion in Chapter 3 may have implied. Rather, it varies with both latitude and time of year.

Measurements of Ozone Column Depth

Several methods exist for measuring the vertical distribution of ozone. Two of the most common methods are remote sensing from satellites and in-situ measurements from **ozonesondes,** balloon-borne instruments that measure the concentration of stratospheric ozone. A comparison of the results of these two methods is shown in Figure 14-3. Ozone concentrations are given in units of **number density,** or number of molecules per cubic cen-

FIGURE 14-3

Ozone vertical profile measured in number density (molecules per cubic centimeter) by ozonesonde (solid curve; data from World Ozone Data Center) and by satellite instrument at sunset over Wallops Island, Virginia.

timeter. At its peak in the stratosphere, near 20 to 25 km altitude, the ozone number density is typically about 5×10^{12} molecules/cm^3. In terms of relative concentration, this amounts to a few parts per million. In the example shown here, the two methods for measuring ozone agree to within a few percent at all altitudes above about 16 km, which is where most of the ozone resides. The satellite does not produce accurate results at low altitudes, because it is looking down through the bulk of the ozone layer.

The flux of solar UV radiation that reaches Earth's surface depends on the **solar zenith angle**—that is, the angle of the Sun from the vertical (Fig. 14-4)—and on the vertical column depth of ozone. The **column depth** is the total amount of ozone per unit area above a certain location at the surface. It is measured in several ways. The simplest unit is molecules of O$_3$ per square centimeter. In this unit, the sum of the ozone number densities at all altitudes multiplied by the height of the atmosphere is equal to the ozone column depth. A typical, midlatitude ozone column depth is about 8×10^{18} O$_3$ molecules/cm^2, meaning that there are 8×10^{18} molecules over 1 cm^2 of Earth's surface.

A second way of measuring column depth is to express it as the thickness that a layer of pure ozone would have at 1 atm pressure. One *atmosphere-centimeter* (1 atm-cm) is equal to 2.687×10^{19} molecules/cm^2. Thus, a typical midlatitude ozone profile would have a column depth of about 0.3 atm-cm. Physically, this means that the ozone in Earth's atmosphere is equivalent to a 0.3-cm-thick layer of pure ozone at the surface. The small magnitude of this number gives a preliminary indication of why the ozone layer is so fragile.

The unit atmospheric chemists use to measure ozone column depth is called the **Dobson unit** (DU). One Dobson unit is equivalent to a layer of pure ozone 0.001 cm

FIGURE 14-4

The pathlength through the ozone layer depends on the solar zenith angle, θ (theta).

(a) Sun directly overhead   (b) Sun near the horizon

thick at 1 atm pressure. Because 1 atm-cm is equal to 1000 DU, a typical midlatitude ozone column depth is about 300 DU. The Dobson unit is named after George Dobson, an English physicist who in 1931 became the first person to measure ozone column depth accurately. The ground-based device that he developed has become known as a *Dobson spectrophotometer*. The instrument compares the different amounts of ozone absorption at two wavelengths of sunlight. By measuring the solar UV flux at both wavelengths and at two or more solar zenith angles, it is possible to calculate the ozone column depth directly.

Spatial Distribution of Ozone

Satellite measurements represent a tremendous advance over ground-based measurements, because satellites make it possible to determine ozone column depth at all points over the globe during all seasons of the year. The results from a composite of several years of such mapping by the TOMS instrument (see Chapter 1) aboard the *Nimbus-7* satellite are shown in Figure 14-5. Ozone column depths are highest during springtime at mid-to-high latitudes and lowest over the equator. These data were collected during the early years of TOMS, before the Antarctic ozone hole had become as pronounced as it is today; that is why the ozone hole is not visible in the figure.

The reason ozone is distributed in this manner is complex. As we shall soon see, the greatest production of ozone occurs in the tropics, where the solar UV flux is the highest. Both the mid-to-high latitude peak and the equatorial minimum in column depth are caused by stratospheric circulation: Ozone-rich air from the tropical upper stratosphere is transported poleward by north-south winds. The details of this stratospheric circulation are beyond the scope of our discussion. Figure 14-5 does explain, however, why it is easier to get sunburned in Florida (approximately 30° N) than in New York (about 43° N). Not only is the Sun higher in the sky at low latitudes than it is farther north, but the ozone column depth is also smaller than at higher latitudes. The combination of these two effects allows a much higher percentage of the incident UV photons to make it to the surface at low latitudes.

Tropospheric Ozone

Not all of Earth's ozone is confined to the stratosphere. Roughly 10% is in the troposphere, where it plays a key role in tropospheric chemistry. Its most important function is to provide a source of oxidizing radicals (highly reactive molecules) that help cleanse the lower atmosphere of pollutants such as carbon monoxide (CO) and sulfur dioxide (SO_2). The radicals are produced when the ozone molecule is split by short-wavelength UV radiation. So, a certain amount of ozone in the troposphere helps keep our air fit to breathe.

Too much ozone at ground level, conversely, is a bad thing. Ozone in high concentrations is toxic to plants and

Ozone column depth (DU)

FIGURE 14-5

Latitudinal and seasonal distribution of ozone column depth as measured by satellite, composited over several years. Contours represent column depths in Dobson units.

irritates both our eyes and lungs. Indeed, ozone is a key component of the photochemical smog that forms over Los Angeles, Denver, and many other large cities. As with many other things in life, ozone has both a good side and a bad side. Ozone that is found in the stratosphere, however, is always good from our human perspective.

The Chapman Mechanism

To understand why the ozone layer is susceptible to alteration by anthropogenic processes, we must examine the chemistry by which ozone is formed and destroyed. The first person to describe these processes was Sydney Chapman, an atmospheric chemist who published a ground-breaking paper on ozone photochemistry in 1930.

TABLE 14-2

The Chapman Mechanism of Ozone Production and Destruction		
*Reaction**		*Rate*
1) $O_2 + UV\ photon \rightarrow O + O$ ⎫ production		Slow
2) $O + O_2 + M \rightarrow O_3 + M$ ⎭		Fast
3) $O_3 + photon \rightarrow O_2 + O$ ⎫ destruction		Fast
4) $O + O_3 \rightarrow 2\ O_2$ ⎭		Slow

* The symbol "M" represents a third molecule necessary to carry off the excess energy of the collision between an O atom and an O_2 molecule.

The four chemical reactions that he proposed for a simplified (pure oxygen–nitrogen) atmosphere have become known as the **Chapman mechanism** (Table 14-2).

Production of Ozone

The first step in the Chapman mechanism is the splitting apart of molecular oxygen, O_2, by an ultraviolet photon to form two atomic oxygen (O) atoms. The splitting of a molecule by the absorption of light or by UV radiation is called **photolysis,** or **photodissociation.** UV radiation at wavelengths shorter than 240 nm is required to photolyze O_2. One UV photon in this wavelength range will split apart one O_2 molecule. Here is where the particulate behavior of light, mentioned in Chapter 3, becomes evident. Two visible-light photons with the same combined energy as one UV photon will not be able to split O_2, because photolysis reactions are discrete events.

Once it has formed in the atmosphere, an atomic oxygen atom reacts quickly with another O_2 molecule to form ozone (reaction 2 of the Chapman mechanism). This process cannot occur in isolation, however, because the colliding molecules have too much energy to stick together unless some third molecule, represented by "M," is available to carry it away. Molecule M can be N_2, O_2, ^{40}Ar, or any of the other molecules in Earth's atmosphere.

The situation is analogous to a billiards shot (Fig. 14-6). If you shoot at a single, isolated ball (Fig. 14-6a), chances are that the cue ball and target ball will roll away in different directions. At best, if you hit the target ball straight on with the proper amount of backspin, the cue ball will remain motionless after contact, while the target

(a)

(b)

FIGURE 14-6

The ozone formation reaction (Chapman reaction 2) is analogous to hitting a billiards shot. (a) Hitting a target ball with the cue ball will result in the movement of both in different directions. (b) A third ball is required to allow the cue ball and the target ball to stick together.

ball will roll in the forward direction. If you place a third ball in contact with the target ball, however, and hit the pair dead on, the third ball will roll away and the cue ball and the target ball will remain together (Fig. 14-6b). This, essentially, is what happens in the atmosphere when an ozone molecule is formed by the collision of an O atom with an O_2 molecule.

Destruction of Ozone

Once formed, an ozone molecule can be photolyzed by the absorption of another photon (reaction 3 of the Chapman mechanism). Photolysis reverses the process that occurred in reaction 2. Unlike O_2, O_3 can be split by radiation in the visible-light range. Because many more visible photons than UV photons are available, O_3 is photolyzed much faster than O_2. Furthermore, O_3 can be photolyzed all the way down to Earth's surface, whereas O_2 can be photolyzed only above about 20 km. All of the short-wavelength UV radiation required to split O_2 is absorbed above this height. That is why the ozone layer is located in the stratosphere and not near Earth's surface.

The fourth reaction in the Chapman mechanism is the reaction of an O atom and an O_3 molecule to yield two O_2 molecules. This is a slow reaction, as measured in the laboratory, but as we will see, it is the key to understanding ozone photochemistry. The reason is that when ozone reacts with atomic oxygen (reaction 4), the ozone is destroyed permanently, whereas when ozone is photolyzed (reaction 3), the resulting O atom is free to combine with another O_2 molecule, re-forming O_3.

Odd Oxygen

We can express the difference in reactions 3 and 4 of the Chapman mechanism more precisely by defining a quantity called **odd oxygen,** or O_x. As its name implies, odd oxygen includes all pure, oxygen-containing atoms or molecules that have an odd number of oxygen atoms. Thus, the concentration of odd oxygen, denoted by $[O_x]$, is equal to the sum of the concentrations of atomic oxygen and ozone: $[O_x]$ = $[O] + [O_3]$. Ordinary molecular oxygen, O_2, contains an even number of oxygen atoms; it is not counted as O_x.

With the concept of odd oxygen, we can analyze the Chapman mechanism at a deeper level. Table 14-3 shows

the change in odd oxygen for each of the four reactions. Reaction 1 produces two O atoms from one O_2 molecule, so the change in O_x is +2. Reactions 2 and 3 merely interconvert two forms of odd oxygen, O and O_3, so the change in O_x is 0 for each reaction. Reaction 4 destroys both an O atom and an O_3 molecule, so the change in O_x is −2.

The significance is the following: The two slow reactions, 1 and 4, are the most important, because they control the abundance of odd oxygen. The two fast reactions, 2 and 3, determine the ratio of the O abundance to the O_3 abundance, but they have no direct effect on the concentration of odd oxygen. As we shall see in the next section, it is the destruction of odd oxygen that really matters in trace element chemistry.

Catalytic Cycles of Nitrogen, Chlorine, and Bromine

The Chapman mechanism provides the basis for understanding ozone chemistry, but it does not by itself provide an accurate description of Earth's stratosphere. When the Chapman mechanism is included in a computer model of the stratosphere, the model predicts about 30% more ozone than is actually present. Other processes must be destroying ozone as well or, equivalently, destroying odd oxygen.

The shortcoming of the Chapman mechanism is that it ignores the effects of atmospheric trace constituents such as nitrous oxide (N_2O), water vapor, and freons. These trace gases can be photolyzed, producing highly reactive radicals that keep ozone abundances lower than they would otherwise be.

The Nitrogen Catalytic Cycle

One such radical is nitric oxide, NO. In Chapter 9 we noted that NO produced by lightning discharges constitutes a natural source of fixed nitrogen in the oceans. Lightning does not occur in the stratosphere, but NO can be produced from nitrous oxide, N_2O (as we shall see in the next section). Nitric oxide is one of several radicals that can facilitate the destruction of ozone. The destruction process consists of the following two reactions:

TABLE 14-3

The Chapman Mechanism and Odd Oxygen		
*Reaction**	*Rate*	ΔO_x
1) $\quad O_2 + UV\ photon \rightarrow O + O$ } production	Slow	+2
2) $\quad O + O_2 + M \rightarrow O_3 + M$	Fast	0
3) $\quad O_3 + photon \rightarrow O_2 + O$ } destruction	Fast	0
4) $\quad O + O_3 \rightarrow 2\,O_2$	Slow	−2

* The symbol "M" represents a third molecule necessary to carry off the excess energy of the collision between an O atom and an O_2 molecule.

$$NO + O_3 \longrightarrow NO_2 + O_2 \qquad \textbf{fast}$$
$$NO_2 + O \longrightarrow NO + O_2 \qquad \textbf{fast}$$

$$\text{Net:} \quad O_3 + O \longrightarrow 2\,O_2 \qquad \textbf{fast}$$

In the first reaction, nitric oxide reacts with ozone, forming nitrogen dioxide (NO_2) and molecular oxygen. Nitrogen dioxide is a brownish gas that is a major component of photochemical smog. It is visible in the "brown cloud" that hangs over Denver and many other large cities on weekdays. In the second reaction, nitrogen dioxide reacts with atomic oxygen, re-forming nitric oxide and producing a second O_2 molecule.

We can determine the net effect of the two reactions by adding the reactions together and cancelling out atoms or molecules that occur on both sides of the reaction arrows. When we do this, we find that the net reaction is exactly the same as step 4 in the Chapman mechanism. Thus, this reaction results in the destruction not just of ozone but of odd oxygen as well.

This destruction process is an example of a **catalytic cycle,** a set of chemical reactions facilitated by the presence of a catalyst. A **catalyst** is a substance that increases the rate of a chemical reaction but is itself unchanged by the reaction. In the nitrogen catalytic cycle, the catalyst is the NO molecule. The NO molecule is destroyed in the first step of the cycle, but it is re-formed in the second step and, hence, is free to react again. In the lower stratosphere, one NO molecule can destroy hundreds or thousands of O_3 molecules. Furthermore, both steps in the cycle are fast reactions, as measured by laboratory experiments. As a catalyzed reaction should, be the net reaction is therefore also fast in comparison with the direct reaction between O_3 and O.

The Chlorine Catalytic Cycle

Similar ozone-destroying catalytic cycles can be created by other radicals. One very important cycle involves atomic chlorine, Cl:

$$Cl + O_3 \longrightarrow ClO + O_2 \qquad \textbf{fast}$$
$$ClO + O \longrightarrow Cl + O_2 \qquad \textbf{fast}$$

$$\text{Net:} \quad O_3 + O \longrightarrow 2\,O_2 \qquad \textbf{fast}$$

In the first step of this cycle, atomic chlorine reacts with ozone, forming chlorine monoxide (ClO) and O_2. The chlorine monoxide then reacts with atomic oxygen, forming Cl and another O_2. This cycle is analogous to the nitrogen catalytic cycle, with NO replaced by Cl and NO_2 by ClO. The individual reactions in the chlorine catalytic cycle are faster than those in the nitrogen cycle. Hence, this cycle is even more effective at destroying ozone. Or, to look at it in another way, less chlorine is needed than nitric oxide before the loss rate of ozone becomes significant.

Other Important Catalytic Cycles

Nitrogen and chlorine compounds are not the only compounds that affect stratospheric ozone. Ozone can also be destroyed by catalytic cycles involving bromine (Br) radicals and **hydroxyl** (OH) **radicals.** Bromine radicals have both natural and anthropogenic sources (as we shall see next), but hydroxyl radicals are entirely natural. Thus, the photochemistry of even the unperturbed stratosphere is, in reality, quite complex. Elaborate computer models are needed to simulate all of the possible catalytic cycles that affect the ozone concentration.

Sources and Sinks of Ozone-Depleting Compounds

The Odd Nitrogen Cycle

The NO and NO_2 that participate in the nitrogen catalytic cycle are referred to as **odd nitrogen** (NO_x) compounds, which contain an odd number of nitrogen atoms, to distinguish them from N_2. Odd nitrogen is similar in concept to the *fixed nitrogen* discussed in Chapter 9. The first term is used by atmospheric chemists, the second by biologists. In both cases, the important characteristic of the molecules is that the strong bond between the two nitrogen atoms in N_2 has been split. Hence, odd nitrogen molecules are much more reactive than is N_2.

Stratospheric odd nitrogen derives primarily from nitrous oxide, N_2O. The reaction that produces NO is

$$N_2O + O^* \longrightarrow 2\,NO.$$

Here, O* is an electronically excited atomic oxygen atom produced by the UV photolysis of ozone. It is much more reactive than normal, ground-state, atomic oxygen. (Normal O atoms are incapable of reacting with N_2O.) The N_2O itself comes from Earth's surface, where it is produced by microbial activity in soils and in the ocean (Fig. 14-7.) This activity is enhanced in places by the addition of nitrate fertilizers to the soil. Much of the nitrate in such fertilizers is taken up by growing plants. However, a substantial fraction undergoes bacterial denitrification (Chapter 9) and is subsequently released as either N_2 or N_2O.

The nitrous oxide makes its way to the stratosphere, where some of it reacts to form NO. The rest is photolyzed back to N_2 and O. The NO so produced participates in the ozone-destroying nitrogen catalytic cycle, forming NO_2 in the process. Every once in awhile, however, the resulting NO_2 molecule, instead of reacting with atomic oxygen, encounters a hydroxyl radical instead. The hydroxyl radical combines with the NO_2 molecule, producing nitric acid, HNO_3:

FIGURE 14-7

The atmospheric odd nitrogen cycle.

$$NO_2 + OH + M \longrightarrow HNO_3 + M.$$

The "M" in this reaction is a third molecule that carries off the excess energy of the collision. Nitric acid then diffuses down into the troposphere, where it dissolves in cloud droplets and is removed by precipitation. The nitric acid makes the rain slightly more acidic, but the amount of acid formed is so small that it poses no environmental problem. Acid rain itself is a problem in some areas, but it is caused by nitric and sulfuric acids formed within the troposphere.

Although N_2O is currently the largest source of stratospheric odd nitrogen, most of the concern about these compounds stems from potential increases in their abundance caused by high-flying, *supersonic transport* airplanes, or SSTs. Jet airplanes produce nitric oxide during the process of combustion. The reaction is similar to the production of NO by lightning. In each case, the high temperatures that are generated cause N_2 and O_2 to react with each other to form two NO molecules. Conventional jets fly in the upper troposphere and, hence, are not a threat to the ozone layer. The SSTs, however, inject nitric oxide and other exhaust products directly into the stratosphere. The French and British have been flying an SST called the *Concorde* since 1977, mostly on trans-Atlantic flights. So far, the frequency with which the *Concorde* has flown has not been high enough to affect stratospheric ozone seriously. Other governments, however, including the United States, are currently considering developing this type of aircraft. Because of the possible impact on ozone, careful environmental impact studies should be performed before large fleets of SSTs are built and flown.

The Chlorine Cycle

The attention of most stratospheric chemists is currently on chlorine. Chlorine is introduced into the stratosphere by several different gases produced at Earth's surface. The chlorine-containing gases that occur naturally are methyl chloride (CH_3Cl), and hydrogen chloride (HCl). Methyl chloride is produced in large quantities by marine plankton. Most of the methyl chloride released at the surface reacts in the troposphere, however. The amount that makes its way up to the stratosphere is enough to produce a stratospheric chlorine concentration of only about 0.6 ppb. By comparison, the current stratospheric chlorine concentration is approximately 3.3 ppb. So, most of this chlorine must derive from other sources.

Hydrogen chloride has received a great deal of attention because it is released in large quantities during volcanic eruptions. Particularly violent eruptions inject gases directly into the stratosphere. This fact has led some skeptics to suggest that most stratospheric chlorine derives from volcanoes and that we need not worry about anthropogenic sources. However, detailed studies of the El Chichón eruption in Mexico in 1982 and of the Mt. Pinatubo eruption in the Philippines in 1991 have led to the conclusion that very little of the HCl released in such events reaches the stratosphere. Most of the emitted HCl dissolves in water droplets that condense out of the volcanic plume. It is removed when these droplets fall out as rain.

Hydrogen chloride is also emitted by less violent volcanic eruptions and by the evaporation of sea spray. Seawater contains chloride ion, Cl^-, which forms HCl when the water evaporates. However, the HCl from these sources typically does not reach the stratosphere, because it, too, is removed by precipitation before it can leave the troposphere. The same is true of the chlorine released from swimming pools. This chlorine comes off the water surface as molecular chlorine, Cl_2, but is quickly converted to HCl by photochemical reactions in the lower troposphere. Hence, nearly all of it is removed from the atmosphere before it can damage the ozone layer.

The largest sources of stratospheric chlorine today are *chlorofluorocarbons* (CFCs), or *freons,* which are anthropogenic compounds. The two most common of these are freon-11 (CCl_3F) and freon-12 (CCl_2F_2). These compounds are also known as F-11 and F-12, respectively. Their atmospheric concentrations have increased since first measured in 1977 by about a factor of 2 since that time (Fig. 14-8). We mentioned these gases in Chapter 3 and again in Chapter 13 because they contribute to Earth's greenhouse effect. Freon-11 has been used as a propellant in spray cans and as a blowing agent for producing foams (Styrofoam®, for example). It is also used to clean semiconductor chips for computers and other electronic devices. Freon-12 is used as a refrigerant and until recently was the working fluid in most car air conditioners. The uses of freons are changing rapidly, though. Freon-11 has been banned from use in spray cans in the United States since 1978. Freon-12 has already been replaced by more ozone-friendly compounds in most new cars. This decrease in freon usage has caused the atmospheric growth rates of these compounds to decrease substantially in the 1990s. Indeed, freon-11 appears to have peaked in 1993–1994 and may be beginning a slow decline (see Fig. 14-8a). Freon-12 concentrations were still increasing in 1994 but not as rapidly as they had been earlier.

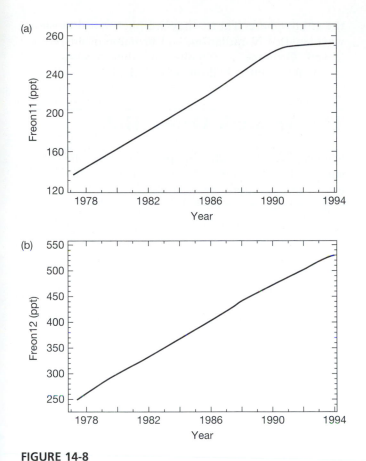

FIGURE 14-8

Atmospheric concentrations of (a) freon-11 and (b) freon-12 since 1977. (After the National Oceanic and Atmospheric Administration.)

FIGURE 14-9

The atmospheric chlorine cycle.

When freons were first introduced in the 1930s, they were considered to be wonder chemicals. Not only did they have useful thermodynamic properties, but also they were inert and nontoxic to humans. In contrast, early refrigerants such as ammonia and sulfur dioxide can be quite dangerous if they leak. It is this very property of being inert, however, that makes freons dangerous to the ozone layer (Fig. 14-9). Freons do not react in either the troposphere or the lower stratosphere. Hence, freon gases

How the Link Between Freons and Ozone Depletion Was Discovered

The first step in establishing the link between freons and ozone depletion was made by James Lovelock, the same Lovelock who formulated the Gaia Hypothesis (Chapter 1). Long before he became known as a theorist, Lovelock established his scientific reputation by inventing a device called the *electron capture detector*. When mounted on an instrument called a *gas chromatograph*, this device can be used to measure the concentrations of extremely dilute atmospheric gases. Lovelock used his detector to measure the concentration of freon-11 and other chlorine-containing gases. He published his measurements in 1970.

By comparing his measured concentrations with estimates of freon emission rates, Lovelock was able to show that most of the freon-11 that had been produced up until that time was still pre-

sent in the atmosphere. This meant that freon-11 must have a very long atmospheric lifetime. The lifetimes of freon-11 and freon-12 have since been determined to be about 60 years and 130 years, respectively. Ironically, Lovelock was not the least bit concerned about this finding, because freon gases were at that time considered to be totally harmless. Indeed, he wrote that "The presence of these compounds [CFCs] in the atmosphere constitutes no conceivable hazard." This sentence was added to ward off environmentalists, who were at that time quick to warn about the health effects of newly discovered chemicals.

Lovelock, though, was wrong on this last count, for reasons that were entirely different from what he had in mind. Chemists had known since the 1930s that Cl and ClO were capable of destroying ozone. It did not take long after Love-

lock's discovery for atmospheric scientists to put two and two together. The critical paper was published in 1974 by Sherwood Rowland and Mario Molina, two atmospheric chemists at the University of California at Irvine. Rowland and Molina pointed out that the long lifetimes of freon gases would allow them to diffuse up into the stratosphere, where the chlorine released by their decomposition could wreak havoc on the ozone layer. This idea, obvious in retrospect but quite novel at the time, caught on quickly in the scientific community. More than 10 years passed and many pitched battles were fought, however, before it was finally accepted by the chemical industry. These battles have been referred to as the "ozone wars." We owe a debt of gratitude to all the scientists who participated in those wars and helped bring us to the understanding that we have today.

released at the surface diffuse all the way to the upper stratosphere (above 40 km altitude). Once there, they are photolyzed by short-wavelength UV radiation, which breaks the molecules apart and releases their chlorine as atomic Cl. The Cl then proceeds to destroy ozone by way of the chlorine catalytic cycle.

Skeptics of the prospect that ozone depletion is anthropogenic occasionally suggest that freon gases, being heavier than air, should tend to stay in the lower atmosphere rather than diffusing upward to the stratosphere. (The molecular weight of freon-11, for example, is 136, whereas that of air is only about 29.) This idea, however, is based on a misconception about how the atmosphere mixes. In actuality, winds and eddies cause the atmosphere to be well mixed up to an altitude of about 100 km. Only above that height do heavier gases begin to separate from lighter ones. So, freons are able to deliver ozone-destroying chlorine atoms to the stratosphere despite their high molecular weights.

As in the case of the odd nitrogen cycle, the chlorine cycle in the stratosphere is eventually broken. In this case, Cl reacts with methane (or with H_2), forming hydrogen chloride:

$$Cl + CH_4 \longrightarrow HCl + CH_3.$$

Hydrogen chloride, like nitric acid, is relatively unreactive and diffuses downward into the troposphere, where it is removed by precipitation. Dissolved in water, HCl is the very strong acid hydrochloric acid. However, the actual amount of hydrochloric acid formed is relatively small, so it has little effect on the acidity of rainwater.

The Bromine Cycle

Atmospheric bromine, like chlorine, has several different sources. Methyl bromide (CH_3Br) like methyl chloride, is produced naturally as a byproduct of biological activity in the oceans. However, it is also used as a fumigating agent for soil pests, including termites. Some 15% of manufactured methyl bromide is used in California for agricultural fumigation. The other major source of atmospheric bromine involves artificial chemical compounds known as **halons,** which are used in certain types of fire extinguishers. The two most common of these are halon-1211 (CF_2ClBr) and halon-1301 (CF_3Br).

Some types of fire extinguishers use CO_2 instead of halons. Both gases are inflammable and heavier than air, so they are relatively efficient at keeping oxygen away from a flame. Being heavier than air does make a difference in this case, because the gas released from a fire extinguisher does not have time to disperse before it reaches the flame.

The atmospheric bromine cycle is analogous to the chlorine cycle: Most of the methyl bromide reacts in the troposphere, whereas the halons diffuse up into

the stratosphere. There, they are photolyzed by short-wavelength UV radiation, and bromine atoms are released. Bromine is removed by the formation and eventual rainout of hydrogen bromide, HBr.

The Antarctic Ozone Hole

Science, like life, is full of surprises. Perhaps the biggest surprise for most atmospheric chemists was the discovery of the Antarctic ozone hole in 1985. Recall from Chapter 1 that the ozone hole was overlooked for 6 years in the *Nimbus-7* satellite measurements because the observed column depths were considered too low to be real. Computer models also failed to predict its occurrence until long after it was observed. By taking a closer look at what causes the ozone hole, we can see why it caught atmospheric scientists off guard.

To begin, recall that the ozone hole occurs only during the month of October; it apparently did not occur at all prior to about 1976; and it is found only over Antarctica. No corresponding Arctic ozone hole has been seen, although scientists have been looking very carefully to see if one might develop. So, perhaps we should not be surprised that the explanation of the ozone hole turns out to be rather complicated.

Homogeneous and Heterogeneous Reactions

The chemistry that we have touched on in the last few sections is reasonably well understood. The chemical reactions that we have discussed are all **homogeneous reactions**—that is, reactions between molecules that are in the gas phase. We can study gas-phase reactions in the laboratory if we keep the reacting molecules away from the walls of their container. In contrast, the chemistry that causes the ozone hole involves **heterogeneous reactions,** such as reactions that occur on solid surfaces, such as particles. These reactions are more difficult to study experimentally; consequently, much less is known about them.

The particles involved in the formation of the Antarctic ozone hole are collections of droplets called **polar stratospheric clouds** (PSCs). These clouds were first discovered by high-flying spy planes. Over most of the globe, clouds form only in the troposphere; the stratosphere is too dry for condensation to occur. In winter the polar stratosphere, however, is so cold ($-80°C$ or below) that certain trace atmospheric constituents can condense. The particles that form typically consist of a mixture of water and nitric acid. Although they are very tenuous by comparison with normal, tropospheric clouds, these PSCs alter the chemistry of the lower stratosphere in two fundamental ways: (1) by coupling between the odd nitrogen and chlorine cycles and (2) by providing surfaces on which heterogeneous reactions can occur, as we shall see next.

Coupling between the Odd Nitrogen and Chlorine Cycles

Throughout most of the lower stratosphere, the nitrogen and chlorine cycles are coupled by way of the reaction

$$ClO + NO_2 + M \longrightarrow ClONO_2 + M.$$

The product formed in this reaction, chlorine nitrate ($ClONO_2$), does not react directly with either ozone or atomic oxygen and can be converted back to ClO only with difficulty. Thus, it serves as a relatively inert storage reservoir for chlorine, keeping it out of the more reactive forms, Cl and ClO. Because they can directly catalyze ozone destruction, Cl and ClO, along with compounds that can be easily converted into them by photolysis, are collectively termed **reactive chlorine.**

Under normal conditions in the polar stratosphere, NO_2 is always sufficiently abundant to tie up a significant fraction of the available chlorine in the form of chlorine nitrate. In the wintertime Antarctic stratosphere, however, NO_2 concentrations become very low, because most of the odd nitrogen has been converted into HNO_3 and subsequently incorporated into cloud droplets as PSCs. This removal allows reactive chlorine concentrations to increase, because less chlorine is bound up as chlorine nitrate.

The PSC particles also help convert unreactive forms of chlorine into reactive chlorine by providing surfaces on which heterogeneous reactions can occur. For example, one reaction that is thought to be important is

$$ClONO_2 + HCl \longrightarrow Cl_2 + HNO_3.$$

Molecular chlorine, Cl_2, does not react directly with ozone itself but is readily photolyzed to atomic chlorine:

$$Cl_2 + photon \longrightarrow Cl + Cl.$$

Once formed, atomic chlorine can execute its destructive effect on ozone.

The Polar Vortex

We now have the chemistry we need to understand how the ozone hole forms. Some knowledge of stratospheric circulation is required, however, to complete the story. We saw in Chapter 4 that the prevailing winds at polar latitudes blow from east to west. This is true in the stratosphere as well, especially during winter. When viewed from above the South Pole, this wintertime circulation pattern appears as a gigantic whirlpool called the **polar vortex** (Fig. 14-10). The cold, dense air in the middle of the vortex is subsiding in much the same way as water going down a drain. The sinking air carries cloud particles along with it, permanently removing odd nitrogen from the stratosphere. The polar vortex also effectively isolates the Antarctic stratosphere from the rest of the globe. The air inside the vortex is rotating much faster than the air outside, making it difficult for air to pass through the vortex boundary. Some air does get through and replaces the air that is sink-

FIGURE 14-10

Schematic diagram of the Antarctic circumpolar vortex. (From Tom Van Sant/Geosphere Project, Santa Monica/Science Photo Library/Photo Researchers, Inc.)

ing in the center of the vortex, but the amount is small compared with normal rates of latitudinal mixing. Thus, the odd nitrogen removed from the polar stratosphere during winter is not replaced by odd nitrogen from lower latitudes. The same phenomenon also ensures that very little new ozone can be brought in once the ozone hole forms.

The full story of the ozone hole involves an intricate coupling of chemistry, atmospheric circulation, and the availability of sunlight. During the long, Antarctic winter (May through September), the stratosphere becomes cold enough to allow PSCs to form. These PSCs deplete the polar stratosphere of odd nitrogen and help convert unreactive forms of chlorine ($ClONO_2$ and HCl) into more reactive forms, such as Cl_2. The reactive chlorine, however, remains bound to the surfaces of cloud particles until late September, when the Sun first peeks over the Antarctic horizon. The sunlight releases reactive chlorine from the particle surfaces, and the chlorine wreaks havoc with the ozone layer during October. Near the end of the month, the polar vortex breaks down, allowing fresh ozone and odd nitrogen to be brought in from lower latitudes. And, so, the ozone hole disappears until the following October.

Why does this process not occur over the Arctic during Northern Hemisphere spring? The answer seems to lie in the stratospheric circulation. As a result of geography, the Arctic polar vortex is less fully developed than the Antarctic vortex and breaks up earlier in the spring. The Southern Hemisphere contains a roughly symmetric continent (Antarctica) surrounded by a large ocean. The Northern Hemisphere contains a mixture of land and ocean, and it also has large, north–south mountain ranges, such as the Rocky Mountains in North America. The mountains disrupt the east–west tropospheric circulation and create atmospheric waves that propagate up to the stratosphere and deposit energy there. This energy warms the stratosphere and disrupts the smooth, circumpolar flow. If an Arctic ozone hole were to occur, it would form during April. By this time of year, however, the Arctic polar vortex has already disappeared. Thus, the unique combination of conditions that occur in the Southern Hemisphere is never duplicated in the north.

Evidence of Midlatitude Ozone Depletion

The fact that we do not observe an ozone hole over the Arctic does not mean that residents of the Northern Hemisphere should be complacent about ozone depletion. Indeed, there is considerable evidence that ozone is decreasing slowly at midlatitudes in both hemispheres. The data come from three satellite-borne ozone-monitoring instruments.

Sources of Natural Variability in Ozone

Detecting trends in ozone column depths at low- and midlatitudes is difficult, because the rates of change are not nearly as large as those seen over Antarctica. Furthermore, before we can identify any long-term trends, we must first correct for natural variability. Figure 14-11, which shows satellite measurements of ozone column depth over Hohenpeissenberg, Germany, illustrates the problem. The first signal that we see in the data is a strong annual cycle of ±75 Dobson units. For reasons that are related to stratospheric circulation patterns, ozone is most abundant during the spring and least abundant during the fall. (Verify this by referring back to Fig. 14-5.) But there are other, more subtle oscillations in the data as well. These oscillations include the solar cycle—the 11-year cycle in sunspot number—and the *quasi-biennial oscillation,* or QBO, which is a 27-month cycle in the direction of winds in the equatorial lower stratosphere.

The sunspot cycle affects ozone because the amount of ultraviolet radiation emitted by the Sun varies with sunspot activity. When sunspots are abundant (termed **solar maximum**), as they were in 1991, the Sun gives off more UV energy. The added UV radiation comes not from the spots themselves but from bright areas called plages surrounding the spots (see Chapter 12). Higher UV fluxes cause more O_2 to be photolyzed, resulting in increased production of ozone. At times of low sunspot activity **(solar minimum),** as in 1986 and 1997, the solar UV flux is low and ozone production decreases.

The effect of the solar cycle on stratospheric ozone is illustrated in Figure 14-12. The solid curve shows the flux of 10.7-cm radio emission from the Sun, which happens to be well correlated with the solar UV flux. The dashed curve shows the percentage change in average ozone column depth (40° N to 40° S) after the seasonal cycle, QBO, and longer-term trend have been removed. Ozone column depth is about 2 to 3% higher during solar maximum than during solar minimum. As this variation is of the same magnitude as the trend in the satellite data, it, too, must be removed before the trend can be determined.

Long-Term Trends in Ozone

By removing all the known sources of natural variability from the satellite data, it is possible to identify long-term trends in ozone column depths. The trend for the entire region between 60° N and 60° S is shown in Figure 14-13. The average ozone column depth apparently decreased by about 2.9% per decade between 1979 and 1991. The downward trend over the entire dataset (1979 to 1994) is even steeper, but the last 3 years of data were anomalously low because of the eruption of Mt. Pinatubo in

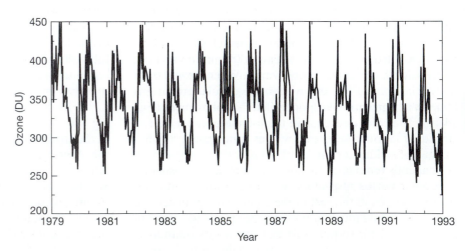

FIGURE 14-11

Weekly averaged satellite data on ozone column depth over Hohenpeissenberg, Germany. A strong annual cycle is observed in the data. (After World Meteorological Organization, 1994.)

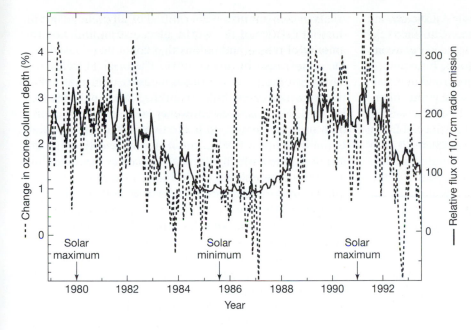

FIGURE 14-12

Response of average ozone column depth ($40°$ N to $40°$ S) to the solar cycle, after the seasonal cycle, quasi-biennial oscillation, and long-term trend have been removed. (After World Meteorological Organization, Scientific Assessment of Ozone Depletion 1994.)

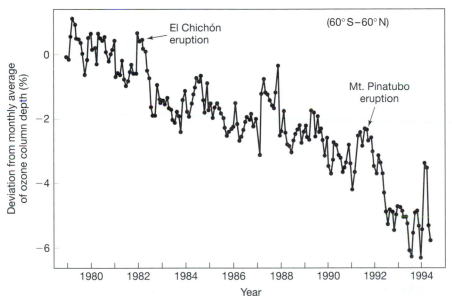

FIGURE 14-13

Percentage deviation of monthly averaged ozone column depth ($60°$ N to $60°$ S) from the 1979 value. (After World Meteorological Organization, Scientific Assessment of Ozone Depletion 1994.)

June 1991. This eruption, and that of the Mexican volcano El Chichón in 1982, injected large amounts of sulfur dioxide, SO_2, into the stratosphere. The SO_2 was oxidized to sulfuric acid, which then condensed, forming a layer of sulfate aerosols. The aerosols depleted the ozone layer by blocking some of the incident solar UV radiation and by providing surfaces on which heterogeneous reactions could take place. In each case, the ozone layer recovered within 2 to 3 years, because most of the aerosol particles had fallen out by this time.

If we look at the ozone data more closely, we see that the downward trend in ozone was not uniform over the region $60°$ N to $60°$ S but, rather, was confined to midlatitudes (Table 14-4). Midlatitude ozone has been decreasing by approximately 4% per decade in both hemispheres, whereas

TABLE 14-4

Percentage Change in Ozone Column Depth per Decade, 1979–1991*	
Latitude Belt	*% Change*
Mid North	-4.0 ± 2.1
Equatorial	$+0.4 \pm 2.1$
Mid South	-4.5 ± 2.1

* *Source: Scientific Assessment of Ozone Depletion: 1994*, World Meteorological Organization

tropical ozone has remained essentially constant. (The data actually indicate a slight upward trend in the tropics, but the magnitude of the trend is smaller than the errors in the measurements.) The trend at midlatitudes is statistically significant and is an obvious cause for concern. Surface UV fluxes would increase substantially if this trend were to continue for even another decade or two into the future.

What could be causing the downward trend in ozone at midlatitudes? The obvious candidates are stratospheric chlorine and bromine, which have been increasing throughout the past few decades as a consequence of freon and halon emissions. However, atmospheric chemists who have tried to simulate this effect with computer models have been able to account for only about one-third of the observed rate of ozone decrease. The models are fairly accurate during summer and fall, but they miss badly during winter and spring, which is when most of the observed depletion seems to be occurring. Either the chemistry of ozone is still not fully understood or the stratosphere is even more complex than we have already discovered.

Mechanisms for Halting Ozone Depletion

We began this chapter by pointing out that ozone depletion is now widely recognized as a global threat and that steps have already been taken to reduce or eliminate the problem. These steps fall into two categories: (1) the ratification by the international community of treaties designed to reduce freon and halon emissions and (2) the development of "ozone-friendly" substitutes for freons by the chemical industry.

International Agreements

The first and most important step in terms of diplomacy was the signing of the **Montreal Protocol** in 1987. This treaty, which has now been ratified by all of the major industrial nations of the world, placed strict limits on the amount of freons and halons that could be released into the atmosphere by any country. The original treaty was agreed on prior to the crucial measurement that linked the Antarctic ozone hole to anthropogenic chlorine (see Fig. 1-7). That discovery, however, solidified international opinion on the issue and led to two subsequent revisions of the treaty (the 1990 *London Accord* and 1992 *Copenhagen Accord*), which ultimately placed even stricter limits on freon emissions (Table 14-5).

The expected effect of these treaties up to A.D. 2075 is shown in Figure 14-14. The figure shows that in the absence of any kind of international agreement ("business as usual"), stratospheric chlorine levels could be expected to rise quite rapidly over the next few decades. Under such circumstances, computer models predict that ozone column depths could decrease precipitously. The Montreal Protocol alone would have slowed this process, but stratospheric chlorine concentrations would still have continued to rise. With the most recent international treaty (the Copenhagen Accord) in place, the outlook is much improved. Tropospheric concentrations of freons and halons peaked in 1995 and are beginning a slow decline. Gas concentrations in the stratosphere lag those in the troposphere by 3 to 5 years, so stratospheric chlorine and bromine abundances will probably continue to increase until about the turn of the century. By A.D. 2060, stratospheric chlorine should be back down to about 2 ppb, its value in the late 1970s, when the Antarctic ozone hole first appeared. Hopefully, the hole will disappear at that time. Within the following century, stratospheric chlorine should reach the background level of 0.6 ppb that results from naturally produced methyl chloride.

TABLE 14-5

International Controls on Ozone-Depleting Substances (ODS) (Montreal Protocol with Adjustment and Amendments)*		
Meeting	*ODS Controlled*	*Control Provisions*
Montreal, 1987	Higher-volume CFCs	50% reduction over 10 years
	Halons	Freeze beginning 1/1/1992
London, 1990	CFCs	Phaseout by 1/1/2000
	Halons	Phaseout by 1/1/2000
	Carbon tetrachloride	Phaseout by 1/1/2000
	Methyl chloroform	Phaseout by 1/1/2005
Copenhagen, 1992	CFCs	Phaseout by 1/1/1996
	Halons	Phaseout by 1/1/1994
	Carbon tetrachloride	Phaseout by 1/1/1996
	Methyl chloroform	Phaseout by 1/1/1996
	HCFCs	Cap with phaseout by 1/1/2030
	HBFCs	Phaseout by 1/1/1996
	Methyl Bromide	Freeze beginning 1/1/1995

* Ten-year delay for developing countries.

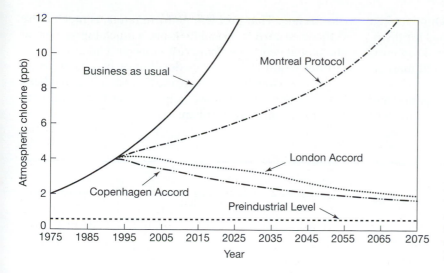

FIGURE 14-14

Projected atmospheric chlorine concentrations under the various international agreements. The horizontal dashed line shows the estimated preindustrial level of chlorine. (After Dr. M. McFarland at DuPont.)

Freon Substitutes

Can the world make do without conventional freon gases? The answer appears to be yes. Substitutes for freons are already being developed by major chemical companies. At least two different strategies are being employed. One of these is to replace one of the chlorine or fluorine atoms with a hydrogen atom to make compounds referred to as *HCFCs*. An example is the compound HCFC-22 ($CHFCl_2$), which can be used as a substitute for freon-12. Attaching a hydrogen atom to a CFC makes the CFC susceptible to chemical attack by the hydroxyl radical, OH (Fig. 14-15). Recall from earlier in the chapter that radicals play an important role in removing pollutants from the atmosphere. Because of its increased reactivity, an HCFC-22 molecule released at the surface has only a 5% probability of reaching the stratosphere, compared with a virtually 100% probability for freon-11 or freon-12. The other 95% of the HCFC-22 is destroyed in the troposphere, where it can do no harm to ozone. So, this compound is 20 times more ozone-friendly than normal freon gases.

A potential downside of this strategy is that the more reactive HCFCs may pose direct health threats. Hence, thorough testing of the biological effects of these compounds is required. On the bright side, HCFC-22 has a short lifetime (about 15 years), which means that it contributes far less greenhouse-enhanced warming than does a long-lived gas such as freon-12. Thus, replacing CFCs with HCFCs can help slow global warming while it reduces the destruction rate of stratospheric ozone.

A second strategy is to eliminate chlorine from the CFC molecule, creating an HFC compound. An example of this type is the compound HFC-134a (CF_3CH_2F), which is already in use as a substitute for freon-12 in some new cars. Because HFCs contain no chlorine, they pose no significant threat to ozone. (The reason is somewhat paradoxical: Fluorine radicals are *more* reactive than Cl radicals. Indeed, they are so reactive that they combine directly with water vapor to form hydrofluoric acid, HF: $F + H_2O \rightarrow HF + OH$. Like HCl, HF is a tightly bound molecule that drifts down to the troposphere, where it is removed by rainout. So, fluorine radicals are removed from the stratosphere before they do much damage to ozone.) Also, HFC-134a is relatively short-lived (about a 16-year lifetime), so it is a good choice in terms of climate as well. As shown in Table 14-5, HCFCs are scheduled to be eliminated by A.D. 2030, when they will likely be replaced by HFCs. Hopefully, this action will eliminate all threat to the ozone layer from these types of compounds.

Lessons Learned from the Ozone Experience

The manner in which the international community and the chemical industry have responded to the threat of ozone depletion shows that we *can* deal successfully with global environmental problems. This success, however, did not come without a fight. In the 1970s, when the danger posed by freons had just been discovered, the chemical industry vigorously opposed the idea that their products might be harmful. A concerted effort by atmospheric scientists associated with universities and national laboratories caused the industry to look more closely at

Hydroxyl radical
OH

Cl–C–H ⟶ Cl–C· + H–O–H

HCFC-22   Unstable radical   Water

FIGURE 14-15

Hydroxyl (OH) radicals can react with hydrogen atoms in HCFCs to form unstable radicals plus water.

the problem. Eventually, companies such as Dupont hired their own atmospheric chemists and confirmed for themselves that action needed to be taken. Now, Dupont is leading the chemical industry in the design and marketing of freon substitutes.

It remains to be seen whether energy companies and consumers will follow a similar path with regard to global warming. That problem, unfortunately, is much more difficult to correct. Fossil fuels are a much bigger part of the global economy than are freons, and adequate substitutes might be harder to develop or be prohibitively expensive. Hopefully, though, the precedent that has been set on ozone depletion will help guide us to a solution to the global warming problem.

Chapter Summary

1. Solar ultraviolet radiation between 200 nm and 320 nm poses significant dangers to humans and other organisms if not effectively blocked by stratospheric ozone. The wavelength region of greatest concern is the UVB region, between 290 nm and 320 nm, where the solar flux is relatively high and the ozone absorption coefficient is relatively low.

2. The flux of solar UV radiation that reaches Earth's surface depends on the ozone column depth, which is a measure of the total amount of ozone overhead. The column depth varies regionally; it is generally highest near the poles and lowest near the equator. This distribution, combined with the fact that the Sun is generally higher in the sky near the equator, produces much higher ground-level UV fluxes in the tropics than at high latitudes.

3. Ozone chemistry in a pure oxygen–nitrogen atmosphere can be described by a set of four reactions known as the Chapman mechanism.
   a. The first and last of the Chapman reactions affect the abundance of odd oxygen (O_3 and O); the middle two reactions cause the different forms of odd oxygen to interconvert.
   b. In the real atmosphere, catalytic cycles involving nitrogen, chlorine, and bromine provide alternative, faster ways of destroying odd oxygen. Both chlorine and bromine have large anthropogenic sources (freons and

halons, respectively) that are considered to be a threat to the ozone layer today.

4. The most striking evidence of ozone depletion comes from Antarctica, where an ozone hole forms each year during October.
   a. The ozone hole is apparently caused by a complex series of reactions that occur on the surfaces of polar stratospheric cloud particles. These same particles remove odd nitrogen from the polar stratosphere, breaking the coupling between the odd nitrogen and chlorine cycles and allowing free chlorine to wreak havoc on the ozone layer.
   b. The Arctic stratosphere behaves differently than the Antarctic stratosphere, because the wintertime polar vortex is not as well developed or long-lasting in the Arctic.
   c. Ozone has also been decreasing at a lesser rate at midlatitudes in both hemispheres for reasons that are not fully understood.

5. International agreements, namely the Montreal Protocol and the subsequent London and Copenhagen Accords, have placed strict limits on future freon and halon emissions. If these accords are enforced and obeyed, stratospheric chlorine concentrations should decline to their natural levels during the course of the next century, and the ozone hole should disappear. It is important that we continue to recognize the wisdom of these agreements and the reasons why they were established.

Key Terms

absorption coefficient
action spectrum
catalyst
catalytic cycle
Chapman mechanism
column depth
Dobson unit
dose rate
erythemal action spectrum

halons
heterogeneous reaction
homogeneous reaction
hydroxyl radical
Montreal Protocol
number density
odd nitrogen
odd oxygen
ozonesondes

photodissociation
photolysis
polar stratospheric clouds
polar vortex
reactive chlorine
solar maximum
solar minimum

Review Questions

1. What are the three categories of UV radiation? Which of these are considered to be biologically harmful?
2. What is ozone column depth? In what units is it measured?
3. What are the four chemical reactions that comprise the Chapman mechanism? Which ones affect odd oxygen?
4. What is a catalyst?

5. How do nitrogen and chlorine catalyze the destruction of ozone?
6. What role do polar stratospheric clouds play in the formation of the Antarctic ozone hole?
7. Why is a springtime ozone hole observed over the Antarctic but not over the Arctic?

8. What is the long-term trend in midlatitude ozone column depths? Can it be explained by the observed increase in stratospheric chlorine and bromine?

9. What strategies have been adopted for reducing or eliminating the use of freons?

Critical-Thinking Problems

1. a. According to Figure 14-5, the maximum ozone column-depth occurs at high northern latitudes during late winter. The column depth there is 460 DU. How many molecules of ozone per square centimeter does this correspond to?
   b. The minimum ozone column depth, which occurs in the tropics, is 240 DU. How many ozone molecules are in a 1-cm^2 vertical column there?
   c. What was the approximate ozone column depth (in Dobson units) over Wallops Island, Virginia, at the time of the measurements shown in Figure 14-3? [*Hint:* Ozone column depth is ozone concentration (in molecules/cubic centimeter) multiplied by the height of the column. You may want to use a different average ozone concentration for the troposphere and the stratosphere.]

2. The absorption of solar ultraviolet radiation of a given wavelength λ by atmospheric ozone follows Beer's Law:

$$\frac{F}{F_0} = \exp\left[\frac{-kN}{\cos\theta}\right],$$

where F = the UV flux at altitude z; F_0 = the UV flux at the top of the atmosphere; k = the absorption coefficient at wavelength λ; N = the ozone column depth above height z; and θ = the solar zenith angle.

The function $\exp(x)$ (or e^x on some calculators) is the exponential function, which also describes radioactive decay, population growth, and other interesting phenomena.
   a. The wavelength region where changes in the solar UV flux have the most potential to do harm is around 290 nm, in the UVB range. The absorption coefficient of ozone at this wavelength is about 10 atm^{-1}-cm^{-1}. The average column depth of ozone from the ground up to the top of the atmosphere is about 0.3 atm-cm. By what factor is the incident solar UV flux at 290 nm attenuated today—that is, what is the current value of F/F_0 at ground level? Evaluate your answer for a solar zenith angle of 45°.
   b. If the ozone column depth were to be reduced by 1% as a consequence of increasing concentrations of chlorofluorocarbons, what would be the resulting percent increase in the ground-level UV flux at 290 nm? What about for an ozone decrease of 10%? 50%? Assume a solar zenith angle of 45° in each case.
   c. Suppose that we decided that we could tolerate a 10% increase in the UV flux at 290 nm for $\theta = 45°$, but no more. What would be the maximum percentage decrease in ozone that we could allow? [*Hint:* The inverse of the exponential function is called the natural logarithm, abbreviated as *ln*. If $y = \exp(x)$, then $x = \ln(y)$.]

Further Reading

General

L. Dotto and H. Schiff, *The Ozone War,* Doubleday, Garden City, NY, 1978.

R. P. Turco, *Earth Under Siege* (Ch. 13), Oxford University Press, New York, 1997.

Advanced

D. L. Albritton, R. T. Watson, and P. J. Aucamp, *Scientific Assessment of Ozone Depletion: 1994,* World Meteorological Association, Geneva, Switzerland, 1994.

Chapter 15
Human Threats to Biodiversity

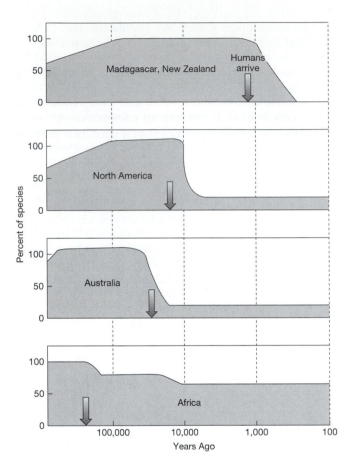

The extinction of large mammals and birds corresponds to the spread of human populations. Humans evolved first in Africa; in fact, they coevolved with other animals there over millions of years. As a result, the curve for Africa stands out from the others. (After E. O. Wilson, 1992. *The Diversity of Life,* New York: W. W. Norton and Company)

Key Questions

- Where is species loss occurring, and why?
- What are ecosystems, how do they work, and how do they relate to bidiversity?
- Why should we care about species loss?
- Could the loss of biodiversity have important consequences for humans and for the health of the Earth system?

Chapter Overview

Today, humans are responsible for an extinction rate of species that rivals any of the mass extinctions Earth has experienced in the past. Most of this species loss is occurring as a result of human land-use practices that destroy natural habitats, and we may soon reach the point at which not only are species becoming extinct, but also we are in danger of destroying whole ecosystems. This species loss is particularly severe in tropical rainforests. Focusing on ecosystems rather than simply considering individual species is important. The loss of biodiversity could affect the health or stability of the planet and may well reduce world food supplies.

Introduction

In Chapters 13 and 14 we saw two ways in which human activity is having, or has the potential to have, a significant impact on the Earth system. In this chapter we will address a third issue, that of biodiversity and the loss of species. In Chapter 10 we introduced the concept of biodiversity and discussed the mass extinctions that have occurred in the past. We also illustrated the dramatic changes that took place in the biosphere as a result of these extinction events. What many people fail to realize is that the rate of species loss due to human activity in the present is equal to, or greater than, the rate of species loss that marked these past mass extinctions.

The implications of species loss can be harder to grasp than are those of other environmental threats we have discussed previously. It is relatively easy to see the impact of the loss of stratospheric ozone or of a change in climate. Most of us have experienced extreme summers or winters in the past and can imagine what it might be like if those extremes were to become the norm in the future. Few of us, however, have any direct experience of species loss or of the consequences that it might entail.

On evolutionary time scales, new species continuously develop and others go extinct; in fact, the average life-time of most species is between 1 million and 10 million years. When we consider that life has been present on Earth for the past 3.8 billion years, it is apparent that many species have evolved and disappeared over that time period. Indeed, most of the species that have ever lived are now extinct. However, the fossil record shows that the rate of **speciation** (the origination of new species) is slightly greater on average than the rate of extinction. Consequently, there are more species today than there have ever been at any single time in the past.

- How many species are living today? We do not know for certain. Approximately 1.4 million species have been described, and new species are being found faster than they can be catalogued.

- How many more species have yet to be discovered? Estimates range from 10 million to 100 million.

- How many species go extinct each year? Again, we do not know. The estimates are based on the proportion of species that are lost; hence they also depend on the estimates used for the existing number of species.

Some estimates suggest that at the present rate of extinction, one-quarter of all species on Earth may be lost within the next 50 years. If we accept 100 million as the number of existing species, then the extinction rate is half a million species per year. If 10 million is closer to reality, then the extinction rate is approximately 50,000 species each year. Either way, we do know that species disappear every day as a direct consequence of human activity.

Why is this massive species loss occurring, and does it make a difference to us? These are the questions that we will address in this chapter. Using some of the same systems concepts developed in previous chapters, we will examine processes in the biosphere at various scales. Recall from Chapter 1 that the *biosphere* comprises that part of Earth inhabited by organisms; hence the term includes both living and nonliving components. A simple hierarchy has been developed that subdivides the biosphere (Fig. 15-1). The smallest subunit is the *species,* which consists of all closely related organisms that can potentially interbreed. All the members of a single species that live in a given area make up a **population.** In any area you will tend to find a characteristic assemblage of two or more groups of interacting species, known as a **community.** A community may include any combination of animals, plants, fungi, and microbes. A region with a characteristic plant community (such as a desert or tropical rainfor-

est) is called a **biome.** A community of animals, plants, fungi, and microbes, together with the physical environment that supports it, is referred to as an **ecosystem.** All the ecosystems on Earth in turn make up the biosphere. Although it is usual to discuss biodiversity and extinction in terms of species, it is important to recognize that one species does not exist independent of the other species around it. Species coexist and interact with a specific assemblage of other species in ecosystems.

Much of our discussion will revolve around ecosystems, and it is possible to analyze individual ecosystems by using the same systems approach that we have used in previous chapters. Note, however, that we know very little about ecosystems from an Earth systems perspective. Although some studies describe the cycling of a particular nutrient through a given ecosystem or the transfer of energy or water through that ecosystem, there are very few studies of how any particular ecosystem functions in its totality. Furthermore, we have little idea of how a particular assemblage or distribution of ecosystems interacts at the global scale. In earlier chapters we were often able to reduce the complexity of the Earth system to a few relatively simple concepts and systems diagrams, but that is not really possible here. Ecologists have been studying ecosystems for some time and have made extensive use of systems theory. We could fill this chapter with diagrams illustrating the different functions of numerous ecosystems, but none of them would answer the questions posed earlier: Why is the species loss occurring? and Should we care about it?

In Chapter 10 we introduced the idea of a possible relationship among diversity, community stability, and environmental stability; however, very little work has actually been done along these lines. In the remainder of this chapter, we will describe some of the ecosystems in which species are being lost some of the reasons why this loss is occurring. We will then talk about the value of species, revisiting the idea that biodiversity is vital for the long-term health of our planet.

Ecosystems

What Are Ecosystems?

As we have said, ecosystems are subsets of the (global) biosphere, assemblages of animal, plant, fungal, and microbial species that interact with each other and their surrounding environment (see "Species Interactions," below). For terrestrial ecosystems the environment includes the topography, soils, atmosphere, and climate. For aquatic ecosystems the environment includes the physical and chemical characteristics of the particular freshwater or saltwater body concerned. Each ecosystem is located in a slightly different physical environment; consequently, there is a tendency to think of the environment as controlling the type of ecosystem that develops. This is

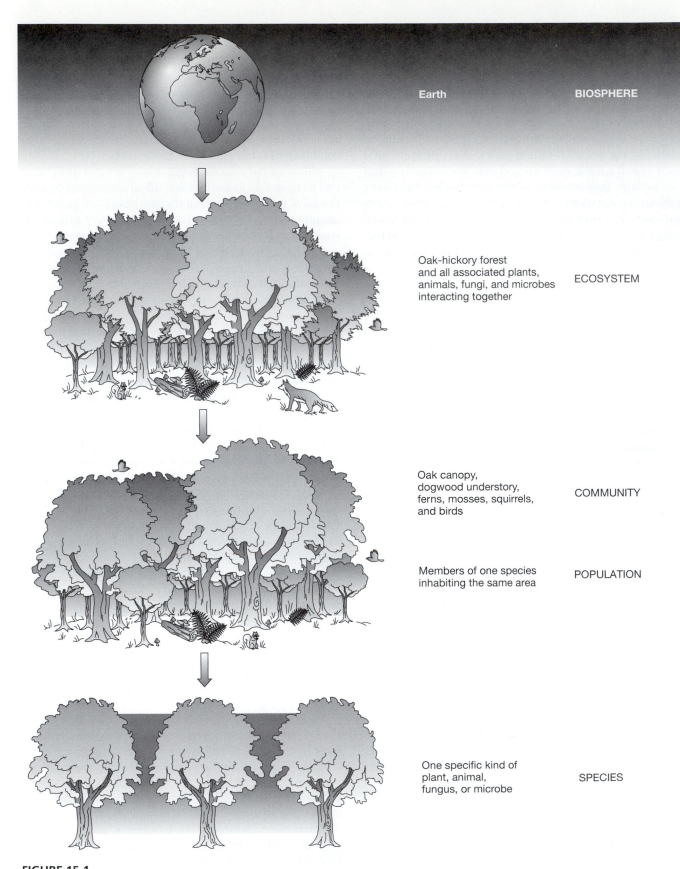

Earth   BIOSPHERE

Oak-hickory forest
and all associated plants,
animals, fungi, and microbes
interacting together   ECOSYSTEM

Oak canopy,
dogwood understory,
ferns, mosses, squirrels,
and birds   COMMUNITY

Members of one species
inhabiting the same area   POPULATION

One specific kind of
plant, animal,
fungus, or microbe   SPECIES

FIGURE 15-1

Division of the biosphere into ecosystems, communities, populations, and species. (From Nebel and
Wright: *Environmental Science,* 4/e, 1993. Reprinted by permission of Prentice Hall, Upper Saddle
River, N.J.)

true to some degree, but it is not the whole story: The organisms within a particular ecosystem interact with their environment. For example, the type of soil present may help determine which plants will grow, but the plants themselves add organic matter to the soils that changes the soil chemistry, possibly allowing different species to grow there. Certain species might tolerate only particular temperature and precipitation regimes, but those species may be capable of altering local climate, thereby promoting their own growth or the growth of other species.

We indicated this possibility with James Lovelock's Daisyworld example in Chapter 2. We also illustrate it here with an example from the boreal forests of North America and Asia. The coniferous trees in the boreal forest hide the snow-covered ground in winter, reducing the albedo of the forested region (the albedo of the forest canopy is much lower than the albedo of the snow cover). The reduced albedo should result in higher winter temperatures than would occur if the forest were not present. By modifying its local environment at its northern edge, Lovelock suggests that the forest may be able to push farther north beyond the point where temperatures would otherwise be too cold to allow the trees to grow. Is there any indication that this might, in fact, be happening? Scientists at the National Center for Atmospheric Research in Boulder, Colorado, conducted a general circulation model experiment in which they changed all of the forest north of 45° N to bare ground. This is equivalent to moving the border between the boreal forest and the treeless tundra southward. The effect of this change was to produce a large increase in the wintertime albedo, because the snow cover was revealed by the removal of the forest. In the model, as in the real world, where snow occurs in forested regions (particularly evergreen forests), it is covered by the forest canopy, and the albedo is much less than that of snow-covered bare ground. With the removal of the forest, the increased albedo caused a large drop in air temperatures, the greatest change being in April—up

to a 12°C (21.6°F) decrease over the land surface. The colder winter temperatures increased the sea-ice cover, and the higher albedo further enhanced the cooling effect. The colder temperatures were maintained through the summer, with July being as much as 5°C (9°F) colder than before the removal of the forest. These lower summer temperatures were caused by the earlier winter cooling, which reduced sea-surface temperatures in the Arctic Ocean. With less energy available in the ocean, warming was inhibited during the summer.

These processes are illustrated in the systems diagram in Figure 15-2. The solid lines indicate the interactions simulated by the model, including the ice–albedo feedback discussed in several earlier chapters. The dashed line completes another positive feedback loop implied by the model results. The forest cover is not an interactive part of the model, so we cannot see the feedback from the change in temperature to the change in forest cover. However, the model does show that the July 18°C (64.4°F) isotherm (which correlates well with the present northern limit of the forest) shifts southward far enough to prevent forest regrowth. Although the interactions are a little more complex than those suggested by Lovelock, we do see that, at least in terms of the model, the forest has a significant impact on the climate system. By keeping high-latitude temperatures from being as cold as they would otherwise be, the forest helps perpetuate an environment conducive to their own growth.

It is very apparent, therefore, that ecosystems are not divorced from their environment. As the environment changes, the ecosystem changes; as the ecosystem changes, it can in turn change the local environment. The obvious conclusion is that ecosystems are not static. Changes in climate can cause ecosystems to move gradually to new locations, such as arctic tundra and its biologic community spreading equatorward during glacial periods and retreating poleward during interglacials. More interestingly, new environmental conditions could give rise to a totally new assemblage of species: an ecosystem that has not been seen

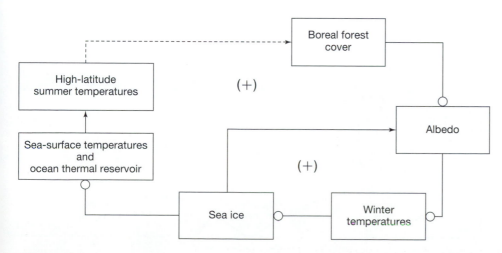

FIGURE 15-2

Possible feedbacks between the boreal forest and climate.

before. Furthermore, because ecosystems and the environment interact, it may be possible for a new ecosystem to evolve without any large-scale change in the environment. In this case, the initial environmental conditions support one ecosystem but the interactions change the local environment, so the ecosystem evolves into something new.

Although Figure 15-1 implies an ordered hierarchy, the levels are not discrete and ecosystems themselves are not discrete units. Each level in the hierarchy interacts with all other levels, and ecosystems overlap each other. One ecosystem gradually merges into another geograph-

ically at a diffuse boundary called an **ecotone** (Fig. 15-3). An ecotone may include an entirely different assemblage of species that do not match those in the ecosystems on either side of the boundary (Fig. 15-4). There is also considerable overlap between ecosystems in a structural sense: Several ecosystems may share many common physical attributes. According to the plant communities ecosystems suport, we can identify various terrestrial biomes; the seven most important are shown in Figure 15-5 (Plate 16).

Note that whereas the term *ecosystem* applies to the total assemblage of species in a particular environment,

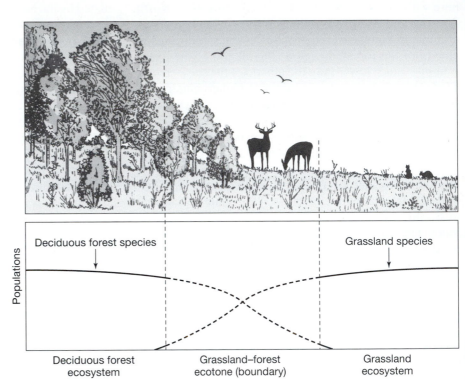

FIGURE 15-3

Ecosystems and ecotones. (From Nebel and Wright: *Environmental Science,* 6/e, 1998. Reprinted by permission of Prentice Hall, Upper Saddle River, N.J.)

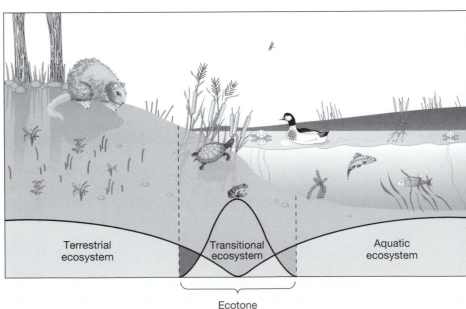

FIGURE 15-4

An ecotone may create a unique habitat of its own. (From Nebel and Wright: *Environmental Science,* 6/e, 1998. Reprinted by permission of Prentice Hall, Upper Saddle River, N.J.)

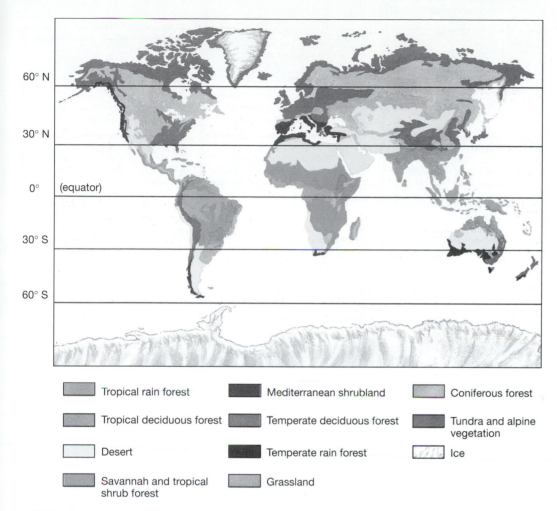

FIGURE 15-5

World distribution of the major terrestrial biomes. (From Audesirk and Audesirk: *Biology: Life on Earth*, 5/e, 1999. Reprinted by permission of Prentice Hall, Upper Saddle River, N.J.) [See color section]

habitat refers to the plant community and physical environment that supports a single species. Although some species, such as the dodo, have been hunted to extinction, most species loss occurs because the habitat in which a species lived has been destroyed.

Species Interactions

Although ecosystems may appear to be very different from one another, they all exhibit a common biotic structure. All ecosystems include *autotrophs,* organisms (such as plants), that use inorganic nutrients and an external energy source to produce their own organic matter, and *heterotrophs,* organisms (such as animals and fungi) that obtain energy by feeding on organic matter—that is, they feed on autotrophs. Autotrophs are classified as *producers,* whereas heterotrophs are further divided into *consumers, decomposers,* and *detritus feeders* (Refer back to the discussion of the organic carbon cycle in Chapter 5.)

In this type of organization, primary consumers live off the producers, secondary and higher-order consumers feed on lower-order consumers, and the decomposers and detritus feeders feed on dead organic matter of both producers and consumers (Fig. 15-6). In assisting the chemical breakdown of organic matter, the decomposers and detritus feeders return nutrients to the system that are then re-used by the producers.

We can represent which organisms feed on which by means of a **food chain** that links particular organisms in an ecosystem (Fig. 15-7). Because one organism may feed on several other types or may be eaten by several different types of organisms, food chains are usually interconnected into *food webs.* Despite the potential complexity of these webs, the overall structure is very simple: Each web passes through a series of feeding levels called *trophic levels.* For example, the following trophic levels range from the bottom up: producers > primary consumers > secondary consumers > higher-order consumers. There are normally no more than four trophic levels in any system.

How many organisms are there at each level? Rather than looking at the number of individuals, we can simplify the comparison among levels if we talk in

Autotrophs Make their own organic matter from inorganic nutrients and an environmental energy source	Heterotrophs Must feed on organic matter for energy	
Producers	**Consumers**	**Decomposers and detritus feeders:** Organisms that feed on dead organic material
Photosynthetic green plants: Use chlorophyll to absorb light energy	Primary consumers/herbivores: Animals that feed exclusively on plants	Decomposers: Fungi and bacteria that cause rotting
Photosynthetic bacteria and algae: Use a variety of pigments to absorb light energy	Omnivores: Animals that feed on both plants and animals	Primary detritus feeders: Animals and microbes that feed directly on detritus
Chemosynthetic bacteria: Use high-energy inorganic chemicals such as hydrogen sulfide	Secondary consumers/carnivores: Animals that feed on primary consumers	Secondary and higher orders of detritus feeders: Animals and microbes that feed on primary detritus feeders
	Higher orders of consumers/carnivores: Animals that feed on other carnivores	
	Parasites: Plants, animals, fungi, bacteria, and protozoa that become associated with another organism and feed on it over an extended period of time	

FIGURE 15-6

Autotrophs and heterotrophs. (From Nebel and Wright: *Environmental Science,* 6/e, 1998. Reprinted by permission of Prentice Hall, Upper Saddle River, N.J.)

FIGURE 15-7

A simple food web. Food (energy and nutrients) is transferred from one organism to another along these pathways. (From Nebel and Wright: *Environmental Science,* 6/e, 1998. Reprinted by permission of Prentice Hall, Upper Saddle River, N.J.)

terms of **biomass:** The total combined weight of organic material in each trophic level. Each trophic level, except for the producers, uses organic matter from the level below. The organisms use that organic matter for growth and to produce energy. In converting organic matter to energy, some organic matter is lost as we move from lower-order to higher-order trophic levels. In terrestrial ecosystems, the biomass is reduced by 90 to 99% at each higher level.

Ecosystems are not organized entirely according to which species is feeding on whom; other forms of interaction are also found. These may include mutually supportive relationships, such as the relationship between flowering plants and insects: Insects feed off of the nectar or pollen from the flowers, which are then pollinated in the process. The extreme example of this mutualism is **symbiosis,** a relationship in which two species are dependent on each other to the degree that neither can live alone.

At the other extreme are species that coexist in a competitive relationship, although this competition tends to occur less frequently than we might suppose. Most species tend to adapt to a particular habitat and, even where potentially competitive species occupy the same habitat, each tends to develop its own particular *niche.* An animal's niche describes not only the food it eats, but also where and when it eats, where it lives, where it nests, and so on. Specialization to this degree, in which different species occupy the same geographic location but have different living habits, reduces potential contact and help reduce competition among species at the same trophic level.

The species in an ecosystem interact with their environment as well as with each other, and different species thrive under different physical conditions. There is usually some optimal range of conditions over which each species is best adapted. The species comes under stress as the environment moves away from that range—for example, as it becomes wetter or drier, warmer or colder, shadier or sunnier, or more or less acidic. At some point, the stress may be large enough that the organism reaches its limit of tolerance for those conditions, and death occurs.

When we throw human beings into the mix, the level of complexity and the nature of the interactions increases substantially. We then must take into account the social, political, and economic interactions among different human societies and also the dramatic effects that these societies can have on the physical environment.

The Importance of Ecosystems

Why is the concept of an ecosystem important in our discussion of biodiversity and species loss? When we deal with systems, a perturbation to one part of the system can have impacts throughout the system. Because there is an interdependency among species in an ecosystem, if one species is removed, the loss of other species may fol-

low. How significant a loss will depend on the role that the removed species plays in the ecosystem. In Chapter 10 we suggested that biodiversity should not be measured simply in terms of the number of species but should also describe the interactions among those species. We can suggest further that to know the number of interactions might not be sufficient; it is quite likely that interactions among some species will be more important than others. Ecologists use the term **keystone species** to describe a species that plays a vital role in the operation of an ecosystem.

The El Niño–Southern Oscillation (ENSO) events described in Chapter 12 provide a graphic illustration of the importance of ecosystems. When the westward-flowing ocean current in the tropical Pacific Ocean slows, warm water flows back toward the South American coast, preventing the upwelling of cold, nutrient-rich water off the coasts of Peru and Ecuador. The loss of these nutrients prevents the growth of the phytoplankton that form the basic food source for the local fish population. The reduction of the fish population in turn results in the loss of the birds that feed on those fish. The end result is a massive drop-off in the marine and bird populations and a drastic change in the local ecosystem.

These losses are not permanent: After several years the populations gradually grow back to their previous levels. If there were a drop-off of species found only in this locality, however, such a disturbance could easily lead to their extinction. In fact, the warmer waters that appeared off the western coast of South America during the 1982 ENSO event also killed large areas of coral, and three coral species that lived only in this area went extinct.

The Modern Extinction

The Beginnings

The modern mass extinction episode began when the first humans evolved and spread to colonize larger and larger areas (Fig. 15-8). Seventy-three percent of the large mammal genera of the late Pleistocene are extinct. A similar percentage of the large bird genera has also been lost. The archeological evidence shows that the losses coincide with the spread of human populations:

- When the first humans arrived in Madagascar in about A.D. 500 there were 6 to 12 species of large flightless birds, 17 genera of lemurs, and numerous species of other large birds and mammals. Within a very short time the flightless birds were gone, as were seven of the lemur genera, plus one species of aardvark, a pygmy hippopotamus, and two species of large land tortoises.

- Humans arrived in New Zealand around A.D. 1000. Since then, 20 species of large land birds and 22

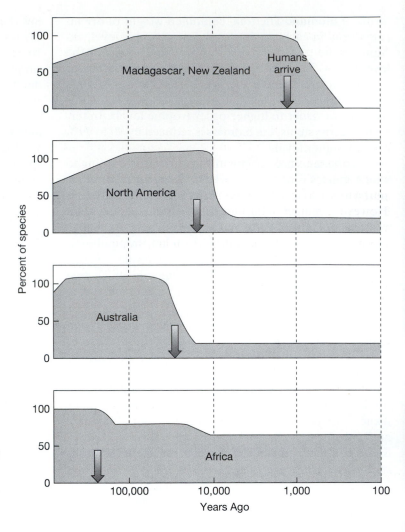

FIGURE 15-8

The extinction of large mammals and birds corresponds to the spread of human populations. Humans evolved first in Africa; in fact, they coevolved with other animals there over millions of years. As a result, the curve for Africa stands out from the others. (After E. O. Wilson, 1992. *The Diversity of Life,* New York: W. W. Norton and Company)

other species of flightless birds have been hunted to extinction.

- As they spread through the western Pacific, the Polynesians killed off more than half of the native species they found. Species that are indigenous, or native, to a particular location are called **endemic species.**

- The native Hawaiians extinguished 35 to 55 species of land birds. Fifty species remained when Captain Cook arrived in 1778, and one-third of those are now extinct.

There is evidence of a similar loss of species in North America. The first humans who crossed from Siberia into North America found vast grasslands with herds of large mammals, including bison, antelope, and mammoth, together with numerous large birds and other mammals that are now extinct. However, there is an additional explanation for the North American extinction besides hunting. The Pleistocene was a period of large glacial advances and retreats, which accompanied significant swings in mid- and high-latitude climate. Fossil evidence suggests that anomalously high extinction rates coincided with large climate changes. Humans arrived in Australia 30,000

years ago, and there is fossil evidence of the existence of a large number of species that are now extinct. However, the extinction was more gradual than elsewhere, and Australia suffered a long period of drought between 26,000 and 15,000 years ago, when the most rapid species loss also occurred.

When attributing blame for the extinctions in North America and Australia, a good case can be made for both climate change and over-hunting. In other parts of the world, such as New Zealand, the Pacific islands, and Madagascar, however, there is no indication of a significant climate change. In those areas, the evidence supporting over-hunting appears overwhelming.

The Present Day

As we move forward in time, the predominant agent of destruction has changed from over-hunting to habitat destruction, and the pace of species loss has increased dramatically. Across Europe, from the hedgerows of southern Britain to the vine-covered hills of Cyprus, are widely diverse landscapes and ecosystems; all of them, except for the high peaks of the Alps and the Pyrenees, have been created by human activity. It is becoming increasingly difficult, if not impossible, to find landscapes anywhere in the world that have not been modified in some way by human actions. Today this activity almost always involves changes in land use, resulting in habitat destruction and species loss. With few exceptions, human land use leads to a reduction in biological complexity and reduced biodiversity. One interesting exception is the hedgerows of southern England. The original forest was cleared and the land used for cultivation. Later, the large medieval fields were subdivided, and hedgerows were planted to separate the individual fields. These hedgerows actually support a more diverse ecosystem than did the original forest. However, changes in farming practices are leading to the removal of many of these hedgerows in the latest chapter of human modifications to the British landscape.

Tropical Deforestation.   The greatest rate of species loss today is found in the tropical forests (Fig. 15-9). The climates of these forest regions are characterized by high rainfall (in excess of 2 m/yr), high mean annual temperatures, and low seasonal contrasts. For the most part, the forests are located in the areas of trade-wind convergence associated with the Hadley circulation, as described in Chapter 4. These forests cover only 6% of the land surface, yet they are thought to contain three-quarters of all the living things on Earth, or two-thirds of all plant and animal species. Great Britain has 1430 species of flowering plants and 35 native tree species, whereas the Malay Penin-

sula, with only half the area of Great Britain, has 7900 flowering plants and 2500 native tree species. Of the 9040 bird species in the world, almost half live in the Amazon Basin or in Indonesia. Studies in Peru found 300 tree species in a 2.5-acre plot (approximately 10,000 m^2), and 1000 tree species were found in a combined census of ten 2.5-acre plots in Borneo. In contrast, there are only 700 native tree species in the whole of the United States and Canada. Although this chapter focuses on loss of biodiversity, it is important to recognize that tropical deforestation has other regional and global consequences (see the box "Other Consequences of Tropical Deforestation").

IMPACT OF FOREST CLEARANCE.   There are no hard and fast rules, but in general the number of species living in a tropical rainforest is related to the area of forest coverage: A tenfold increase in area results in double the number of species. Applying these numbers to the Atlantic coast of Brazil, where the forest has been reduced to less than 1% of its original cover, suggests that the forest biota may have already decreased by 75%. Similar losses are occurring throughout the tropics. The Harvard University biologist Edward O. Wilson concludes that, even using the most optimistic and conservative estimates, 27,000 species are lost each year—that is, 74 species per day or 3 species every hour—simply as a consequence of tropical deforestation.

If an area of temperate forest is cleared and then abandoned, something very similar to the original forest

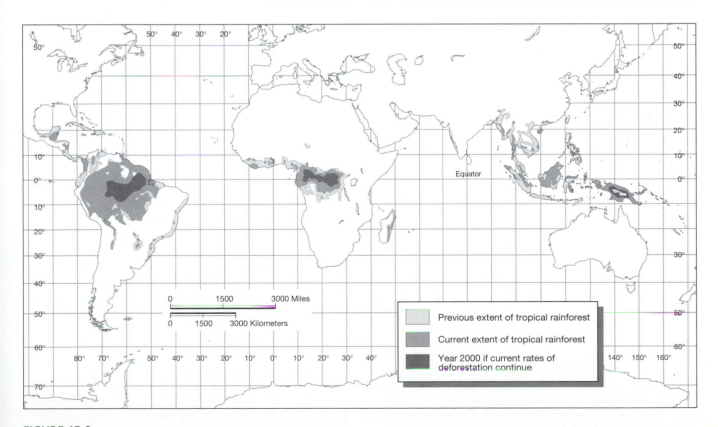

FIGURE 15-9

Distribution of tropical rainforest and extent of tropical deforestation.

Other Consequences of Tropical Deforestation

Clearing tropical forests and converting the land to other purposes, such as agriculture and grazing, has other consequences besides contributing to an overall loss of biodiversity. At the local scale, deforestation tends to result in soil degradation and increased erosion. Without the continuous addition of fertilizer, the nutrient-poor soils rapidly lose their usefulness for agriculture. Where before much of the rainfall was intercepted by the tree canopy, now more of it reaches the surface directly, leading to erosion, which increases as agricultural productivity and surface cover decrease. The result is very rapid soil loss.

Deforestation also has an effect on local rainfall. Estimates from the Amazon Basin indicate that approximately half its precipitation is derived from recycled water. In other words, rainfall occurs and the water is intercepted by the canopy; some is used by the trees and some is returned to the atmosphere through *evapotranspiration,* the combined effect of water loss by evaporation and by transpiration (the transfer of water, in vapor form, from plants to the atmosphere through pores in the leaves). The water that is returned to the atmosphere falls again as rain elsewhere in the basin. When large areas of forest are cleared, a large proportion of the rainfall is lost as runoff, evapotranspiration decreases, and rainfall over the basin as a whole decreases. Several groups of researchers have used atmospheric general circulation models to examine these processes. The models suggest that deforestation has its greatest impact on the local climate by changing the albedo, the surface hydrology, and the surface roughness. (Surface roughness is important because it affects the windflow, which, in turn, affects evaporation and the transfer of latent and sensible heat.)

One such study was carried out by J. Lean and D. Warrilow at the British Meteorological Office. They simulated the replacement of all the tropical forests and savannahs in South

TABLE 15-1

Impacts of Deforestation on Local Climate			
Surface Variable	*Observed*	*Control**	*Deforested**
Evaporation (mm/d)	3.34	3.12	2.27 (-27.2%)
Precipitation (mm/d)	5.26	6.60	5.26 (-20.3%)
Soil moisture (cm)		16.13	6.66 (-58.7%)
Runoff (mm/d)	2.76	3.40	3.00 (-11.9%)
Net radiation (W/m^2)		147.3	126.0 (-14.5%)
Temperature (°C)	24.0	23.6	26.0 ($+2.4$°C)

* Model simulations are averaged over 3 years.
SOURCE: Adapted from J. Lean and D. Warrilow: Simulation of the regional climatic impact of Amazon deforestation. *Nature* **342**:411–413, 1989.

America (north of 30° S) by tropical pastures and examined the effect that this simulation had on the climate model. Table 15-1 presents some of the results, comparing observational data with the model simulation before (the control run) and after the deforestation. We can see from the table that the control run produces a little too much rainfall and runoff and slightly underestimates evaporation, but it generally produces a climate very similar to the observed climate.

More interestingly, when we compare the deforestation experiment with the control run, we see some major differences and some interesting contradictions. Overall, the hydrologic cycle is weakened, which is what we would expect, except that runoff decreases instead of increases. Also, the net radiation (the total amount of available energy) decreases, but temperature goes up. The reasons for this lie in the multiple interactions that occur in the climate system. The pasture has a higher albedo than that of the forest or the savannah, so deforestation

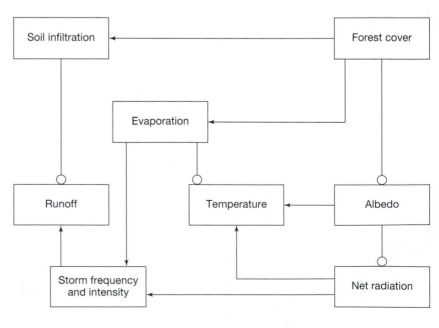

BOX FIGURE 15-1

Systems diagram linking forest cover with local climate and hydrologic parameters.

(Continued)

Other Consequences of Tropical Deforestation   *(Continued)*

increases the albedo and reduces the incoming solar energy. These effects contributed to the decrease in net radiation. Temperatures go up instead of down, however, because, prior to deforestation, evaporation had a cooling effect on the forest; when the forest is removed and evaporation decreases, the cooling effect is reduced and temperatures increase despite the reduction in net radiation.

The surface runoff is determined by several factors, including the rate at which water infiltrates the soil (which is the only soil parameter changed in this model for the deforested case). The infiltration rate is reduced for the deforested soil, so runoff should increase. However, runoff is also affected by the intensity and frequency of the precipitation events. Infrequent, high-intensity events deliver rain to the surface faster than the soil can remove it, and there is high runoff. If the same amount of rain falls in more-frequent, less-intense events, water is delivered to the surface at a slower rate, more infiltrates, and runoff is reduced. In the deforestation case, runoff is reduced because, although less water infiltrates the soil and is held by the vegetation, there is

also a decrease in frequency and intensity of rainfall events. The soil moisture shown in Table 15-1 is not the total amount stored in the soil but actually the amount that is available to the vegetation. The decrease that the model shows in available soil moisture is due to the rooting depth of the vegetation: The shallower roots of the pasture cannot reach down far enough to tap the water that exists at greater depth. These interactions are illustrated in Box Figure 15-1, a systems diagram. There we see that runoff and temperature have both positive and negative couplings from other system components, thus allowing for their apparent contradictory response to what we might, at first, have expected.

In addition to these local effects, deforestation has an impact on the global climate. The tropical forests represent a large store of organic carbon. When the forests are cut and burned to clear the land for agricultural use, this carbon is returned to the atmosphere as carbon dioxide, which contributes to the buildup of atmospheric greenhouse gases. Approximately half of the postindustrial increase in atmospheric CO_2 has come from tropical deforestation.

---

will grow back within 100 years, and it will have similar levels of biodiversity as the original forest. There is little information available on the regeneration time of tropical rainforests, but the indications are that when large areas are cleared the forest takes hundreds of years to grow back to its original state. In some cases it is likely that the original rainforest will never return. Why are tropical rainforests so sensitive to change?

When a large area of temperate forest is cleared and then abandoned, grasses will spread, shrubs and a few trees will appear, and eventually the spreading vegetation cover will produce the temperature, moisture, and shade conditions that promote the return of the original forest species. This sequence of events is facilitated by the nature of the trees' seeds, which tend to be fairly resistant to stress. They are able to lie dormant for long periods, ready to germinate when the right environmental conditions return. Furthermore, a large fraction of the organic material in the forest is returned to the soil through the fall of litter and the decay of dead organic matter. Over time, this recycling produces a nutrient-rich soil layer, which, after the forest has been cleared, is still available to help promote new forest growth. Consequently, it is typically possible for temperate forests to return to something similar to the original forest cover.

The situation in tropical forests is very different. Figure 15-10 illustrates the nutrient flow through a forest–soil system. In simple terms, there are three primary nutrient reservoirs: the trees, the litter, and the soil. The transfer of nutrients to the litter is controlled by the death rate, the transfer to the soil is determined by the rate of decomposition, and the transfer back to the trees depends on the photosynthetic uptake. Tropical forests return lit-

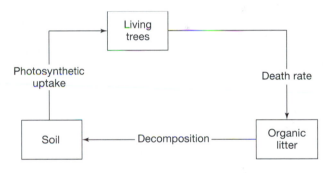

FIGURE 15-10
Nutrient flow through a forest–soil system.

tle organic matter and few nutrients to the soil. When trees die and decompose, the nutrients in them are returned to the living biomass very quickly. This scenario implies a very efficient nutrient recycling system: higher temperatures and moisture conditions favor rapid decomposition, whereas extensive root systems (roots constitute 60% of the biomass) extract the nutrients from the soil very quickly. In tropical forests, therefore, high rates of decomposition and photosynthetic uptake cycle the nutrients rapidly through the soil and back to the trees, making the living biomass the major nutrient reservoir. In temperate forests the rate of decomposition is lower, as is the rate of photosynthetic uptake, and nutrients tend to accumulate in the organic litter and in the soil.

There is a further important difference between tropical and temperate forests. After rainforests have been cleared, heavy rains, characteristic of the tropics, wash away much of the organic material that does enter the soil, leaving behind soils that are very acidic and nutrient deficient. In contrast to the seeds of temperate forest plants, the seeds of the tropical forest plants tend to be less resistant to stress and typically germinate within a few weeks. For small clearings this is not a problem: Dead organic matter is rapidly broken down, releasing nutrients that are used for new growth. Where the clearing is extensive, however, the nutrients are removed very quickly, the cleared areas are usually very hot (because there is no longer a forest canopy to shade the surface), and the germinating seeds cannot survive. Eventually vegetation and a forest cover return, but the forest composition is normally very different from what was there before the land was cleared.

In terms of species loss, there is yet another major difference between temperate and tropical forests. Whereas some temperate forest species are confined to relatively small geographic locations, many of the plant and animal species have a wide geographic distribution. If the forest is cleared in one area, many of the species will continue to thrive elsewhere and can eventually return to recolonize the cleared regions. Some tropical forest species also have extensive distributions, but a significant proportion of the plant and animal species live only in very small geographic areas. Over time, the overall forest structure might not change and the same families of plant and animal life will be present, but the actual species composition can change very rapidly. Thus, when a large area of tropical forest is cleared, all the species that were found only in that area go extinct.

There was about 8 million km^2 of tropical rainforest in 1989—half of what existed in prehistoric times. In the late 1980s, this forest was being removed at a rate of 1.8% (140,000 km^2) per year. The total global cover of tropical rainforest in 1989 was approximately equal to the area of the contiguous 48 states of the United States; at a clearance rate of 1.8%, an area of tropical forest the size of Florida was being removed each year. What effect does this have on biodiversity and species loss? If a million years is the typical lifetime of a species, then the normal extinction rate is one in 1 million species per year. If we use 10 million as a conservative estimate of the number of species that live in the rainforests, then the background extinction rate should be 10 species per year. What is the actual extinction rate? No one knows for certain, but we presented Wilson's estimates earlier: 27,000 species per year due to forest clearance alone. Even with these very conservative estimates, the extinction rate due to human activity is 3000 times greater than the background rate of extinction.

REASONS FOR FOREST CLEARANCE. Why is so much forest being destroyed? The superficial response is to say that the forests are cleared for logging; to open new areas for agriculture and cattle ranching; and for large-scale development, which includes hydroelectric projects, exploitation of mineral resources, resettlement projects, and so on. Approximately 8 million hectares (800,000 km^2) of

forest are cleared for agriculture each year. The normal procedure is to cut and burn the forest and then use the cleared area to grow crops—a practice known as *slash-and-burn* agriculture. The burning releases nutrients to the soil and provides fertilizer for the crops. After several seasons the nutrients are depleted, and the farmer moves on to slash and burn a new area. This is not necessarily as wasteful as it sounds: If the areas are small and the old fields are left fallow for a sufficient time (up to 30 years), enough vegetation grows back that nutrient levels are replenished and the area can be burned and cultivated again. Problems occur when large areas are cleared or when the vegetation is not given enough time to regenerate. In these cases the soil loses its fertility, crop yields decrease, and permanent soil degradation is commonly the end result. This type of damage is normally caused by population increase; however, in some areas extensive deforestation is caused by the production of commercial crops (cash crops), which are typically used for export, by large agricultural plantations.

Apart from agriculture, approximately 50,000 km^2 of forest is cleared for timber, pulp, and other wood products each year. The most extensive logging is taking place in Asia and West Africa. Approximately half of what is cut is exported, with Japan taking the largest amount and the United States being the next largest importer. Wood is also cut for fuel use: For almost half the world's population, wood is the principal energy source for cooking and heating. Large areas are also being cleared for cattle ranching—approximately 20,000 km^2 per year in Latin America. Much of the beef once went to the United States, but most Latin American beef is now consumed in Latin America or in Europe. Cattle ranching is particularly destructive, because once large areas are cleared for pasture, soil degradation, erosion, and compaction from hooves and vehicles soon makes the land unusable, requiring more forest to be cleared.

The above description reflects the obvious transition from one land cover, the forest, to something else, as the human use of the land changes. What it hides is a multitude of social, economic, and political forces that are causing the land use to change. The particular combination of factors varies from place to place and through time, but examples might include the following:

- Growing population pressures that force an increasing number of people into the forested areas, requiring an expansion of food crops.

- Unequal division of land that results in a very small proportion of the population owning a very large fraction of the existing developed land. This arrangement forces people who do not own land to develop new areas, encroaching farther and farther into the forest.

- Deliberate resettlement programs in which governments encourage migration into the forests to reduce population pressures elsewhere. Such programs themselves may be driven by the hope of solving (or hiding) political and economic problems in other areas—poverty and unemployment in large cities, for example.

- A growing need in the industrializing nations to produce cash crops and to develop mineral resources to pay ever-increasing international debts.

Understanding the human factors that are causing forest clearance in a particular location is essential if efforts are to be made to stem the tide of tropical deforestation. Preaching conservation and offering plans for the sustainable use of forest resources do little good if the plans do not address the problems that led to the deforestation in the first place.

Much of the world's attention has focused on the Amazon forest, the world's largest remaining area of undisturbed rainforest. In the 1980s, the Brazilian government was very active in promoting the development of the forest. It granted 250 acres of land to people willing to migrate to the forest, provided subsidies for cattle ranchers, and promoted two huge development projects: the Grande Carajas project, a $62 billion industrial enterprise in eastern Brazil expected to involve an area the size of France and Great Britain; and Polonoreste, a development scheme built around a highway that has been ex-

tended through the forest. The Brazilian government has since recognized some of the dangers involved with this type of development. It has halted the subsidies to cattle ranchers, and it has placed 7.5 million acres of rainforest into extractive reserves—land that is protected and where the two major rainforest products, rubber and brazil nuts, can be extracted in a sustainable fashion. Unfortunately, although there have been some recent efforts to slow deforestation in the Amazon, the pace has accelerated elsewhere. Suriname and Guyana in South America, for example, which contain one of the world's largest unbroken tracts of tropical forest, signed agreements that could lead to massive deforestation. Suriname sold the rights to 375,000 acres near the Guyana border to an Indonesian investment group; Guyana granted a joint Malaysian–Korean group the rights to log 4.2 million acres, almost 10% of the country.

Hot Spots. The tropical rainforests have the largest extinction rates, but they are not the only places where species are at risk. Around the world, numerous areas have large numbers of endemic species threatened by the loss of habitat. Norman Myers of Oxford University identified 18 such areas. In each case the habitat has been reduced to less than 10% of its original cover or is expected to be reduced to that amount within the next few decades (Fig. 15-11). The descriptions given here are summarized

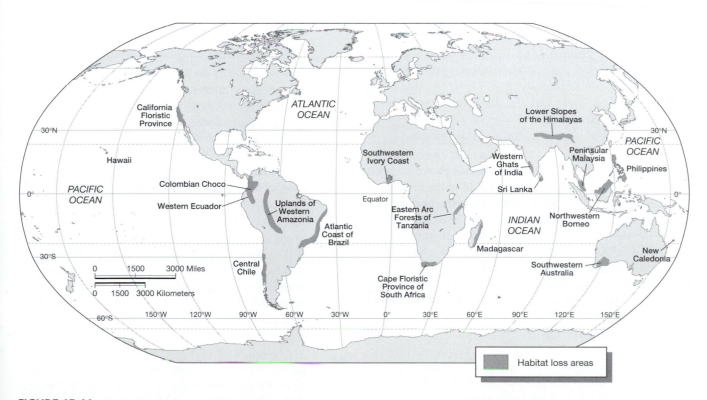

FIGURE 15-11

Hot spots of habitat loss—areas with many species that live nowhere else and that are in greatest danger of extinction as a result of human activities. (After E. O. Wilson, 1992. *The Diversity of Life,* New York: W. W. Norton and Company)

from *The Diversity of Life* by Edward O. Wilson (see "Further Readings" at the end of this chapter). These regions account for only 0.5% of Earth's land surface, but they are home to 20% of the world's plant species that are found nowhere else on Earth and are in danger of extinction:

- *California floristic province.* California has a climate and vegetation similar to that of the Mediterranean. The California floristic province contains one-quarter of all plant species of the United States and Canada. Half of its 2140 species are found nowhere else in the world. This ecosystem is being destroyed by urban expansion and by agricultural development.

- *Central Chile.* This Mediterranean-type ecosystem has 3000 plant species. Covering 6% of Chile, this area has more than half of the entire Chilean flora. Only one-third of the original tree cover remains, and the area is the most densely populated (by humans) in the country.

- *The Colombian Choco.* These coastal plain forests extend up the lower mountain slopes along the entire length of Colombia. They house one of the richest but least explored floras in the world, with 3500 known plant species. Many more unknown species are thought to live in the forest, and one-quarter are thought to be endemic to the area. One-quarter of this forest has already been destroyed by logging or has been cleared for agriculture.

- *Western Ecuador.* The foothills on the western slopes of the Andes Mountains once contained 10,000 plant species, with one-quarter endemic to the region. Very little of the forest now remains, and the residual habitat may be too small to allow the remaining plant populations to survive.

- *Uplands of Western Amazonia.* The most extensive flora and fauna of any place on Earth are found in a narrow arc stretching from Colombia to Bolivia. This 50-km-wide belt forms the western edge of the Amazon Basin and occupies the eastern slopes of the Andes Mountains at an elevation between 500 and 1500 m. Each mountain ridge has its own unique collection of local plants and animals, most of which have never been studied. This belt is rapidly being settled; about 65% has already been cleared or converted to palm-oil plantations.

- *Atlantic coast of Brazil.* A unique rainforest once covered about 1 million km² of this coastal region. Now it is the most densely populated and agriculturally productive part of Brazil. Less than 5% of the original forest cover remains.

- *Southwestern Ivory Coast.* This rainforest, originally 160,000 km² in area, is a distinct botanical province of West Africa. Logging and slash-and-burn agriculture have reduced the tree cover to 16,000 km², and the remaining cover is disappearing at a rate of about 2000 km² each year.

- *Eastern arc forests of Tanzania.* Nine sections of *montane* forest (characterized by cool, moist conditions and dominated by evergreen trees) extend in an arc across eastern Tanzania. Each has been virtually isolated since prehuman times, and each has developed a very distinct biota. Deforestation has reduced these forests to half of their original cover.

- *Cape floristic province of South Africa.* The southern tip of Africa is home to a distinct type of heathland called *fynbos.* Agriculture, development, and the introduction of *exotic* (nonnative) species has reduced the fynbos by one-third of its original cover. In the 89,000 km² that remain, 8600 plant species live, 73% of which are found nowhere else in the world. What is left of the fynbos is being rapidly fragmented and degraded.

- *Madagascar.* The initial tectonic rifting that separated Madagascar from the rest of Africa began about 150 million years ago, and Madagascar has been effectively isolated from the African continent for the past 10 million years. Madagascar's flora and fauna are unique; there are 10,000 plant species, 80% of which are endemic. By 1985, slash-and-burn agriculture had reduced the forest to one-third of the original cover. Most of this loss occurred since 1950, and the pace has not slowed.

- *Lower Slopes of the Himalayas:* An unusual mountain forest of mixed tropical and temperate species is found along the lower slopes of the Himalayas from northern India, across Nepal, and to the western provinces of China. It houses 9000 plant species, with 39% being endemic to the region. More than two-thirds of the forest has been lost through logging, conversion to farmland, or harvesting for fuel.

- *Western Ghats of India:* The seaward-facing slopes of the Western Ghats receive large amounts of monsoon rainfall that supports a tropical forest covering an area of about 17,000 km². There are 4000 known plant species, 40% of which are endemic. Logging and conversion to farmland have reduced the forest by one-third. The remainder is being cleared at a rate of 2 to 3% per year.

- *Sri Lanka:* The Sri Lankan forest is a remnant of a forest that once covered most of the Indian continent. Of the 1000 plant species there, half are endemic, but timber and agricultural demands have reduced the forest to less than 10% of the original island cover.

• *Peninsular Malaysia:* Most of the peninsula was forest covered. One-third of its 8500 plant species are endemic to the region. By the mid-1980s, half the forest had been cleared; about half of the endemic tree species are classified as endangered or extinct.

• *Northwestern Borneo:* The forest that once covered almost the entire island is rapidly disappearing. The northern third of the island, where 40% of the plant species are endemic, has been extensively cleared by logging.

• *The Philippines:* The Philippines consists of 7100 islands close enough to each other and to the Asian mainland to receive many plant and animal colonists. However, each island evolved a slightly different flora and fauna, many of which are endemic. Two-thirds of the forest has been cleared in the past 50 years, whereas logging and agricultural settlement are rapidly clearing the remainder. There are plans to save 6450 km^2 in forest preserves, but that is merely 2% of the land surface.

• *New Caledonia:* The forests of New Caledonia are home to 1575 species of plants, 89% of which are endemic. After many years of logging, mining, and burning, only 1500 km^2 of undisturbed forest remains—9% of the island.

• *Southwestern Australia:* This Mediterranean-type ecosystem is similar in form to the Cape floristic province of South Africa and no less diverse. Of its 3630 plant species, 78% are endemic. Agricultural conversions since the mid-1950s, together with mining operations and invasion by exotic weeds, have reduced the tree cover by half. One-quarter of the species of Southwestern Australia are now classified as rare or threatened.

Although these are all 18 hot spots originally identified by Norman Myers, the list does not end here; many more places could be included. One ecosystem that is worth mentioning is the coral reef. Despite its appearance, coral is actually an animal. Found in shallow tropical seas, coral reefs are the most productive and diverse of ocean ecosystems. The reef is dynamic, continuously changing, and very fragile. It is easily damaged by shifting sand and storms; left alone, it normally recovers quickly. However, most coral reefs are located close to shore, so any natural damage is now augmented by human activities, notably pollution, mining for coral rock, collection of coral specimens, overfishing, and damage from accidental grounding by ships.

Coral reefs all around the world are affected, and in some cases the reduction in reef area is about 10%. There are also increasing episodes of coral bleaching on many reefs. Bleaching is caused by the loss of symbiotic *zooxanthellae* (single-celled algae that live in the tissue of the coral animals). This loss is considered to be a general response of the coral to stress. Stress can result from numerous causes, including excessive hot or cold temperatures, chemical pollution, or dilution by freshwater. Many areas have attempted to reverse the damage by establishing marine parks and by controlling the human activities that take place on or near the reefs. These measures help prevent accidental damage to the reef but offer little protection from the more pervasive damage of environmental pollution.

Why We Should Care About Biodiversity

You probably live in or close to a city; approximately half the world's population lives in urban areas. Although you may have seen one city building knocked down to be replaced by another or an open field turned into a shopping mall, you probably do not have much first-hand experience in large-scale habitat destruction. Unless you are a tropical ecologist or a scuba diver, the closest you may ever get to the backwaters of the Amazon or to a coral reef is a nature program on television. In the 1600s and 1700s, forests were cut down in Europe and in what would become the United States to power steam engines and to expand agriculture to support growing populations. You could say that, in doing so, we were able to support a workforce and develop the technology that eventually enabled us to see the remaining forests or the reefs on those television sets. You might also argue that if the industrializing nations need or want to do the same now, or if Californians prefer suburbs to chaparral, that is their decision and why should we get excited about it? These are valid questions. We are going to try to answer them in several ways—first by looking at why individual species may be important, and then by stepping back and looking at the importance of biodiversity to the Earth system as a whole.

Instrumental and Intrinsic Values of Species

Ecologists tend to discuss the value of a species in terms of its *instrumental* or *intrinsic value*. A species' **instrumental value** is the degree to which the existence of the species benefits another species in some way. We are normally the other species, and the value or benefit is typically economic. A species' **intrinsic value,** conversely, is its value for its own sake, regardless of whether we benefit by it or not. Some argue, for example, that all species and all individual organisms have the same rights to exist that humans do—that humans have no greater value than any other species. Others argue that humans are unique: We dominate over nature, nature is ours to exploit, and other species have no individual rights per se. Still others argue that this very uniqueness, in fact, imposes a special responsibility: We are capable of moral and ethical judgments, and these should include concern for other species. Assigning an economic value is relatively easy; assigning intrinsic value, however, is a moral and ethical issue that is more subjective. The assignment of intrinsic value varies from society to society and from individual to individual, according to their outlook on nature.

Instrumental Value. We can look at instrumental value from several perspectives. We can view species as being resources:

- *Medicine:* Of all pharmaceuticals produced in the United States, 25% contain ingredients originally derived from native plants. The often-quoted example is a plant called the rosy periwinkle, which grows only on Madagascar and fortunately was not a victim of the large species loss that has occurred on the island. In the 1960s scientists extracted two substances from the plant: vincristine and vinblastine, which became our primary defense against childhood leukemia and Hodgkin's disease. Before the discovery of these drugs, childhood leukemia was almost always fatal; since the 1960s, with the use of vincristine, there is now a 95% chance of remission. These two drugs, developed from a single plant species, now represent a $100-million-a-year industry—just a small part of the $8 billion a year derived from pharmaceuticals based on "natural" products. How many other rosy periwinkles are waiting to be discovered, and how many disappear every year without us even knowing they existed?

- *Scientific value:* Although a considerable amount of discovery takes place in scientific laboratories, much of what we know about evolution and ecology, and our understanding of how natural systems work, has been derived from the study of species and ecosystems in their natural state. However, our understanding of these systems and how they interact, particularly at the global scale, is far from complete. At the present rate of habitat destruction, we are in danger of wiping out whole ecosystems without ever understanding how they operate or what role they may have played in regional or larger-scale ecology.

- *Recreational and aesthetic value:* It is difficult to assign a monetary amount to the scientific value of species and ecosystems. However, it is much easier to assign such a number to their recreational value. Each year, more than 50 million people in the United States spend a total of $38 billion on recreational hunting or fishing. There are also the millions of people who escape the cities to spend their weekends or vacations walking, climbing, and photographing the mountains, forests, and beaches of the more "natural" parts of the world. They may sail on rivers, lakes, or the ocean, sit on shore and admire the view, or simply drive by and look. The continuing popularity of television nature shows also attests to the aesthetic value we place on the natural world, whether we ever get to see it in person or not.

- *Commerce:* There are both direct and indirect commercial benefits to be gained from the natural environment. Direct commercial interests revolve around activities such as logging, commercial fishing, and the collection and sale of exotic plant and animal species. Indirect benefits arise from recreational activities that support commercial interests, for example, in travel, accommodations, restaurants, sporting goods stores, and so on. The business of arranging tours to remote locations, such as trekking in the Himalayas and hunting and photographic safaris in East Africa, is thriving. There is also now a growing business in ecotourism, where tourists visit places specifically to observe rare or endangered species or ecosystems.

- *Agriculture, forestry, aquaculture, and animal husbandry:* Those of us who live in the developed world, in particular, tend to regard food production as simply one component of our industrialized society; we see little connection between miles of uniform crops and mechanized production, and the grasslands or forests that would otherwise be there. In reality, there is a very close connection, and the existence and maintenance of our agricultural food supply is much more dependent on natural species than many people realize. We will return to this connection when we look at the value of species from the perspective of a global system.

The Loss of Biodiversity

Now we shall step back from individual species and consider the whole diversity of life. What are the consequences involved in the loss of that biodiversity?

Biodiversity and Food Supply. Biodiversity is of critical importance to the modern agricultural techniques we employ to feed our ever-growing global population. Most people are well aware of many of the local environmental problems that can arise from our attempts to increase productivity: soil degradation and erosion, salinization, groundwater pollution from fertilizer and pesticides, and so on. From our larger-scale or global perspective, however, a more serious problem is developing with respect to the basic ingredients we need for modern agriculture—the seeds themselves.

Early farmers and agronomists noted that some varieties of a crop are more productive more disease-resistant, or more drought-tolerant than others. The selective use of different strains or varieties resulted in increased productivity. Today biotechnology and gene splicing can produce new crop varieties to order. At first glance it is hard to see the problem: With modern techniques we can produce seeds with high yields and *uniform* characteristics. Crops can be planted, harvested, processed, packaged, and transported all with minimum human intervention—just what we need to keep the world fed. Unfortunately, we never quite manage to feed the world. The lack of adequate food supplies is a result of social, political, and economic factors that determine the global distribution of resources, not an inability to produce enough food for all. The potential problem with modern agriculture is not that it is not productive enough but that it is *uniform.*

We have already seen that the ability of an ecosystem to resist change is partly determined by its biodiversity. The same goes for individual species. A food crop such as corn or rice in its *natural* state is likely to consist of many different strains growing together. Some are more productive than others, some more drought-resistant; some would be resistant to one pest, others more resistant to a different pest. In any year, whatever the conditions, whatever pests flourish, there is a good chance that some corn or rice will grow-whichever strain is best adapted to that particular set of conditions.

Over time the environmental conditions change, and the pests and diseases that attack a particular plant evolve. At the same time the crop plant also changes, and new strains evolve that are resistant to the new conditions. Every strain carries its own particular genes. *Genes* are units of heredity; they control the life processes of all cells and contain the information that controls how the genes work—in other words, the information that produces exactly that strain of corn or rice. With thousands of subtly different strains, there is a vast natural reservoir of genetic material available that ensures the long-term health and survival of the species. Vitality depends on diversity.

However, the whole approach to modern agriculture is uniformity. The object is to reduce diversity as much as possible, to make a high-yield variety of a crop that is resistant to drought and a particular assemblage of pests and diseases. This variety flourishes for a few years until a new strain of pest or disease evolves, against which it has no defense. At that point it is back to the drawing board, or the biotechnology lab, where the scientists mix-and-match until they produce a corn variety that is resistant to the new conditions. Everything is fine for a few more years. So, where does the genetic mix-and-match material come from? It comes from very specialized seed banks that have been established around the world. We have not actually destroyed the diversity of our food crops, we have simply concentrated it in a few locations. The International Storage Center for Rice Genes in the Philippines, for example, keeps more than 86,000 varieties of rice. The geneticist needing some new rice strains

to work with simply goes to the seed bank and makes a withdrawal.

But the genes in the seed bank are, in a sense, frozen in time. They represent the state of genetic diversity that the rice had achieved when the bank was established. The diseases and pests, conversely, do not live in seed banks—they exist in the wild and mutate and evolve, continuously producing new strains. In some cases they produce something that the seed bank cannot deal with. Then our only recourse is to return to the wild. The geneticist goes back to the "genetic home" of the species and searches for a wild variety that has the traits needed to solve this particular problem. However, if the natural habitat is destroyed, the only repository of genetic diversity is in seed banks, which are more vulnerable to accidents (such as fire), war, or sabotage.

Despite our modern techniques, despite seed banks and biotechnology, ultimately the health of our food crops depends on the continued survival of a diverse population of wild strains. This is where our real problem emerges.

Humans make use of a huge number of different food crops. Yet, the bulk of the world's food supply and animal feed comes from just 130 species of plants. The ultimate storehouses of the genetic material we need to ensure long-term health and high productivity are restricted to only 12 locations around the world. These 12 centers of genetic diversity, called *Vavilovian centers* in honor of the Russian geneticist N. I. Vavilov, who first described them, are shown in Figure 15-12. Each of these centers is located in an area where population pressures or development are placing an increasing stress on the existing natural habitat.

Just how vulnerable we might be is illustrated by an episode in the late 1970s that threatened the rice crop of Southeast Asia. The rice, a basic food source for hundreds of millions of people in the region, was threatened by a disease called grassy stunt virus. Scientists searched through the gene banks—through 47,000 varieties of rice—and eventually found the genetic material that would allow the crop to resist the disease, in a single wild species from a valley in India. Soon after the retrieval of the material, the valley was flooded in the construction of a new hydroelectric project.

Biodiversity and Stability. As an extreme case of loss of biodiversity, imagine what would happen if we were suddenly to cut down all the world's forests. Would we suffocate? No, we would not quickly run out of oxygen, even though green plants are responsible for the buildup of oxygen that gave us the atmosphere we breathe today. Even if we removed all the photosynthesizing organisms on the land and the phytoplankton in the sea, it would still take us millions of years to suffocate. (You can calculate just how long in Critical-Thinking Problem 3 at the end of this chapter.)

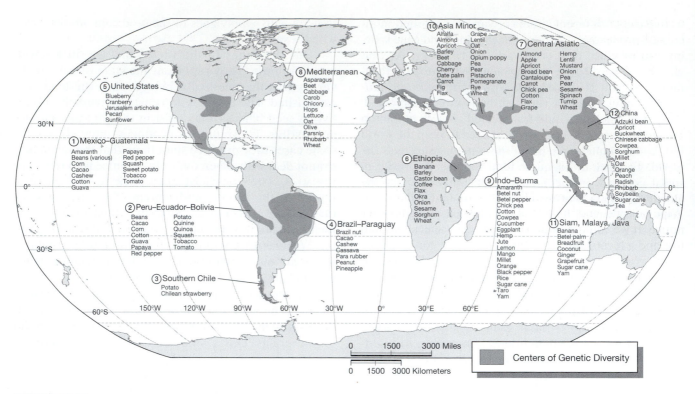

FIGURE 15-12

The 12 centers of genetic diversity. All of the genetic diversity essential for maintaining the world's food supply is limited to just these 12 areas. (After S. C. Witt, 1985. *BriefBook: Biotechnology and Genetic Diversity,* San Francisco: CSI)

We would not suffocate if we lost all the photosynthesizing organisms, but would we starve? Yes. Without these primary producers at the base of the food chain, all higher-order consumers, including humans, would quickly run out of food. We are not separate from the rest of Earth's biota; we are an integral part of the Earth system. Without the rest of the system in place, we could not survive. Our existence depends on the continued presence of a flourishing biota, and, as we have seen, diversity enhances the health and vitality of the biota. The greater the number of species in an ecosystem, the healthier the ecosystem will be. The more diverse the ecosystem, the greater chance it can survive disruption: If one species is lost from a diverse ecosystem, some other species is likely to continue in a similar role, performing a similar function in the ecosystem. For the ecosystem to be healthy or productive, how diverse does it have to be?

Studies of productivity in different ecosystems have shown that productivity increases as biodiversity increases. Other studies have shown that the resilience of an ecosystem—how well it is able to withstand different types of stress—also increases as biodiversity increases. Both the productivity and the stability of an ecosystem are strongly affected by biodiversity. However, these studies also suggest that beyond a certain level of biodiversity, there is little further increase in either productivity or stability. Although we can document this threshold for individual ecosystems in particular locations, however, we still do not fully understand what this means at the global scale.

Beyond the moral or ethical issue and beyond the potential medical or economic value of individual species, we can still ask the following question: If we kill off a large number of the species existing on Earth, will it matter?

At present there are two very different views on the importance of biodiversity. One, put forward by the Stanford University ecologists Paul and Anne Ehrlich, likens biodiversity to the rivets on an airplane. Each species has a small but important role to play in the overall functioning of the ecosystem, just like the rivets on the airplane. If you remove the rivets (or species) one by one, sooner or later the stress on the system becomes so great that it fails: The airplane falls out of the sky, or the ecosystem collapses. The second hypothesis was proposed by Brian Walker, an ecologist for the Australian Commonwealth Scientific and Industrial Research Organization. He suggested that, in fact, most species are superfluous and that the system is maintained by a few keystone species. In this case, loss of species is not a problem as long as you do not lose the keystone species.

Where does this leave us? If we wiped out all tropical rainforests but replanted an equal area of temperate forest, the new forest would continue to release oxygen and take up CO_2. If we were to plant a large enough area, we could replace the biomass of the tropical forests and even begin to take up some of the CO_2 released by the burning of fossil fuels. (To take up all of the CO_2 that could be released by the burning of fossil fuels, we would need to more than double the organic carbon content of living biomass and soils.) However, we would have exterminated more than half the species on Earth. We cannot tell what repercussions this might have throughout the Earth system. If we accept the rivet hypothesis, then we might imagine a loss of half the species on Earth as catastrophic. If we assume that only the keystone species are really important or that, for any ecosystem, there will be redundancy in the roles played by different species, then we can accept the loss of some species. Multiple species of insects, for example, might fill similar niches and interact with the rest of the ecosystem in similar ways. Wipe out one, and the rest of the ecosystem might never notice. If we could pick and choose which species should live and which should die, then maybe we could reduce the number of species in the world by half, and the rest of the Earth system would continue along quite happily.

But extinction does not happen that way. As the human population grows, so do our demands on natural resources. The loss of habitat caused by our building of cities, extraction of minerals, logging of forests, and conversion of forests and grasslands to agriculture and graz-

ing lands kills many species and can destroy whole ecosystems. The destruction is not selective; neither are the effects. From a long-term perspective (Chapter 10), we can judge that the Earth system is very resilient. It can recover from very large perturbations, and no doubt Earth will recover from this one, although it may take tens of millions of years. In terms of individual species, it is a very different matter. Although past extinction events were different in cause and structure to the present one, it might be worth considering that whenever such a large extinction event occurred in the past, the species at the top of the food chain before the event were no longer there once it was all over. Dinosaurs ruled the Mesozoic Earth and are now extinct. Humans could be next.

We do not know enough about how most ecosystems operate or about how important different ecosystems are for the overall global system. For purely pragmatic reasons, it makes sense to slow the rate of habitat destruction and the extinction of species. Pragmatism aside, we pose the following ethical question: Does any species have the right to exterminate another?

Chapter Summary

1. A mass extinction episode as large as those that have occurred periodically throughout Earth's history is taking place today. It is a result of the spread of human populations and the destruction of much of the world's most productive natural habitats.
2. Much of this species loss is taking place in the tropical rainforests of the world, which are also the most diverse of the terrestrial ecosystems.
   a. The tropical rainforests are thought to contain three-quarters of all the living things on Earth and two-thirds of all plant and animal species.
   b. Deforestation in the tropics is wiping out species faster than they can be catalogued and much faster than they can be studied. Every day, species are going extinct that we do not even know exist.
3. The instrumental value of a species can be viewed in different ways—for example, their potential value as pharmaceuticals, their recreational value, their scientific value, or their commercial value.
   a. The diversity of species has value in terms of the productivity and stability of ecosystems and plays an important role in the stability, or health, of the Earth system as a whole.
   b. The destruction of tropical forests has other impacts besides the reduction of global biodiversity. It also results in large changes in local climate and the hydrologic cycle,

and, through the release of CO_2, it has an impact on the global-scale climate.
4. Biodiversity, or the loss of biodiversity, is of critical importance to modern agricultural production. Current practices, particularly in industrialized nations, are to increase productivity through increased specialization and the use of a limited number of selectively developed varieties of crops.
   a. When only a few varieties of a particular species exist in a location, they become particularly vulnerable to new strains of diseases or pests that may evolve.
   b. Diversity is essential for maintaining the long-term health and survival of a species, but modern agriculture is dependent on reducing that diversity.
   c. In the case of particular food crops, we compromise by building large reservoirs of genetic diversity in specialized seed banks.
   d. As new diseases arise or as new pests evolve, the genetic material contained in these seed banks might not be sufficient to develop a resistant variety of the crop. We then have to turn to the region in the world where the plant first evolved (the genetic home of the plant) to look for a wild strain that might be resistant to that particular threat.
   e. The vast majority of the world's food and animal feed crops come from just 12 regions-each located in an area where population pressures or development are placing an increasing stress on the existing natural habitat.

Key Terms

biomass
biome

community
ecosystem

ecotone
endemic species

food chain	intrinsic value	speciation
habitat	keystone species	symbiosis
instrumental value	population	

Review Questions

1. Explain why it is difficult to give an accurate count of the number of species that go extinct each year.
2. Use a diagram to help explain the relationships among *ecosystems, communities,* and *species.*
3. What is meant by the term *habitat?* Explain its importance to issues of biodiversity and extinction.
4. What is meant by the term *keystone species?* Explain why it is important to issues of biodiversity and extinction.
5. Where does the greatest rate of species loss occur today?
6. If a large area of temperate forest is cleared, it can commonly grow back fairly rapidly to something that closely resembles the original forest. If a large area of tropical forest is cleared, regrowth is slow and the original forest cover may never return. Explain why these forests respond differently to clearing.
7. Describe three consequences of tropical deforestation other than loss of biodiversity.
8. List four social and economic factors that contribute to tropical deforestation.
9. What characterizes the 18 hot spots identified by Norman Myers?
10. What is meant by the instrumental and intrinsic values of species?
11. Explain why lack of diversity in food crops might present problems for the world food supply.
12. Why is there a relationship between biodiversity and the long-term health of an ecosystem?
13. Explain the difference between the Ehrlichs' rivet theory and Walker's theory of the importance of biodiversity.

Critical-Thinking Problems

1. Figure 15-2 presents a systems diagram of the feedbacks involving boreal forest cover, albedo, temperatures, sea ice, and the oceans. We used this diagram to show that it is possible for the northern boreal forest to have a significant impact on the larger-scale climate. Using the information you now have about the possible impacts of anthropogenically induced greenhouse climate change, expand on this diagram and discuss the implications in terms of climate and forest cover.
2. Consider the following set of statements for a situation in the tropical forest where forest is cleared for crops:
   - Forest clearance requires access to the forest (roads).
   - An increasing population requires access to services and commodities (a town or city).
   - Settlements require roads.
   - Construction and forest clearance are easier (may increase) if roads are present.
   - The presence of roads and settlements attracts more people into the region.
   - Forest clearings rapidly lose productivity, requiring more clearing.
   - Clearing results in loss of biodiversity.

     Take these statements as a minimum; you might think of some other conditions or relationships that could also be included. Construct a systems diagram showing how the statements all interact. Describe what the diagram shows. Are any feedback loops present? If so, are they positive or negative? Is this a useful way of looking at some of the forces that promote deforestation? Is anything missing that you think should also be taken into account?
3. What would happen to atmospheric oxygen if photosynthesis were to shut off suddenly? Photosynthesis is the source of most of the O_2 in the atmosphere, so it is reasonable to guess that O_2 would decline. Analyze how it would do so by performing the following calculations:
   a. Forests and soils together contain about 2200 Gton of carbon. The amount of carbon stored in the living marine biosphere is negligible in comparison. The atmosphere contains 3.6×10^{19} mol of O_2. By what percentage would atmospheric O_2 decrease if Earth's biota were to die suddenly and if the forest and soil organic carbon were to be oxidized? (*Hint:* Remember that in the decay of organic matter, 1 mol of carbon reacts with 1 mol of O_2 according to the formula $CH_2O + O_2 \rightarrow CO_2 + H_2O$.)
   b. Approximately how long would this process take? Assume the decay proceeded at the modern rate of 30 Gton(C)/yr.
   c. The remaining atmospheric O_2 would have to be removed by the weathering of kerogen and other reduced minerals (iron and sulfides) in rocks. Kerogen is weathered at a rate of 0.05 Gton(C)/yr. Express this rate in terms of moles of carbon per year. If the oxidation of iron and sulfides were to use up oxygen at this same rate, how long would it take for atmospheric O_2 to disappear?

Further Reading

General

Baskin, Y., 1997. *The Work of Nature: How the Diversity of Life Sustains Us.* Washington, D.C.: Island Press.

Eldredge, N., 1998. *Life in the Balance: Humanity and the Biodiversity Crisis.* Princeton, New Jersey: Princeton University Press.

Wilson, E. O., 1992. *The Diversity of Life.* New York: W. W. Norton and Company.

Advanced

Balich, M. J., E. Elizabetsky, S. A. Laird (Eds.), 1996. *Medical Resources of the Tropical Forest.* New York: Columbia University Press.

Chapter 16
Climate Stability on Earth and Earthlike Planets

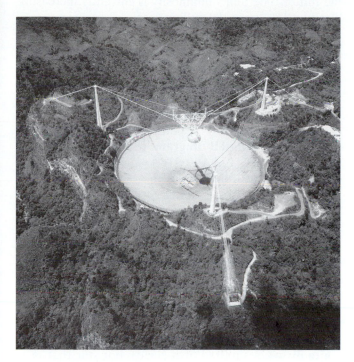

The Arecibo radio telescope. (From David Parker/Science Photo Library, Photo Researchers, Inc.)

Key Questions

- How will Earth's climate evolve in the distant future as the Sun continues to brighten?
- Why are Venus and Mars so different from Earth?
- What determines the width of the habitable zone around our Sun and other stars?
- What are the chances that life exists elsewhere in the universe, and can we detect it if it is there?

Chapter Overview

The Sun will continue to brighten in the future, as it has done throughout Earth's past. As Earth's climate warms, atmospheric CO_2 concentrations should decline in response to the negative feedback provided by the carbonate–silicate cycle. Eventually, this decline may pose a problem for life, because there will not be enough CO_2 in the atmosphere to support photosynthesis; thus, the lifespan of Earth's biota is limited.

Although the CO_2-climate feedback may have stabilized Earth's climate over billions of years, it evidently did not operate on our neighboring planets, Venus and Mars. Venus is clearly outside the habitable zone around the Sun—that is, the region in which a planet can support liquid water at its surface. Mars appears to be outside the habitable zone as well, though perhaps as a result of the planet's small size, which prevents it from recycling CO_2 back into its atmosphere, rather than its distance from the Sun. Habitable, Earthlike planets may exist around other nearby stars. Some of these planets may harbor life; of those, a few may harbor intelligent beings like ourselves. Space-based infrared telescopes and ground-based radio telescopes may eventually tell us whether extraterrestrial life does indeed exist.

Introduction

A recurrent theme of ours has been the interaction between atmospheric CO_2 levels and climate. On long time scales, we have argued that these variables are connected by way of a strong negative feedback loop. A warmer climate increases the amount of precipitation, which, in turn, increases the rates at which silicate rocks are weathered and carbonate sediments are formed. As a result, the atmospheric CO_2 concentration drops, thereby decreasing the magnitude of the greenhouse effect and causing the climate to cool. The negative feedback produced by this mechanism was probably a major reason why Earth's climate has remained relatively stable for at least 3.5 billion years (the time during which we know that life has existed).

Climate, of course, is not fixed. On shorter time scales, Earth's climate exhibits substantial fluctuations associated with glacial–interglacial cycles; atmospheric CO_2 seems to have provided positive feedback in this case. But from a broader viewpoint, Earth's climate is stable in the sense that it has always remained within the temperature regime where liquid water can exist on at least part of the planet's surface. This is the fundamental criterion for a habitable planet, because all known organisms require liquid water during at least part of their life cycle. And most advanced organisms, such as humans, require liquid water on a day-to-day basis.

Could life exist elsewhere in the universe? To answer this question, we need to determine the width of the liquid water *habitable zone* around the Sun and other stars. We know that within our own Solar System only one planet, Earth, has liquid water at its surface. As we pointed out in Chapter 3, Venus is much too hot to have liquid water, and Mars is much too cold. Liquid water could be present on Mars several kilometers beneath its surface, where heat from Mars' interior could help keep temperatures above the freezing point. Jupiter's moon Europa may also have liquid water beneath its icy crust. In this case, the heat source is thought to be *tidal flexing* as Europa follows its slightly eccentric orbit around Jupiter. (Jupiter's gravity raises tides on Europa, just as the Moon raises tides on Earth. The same side of Europa always faces Jupiter, so these tides do not move across the surface as they do on Earth. However, because Europa's orbit is noncircular, it wobbles back and forth a little; this wobble causes heat to be released within Europa's crust.) But both of these environments are highly specialized. If life does exist in either place, it will be extremely difficult to detect—and it will not be like us. If we hope to find advanced life, we need to look for a planet not too different from Earth.

What went wrong with Venus and Mars? Why were their climates not stabilized by the same silicate-weathering feedback that appears to have operated on Earth? In this chapter, we shall attempt to answer that question and to determine its implications for the possibility of life on planets around other stars. But first let us consider what the silicate-weathering feedback may mean to Earth's own climatic future.

Climate Evolution in the Distant Future

As we noted in earlier chapters, the Sun has gotten brighter throughout its history as a result of the gradual conversion of hydrogen to helium. This process is ongoing, so the Sun is expected to continue to get brighter in the future. The Sun's luminosity is currently increasing by about 1% every 100 million years.

How long will Earth continue to be habitable? We know that the CO_2–weathering feedback must break down at some point. The Sun is currently a **main-sequence star**—that is, a normal, middle-aged star that falls along a band running from the upper left to the lower right on a plot of luminosity versus effective radiating temperature. Such a plot is called a **Hertzsprung–Russell diagram.** (See the box "Main-Sequence Stars and the Hertzsprung–Russell diagram.") About 5 billion years hence, the Sun will reach the end of its main-sequence lifetime. Over the next few hundreds of millions of years, it will expand into a **red giant,** a large, reddish star that is intermediate in the post-main-sequence phase of its evolution. Its luminosity will increase by a factor of more than 1000, and its outer envelope will extend to somewhere between the orbits of Mercury and Venus. Any life form existing on Earth at that time will be fried to a crisp, regardless of the atmospheric CO_2 concentration. After that event, the Sun will contract and become a **white dwarf,** a small (Earth-sized), hot star that is at the final stage of evolution of a star such as our Sun. Earth itself will then become very cold, but this change will not matter to anyone because all life will have been extinguished during the prior, red-giant phase of the Sun's evolution.

In actuality, Earth's climate regulation system could begin to fail long before the Sun's red-giant stage is reached. Figure 16-1 shows the result of a computer simulation of Earth's climate over the next 1.6 billion years. During this time, solar luminosity should increase by about 16% (Fig. 16-1a). Earth's surface temperature, T_S, is predicted to increase slowly at first, then more rapidly starting about 1 billion years hence (Fig. 16-1b). By 1.6 billion years from now, T_S could exceed 100°C (212°F), and Earth's surface could become uninhabitable for all but the most thermophilic (heat-loving) of microbes. Furthermore, the computer models indicate that another, irreversible change will take place, starting in about 1 billion years when solar luminosity has increased by about 10%: The stratosphere will become wet, and water will be lost rapidly by photodissociation followed by the escape of hydrogen to space. Eventually, Earth should evolve into a planet similar to Venus.

Main-Sequence Stars and the Hertzsprung–Russell Diagram

The Hertzsprung–Russell diagram, or H-R diagram for short, is a standard means of categorizing different types of stars (Box Fig. 16-1). The horizontal axis represents the effective radiating temperature of the star, as determined by Wien's law (Chapter 3) or by some equivalent method of analyzing the star's spectrum. For historical reasons, temperature is always shown increasing to the left. The vertical axis represents the luminosity of the star relative to that of the Sun.

As Box Figure 16.1 shows, most stars fall along a well-defined band that runs from the upper left of the diagram to the lower right. This band is referred to as the main sequence. It consists of "normal" stars—that is, stars that are in the slowly evolving, middle phase of their existence. Stars are further grouped into seven classes on the basis of their spectra as O, B, A, F, G, K, or M. The brightest, bluest, and most massive (O and B) stars are referred to as *early type stars;* the dimmest, reddest, and least massive (K and M) stars are called *late-type* stars. Within these categories, stars are further assigned numbers ranging from 0 (early) to 9 (late).

Our Sun is an unremarkable G2 star that occupies a spot near the middle of the main sequence. In about 5 b.y., the Sun will evolve off the main sequence and become first a red giant (at the upper right in the H-R diagram) and eventually a white dwarf (at the lower left). The Sun is drifting slightly upward on the H-R diagram during its main-sequence lifetime as a consequence of the conversion of hydrogen to helium. This effect is small compared with the changes that occur before and afterward, but it has important effects on planetary climates. It is this slow, main sequence evolution that gives rise to the faint young Sun paradox and that may limit the lifetime of Earth's biota in the future.

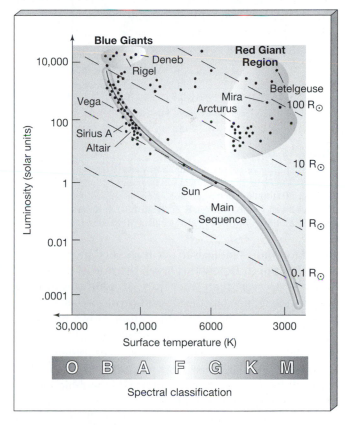

BOX FIGURE 16-1

The Hertzsprung–Russell diagram. (From E. Chaisson and S. McMillan: *Astronomy: A Beginner's Guide to the Universe,* 2/e, 1998. Reprinted by permission of Prentice Hall, Upper Saddle River, N.J.)

The CO_2 Compensation Point

Earth's biota could run into trouble long before the temperature becomes high and the oceans disappear. The reason that T_S remains low for the next billion years in this simulation is that atmospheric CO_2 is predicted to decrease rather rapidly (Fig. 16-1b). (The current fossil-fuel CO_2 pulse, which will last for only a million years or so, is neglected in this calculation.) The calculation assumes that the silicate weathering rate is controlled by the partial pressure of CO_2 in the soil, P_{soil}, which is higher than that in the atmosphere, P_{atm}, as a result of the microbial decomposition of organic matter in soils (Chapter 7).

This pumping is assumed to be provided by C_4 plants, which are capable of photosynthesizing down to CO_2 levels of about 10 ppm. The critical concentration of CO_2 re-

FIGURE 16-1

Long-term projections of (a) solar luminosity and (b) surface temperature (T_S) and CO_2. Here S_0 is the present solar flux and P_{atm} and P_{soil} represent the CO_2 concentrations in the atmosphere and in the soil, respectively. (Reprinted with permission from K. Caldeira and J.F. Kasting, Nature 300: 721-723, 1992. Copyright 1992 Macmillan Magazines, Ltd.)

quired for photosynthesis to occur is called the **CO_2 compensation point.** Below this level, the rate of photosynthesis is slower than the rate of **photorespiration** (respiration induced by the absorption of sunlight), so the plants cannot grow. In the simulation shown in Figure 16-1b, C_4 plants are predicted to become extinct about 900 million years from now, when the atmospheric CO_2 concentration falls below the CO_2 compensation point. At this time, P_{soil} becomes equal to P_{atm} because the biological pumping action disappears.

The C_3 plants, which fix carbon by means of the Calvin cycle (Chapter 13) and constitute about 95% of the current terrestrial vegetation, could go extinct even earlier. Their CO_2 compensation point is about 150 ppm, approximately half of today's atmospheric concentration. A similar simulation (not shown) in which C_3 plants control the CO_2 partial pressure in soils indicates that C_3 plants should survive for only about another 500 million years. As their CO_2 compensation point is approached, they should gradually be replaced by C_4 plants (which photosynthesize at much lower CO_2 concentrations than do C_3 plants) or by new species of plants that evolve mechanisms for photosynthesizing at low atmospheric CO_2 concentrations.

Although the predicted demise of the biota is not exactly imminent, the results of this simulation are in some sense disturbing. It took almost 4 billion years to evolve multicellular life and another half billion years to produce humans. After all of this evolutionary hard work, less than 1 billion years may remain before the planet becomes uninhabitable. Evidently, even a planet as well-suited for life as Earth is cannot be expected to harbor advanced life for more than a small fraction of its lifetime.

Climate Evolution on Venus and Mars

Why are the climates of Venus and Mars so different from that of Earth? Both of these planets probably started out with the necessary ingredients (water, carbon, and silicate rocks) to allow the carbonate–silicate cycle to operate. Thus, the CO_2–climate feedback mechanism could in principle have operated on these planets as well. But this mechanism was clearly incapable of keeping their surface temperatures within the liquid-water regime.

The Runaway Greenhouse on Venus

The answer to our question about the climates of Venus and Mars is that even systems containing negative feedback loops commonly have limits beyond which they become unstable (see Chapter 2). In the case of Venus, the negative feedback loop between the surface temperature and the outgoing infrared flux (see Fig. 3-22) breaks down when the solar heating is too strong. Venus orbits the Sun at about 0.72 AU, so the inverse-square law (Chapter 3) shows that the solar flux at the surface of Venus is about $1.9 (=1/0.72)^2$ times that at Earth's surface. Even early in the Solar System's history when the Sun was 25 to 30% dimmer, the solar flux at Venus' orbit would still have been some 40% higher than the present flux at Earth's orbit.

Under these circumstances, climate models predict that the positive coupling between surface temperature and the outgoing IR flux disappears (Fig. 16-2). Recall from Chapter 3 that this coupling is part of the funda-

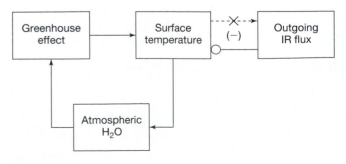

FIGURE 16-2

Systems diagram illustrating the runaway greenhouse on Venus.

mental feedback loop that keeps Earth's climate stable. On Venus, the atmosphere becomes so warm and so full of water vapor that virtually no infrared radiation from the surface is able to escape to space. Only the radiation from the cold upper troposphere and stratosphere is able to escape, and the flux of this radiation does not change as the surface becomes warmer and warmer. The resulting out-of-control climate system is called a **runaway greenhouse.** With the stabilizing feedback loop gone, the surface temperature "runs away" to values as high as 1500 K, the oceans evaporate, and all the water on the planet's surface turns to vapor. Even the stratosphere becomes filled with water vapor.

Once the stratosphere of Venus became wet, water would have been lost by photodissociation, followed by the escape of hydrogen to space. The oxygen left behind would have reacted with reduced minerals, such as ferrous iron, present at the planet's surface and with any reduced gases remaining in its atmosphere. With no liquid water remaining to facilitate silicate weathering, volcanic CO_2 would have accumulated, forming the dense atmosphere that exists today. The loss of water would also have removed the sink for sulfur-containing volcanic gases. These gases would likewise have accumulated in the atmosphere, giving rise to the dense sulfuric acid clouds that presently obscure the Venusian surface. The end result of this process is the hot, dry planet that we see today (Fig. 16-3).

How do we know that this is actually what happened on Venus? Could Venus have simply started out with less water than Earth did as a consequence of being closer to the Sun? We cannot be sure because we cannot look back in time and because Venus is so hot that it would be extremely difficult to go there and look for the oxygen left behind from the loss of water. An important clue has been left behind in the Venusian atmosphere, however: The ratio of deuterium to hydrogen in the little water vapor that remains is more than 150 times higher than that ratio in Earth's oceans. Recall from Chapter 1 that deuterium, D, is an isotope of hydrogen that has an extra neutron in its nucleus. If Venus initially had a significant amount of water and if this water was lost by photodissociation followed by escape of hydrogen to space, we would expect that the lighter hydrogen isotope, H, would escape more quickly and that the heavier one, D, would become enriched in the water vapor that was left behind. The fact that this is exactly what we observe supports the idea that Venus was once a water-rich planet and that the runaway greenhouse hypothesis is correct.

Martian Climate Evolution

Mars, in contrast, faced just the opposite of Venus's climate problem. The solar flux reaching Mars is only 43% of that reaching Earth and would have been even lower early in the Solar System's history. The solar heating was so low that CO_2 itself should have condensed, forming clouds and polar ice caps. We know CO_2 ice by the name *dry ice,* and it has many familiar uses, such as keeping ice cream cold. The present Martian polar caps (see Fig. 3-1) contain a mixture of CO_2 ice and water ice.

The fact that CO_2 can condense on Mars makes that planet's climate evolution very different from that of Earth. On Mars, volcanic CO_2 would not simply have accumulated in the atmosphere if the surface temperature fell below freezing. Thus, the stabilizing feedback provided by the carbonate–silicate cycle would not have operated in the same way as on Earth.

The surprising thing about Mars is that, despite this problem of CO_2 condensation, the Martian surface was once warm enough to allow liquid water to flow on its surface (Fig. 16-4). Most of the Martian channels, or *valleys,* appear to have formed prior to about 3.8 b.y ago. (On Earth, a valley is the relatively wide region carved by a stream and a channel is the narrow stream that runs through it.) We infer this because the terrain on which most of the valleys are located is covered with lots of impact craters. By analogy with the cratering record of the Moon, this cratering occurred during the first 700 million years of the Solar System's history (see Chapter 8). Evidently, Mars had a denser atmosphere at that time with enough greenhouse effect to keep its surface warm despite the low solar luminosity. Climate modelers are still not sure how to explain this warm climate, although greenhouse warming by CO_2 was almost certainly involved. High-altitude CO_2 clouds, which act somewhat like cirrus clouds on Earth, may also have played a role. (Recall from

FIGURE 16-3

The surface of Venus, as observed by radar from the *Magellan* spacecraft. (From Jet Propulsion Laboratory, NASA Headquarters)

FIGURE 16-4
A Martian valley, seen from the *Mars Global Surveyor* spacecraft. (Courtesy of NASA)

Chapter 3 that high-altitude clouds warm a planet's surface whereas low-altitude clouds tend to cool it.)

Why, then, did Mars become so cold? The answer may lie in the planet's small size. Mars has about half Earth's diameter and only about one-tenth its mass. A planet that size should have cooled off relatively quickly after its formation and would have a smaller radioactive heat source. (A planet's internal heat energy is proportional to its volume, which decreases more rapidly with radius than does the planet's surface area.) With less geothermal energy available to drive plate tectonics, the Martian crust might have solidified into a single unit early in the planet's history. This solidification would have shut down the carbonate–silicate cycle, because carbonate rocks would no longer have undergone metamorphism. With most of the planet's CO_2 trapped in carbonates or stored as dry ice in the Martian soil, the greenhouse effect would have diminished and Mars would have approached its present, frozen state.

Habitable Planets around Other Stars

The question of what happened to Venus and Mars bears on another, even more interesting question: What is the chance that habitable planets exist elsewhere in our galaxy? We could broaden this to include the rest of the universe, but the question would then become entirely philosophical because we would have no way of verifying our predictions. As we shall soon see, within the next 20 to 30 years we may be able to determine whether habitable (or even inhabited) planets exist within at least part of our own galaxy. Thus, it is not too soon to start thinking about whether they should be expected.

We know now that planets exist around other stars. At least 11 stars other than the Sun have been shown to

have planetary companions. All these planets are large, gas-giant planets like Jupiter (see Chapter 8). (At least, we assume they are gas giants because they are so massive.) Large planets are easier to detect than small ones, because their gravitational pull on their parent star is strong; it is the motion of the stars themselves that has been observed. But gas-giant planets are not likely abodes for life. They have no solid surface, and their atmospheres are highly convective. Thus, even though organisms could theoretically exist at some level in their atmospheres where liquid water is stable, the organisms would periodically be lofted to great heights, where the temperature is very cold, or carried to great depths, where they would be incinerated. Life *as we know it* requires a much more stable environment that is best provided by small, rocky planets like our own.

Formation of Earthlike Planets

What are the chances that we will find Earthlike planets around other stars? The answer to this question depends partly on how planets form and partly on what their climates are like later on. How a planet's climate evolves depends on a number of factors that we have discussed in previous chapters.

The formation of terrestrial-type planets is probably a common process. The rock-forming elements (silicon, oxygen, iron, and magnesium) that make up the bulk of Earth have been identified in the spectra of other stars and are thought to be abundant throughout the universe. The volatile elements on which life depends—carbon, nitrogen, hydrogen (as water), phosphorus, and sulfur—are seen in interstellar clouds and are likewise thought to be abundant elsewhere. Disks of dust and gas are known to exist around many young stars, so the general conditions that are believed to lead to planet formation are commonplace. If the formation of other planetary systems has followed the same general pattern as that of our Solar System, it is likely that planets formed in their inner parts would have been chemically much like Earth. In particular, they are likely to have had enough water to form an ocean and enough carbon to have a carbon cycle. A planet that is similar in size to Earth might also be expected to have enough internal heat to power Earthlike plate tectonics.

The Habitable Zone around the Sun

Whether another planet could have an Earthlike climate depends on how far from its star the planet forms and on how fast the parent star evolves in luminosity. As mentioned earlier, a fundamental requirement for life as we know it is that the planet's surface temperature remain in the range at which liquid water can exist. The region in space around a star where this condition is satisfied is called the **habitable zone** (HZ), or **ecosphere.** Because stars increase in luminosity as they age, their habitable zone moves outward with time. The region where a plan-

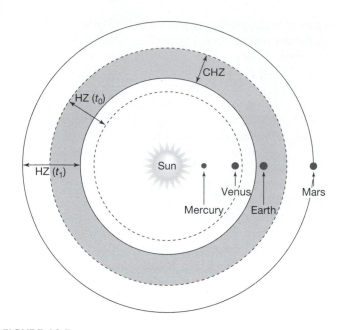

FIGURE 16-5

The relationship between the habitable zone (HZ) and the continuously habitable zone (CHZ). The CHZ is defined for some particular time interval $\Delta t = t_1 - t_0$, taken here to be 4.6 b.y.

et can remain habitable for some finite time interval is called the **continuously habitable zone,** or CHZ. The CHZ represents the overlap between the HZ at some initial time, t_0, and at some later time, t_1 (Fig. 16-5). It is defined for some specific interval, $\Delta t = t_1 - t_0$. For our Sun, we usually take t_0 to be the time the Sun formed (4.6 b.y. ago) and t_1 to be the present, so $\Delta t = 4.6$ b.y.

How wide is the current habitable zone around our own Sun? Climate models can be used to help answer this question. So far, only one-dimensional, radiative–convective models have been used, but ultimately this question could be investigated with three-dimensional general circulation models. The one-dimensional models predict that the inner edge of the HZ around our own Sun is at about 0.95 AU. Inside this distance, a planet's stratosphere is predicted to become wet, and water is lost rapidly by photodissociation followed by the escape of hydrogen to space, as is thought to have occurred on Venus. Note that this result is consistent with the predictions for Earth's future climate discussed earlier in this chapter. Figure 16-1 indicates that Earth's surface temperature should start to increase rapidly beginning about 1 billion years in the future, when solar luminosity has increased by about 10%. If Earth were located at 0.95 AU rather than at 1 AU, the inverse-square law (Chapter 3) predicts that the solar flux would be higher by a factor of $1/(0.95)^2$, or 1.11. In either case, an increase of approximately 10% in the solar flux would be enough to drive off Earth's water.

Identifying the outer edge of the habitable zone is a little trickier. Early calculations by some climate modelers suggested that the outer edge is very close to Earth's

current orbital distance of 1 AU. Moving Earth out to even 1.01 AU is enough to produce **runaway glaciation** in some models, in which the planet's surface becomes entirely covered by snow and ice. The reason was the positive feedback provided by snow and ice (see Fig. 3-21). But we now believe that this positive feedback loop is countered by a very strong negative feedback loop involving the carbonate–silicate cycle. If Earth were somehow to be moved farther from the Sun, Earth's surface temperature would drop and CO_2 would accumulate in its atmosphere, just as it is thought to have done in the distant past when solar luminosity was lower. When this positive feedback loop is included, the climate models predict that the outer edge of the current HZ is somewhere near the orbit of Mars, around 1.5 AU. The exact distance is difficult to determine, because the effect of CO_2 clouds cannot be estimated reliably with one-dimensional climate models and because other greenhouse gases, such as methane, might extend the HZ even farther out if they were present in appreciable concentrations in a planet's atmosphere. As we mentioned earlier, Mars had liquid water on its surface early in its history even though it appears to have been outside the HZ at that time, according to Figure 16-5. Perhaps the HZ is wider than we think.

Because the CHZ represents the overlap between two HZs at different times in the Sun's history, it is narrower than the HZ. A planet near the outer edge of the current HZ would have been outside the HZ 4.6 b.y. ago when the Sun was only 70% as bright as it is today. When we take into account the change in solar luminosity, the outer edge of the 4.6-b.y.-CHZ is estimated to lie in the vicinity of 1.25 AU. (Note again that the CHZ is defined only for some specific time interval.) This estimate is conservative, because it does not take into account the possible warming caused by CO_2 clouds on planets near the outer edge of the HZ. So, the actual width of the CHZ may be even greater. Evidently, the habitable region around our own Sun is much wider than early predictions had suggested.

Habitable Zones around Other Stars

The same type of climate stability calculations described above can be performed for stars other than our Sun. The results are summarized in Figure 16-6, which shows the instantaneous HZ midway through a star's main-sequence lifetime. (It is difficult to show CHZs on such a plot, because stars of different masses evolve at different rates.) The vertical scale shows stellar mass relative to the Sun's mass; the horizontal scale shows distance in astronomical units. The basic result is very simple: To have a surface temperature similar to Earth's, a planet orbiting a dim, red star would have to be closer than 1 AU to that star, whereas a planet orbiting a bright, blue star would have to be farther than 1 AU from that star.

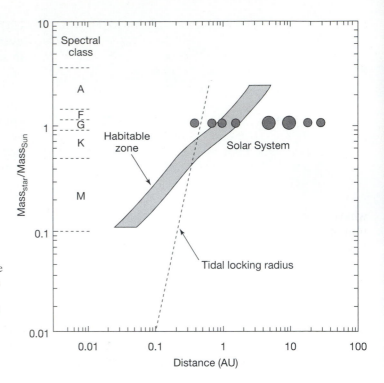

FIGURE 16-6

Habitable zones for other main-sequence stars. The HZ is plotted midway through the star's main-sequence lifetime. For a discussion of spectral classes of stars, see the box "Main-Sequence Stars and the Hertzsprung–Russell Diagram." (After J.F. Kasting et al., *Icarus*, v. 101, p. 108-128, 1993.)

Problems occur, however, for stars much earlier or later in the main sequence than our own Sun. For dim, red stars, the habitable zone falls within the **tidal locking radius** of the star—that is, the distance from a star within which a planet is likely to rotate such that it always faces the star. The Moon shows only one side to Earth; the reason is that it is close enough to Earth for its rotation to have been damped by tides. An astronomer would say that the Moon exhibits **captured rotation.** Close to the star, the tidal forces increase faster than the amount of available starlight. Thus, a planet that is close enough to a dim, red star to be within its habitable zone is likely to develop a captured rotation. The planet would then have one hot, permanently sunlit side and one cold, perpetually dark side. Its atmosphere might freeze out on the dark side, making the entire planet uninhabitable.

Bright, bluish stars pose other problems for planetary habitability. The most serious is that they have relatively short main-sequence lifetimes. The rate of energy generation in a star increases with a star's mass much faster than the mass itself does. The result is that hot, massive stars exhaust the hydrogen in their cores much faster than does a star like the Sun. The hottest, most massive type of star stays on the main sequence for only a few tens of millions of years, compared with 10 billion years for our Sun. Thus, if life were to originate on a planet circling a bright star, it would have much less time to evolve than it did here on Earth.

Hot, blue stars pose another potential problem for life in that they emit a significant fraction of their radiation at ultraviolet wavelengths. According to Wien's law

(Chapter 3), as the temperature of the star's surface increases, the peak wavelength of its emitted radiation decreases. The Sun's effective surface temperature is 5780 K; hence, its radiation peaks at about 0.5 μm, near the middle of the visible part of the electromagnetic spectrum. Hot, blue stars can have surface temperatures of up to 25,000 K, so their radiation peaks well into the ultraviolet. The relative amount of UVB and UVC radiation emitted by such a star is orders of magnitude greater than that emitted by our own Sun. Hence, a planet orbiting a hot, blue star would need to have an extremely well developed ozone screen in order for its surface to be habitable by advanced life. The good news is that the shorter UV wavelengths that split O_2 and make ozone are even more strongly enhanced in these stars, so such "super" ozone screens might actually exist.

We conclude that stars not too different from our Sun are the ones most likely to harbor habitable planets. Among such stars, however, the chances of there being an Earthlike planet appear to be pretty good.

The Drake Equation

The last chapter of many introductory astronomy textbooks is devoted to speculation about the possibility of life elsewhere in the universe. The discussion is generally framed in terms of the **Drake equation,** a relatively innocuous-looking formula used to estimate the number of other intelligent civilizations in our galaxy with which we might one day establish radio communication. The equation was proposed by Frank Drake, an astronomer then at Cornell University. The late astronomer Carl Sagan of Cornell University also had a hand in crafting the equation and wrote about it extensively in his books *Cosmos* and *Intelligent Life in the Universe* (with I. S. Shklovskii).

The equation can be written as follows: N, the number of advanced, communicating civilizations in the galaxy, is expressed as the product of seven terms:

$$N = N_g f_p n_e f_l f_i f_c f_L,$$

where N_g = number of stars in our galaxy; f_p = the fraction of stars that have planets; n_e = the number of Earth-like planets per planetary system; f_l = the fraction of habitable planets on which life evolves; f_i = the probability that life will evolve to an intelligent state; f_c = the probability that intelligent life will develop the capacity to communicate over long distances (for example, by radio telescope); and f_L = the fraction of a planet's lifetime during which it supports a technological civilization.

An alternative form of the Drake equation replaces the factors N_g and f_L with R (the average rate of star formation in the galaxy) and L (the lifetime of a technological civilization). If the number of stars in the galaxy is in steady state, which is probably approximately true, then R is equal to the number of stars in the galaxy, N_g, divided by the average lifetime of a star, L_s. The product RL is thus $(N_g/L_s)L = N_g(L/L_s)$. But L/L_s is essentially the same as f_L, so the two forms of the equation are equivalent.

Despite its apparent simplicity, the Drake equation cannot be solved, because many of the individual terms are difficult to estimate. It is nonetheless useful, because it helps us identify the factors that are important to biological evolution, and to human evolution in particular. It is also an appropriate way to conclude our study of Earth system science, because many of the issues that we have addressed bear on terms in the equation. Hence, we offer the following thoughts as a way of summing up our thoughts about the Earth system and its degree of stability.

N_g—The Number of Stars in a Galaxy

We can estimate the numer of stars in a galaxy by looking at small but representative patches of sky and counting the individual stars. The current estimate for N_g is about 4×10^{11}, or 400 billion, stars. Other galaxies, of which there are billions, have comparable numbers of stars, so the total number of stars in the universe is very much larger than this. But the distances separating galaxies are so large (2 million light-years or greater) that the possibility of radio communication with another civilization is too remote to consider.

f_p—The Fraction of Stars That Have Planets

In the previous section, we discussed briefly the fraction of stars that have planets. Astronomers believe that planetary formation is a natural accompaniment to the formation of stars. One problem that we did not mention is that as many as 90% of the stars in our galaxy are members of binary (two-star) systems or other multiple-star systems. Such stars may not form planets, but theoreticians are not certain. Even if they do, it seems unlikely that many of these planets would be located within the habitable zone of one of the stars. So, let us assume that only single stars have planets; hence we assign f_p a value of about 0.1.

For now, this value for f_p is only a guess. Within the next few years, however, it should be possible to measure this number directly. As mentioned earlier, planets have now been discovered around 10 main-sequence stars. So far, only the largest and closest-in planets have been discovered, because these are the easiest to detect. As observational technology improves, however, and as astronomers accumulate data on stellar motions over longer periods, it should be possible to determine whether planetary systems like our own are commonplace.

n_e—The Number of Earthlike Planets per Planetary System

We have already addressed the number of Earthlike planets per planetary system in our earlier discussion of habitable zones around stars. On the basis of our relatively optimistic estimate of the width of the CHZ around our own Sun, we predict that solar-type stars are very likely to harbor habitable planets. Our estimate of the 4.6-b.y.-CHZ width, about 0.3 AU, is roughly equal to the mean spacing between the four innermost planets in our own Solar System, which lie between 0.4 AU and 1.5 AU. Thus, if planets are spaced similarly around other solar-type stars, there is a very good chance that one of them will be within the HZ. To be conservative, we will estimate this probability at 50%. Only stars of spectral type close to our own sun look to be good candidates, however. This cuts down the number of available stars by a factor of about 5. So, a preliminary estimate of n_e would be 0.5×0.2, or 0.1 Earthlike planets per star.

f_l—The Fraction of Habitable Planets on Which Life Evolves

The fraction of habitable planets on which life evolves is one of the most poorly understood terms in the Drake equation. We discussed the question of life's origin in Chapter 9. Chemists and biologists are uncertain as to how it occurred, so it is impossible to say whether it was a likely or an unlikely event here on Earth.

Two experimental approaches could eventually shed light on the origin-of-life question. One would be to create life in a test tube. The other would be to detect life elsewhere in the universe. Of these, the approach that appears most feasible at present is the detection of life elsewhere. We have already sent two *Viking* spacecraft out from Earth in the late 1970s with the express intent of searching for life on Mars. Neither spacecraft found life on the martian surface; indeed, they all but proved that it could not exist. (The martian soil is so highly oxidized that organic material cannot survive very long.) It now seems unlikely, though not impossible, that life will be found on Mars or anywhere else in our own Solar System.

Other planetary systems are a different matter, however. Life around other stars may be detectable by at least two methods. One method is radio communication. The giant radio telescope at Arecibo, Puerto Rico, is capable of communicating with another, similarly equipped civilization anywhere in our galaxy (Fig. 16-7). If we manage to communicate, however, we will no longer need the Drake equation. The second method of detecting life is to build large telescopes in space and use them to view planets around other stars directly. If connected electronically, these telescopes could act as an **interferometer,** an instrument consisting of two or more telescopes with their images combined. They act effectively as one big telescope. In principle, such an instrument could be used to cancel out the light from the star and to let through only the light from the planets around it. NASA has long-range plans for a four-telescope interferometer that could image Earth-sized planets out to a distance of about 50 light years. This instrument would allow us to look for planets around more than 100 single, solar-type stars.

Such an instrument may be able to perform **spectroscopy** on a planet's atmosphere—a technique in which the composition of a material, such as a gas, is analyzed according to the details of its electromagnetic spectrum. By doing so, an interferometer may also be able to look for the presence of CO_2, H_2O, and O_3. The first two compounds are required for life as we know it. The third, ozone, is formed photochemically from O_2, as we discussed in Chapter 14. Almost all of Earth's O_2 was generated by photosynthesis. The atmospheres of Venus and Mars have only small amounts of O_2 and, hence, do not show evidence of O_3 (Fig. 16-8). So, the detection of ozone in the atmosphere of an extrasolar planet would be fairly strong evidence that the planet was inhabited. Note that this method of detecting extraterrestrial life is much more general than radio communication, because it requires evolution only to the bacterial level of complexity.

FIGURE 16-8

Infrared spectra of the atmosphere of Earth, Venus, and Mars. Only Earth shows the presence of H_2O and O_3. (After R. Hanel, Goddard Space Flight Center)

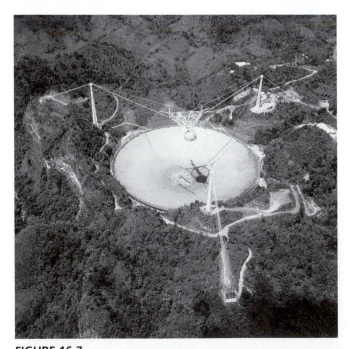

FIGURE 16-7

The Arecibo radio telescope. (From David Parker/Science Photo Library, Photo Researchers, Inc.)

In the meantime, as we do not know how to estimate f_l, we shall for the moment leave its value unassigned.

f_i—The Probability That Life Will Evolve to an Intelligent State

The probability that life will evolve to an intelligent state is another poorly understood term in the Drake equation, in part because it is not entirely obvious what intelligence actually is. Are humans the only intelligent species on this planet? Or should we include dolphins and perhaps a few other species that are not that far behind in terms of mental capabilities?

If we confine our discussion of intelligence to the Drake equation, the appropriate definition becomes clear: To be detectable by radio communication, a species must have the type of intelligence to develop electronic equipment and other aspects of a technological civilization. For this ability, they probably have to have fingers instead of fins. So, in terms of the Drake equation, an intelligent species needs to be fairly similar to our own.

We will not attempt to elucidate here the many factors involved in the evolution of human intelligence. We can, however, point out two topics that we have discussed in previous chapters that bear on this question. The first is the rise of atmospheric oxygen about 2 billion years ago. The rise of oxygen was an absolute necessity not only for human evolution, but also for the evolution of all multicellular life. Indeed, its importance is even more fundamental: Even single-celled eukaryotes (cells with nuclei) require oxygen for their metabolism. Although we cannot be certain that biological evolution would have proceeded elsewhere as it did on Earth, chances are that oxygen is a metabolic requirement for all advanced organisms.

Was the rise of oxygen a likely event? On the basis of the cyanobacteria-like microfossils found in the Warrawoona Formation of Australia (Chapter 9), oxygenic photosynthesis was an early biological invention. Thus, the probability of evolving it must have been relatively high. Once photosynthesis had been invented, the atmosphere was almost bound to become oxidizing, sooner or later. If the hypothesis outlined in Chapter 9 is correct, the key factor was the gradual oxidation of Earth's uppermost layers caused by the recycling of seawater through the mantle and the escape of hydrogen to space. A similar type of process would presumably operate on any planet that had a surface gravity similar to Earth's (so hydrogen could escape) and that had enough internal heat to support plate tectonics. Thus, the rise of oxygen in an Earthlike planet's atmosphere may be a relatively commonplace event.

A second important factor that must influence biological evolution is the frequency of large impacts. Energetic collisions with asteroids or comets have evidently caused mass extinctions of organisms on our planet and have paved the way for the evolution of new life forms (Chapter 10). Humans might not have evolved if the Cretaceous–Tertiary impact event had not eliminated the dinosaurs. Impacts that were either too large or too frequent would have a detrimental effect on biological evolution, however. The lunar cratering record indicates that large (100-km diameter) impactors probably hit Earth fairly often until about 3.8 b.y. ago. These events would have sterilized most of Earth's surface. In our own Solar System, the flux of these large impactors decreased to close to zero after the first few hundred million years. The same, however, would not necessarily be true of other planetary systems.

The key factor in our Solar System that controls the impact flux that hits Earth is the planet Jupiter. Jupiter is directly responsible for at least some of the impactors that have hit Earth over the past few billion years, because that giant planet perturbs asteroids from certain regions in the asteroid belt into Earth-crossing orbits. But Jupiter may also protect us from cometary impacts by deflecting incoming comets before they can enter Earth-crossing orbits. If other planetary systems have similar cometary fluxes, the presence of a giant planet like Jupiter may be essential to ensuring environmental stability on the inner planets.

Like f_l, we do not know what f_i should be, so we will leave it undefined as well.

f_c—The Probability That Intelligent Life Will Develop the Capacity to Communicate over Long Distances

We cannot assess with confidence the numerical probability that an extraterrestrial civilization will develop radio communication. But we may be able to say something about how the planetary environment influenced the emergence of a technologically advanced civilization on Earth. The Yale anthropologist Elizabeth Vrba has suggested that our prehuman ancestors were forced out of their forested environments in Africa around 2.5 m.y. ago by ecological changes associated with the Pleistocene glacial cooling. According to Vrba, these changes led to the replacement of forests by vast areas of savannah (grassland). Our ancestors adapted to their changed environment by developing a more upright manner of walking and by developing weapons with which to hunt the large herbivores that grazed on the savannah.

These changes in lifestyle may also have accelerated intellectual development in humans. Whether the same evolutionary changes would have taken place in the absence of climate change is not clear. But we can see how mildly destabilizing events such as climate change may have helped guide, and perhaps accelerate, both the evolution of humans and the gradual emergence of civilization.

A second, somewhat better documented correlation between climate change and the development of civilization is seen at the end of the last ice age, about 10,000 years ago. At about this time, agricultural societies emerged spontaneously in several different regions of the world, including Mesopotamia, China, and the Indus River valley in India. Very likely, the switch from hunter–gatherer societies to agriculture-based societies was facilitated by the improved climatic conditions. Again, this switch could have occurred without any forcing from climate. But, Milankovitch cycles do seem to have played some role in the development of our current civilization.

Imagine what might happen, though, on an Earthlike planet without a large moon. The Milankovitch climate cycles might then be driven by a chaotically varying planetary obliquity (see Chapter 11). They would probably appear to be regular and to have relatively small amplitudes on a time scale of tens to hundreds of thousands of years, but on longer time scales the planetary obliquity might reach extreme values. What this would do to the emergence of agriculture (and to the soil itself) is unclear. We suspect, however, that it would make the emergence of civilization much more difficult.

Collecting Terms to This Point

Let us stop for a moment to take stock of what we have learned so far. If we combine what we know with what we do not know about the terms in the Drake equation, we are left with

$$N = N_g f_p n_e f_l f_i f_c f_L,$$
$$\approx (4 \times 10^{11}) \times (0.1) \times (0.1) f_l f_i f_c f_L$$
$$= 4 \times 10^9 f_l f_i f_c f_L \text{ yr}^{-1},$$

where we have inserted approximate values for the three leading terms, N_g, f_p, and n_e, and where the symbol $\approx$ means "approximately equals." Our point in writing the equation in this manner is the following: Only the physical science terms in the Drake equation can be assigned numerical values with any degree of certainty. (Even these terms might be subject to considerable debate.) The last four terms deal with biological and sociological evolution and are much more speculative.

Suppose, however, that we are optimistic about the chances of life originating elsewhere and of the evolution of intelligence and civilization. Carl Sagan, for example, was always an optimist, and he estimated (in *Cosmos*) that the product $f_l f_i f_c$ is equal to about 1/300. If we accept his estimate and round off our numbers, the Drake equation boils down to

$$N \approx 10^7 f_L.$$

Under these assumptions, whether N is large or small depends exclusively on f_L, the fraction of a planet's lifetime during which it supports a technological civilization. So, let us speculate for a moment about the magnitude of that factor.

f_L—The Fraction of a Planet's Lifetime during Which It Supports a Technological Civilization

The largest uncertainty in the Drake equation is apparently the fraction of a planet's lifetime during which it supports a technological civilization. Let us assume that the average lifetime of planets themselves is about 10 b.y., the approximate main-sequence lifetime of the Sun. The question then becomes: How long does an average technological civilization last? How long will ours last? The answer clearly depends on what eventually causes our demise. If we do ourselves in by a nuclear war, the answer could be as short as 100 years. If other civilizations annihilated themselves in the same way, the factor f_L would be equal to 10^2 years/10^{10} years $= 1 \times 10^{-8}$, and N would be a paltry 0.1. A value of N below one implies that we might well be the only technological civilization in the entire galaxy.

If left to natural causes, however, the lifetime of our civilization could be much longer. The average longevity of a species in the geologic record is about 1 million years (Chapter 10). If we use this value for civilized species, we get $f_L = 10^6/10^{10} = 1 \times 10^{-4}$ and $N = 1000$. But civilized species are much smarter than average and, thus, might survive much longer. If we are able to deal successfully with all our other problems, humans should be able to survive for at least another 500 m.y., the time scale for the loss of C_3 photosynthesis. If technical civilizations in general were this long-lived, f_L could be as high as $5 \times 10^8/10^{10} = 0.05$ and N would be 500,000.

Future technological innovations, such as interstellar space flight or planet-sized solar shields, could allow us to outlive even this disaster. But the basic point has already been made: If technological civilizations can survive for an appreciable fraction of a planetary lifetime, then the number of such civilizations in the galaxy could be quite large. For $N = 500,000$, the average distance between civilizations should be a few hundred light years. We can imagine that a slow dialogue might take place by radio communication were two such civilizations to discover each other's existence.

Ensuring Our Long-Term Survival

What are the chances that we can overcome all our potential problems and survive for hundreds of millions years? We have discussed a number of current environmental problems in previous chapters. Could any of them could trigger our downfall?

Global warming might lead to drastic environmental changes and even to the relocation of large numbers of people, but it is not likely to result in the demise of all of civilization. Ozone depletion is potentially more dangerous, because UV radiation is directly harmful to almost all forms of life. However, we are already dealing with it in what looks to be an effective manner. Loss of biodiversity might pose a significant threat to global food production (Chapter 15), but it does not seem likely by itself to wipe out the human population. Human population growth itself might be the most significant problem, but even this will ultimately be self-limiting. Either the availability of food or competition for other resources will eventually cause the population to stabilize, hopefully at a comfortable and sustainable level.

The most significant threat to our continued existence over very long time scales might come from large impacts. As we pointed out in Chapter 10, the expected time between impact events the size of that at the Cretaceous–Tertiary is about 10^8 years. Would a 10-km-diameter impactor destroy civilization? It is difficult to say. On the basis of the amount of energy released, the effects would be far more devastating than those of an all-out nuclear war (see Chapter 1). The initial firestorm ignited by the reentry of impact ejecta into the atmosphere could destroy agricultural crops around the world within hours. The dust generated by the impact would reach the stratosphere and would linger there for 6 months to a year. Photosynthesis might be shut off during much of this time, and continenal interiors could become quite cold. (Coastal areas would remain relatively warm because of the large heat capacity of the oceans.) Some of us might be able to survive such a catastrophe by taking advantage of our already fairly advanced technology, but the number of people killed would still likely be in the billions.

Fortunately, there is no reason why such a catastrophe need ever happen. We are already aware of the existence and approximate orbital parameters of many of the Earth-crossing asteroids, and we have the technology to improve considerably on this database. Within a few years, a space-based telescope dedicated to searching for asteroids and comets could map out all of the potentially dangerous objects in the Solar System. Indeed, the construction of such a telescope has already been suggested and seems likely to happen in the wake of the Comet Shoemaker–Levy impact with Jupiter, which helped alert lawmakers to the potential danger of such impacts. With sufficient advance warning of an impact, we may already be capable of avoiding it. A spacecraft for example, could be sent to rendezvous with the object and then to detonate a small nuclear device on one side of the object to deflect its orbit. As time passes and our space technology improves, there is little doubt that we will be able to construct an asteroid defense system, if we decide to do so. So, there is no reason why we need go the way of the dinosaurs. We know enough about the dangers of the Earth system to survive for longer than they did, if we can survive the danger that we pose to ourselves.

Chapter Summary

1. The Sun continues to brighten at a rate of about 1% per hundred million years.
   a. The resulting increase in Earth's surface temperature could lead to rapid loss of water in about a billion and a half years.
   b. Long before this occurs, first C_3 and then C_4 plants might die off as atmospheric CO_2 concentrations fall below the level required to sustain photosynthesis. The CO_2 decrease is predicted to result from the same feedback of surface temperature–silicate weathering rate that is thought to have stabilized Earth's climate throughout the planet's history.
2. Venus and Mars may have been more Earthlike in the distant past, but they evolved along very different climatic paths.
   a. Venus lost its water because it was too close to the Sun. After that, CO_2 and sulfur gases accumulated, forming the hot, dense atmosphere that exists today.
   b. Mars had liquid water flowing on its surface early in its history, despite the lower solar luminosity at that time. A dense CO_2 atmosphere filled with clouds of CO_2 ice may have provided the necessary greenhouse warming. Mars may have cooled down because it was too small to maintain plate tectonics and to keep recycling CO_2 back into its atmosphere.

3. Jupiter-sized planets have been detected around 10 other main-sequence stars. Earth-sized planets have not yet been observed, but it should eventually be possible to look for them. Climate calculations predict that at least some of these planets should lie within the habitable zone around their parent star. Stars not too different from our Sun appear to be the best candidates for harboring habitable planets.
4. The information gleaned about climatic stability on Earth may be fed into the Drake equation to estimate the probability that advanced, technological civilizations exist elsewhere in the galaxy.
   a. Some of the factors in this equation are already known, and other factors may be determined observationally in the foreseeable future. In particular, we may be able to look for the presence of photosynthetic life on planets in other solar systems by analyzing their atmospheres spectroscopically and by looking for the presence of ozone.
   b. The largest single uncertainty in the Drake equation is the lifetime of a technological civilization, a factor that is largely under that civilization's control. If technological civilizations are able to persist for an appreciable fraction of their planet's lifetime, then the chances that other civilizations exist in our galaxy appear to be very good.

Key Terms

captured rotation

CO_2 compensation point

continuously habitable zone

Drake equation

ecosphere

habitable zone

Hertsprung-Russell diagram

interferometer

main-sequence star

photorspiration

red giant

runaway glaciation

runaway greenhouse

spectroscopy

tidal locking radius

white dwarf

Review Questions

1. How should future solar evolution affect climate and life here on Earth?

2. What is the evidence that Venus once possessed more water than it does today? How did it lose its water, and how did its atmosphere evolve afterward?

3. What is the evidence that Mars was warmer and wetter in the past? Why did Mars cool off over time, even though the Sun has become brighter?

4. How is the habitable zone around a star defined?

5. What is the relationship between the instantaneous habitable zone and the continuously habitable zone around the Sun? How wide are these zones thought to be?

6. Why are bright, blue stars and dim, red stars not good candidates for supporting habitable planets?

7. What factors in the Drake equation are we capable of estimating today?

8. How do we know that planets exist around other stars? How many such planets have been detected to date?

9. How might the presence of life on another planet be inferred from the composition of that planet's atmosphere?

10. What factors might limit the lifetime of a technological civilization? Could any of today's global environmental problems (global warming, ozone depletion, and loss of biodiversity) destroy our technological society?

Critical-Thinking Problems

1. The inner edge of the HZ is currently estimated to be at a distance of 0.95 AU from the Sun. Where would it have been 4.6 b.y. ago when solar luminosity was about 70% of its present value? Would Venus, which orbits the Sun at a distance of 0.72 AU, have been inside or outside this critical distance? (*Hint:* Remember the inverse-square law from Chapter 3.)

2. How long do you think our present, technological civilization will last, and what factor (or combination of factors) do you think will bring it to an end? Write a 1- to 2-page essay defending your opinion.

Further Reading

General

Kasting, J. F. "How climate evolved on the terrestrial planets," *Scientific American,* Feb. 1988, 256: 90–97.

Sagan, C., *Cosmos,* New York: Random House, 1980.

Shklovsii, I. S., and C. Sagan, *Intelligent Life in the Universe,* San Francisco: Holden-Day Publishers, 1966.

Advanced

Caldeira, K., and J. F. Kasting, "The life span of the biosphere revisited," *Nature,* 1992, 360: 721–723.

APPENDIX A
Units and Unit Conversions

Common Unit Equivalences

1 kilometer (km)	= 1000 meters (m)
1 meter (m)	= 100 centimeters (cm)
1 centimeter (cm)	= 0.39 inches (in.)
1 mile (mi)	= 5280 feet (ft)
1 foot (ft)	= 12 inches (in.)
1 inch (in.)	= 2.54 centimeters (cm)
1 square mile (mi^2)	= 640 acres (a)
1 kilogram (kg)	= 1000 grams (g)
1 pound (lb)	= 16 ounces (oz)
1 fathom	= 6 feet (ft)

Unit Conversions

To convert:	Multiply by:	To find:
Length		
inches	2.54	centimeters
centimeters	0.39	inches
feet	0.30	meters
meters	3.28	feet
yards	0.91	meters
meters	1.09	yards
miles	1.61	kilometers
kilometers	0.62	miles
Area		
square inches	6.45	square centimeters
square centimeters	0.15	square inches
square feet	0.09	square meters
square meters	10.76	square feet
square miles	2.59	square kilometers
square kilometers	0.39	square miles
Volume		
cubic inches	16.38	cubic centimeters
cubic centimeters	0.06	cubic inches
cubic feet	0.028	cubic meters
cubic meters	35.3	cubic feet
cubic miles	4.17	cubic kilometers
cubic kilometers	0.24	cubic miles
liters	1.06	quarts
liters	0.26	gallons
gallons	3.78	liters
Masses and Weights		
ounces	20.33	grams
grams	0.035	ounces
pounds	0.45	kilograms
kilograms	2.205	pounds

The International System of Units (SI)

Basic Units

Quantity	Unit	SI Symbol
Length	meter	m
Mass	kilogram	kg
Time	second	s
Thermodynamic temperature	kelvin	K
Amount of substance	mole	mol

Prefixes

Prefix	Multiply Unit by:	Symbol
tera	10^{12}	T
giga	10^9	G
mega	10^6	M
kilo	10^3	k
hecto	10^2	h
deka	10	da
deci	10^{-1}	d
centi	10^{-2}	c
milli	10^{-5}	m
micro	10^{-6}	μ
nano	10^{-9}	n
pico	10^{-12}	p
femto	10^{-15}	τ
atto	10^{-18}	a

Derived Units

Quantity	Unit	Expression
Area	square meter	m^2
Volume	cubic meter	m^3
Frequency	hertz (Hz)	s^{-1}
Density	kilogram per cubic meter	kg/m^3
Velocity	meter per second	m/s
Acceleration	meter per second squared	m/s^2
Force	newton (N)	$kg \cdot m/s^2$
Pressure	newton per square meter	N/m^2
Work, energy, quantity of heat	Joule (J)	$N \cdot m$
Power	watt (W)	J/s

APPENDIX *B*
Temperature Conversions

$T(°F) = T(°C) \times 1.8 + 32$

$T(°C) = \dfrac{T(°F) - 32}{1.8}$

$T(°C) = T(K) - 273.15$

APPENDIX C

Periodic Table

1	2	3	4	5	6	7	8	9	10	11	12	13	14	15	16	17	18
1 **H** 1.00797																	2 **He** 4.0026
3 **Li** 6.939	4 **Be** 9.0122											5 **B** 10.811	6 **C** 12.011	7 **N** 14.0067	8 **O** 15.9994	9 **F** 18.998	10 **Ne** 20.183
11 **Na** 22.990	12 **Mg** 24.312											13 **Al** 26.982	14 **Si** 28.086	15 **P** 30.974	16 **S** 32.064	17 **Cl** 35.453	18 **Ar** 39.948
19 **K** 39.02	20 **Ca** 40.08	21 **Sc** 44.956	22 **Ti** 47.90	23 **V** 50.942	24 **Cr** 51.995	25 **Mn** 54.938	26 **Fe** 55.847	27 **Co** 58.933	28 **Ni** 58.71	29 **Cu** 63.54	30 **Zn** 65.37	31 **Ga** 69.72	32 **Ge** 72.59	33 **As** 74.922	34 **Se** 78.96	35 **Br** 79.909	36 **Kr** 83.80
37 **Rb** 85.47	38 **Sr** 87.63	39 **Y** 88.905	40 **Zr** 91.22	41 **Nb** 92.906	42 **Mo** 95.94	43 **Tc** (99)	44 **Ru** 101.07	45 **Rh** 102.91	46 **Pd** 106.4	47 **Ag** 107.870	48 **Cd** 112.40	49 **In** 114.82	50 **Sn** 118.69	51 **Sb** 121.75	52 **Te** 201.59	53 **I** 200.59	54 **Xe** 200.59
55 **Cs** 132.91	56 **Ba** 137.34	*	72 **Hf** 178.49	73 **Ta** 180.95	74 **W** 183.85	75 **Re** 186.2	76 **Os** 190.2	77 **Ir** 192.2	78 **Pt** 195.09	79 **Au** 196.97	80 **Hg** 200.59	81 **Tl** 204.37	82 **Pb** 207.19	83 **Bi** 208.98	84 **Po** (210)	85 **At** (210)	86 **Rn** (222)
87 **Fr** (233)	88 **Ra** (226)	**															

*Lanthanide Series (Rare Earths)

57 **La** 138.91	58 **Ce** 140.12	59 **Pr** 140.91	60 **Nd** 144.24	61 **Pm** (147)	62 **Sm** 150.35	63 **Eu** 151.96	64 **Gd** 157.25	65 **Tb** 158.92	66 **Dy** 162.50	67 **Ho** 164.93	68 **Er** 167.26	69 **Tm** 168.93	70 **Yb** 173.04	71 **Lu** 174.97

**Actinide Series

89 **Ac** (227)	90 **Th** 232.04	91 **Pa** (231)	92 **U** 238.03	93 **Np** (237)	94 **Pu** (244)	95 **Am** (243)	96 **Cm** (247)	97 **Bk** (247)	98 **Cf** (251)	99 **Es** (254)	100 **Fm** (253)	101 **Md** (256)	102 **No** (254)	103 **Lw** (257)

APPENDIX *D*

Useful Facts

Fundamental Physical Constants

Quantity	Symbol	Value
Speed of light in a vacuum	C	2.998×10^8 m/s
Universal gas constant	R	8.314 J/deg/mol
Stefan-Boltzmann constant	σ	5.670×10^{-8} W/m²/s
Planck's constant	h	6.626×10^{-20} J · s
Avogadro's number	N_a	6.023×10^{23} /mol
Mass of a hydrogen atom	m_H	1.67×10^{-27} kg
Boltzmann's constant	k	1.38×10^{-23} J/K

Physical Properties of Earth

Quantity	Value
Mass	5.974×10^{24} kg
Equatorial radius	6378 km
Polar radius	6357 km
Surface area	5.1006×10^8 km²
Average density	5142 kg/m³
Average Earth–Sun distance	1.496×10^8 km
Average Earth–Moon distance	3.844×10^5 km
Volume of oceans	1.4×10^8 m³
Surface area of oceans	3.6×10^8 km (71% of surface)
Average depth of oceans	3800 m
Mass of atmosphere	5.0×10^{18} kg
Mean sea level atmospheric pressure	1.013×10^5 N/m² ($= 1013$ mbar)
Density of air at sea level	1.225 kg/m³

Glossary

15-μm CO_2 band An absorption band that is a result of a vibration frequency of the CO_2 molecule that allows the molecule to absorb infrared radiation at a wavelength of about 15μm, near the peak of Earth's outgoing radiation. Very little of this radiation is allowed to escape directly to space, and so CO_2 is an important contributor to the greenhouse effect.

Absolute vorticity The sum of the planetary and relative vorticities experienced by a moving fluid.

Absorption coefficient The efficiency with which a molecule absorbs electromagnetic radiation of a given wavelength.

Accretion A process by which small bodies orbiting the Sun (or a planet) collide to form larger ones.

Acid A solution with a high concentration of hydrogen ions—that is, with a pH less than 7.

Acid rain Acidic rainwater produced when various acids, including sulfuric acid produced from SO_2 oxidation, combine with natural rainwater.

Action spectrum The relative efficiency with which UV photons at different wavelengths contribute to a specific biological response in a specific organism.

Adaptation A characteristic that enhances an organism's survival or reproductive success.

Albedo The reflectivity of a surface, usually expressed as a decimal fraction of the total incident sunlight reflected from the surface.

Amino acids Organic compounds containing an amino (NH_2) group and a carboxyl (COOH) group. They are the basic building blocks for proteins.

Anion A negatively charged ion.

Anoxic basin A region of the ocean in which deep water is completely devoid of oxygen; rare in today's ocean.

Anoxic Totally devoid of oxygen.

Anoxygenic photosynthesis A type of photosynthesis, carried out by certain types of bacteria, in which no O_2 is released.

Antarctic Bottom Water (AABW) Bottom water that forms in the Weddell Sea off Antarctica. The AABW circles Antarctica and flows northward as the deepest layer in the Atlantic, Pacific, and Indian ocean basins.

Anthropogenic Induced by humans.

Aphelion The position in a planet's orbit that is farthest from the Sun.

Asteroid belt A region between Mars and Jupiter that has a high concentration of asteroids. The asteroid belt represents the remains of a tenth, inner planet that failed to form or was destroyed by a large meteorite impact early in the history of the solar system.

Asteroids Pieces of rocky material composed of minerals and metallic elements; they range in size from dust size to 1000 km in diameter and most reside in the *asteroid belt,* but some are diverted to planet-crossing orbits by gravitational disturbance.

Asthenosphere The zone of most ductile upper mantle just below the lithosphere; may contain small amounts of molten rock.

Atmosphere The thin envelope of gases that surrounds most planets; on Earth, one of the four components of the Earth system.

Atomic number The number of protons in the nucleus of an atom. All atoms of a given element have the same atomic number.

Banded iron-formations Laminated rocks consisting of alternating layers of iron minerals and chert that were formed almost exclusively during the first half of Earth's history.

Barometric law A relationship stating that atmospheric pressure decreases by about a factor of 10 for each 16-km increase in altitude. More technically, atmospheric pressure decreases exponentially with altitude.

Basalt An igneous rock that reaches Earth's surface as lava and cools rapidly; the major rock type in the oceanic crust.

Base A solution with a low concentration of hydrogen ions—that is, with a pH greater than 7.

Biodiversity The variety of life forms, for example, the number of species in an area.

Biological pump A transfer of CO_2 and nutrients from the surface waters to the deep ocean as a result of photosynthesis in shallow waters, of the settling of organic matter, and of decomposition in deep waters. This process operates on a much faster time scale (days to months) than the thermohaline circulation of the oceans.

Biomass The total combined weight of organic material in each trophic level. In terrestrial ecosystems the biomass is reduced by 90 to 99% at each higher trophic level.

Biomass-based fuels Liquid fuels, such as methanol and ethanol, that are produced from fast-growing plants. Unlike fossil fuels, burning these substances does not contribute to atmospheric CO_2 buildup.

Biome A region with a characteristic plant community.

Biosphere The part of the Earth system that directly supports life, including the oceans, atmosphere, land surface, and soils.

Biota All living organisms.

Bioturbation The stirring of sediments by worms and other burrowing organisms that live at the sea floor.

Black smoker A mid-ocean ridge hydrothermal vent from which clouds of black particles of iron sulfide (FeS) are precipitating.

Blackbody radiation Radiation given off by a blackbody, or a body that emits equally well at all wavelengths. This radiation is characterized by the body's absolute temperature.

Body waves Seismic waves that travel through Earth's interior as they spread outward from the earthquake's focus; categorized as either P waves or S waves.

Bottom water Very dense, cold water that forms along the edges of sea ice in certain areas near the poles and subsides, making up the bottom layer of ocean water as it circulates throughout the world's oceans.

Boyle's law A relationship stating that, with temperature held constant, the pressure and volume of a gas are inversely proportional. In other words, the product of pressure and volume is a constant.

Buffer A dissolved substance that helps maintain a stable pH. Carbonate ion is an important pH buffer in seawater.

Buoyancy A tendency of an object to float, rise, or sink when submerged in a fluid.

C₃ plants Plants that fix carbon into a three-carbon chain during photosynthesis. These plants (which include about 95% of all terrestrial plants) should be positively affected by CO_2 fertilization.

C₄ plants Plants that fix carbon into a four-carbon chain during the initial step of photosynthesis. These plants (which include about 5% of all terrestrial plants, such as corn and sugar cane) experience little CO_2 fertilization at CO_2 levels near 350 ppm.

Calvin cycle The biochemical pathway by which green plants convert CO_2 into organic carbon during photosynthesis, a method of carbon fixation.

Captured rotation A rotation in which a planet (or moon) always shows the same side to a star (or planet). Earth's Moon is a good example.

Carbon fixation The biochemical process that occurs during photosynthesis by which atmospheric CO_2 is converted to organic carbon.

Carbonate metamorphism A chemical reaction occurring at high temperatures and pressures between sedimentary carbonate minerals and silica-rich sediments that forms calcium silicate minerals and releases CO_2.

Catalyst A compound that increases the rate of a chemical reaction and is itself unchanged by the reaction.

Catalytic cycle A set of chemical reactions facilitated by the presence of a catalyst. Ozone can be destroyed by catalytic cycles involving nitrogen oxides and chlorine.

Cation A positively charged ion.

Chapman mechanism A set of four chemical reactions that describe the production and loss of ozone in a simplified (pure N_2-O_2) atmosphere.

Characteristic response time How long a reservoir takes to respond measurably to large imbalances in inflow or outflow, calculated in the same way as residence time.

Charles's law A relationship stating that, with pressure held constant, the volume and temperature of a gas are directly proportional. In other words, the quotient of volume and temperature is a constant.

Chemosynthesis The biological production of organic matter by bacteria that utilize energy stored in chemical compounds (inorganic matter) rather than solar energy.

Chlorofluorocarbons (CFCs) Synthetic compounds containing chlorine, fluorine, and carbon. These gases, also called freons, contribute to the greenhouse effect and are harmful to the ozone layer.

Chrondrites "Primitive" meteorites that have never been melted. They are thought to be remnants of the original material from which the planets formed.

CO₂ compensation point The critical CO_2 concentration below which plants are unable to photosynthesize and hence, grow. Technically, it is the CO_2 level below which the rate of photorespiration exceeds that of photosynthesis.

CO₂ fertilization An increase in the growth rate of plants by the addition of CO_2 to the atmosphere.

CO₂ tax A proposed tax on CO_2 emissions; one mechanism that has been suggested to reduce the rate at which fossil fuels are consumed.

Coal A hydrogen- and carbon-bearing compound produced under high-pressure and high-temperature conditions of burial deep within the solid Earth; a fossil fuel formed from high concentrations of organic matter in terrestrial sediments.

Column depth The total amount of ozone per unit area above Earth's surface. Usually reported in Dobson units.

Comets Balls of frozen gases (water, ammonia, methane, and carbon dioxide) containing rocky and metallic debris, that are mostly in distant orbits around the Sun, beyond Pluto (see *Oort cloud*).

Community A characteristic assemblage of two or more groups of interacting species.

Component An individual part of a system. A component may be a reservoir of matter or energy, a system attribute, or a subsystem.

Condensation The process by which a gas becomes a liquid.

Conduction Transfer of heat energy by direct contact between individual molecules.

Consumers Organisms, such as animals, that are incapable of utilizing solar energy directly but must instead consume plants or other photosynthesizers to utilize the chemical energy stored in plant tissues.

Continental drift Theory stating that the supercontinent Pangea began to break apart at the beginning of the Mesozoic era (200 million years ago) and that the continents then slowly drifted into their current positions; a concept proposed by Alfred Wegener in the early 20th century that has been supplanted by the (related) theory of plate tectonics.

Continental shelves The shallow, submerged part of continental margins, bounded on the landward side by the coastline and on the seaward side by steep slopes falling to great depths.

Continuously habitable zone The region around a star where a planet can remain habitable for some finite time interval.

Convection Transfer of heat energy by the circulating motions of a fluid that is heated from below; one of the three primary mechanisms of heat transfer.

Convergence As used in this book, the inward movement of air or water to a region in the atmosphere or ocean.

Core The central part of a planet or of the Sun. Earth's core—one of the three components of the solid Earth—is dense, composed mostly of metallic iron and nickel, and is part solid, part liquid.

Coriolis effect The apparent tendency for a fluid (air or water) moving across Earth's surface to be deflected from its straight-line path. Fluids are deflected to the right of their initial path in the Northern Hemisphere and to the left in the Southern Hemisphere.

Coupling The links between any two components of a system. Couplings can be positive or negative.

Craton The old, tectonically dormant regions of the continents that have served as the nuclei for continental accretion over the past 3 to 4 billion years.

Crust The thin, outer layer of the solid Earth; consists of light, rocky matter that is in contact with the atmosphere, hydrosphere, and biota.

Cyanobacteria Bacteria capable of both oxygenic and anoxygenic photosynthesis; formerly known as blue-green algae.

Decomposition The breakdown of organic matter, a process typically carried out by certain bacteria and fungi.

Deforestation The clearing of all the trees off an area of land.

Dendrochronology A method of dating trees by counting their annual growth rings. The ring widths can be used to infer environmental conditions during the growing season.

Denitrification The conversion of fixed nitrogen into N_2 or N_2O by bacteria.

Deoxyribonucleic acid (DNA) A double-stranded molecule that is the basis for information storage by all living organisms.

Deserts Areas of low rainfall, generally less than 250 mm per year. Deserts may be warm (such as those located in the subtropics) or cold (as in the central plateau of Antarctica).

Detrital mineral A mineral grain found in a sediment that was never completely dissolved during the weathering process.

Deuterium A stable isotope of hydrogen that has one neutron in its nucleus; denoted as ^2H or D.

Dissociate To come apart. Acids dissociate in solution, forming anions plus hydrogen ions (protons).

Divergence As used in this book, the outward movement of air or water from a region in the atmosphere or ocean.

Dobson unit (DU) A unit for measuring the column depth of ozone. One DU is equivalent to a layer of pure ozone 0.001 cm thick at 1 atmosphere pressure.

Dose rate The number of UV photons per unit time that lead to a specific biological response, such as sunburn or skin cancer.

Double helix The spiral pattern formed by two interlinked strands of DNA.

Downwelling The sinking of surface water caused by convergence and water accumulation at the surface.

Drake equation A formula used to estimate the number of other intelligent civilizations in our galaxy with which we might one day establish radio communication.

Dropstones "Misplaced" chunks of rock dropped into marine sediments by melting icebergs.

Earthquake The sudden release of stored energy as the result of rapid movement between two blocks of rock.

Eccentricity The degree to which a rotating object's orbit is elliptical. Eccentricity is defined for Earth's orbit as the distance from the center of the orbit to its foci (one of which is occupied by the Sun), divided by the average distance between Earth and the Sun.

Ecosphere Synonym for *habitable zone.*

Ecosystem A subset of the biota, consisting of assemblages of plant, animal, and microbial species that interact with each other and with their surrounding environment.

Ecotone The diffuse boundary between two ecosystems.

Effective radiating temperature The temperature a planet such as Earth would have if the planet radiated as a blackbody (or if it had no atmosphere).

Ekman spiral The way in which water changes direction with depth in the ocean, under the influence of the wind-driven circulation, the Coriolis effect, and friction.

Ekman transport The net direction of transport in the water column as a result of the Ekman spiral. The movement is at 90° to the wind direction (to the right in the Northern Hemisphere and to the left in the Southern Hemisphere).

El Niño Originally, a warm ocean current that appears off the coast of Peru and Ecuador shortly after Christmas and flows for only a few weeks; the name now describes a major shift in the oceanic circulation that occurs in this region every 2 to 10 years.

El Niño—Southern Oscillation (ENSO) A climatic event in the tropical Pacific Ocean in which the main area of surface convection moves from the western to the central Pacific. This event is associated with large-scale changes in the ocean circulation, the atmospheric circulation, and tropical precipitation patterns. The effects of an ENSO event may also spread beyond the tropics causing anomalous weather conditions in many midlatitude locations.

Electromagnetic radiation A self-progating electric and magnetic wave such as visible light, ultraviolet, or infrared radiation.

Electromagnetic spectrum The full range of different forms of electromagnetic radiation, which differ by wavelength (or, conversely, by frequency).

Endemic species Species that are indigenous, or native, to a particular location.

Enzymes Proteins that catalyze various biochemical reactions.

Eon One of the four major subdivisions of geologic time.

Equilibrium state A state in which the system is in equilibrium—that is, the state in which the system will remain unless something disturbs it. An equilibrium state can be stable or unstable.

Equilibrium surface warming The surface warming that will occur when Earth's surface achieves full thermal balance.

Erosion The transport of the products of weathering to basins where sediment accumulates; the process whereby crustal materials, decomposed and loosened by weathering, are transported by winds, landslides, and streams.

Erythemal action spectrum The action spectrum for sunburn in humans.

Eukaryotes Organisms whose cells have cell nuclei. All multicellular organisms are eukaryotes.

Evaporation The process by which a liquid is converted to a gas.

Evaporite deposit Mineral deposit formed by the evaporation of seawater from shallow seas. The remaining salts are concentrated and precipitate from solution.

Evolution The descent, with modification, of preexisting life forms. The *natural selection* of favorable genetic mutations is the dominant mechanism of evolution.

Extinction The loss of all individuals within a species.

Feedback factor The ratio of the equilibrium response to forcing (the response with feedback) to the response without feedback. Feedback factors less than 1 are indicative of negative feedback: The equilibrium response (with feedback) is smaller than the response to forcing without feedback. Feedback factors greater than 1 indicate positive feedback: The equilibrium response is larger than the response to the forcing itself.

Feedback loop A linkage of two or more system components that forms a round-trip flow of information. Feedback loops can be positive or negative.

Fermentation An anaerobic form of metabolism in which both oxidized carbon and reduced carbon are produced (often generating methane); a decomposition process typically performed by anaerobic bacteria.

Ferric iron Iron in its highest (3+) oxidation state; insoluble in water.

Ferrous iron Iron in its intermediate (2+) oxidation state; soluble in seawater.

Fixed nitrogen Compounds containing a nitrogen atom bonded to something other than another nitrogen atom.

Flux The amount of energy (or number of photons) in an electromagnetic wave that passes perpendicularly through a unit surface area per unit time.

Food chain A progression of organisms each dependent on the one before for food.

Forcing A persistent disturbance of a system; a longer-term disturbance than a perturbation.

Formaldehyde H_2CO, a chemical that could have formed in the early atmosphere and that can be used to synthesize sugars and amino acids.

Fossil fuels Fuels such as coal, oil, and natural gas that are formed from the non-decomposed organic remains of organisms, concentrated in sedimentary rocks.

Frequency The number of wave crests that pass a fixed point in one second.

Gaia hypothesis A theory suggesting that Earth is a self-regulating system in which the biota play an integral role.

General circulation model (GCM) A three-dimensional computer model of the global atmosphere (or ocean) that simulates winds (currents), moisture transport, and energy balance; also called a global climate model.

General gas law The relationship that combines Boyle's law and Charles's law and states that the product of the pressure and volume of a gas is directly proportional to temperature. In other words, the quotient of the product (pressure) × (volume) and temperature is a constant.

Genetic code The key that determines how sequences of three different nucleotides represent different amino acids. Using this key, cells are able to manufacture various proteins on the basis of instructions stored in the cell DNA.

Geostrophic current A current that flows around oceanic gyres; produced where the Coriolis effect that deflects the flow into the center of a gyre is balanced by the downslope flow from the higher sea-surface elevations in the gyre center. The resulting flow is clockwise in the Northern Hemisphere (counterclockwise in the Southern Hemisphere), approximately parallel to the ocean slope, and in the same direction as the wind-driven flow.

Geothermal heat Heat flowing from Earth's interior up to its surface.

Geothermal power The production of electricity by using temperature gradients within the solid Earth as the energy source.

Giant impact hypothesis A theory of lunar formation in which the Moon forms as the result of a glancing collision between Earth and a Mars-sized body.

Glacial interval An interval of time during the Pleistocene when ice sheets covered much of northern North America and Scandinavia and other parts of northern Europe and Asia (as well as Greenland and Antarctica). Globally averaged surface temperatures during glacial intervals were about 10°C, and atmospheric CO_2 concentrations were about 200 ppm.

Glacial striations Grooves carved into bedrock by rocks frozen to the base of a moving glacier.

Glacial surge The sudden, rapid movement of a glacier.

Global warming A warming of Earth's atmosphere due to an anthropogenic enhancement of the greenhouse effect.

Granite A common igneous silicate rock that solidifies below Earth's surface and forms the cores of many mountain ranges. Granite is less dense than basalt, so when a continental plate collides with an oceanic plate, the oceanic plate is subducted beneath the continental plate.

Greenhouse effect The natural mechanism by which a planet's surface is warmed by infrared-absorbing gases in its atmosphere.

Greenhouse gases Gases such as carbon dioxide, methane, nitrous oxide, and water vapor that warm a planet's surface by absorbing infrared radiation and reradiating some of it back toward the surface. Greenhouse gases—whether natural or anthropogenic—contribute to the atmospheric greenhouse effect.

Groundwater Water that penetrates through soil and rock and collects below the surface.

Gyre A large circular circulation pattern in the ocean. Gyres in the Northern Hemisphere circulate clockwise, whereas those in the Southern Hemisphere circulate counterclockwise.

Habitat The plant community and physical environment that supports a single species.

H_2O rotation band A strong absorption band that is a result of rotational frequencies of the water molecule in vapor form. The H_2O molecule absorbs infrared radiation of wavelengths about 12 μm and longer.

Habitable zone The region around a star in which a life-supporting planet might be found; generally, the region where liquid water could exist on a planet's surface.

Hadley circulation The process by which an air mass undergoes convergence at the tropics and divergence at about 30°N or 30°S latitude in one large convection cell.

Half-life The time it takes for half the initial quantity of radioactive isotope to decay.

Halocline A steep salinity gradient in the pycnocline zone that marks the transition between the surface zone and the deep ocean. Salinity rises rapidly with increasing depth in the halocline.

Halons Artificial chemicals containing the elements bromine, chlorine, fluorine, and carbon, used mostly in fire extinguishers.

Heavy bombardment period The time interval between 4.6 Ga and 3.8 Ga when Earth and the other terrestrial planets were being regularly bombarded by large planetesimals.

Hematite Fe_2O_3, a mineral composed of fully oxidized iron. It has a reddish color when present as small, dispersed grains.

Hertzsprung-Russell diagram (H-R diagram) A type of graph in which stars are displayed in terms of their absolute luminosities (vertical scale) and surface temperatures (horizontal scale).

Heterogeneous reaction A chemical reaction that occurs on a solid surface.

Holocene The geological epoch extending from 10 k.y. to the present; an interglacial interval.

Holocene Climatic Optimum A warm period that occurred during the mid-Holocene.

Homogeneous reaction A chemical reaction between molecules that are in the gas phase.

Hydrogen cyanide HCN, a molecule believed to be an essential building block for life.

Hydrologic cycle The major reservoirs of water in the Earth system and the pattern of water storage and movement throughout that system.

Hydrosphere The component of Earth system that includes the various reservoirs of water and ice on Earth's surface.

Hydrothermal vents Cracks in the sea floor, especially around mid-ocean spreading ridges, through which seawater circulates.

Hydroxyl radical A highly reactive molecule that plays a variety of important roles in atmospheric chemistry; chemical formula OH.

Ice shelves Floating sea ice formed at the margins of continents.

Ice-rafting A process by which chunks of rock are carried to sea by icebergs. The icebergs eventually melt, and the rocks may be preserved as dropstones in sediments.

Igneous rock Rock formed by the cooling and solidification of magma. If the magma solidifies beneath Earth's surface, the rocks are intrusive; if the magma erupts as lava at a volcano, the magma cools rapidly into extrusive rocks.

Impact degassing The venting of water and other volatile compounds directly into a planet's atmosphere during impacts of comets or asteroids onto its surface.

Infrared (IR) radiation Electromagnetic radiation of fairly low energy and wavelengths longer than those of visible light from 0.7 μm to 1000 μm.

Inorganic carbon Carbon not associated with compounds that are typically formed by living organisms and that do not contain carbon–carbon or carbon–hydrogen bonds.

Instrumental value The degree to which a species' existence benefits another species (normally, humans) in some way.

Interferometer An instrument consisting of two or more telescopes connected electronically so that their images can be combined.

Interglacial period An interval during the Pleistocene, such as the Holocene, when continental ice sheets were restricted to Greenland and Antarctica. Globally averaged surface temperatures during interglacial intervals were about 15°C, and atmospheric CO_2 concentrations were about 280 ppm.

Interplanetary dust particles (IDPs) Small particles collected in Earth's stratosphere that originated in space.

Interstellar clouds Clouds of dust and gas from which stars occasionally form.

Intertropical convergence zone (ITCZ) A region of the tropics where surface heating causes uplift in the atmosphere, allowing subtropical air to flow inward to produce a convergence zone. This zone moves north and south of the equator as the seasons change.

Intrinsic value A species' value for its own sake, regardless of whether it benefits humans.

Inverse-square law A relationship describing the rate at which the solar flux decreases with increasing distance.

Iridium An element that is rare at Earth's surface but is concentrated in Earth's interior and in extraterrestrial materials.

Isochron diagram A diagram in which one plots the abundance of two different radiogenic isotopes (for example, ^{206}Pb and ^{207}Pb) measured in a number of different rock samples against each other. If the rocks all have the same age, then the data should fall on a straight line. The slope of the line gives the age of the rocks.

Isotopes Atoms of a given element that have different numbers of neutrons in their nuclei.

K-T boundary The boundary between the Cretaceous (K) and Tertiary (T) periods, about 65 million years ago, when the dinosaurs and many other species went extinct.

Kelvin (absolute) temperature scale A metric temperature scale in which the degree has the same size as a Celsius degree, but in which the zero point is moved downward by 273.15°, to absolute zero.

Kerogen Dispersed organic matter (hydrocarbons) in rocks.

Keystone species A species that plays a vital function in the operation of an ecosystem.

La Niña The opposite phase of the Southern Oscillation from El Niño conditions. It represents a stronger or more extreme version of the "normal" circulation in the tropical Pacific.

Latent heat of fusion The energy required to effect a change of phase between a solid and a liquid. Converting a solid to a liquid requires an addition of energy; converting a liquid to a solid releases energy to the environment.

Latent heat of vaporization The energy required to effect a change of phase between a liquid and a gas. Converting a liquid to a gas requires an addition of energy; converting a gas to a liquid releases energy to the environment.

Latent heat The heat energy released or absorbed during the transition from one phase to another, such as when water evaporates.

Lead–lead dating A radiometric age dating technique that relies on measurements of two different isotopes of lead produced from two uranium isotopes that have different half-lives.

Limestone A sedimentary rock composed largely of calcium carbonate minerals, mainly calcite.

Lithification The transformation of sediments into sedimentary rocks; typically involves compaction and the precipitation of mineral cements between sediment grains.

Lithosphere The uppermost mantle and rigid crust, above the asthenosphere. The lithosphere is divided into plates that move relative to one another in the process of plate tectonics.

Little Ice Age An interval of colder temperatures from the 15th to 19th centuries, interrupted by a warmer interval in the 17th century.

Loess Wind-blown glacial silt.

Logistic growth Growth of a population that increases in size exponentially when the population is small, but then approaches a constant size when the population is large. For logistic growth to occur, as the population size increases, either the population growth rate decreases or the death rate increases.

Luminosity The brightness of a star such as our Sun.

Magma Molten, or liquid, rock that forms igneous rock when cooled.

Magma ocean A layer of molten rock covering the entire surface of a planet. The Moon is thought to have had one early in its history. Earth may have had one as well.

Magnetic dynamo The mechanism whereby convection of the liquid-iron outer core generates Earth's magnetic field.

Main-sequence star A middle-aged star that lies on a band of stars running from the upper left to the lower right of a plot of luminosity versus effective radiating temperature (H-R diagram).

Mantle One of the three layers of the solid Earth; a thick, rocky layer between the core and crust. Composed primarily of silicate minerals.

Mass extinction An extraordinary extinction event in which more than 25% of all extant families are lost.

Mass number   The combined number of protons and neutrons in the nucleus of an atom.

Maunder Minimum   The interval from 1645 to 1715, when very few sunspots were recorded.

Medieval warm period   A period of mild conditions in Northern Europe and North Atlantic that reached a maximum at about A.D. 1100.

Mesosphere   An atmospheric layer that extends from about 50 km to 90 km above the surface; temperature decreases with altitude there.

Metamorphic rocks   Rocks formed from exposure to high temperatures, high pressures, chemically active fluids, or any combination of these agents. The rocks must remain in the solid phase to be classified as metamorphic; if melting occurs, they are called igneous.

Meteorite   A comet or asteroid that strikes Earth. Strictly speaking, the material that remains after such impact.

Methane sulfonic acid (MSA)   An acid that is produced from biogenic dimethyl sulfide and forms cloud condensation nuclei in the atmosphere.

Methanotrophic bacteria   Anaerobic bacteria that convert carbon dioxide and hydrogen into methane.

Methanogens   Methanogenic bacteria.

Methanotropic bacteria   Aerobic bacteria that consume methane and incorporate the carbon from that methane into their body tissues.

Microfossils   The fossilized remains of single-celled organisms.

Midocean ridge   A linear chain of subsea volcanic mountains on the sea floor that is the site of formation of new oceanic lithosphere.

Mixed layer   The surface layer of the ocean that is mixed by wind action.

Moho   The crust/mantle boundary, marked by a sharp increase in seismic wave speeds.

Monsoon   A seasonal reversal in the surface winds caused by large-scale differential heating of land and ocean surfaces. Monsoon circulation is defined by the windfields but usually also has a direct impact on rainfall.

Montreal Protocol   An international treaty signed in 1987 that limits future freon and halon emissions.

Moraine   A ridge of sediment deposited by a glacier.

Mountain glaciers   Ice fields formed on the cold, upper reaches of mountains.

Natural selection   The unequal survival and reproduction of organisms, owing to environmental pressures that result in the preservation of favorable adaptations. The relatively nonselective nature of mass extinctions suggests that natural selection may not play an important role in these large events.

Negative coupling   A link indicating that a change (increase or decrease) in one component leads to a change of the opposite direction (decrease or increase, respectively) in the linked component.

Negative feedback loop   A feedback loop with an odd number of negative couplings. Negative feedback loops tend to diminish the effects of disturbances.

Neutrinos   Massless particles given off in nuclear reactions.

Nitrogen fertilization   The stimulation of plant growth caused by the addition of anthropogenic fixed nitrogen, commonly from automobile exhaust.

Nitrogen fixation   A process by which some organisms (primarily prokaryotes) convert N_2 into fixed nitrogen.

North Atlantic Deep Water (NADW)   Cold, dense water that forms in the Arctic Ocean, sinks, and flows southward at depth into the Atlantic Ocean.

Nuclear fission   The splitting of an atomic nucleus into two fragments, accompanied by the release of energy.

Nuclear fusion   The combining of lightweight atomic nuclei into a heavier nucleus accompanied by the release of energy.

Nucleotides   Organic compounds consisting of a sugar, a base, and a phosphate group. Nucleotides are the basic building blocks of DNA and RNA.

Number density   A measure of the concentration of a gas. The units of number density are molecules per cubic centimeter.

Nutrient   Substances normally obtained in the diet that are essential to organisms.

Obliquity   The angle of a planet's spin axis relative to a line drawn perpendicular to the plane of the planet's orbit around the Sun; also called *tilt*.

Odd nitrogen   Any pure nitrogen-containing compound that has an odd number of nitrogen atoms; NO and NO_2. Odd nitrogen is similar to fixed nitrogen in that the nitrogen atom is coupled to atoms other than another N atom.

Odd oxygen   Any pure oxygen-containing atom or molecule that has an odd number of oxygen atoms; O and O_3.

Oort cloud   A distant region of the solar system, beyond Pluto, with a high concentration of comets.

Organic carbon   Carbon associated with compounds that are typically formed by living organisms and that contain carbon–carbon or carbon–hydrogen bonds.

Oxidation state   The degree of oxidation of an atom, molecule, or compound. Substances with a low oxidation state have a large number of available electrons; substances with a high oxidation state do not.

Oxidative weathering   Reactions between the atmosphere and surface rocks in which minerals react with O_2.

Oxidized carbon   Carbon that, in compounds, is combined with oxygen. The carbon atoms in skeletons composed of $CaCO_3$ and in atmospheric CO_2 are oxidized carbon.

Oxygen minimum zone   A zone at intermediate depths in the ocean—about 1 km from the surface—where dissolved oxygen concentrations reach a minimum (and nutrient concentrations reach a maximum), as a result of high oxygen demand by aerobic decomposers and low oxygen supply from the surface ocean or from below.

Oxygenic photosynthesis   The normal, O_2-generating type of photosynthesis carried out by plants.

Ozone (O_3)   A form of oxygen that is much less abundant than, and chemically unlike, the oxygen that we breathe. The ozone that is dispersed in the stratosphere blocks the Sun's harmful ultraviolet radiation.

Ozone hole   A patch of extremely low ozone concentration in the ozone layer. This hole has appeared near the South Pole each October since about 1976.

Ozone layer   A chemically distinct region of the atmosphere (specifically, the stratosphere) that protects Earth's surface from the Sun's harmful ultraviolet radiation.

Ozonesondes   A balloon-borne instrument that measures the concentration of stratospheric ozone.

P wave Primary wave; a seismic body wave transmitted as a series of compressions and expansions in the overall direction of wave movement through Earth's interior. P waves can travel through fluids or solids.

Paleoclimate Past climate.

Paleoclimatology The study of past climates.

Paleosols Ancient soils which can be used as an indicator of past atmospheric oxygen levels.

Palynology The study of pollen and organic microfossils.

Pangea A supercontinent consisting of all the land masses of Earth until it broke apart about 200 million years ago.

Parameterized Simulated on the basis of experiments or observations, as opposed to being calculated directly. For example, clouds generally need to be parameterized to be included in atmospheric general circulation models.

Partial pressure In a mixture of gases, the pressure a gas would exert if it were the only gas present (that is, the contribution of each individual gas to the total pressure exerted by the mixture).

Perihelion The position in a planet's orbit that is nearest to the Sun.

Period A unit of geologic time, shorter than an era and longer than an epoch. Also, the time required for a planet to go around the sun.

Periodicity A time interval of regular recurrence of a phenomenon.

Perturbation A temporary disturbance of a system; a shorter-term disturbance than a forcing.

Petroleum A hydrogen- and carbon-bearing compound produced under high-temperature conditions of burial deep (3–4 km) within the crust; a fossil fuel formed from high concentrations of organic matter in marine sediments.

pH A measure of acidity, defined as the negative logarithm (to the base 10) of the hydrogen ion concentration (in moles per liter). Acids have pH values less than 7, and bases have pH values greater than 7. A pH of 7 is defined as neutral.

Photic zone The portion of the oceans where there is sufficient sunlight for photosynthesis; about the upper 100m of the water column.

Photochemical models Computer models used to simulate atmospheric chemistry.

Photochemical reactions Chemical reactions initiated by the absorption of a photon.

Photodissociation The splitting of a molecule by the absorption of light or by UV radiation; also called *photolysis*.

Photolysis The splitting of a molecule by the absorption of light or by UV radiation; also called *photodissociation*.

Photolyze To split a molecule apart with visible or ultraviolet radiation.

Photon A single, discrete particle, or pulse, of electromagnetic radiation.

Photorespiration Respiration induced by the absorption of sunlight. (The combining of organic matter with oxygen to yield CO_2, H_2O, and energy.)

Photosphere The surface layer of the Sun from which most of its energy, including visible radiation, is emitted.

Photosynthesis The process by which an organism such as a green plant uses sunlight, carbon dioxide, and water to produce organic matter and oxygen.

Photovoltaic cells Specially designed panels that directly convert sunlight into electricity.

Plages Bright (higher temperature) areas that surround sunspots on the surface of the Sun.

Planetary vorticity The angular rotation about a vertical axis at Earth's surface brought about because of Earth's rotation.

Planetesimals Small proto-planets formed during planetary accretion.

Plate tectonics Theory by which the Earth's surface is divided into rigid plates of sea floor and continent that move relative to one another through time.

Pleistocene Epoch The geological epoch, extending from 1.7 m.y. ago to 10 k.y. ago, that is characterized by oscillations in and out of the glacial state.

Polar front zone A zone of steep temperature gradients formed at approximately 60° N and 60° S latitude, where cold, polar air meets the warm air moving poleward from the subtropics.

Polar stratospheric clouds (PSCs) Collections of droplets, consisting of a mixture of water and nitric acid, that are involved in the formation of the Antarctic ozone hole.

Polar vortex The circular, downward-sinking whirlpool of stratospheric air over the poles in winter. The polar vortex is more pronounced in the Southern Hemisphere than in the Northern Hemisphere.

Polarity The direction of orientation of a magnetic field. Earth's polarity (the geographic location of the North and South Poles) flips irregularly on geological time scales.

Population All of the members of a single species that live in a given area.

Positive coupling A link indicating that a change (increase or decrease) in one component leads to a change of the same direction (increase or decrease, respectively) in the linked component.

Positive feedback loop A feedback loop with an even number of, or zero, negative couplings. Positive feedback loops tend to amplify the effects of disturbances.

Prebiotic synthesis The formation of complex, biologically important molecules from compounds existing in the natural environment.

Precession The rotation of a planet's spin axis around a line drawn perpendicular to its orbital plane.

Primary producer A plant (or other type of photosynthesizer or chemosynthesizer) that provide energy used by consumers.

Primary productivity The amount of organic matter produced by photosynthesis in a unit time over a unit area of Earth's surface; also called productivity.

Prokaryotes Single-celled organisms that lack a cell nucleus.

Proteins Biochemical compounds, composed of amino acids, that perform a wide variety of functions within cells, including assisting with DNA replication.

Proto-Sun The central bulge in the solar nebula that later formed the Sun.

Proxy data Data that cannot be obtained by direct measurement but can be inferred from other evidence.

Pycnocline A steep density gradient (caused by changing temperature, salinity, or both) that marks the transition between the surface zone and the deep ocean. On the order of a kilometer in thickness, it is characterized by a rapid downward increase in density. The steep density gradient in the pycnocline zone makes this layer very stable.

Pyrite FeS_2, an iron-bearing mineral that is normally oxidized during weathering today.

Radiation (of heat energy) The outward transfer of heat energy in the form of electromagnetic rays emitted by a body; one of the three primary mechanisms of heat transfer.

Radiative-convective model (RCM) A one-dimensional computer model of the atmosphere that can be used to simulate the greenhouse effect. In an RCM, the climate system is greatly simplified by averaging the incoming solar and outgoing infrared radiation over Earth's entire surface.

Radical A molecule that is highly reactive because it has an unpaired electron in its outer shell.

Radioactive decay Radioactivity; the spontaneous disintegration of an unstable nucleus of one element, creating a different nucleus of a different element and releasing particles and radiation.

Radiometric dating A method of calculating the age of a sample of material by knowing the half-life of a radioisotope within it. This method has provided absolute dates for the geologic time scale and for other specific events in Earth's history.

Rare earth elements Elements with atomic numbers 58 to 71. The pattern of rare earth elements in banded iron-formations resembles that in modern mid-ocean ridge hydrothermal fluids.

Reactive chlorine Chlorine compounds that either destroy ozone directly, such as Cl and ClO, or are readily converted into these compounds by photolysis.

Red giant A large, reddish star that is in the immediate post-main-sequence phase of its evolution.

Redbeds Sedimentary deposits in which the individual grains are coated with the mineral hematite (Fe_2O_3); thought to have formed under conditions of relatively high atmospheric O_2.

Redfield ratio The ratios of the nutrient elements that marine organisms incorporate into their tissues; these ratios appear to be nearly identical in all species of photoplankton. For these organisms, and for seawater, the atomic ratio of carbon:nitrogen:phosphorus is very nearly 106:16:1.

Reduced carbon Carbon that, in compounds, is combined mainly with other carbon atoms, hydrogen, or nitrogen; organic carbon is a form of reduced carbon. Reduced carbon tends to be reactive in the presence of oxygen gas.

Reduced gases Gases, such as H_2 or CH_4, that can react with oxygen. Such gases typically contain the element hydrogen.

Relative humidity The amount of water vapor contained by a unit volume of air divided by the amount of water vapor that volume would contain if the air were saturated.

Relative vorticity Vorticity produced by processes (other than Earth's rotation) that induce a rotary motion in a fluid (such as clockwise or counterclockwise surface wind patterns acting on the ocean surface, or current shear in the oceans).

Residence time The average length of time a substance spends in a given reservoir that is at a steady state with respect to the processes that add and remove the substance to and from the reservoir. Residence time is calculated as the ratio of the reservoir size to the rate of inflow or outflow (which are equal at steady state).

Ribonucleic acid (RNA) A single-stranded molecule involved in protein synthesis. RNA is capable of copying the genetic information stored in double-stranded DNA.

RNA world A hypothetical period of evolutionary history in which organisms were based entirely on RNA, unlike the modernworld, in which organisms depend on RNA, DNA, and proteins.

Rock cycle The cyclical process of creation and destruction of rocks through tectonic, weathering, metamorphic, and igneous processes.

Rocks Consolidated mixtures of crystalline materials called minerals.

Runaway glaciation An out-of-control climate in which a planet's surface becomes entirely covered by snow and ice.

Runaway greenhouse An out-of-control climate in which all water on a planet's surface is present as vapor. The oceans are completely evaporated.

S wave Secondary or shear wave; a seismic body wave transmitted as displacements perpendicular to the overall direction of wave travel through Earth's interior. S waves can travel only through solids.

Salinity The salt content of a water mass; often expressed in parts per thousand.

Sea-floor spreading The formation of hot, new oceanic crust from magma extruded at mid-ocean ridges. Once it forms, the new sea floor spreads to the sides of the ridges and is replaced at the ridge axis by even younger new sea floor.

Seasonal temperature contrast The difference in average temperatures between summer and winter.

Sedimentary rock Rock formed by the compaction and lithification of sediments or by the chemical or biochemical precipitation of minerals.

Sediments Layers of unconsolidated material that is transported by water or by air.

Seismic wave An earthquake-produced wave that ripples through Earth's interior, away from the earthquake's focus, as a result of the elastic deformation of the solid Earth. Two types of seismic waves are generated: body waves and surface waves.

Shelly fossils The fossilized remains of shelled organisms. The first such fossils occur in the Cambrian period at the beginning of the Phanerozoic eon.

Silicate mineral A mineral rich in silicon and oxygen. A major class of rock-forming minerals, the silicates make up most of Earth's crust.

Slab Downgoing plate; the portion of oceanic lithosphere that is subducted into the mantle at a subduction zone.

Solar maximum A time of high sunspot activity and corresponding high solar UV flux.

Solar minimum A time of low sunspot activity and corresponding low solar UV flux.

Solar nebula The cloud of gas and dust surrounding the Sun shortly after it formed.

Solar thermal power A method of conversion of sunlight into electricity in which the sunlight is first used to heat a fluid that, in turn, drives a turbine.

Solar-power satellites Large satellites whose proposed function is to harvest sunlight in space and beam the energy to Earth.

Solid Earth The component of the Earth system that includes all rocks and all unconsolidated rock fragments. The core, mantle, and crust make up the solid Earth.

Southern Oscillation An oscillation in sea-level pressure between the western and central/eastern portions of the tropical Pacific Ocean.

Southern Oscillation Index A measure of the pressure difference between the western and central eastern parts of the tropical Pacific Ocean. Strong negative values indicate ENSO conditions, whereas strong positive values indicate La Niña (non-ENSO) conditions.

Speciation The origination of new species.

Spectroscopy A technique in which the composition of a material, such as a gas, is analyzed according to the details of its electromagnetic spectrum.

Spörer Minimum An interval of low sunspot activity between 1450 and 1534.

Stable equilibrium A state in which the system will remain if left undisturbed and to which the system will return when disturbed.

Stable isotope An isotope that does not spontaneously change into another isotope or into an atom of another element by radioactive decay.

State The set of important attributes of a system that characterize the system at a particular time.

Steady State A condition in which the state of a system component is unchanging in time. A reservoir is in steady state when the rates of inflow and outflow are equal.

Stefan-Boltzmann law A relationship stating that the flux of radiation emitted by a blackbody is related to the fourth power of the body's absolute temperature; derived from the Planck function.

Stratosphere The stable atmospheric layer between 10-15 km and 50 km above the surface; temperature increases with altitude there. The stratosphere contains most of Earth's ozone.

Stromatolites Laminated or domed structures in rocks produced by layers of single-celled organisms.

Subduction The process whereby one plate of oceanic lithosphere, and at times its sediment cover, is carried underneath another plate (either oceanic or continental) in a convergent margin.

Subsidence The sinking of air from higher levels in the atmosphere down toward the surface. Also the vertical movement of Earth's crust toward the mantle.

Sunspot Dark areas of lower-than-normal temperatures on the surface of the sun.

Surface waves Seismic waves that travel only across Earth's surface as they spread outward from the earthquake's surface.

Symbiosis A relationship between two species in which the two are dependent on each other to the degree that neither can live alone.

System An entity comprised of diverse but interrelated parts (components) that function as a complex whole.

Taxon An individual taxonomic group (species, genus, and so on).

Taxonomy The systematic organization of living or fossil organisms into a hierarchy.

Terrestrial planets The four rocky, inner planets of the Solar System: Mercury, Venus, Earth, and Mars.

Thermal drag The ocean's ability to delay climate warming (or cooling) as a result of its large heat capacity.

Thermal expansion The tendency of a substance (such as water above 4°C) to expand when heated.

Thermocline A steep temperature gradient in the pycnocline zone that marks the transition between the surface zone and the deep ocean. Temperature drops rapidly with increasing depth in the thermocline.

Thermohaline circulation The circulation of the deep oceans; driven by density differences that result from variations in temperature and salinity.

Thermophilic bacteria Bacteria that are adapted to high temperatures.

Thermosphere The atmospheric layer higher than about 90 km above the surface; temperature increases with altitude there.

Tidal locking radius The distance from a star within which a planet is likely to develop a captured rotation, that is, the same side of the planet always faces the star.

Tidal power The production of electricity by using long, floating booms to harness the energy of ocean tides.

Till Sediment transported by glaciers and ice sheets, comprising materials of a large range of grain sizes and compositions.

Tillite Rock composed of lithified till. One of several pieces of evidence used to identify past glaciations.

Trace gases Gases such as carbon dioxide, methane, and nitrous oxide that are present in Earth's atmosphere in very low concentrations.

Tracers Atoms, ions, or chemical compound whose sources and sinks are easily measured and are relatively well understood, so they can be used to determine oceanic (or atmospheric) circulation.

Transform fault A fracture in the lithosphere between two lithospheric plates that, relative to each other, are sliding past one another (parallel to the boundary) in opposite directions.

Transient response The time-dependent (nonequilibrium) change in a quantity, such as temperature, in response to a perturbation.

Tritium An unstable isotope of hydrogen in which two neutrons are present in each atomic nucleus; denoted as ^3H or T.

Troposphere The lowermost, convective layer of Earth's atmosphere between the surface and 10 to 15 km above it; temperature decreases rapidly with altitude there. Weather is confined to the troposphere.

Ultraviolet (UV) radiation Electromagnetic radiation of fairly high energy and wavelengths from 400 nm to about 10 nm, shorter than those of visible radiation.

Uniformitarianism The principle that processes operating today also operated in the past, and that past geologic events can be explained in terms of processes active today.

Unstable equilibrium A state in which the system will remain if left undisturbed, but even slight disturbances will carry the system to some other (stable) equilibrium state.

Unstable isotope An isotope that spontaneously changes into another isotope or into an atom of another element by radioactive decay.

Uplift Any process by which air is forced to rise upward in the atmosphere. Also an upward vertical tectonic movement of Earth's crust.

Upwelling The rising of cooler, nutrient-rich ocean water to the surface to replace warm, divergent surface water.

Uraninite UO_2, a uranium-containing mineral that is normally destroyed during weathering in an O_2-rich atmosphere.

Vascular plants Plants that have a well-developed stem or trunk for transporting water and nutrients from the ground up to their leaves. They also tend to have well-developed root systems, which take up water and nutrients and help support the plant.

Visible radiation Visible light; electromagnetic radiation of moderate energy and a relatively narrow range of wavelengths, from about 400 nm to 700 nm. Within this range, the color of the light depends on its wavelength.

Visible spectrum The range of component wavelengths of visible light; the colors of the rainbow.

Volatile compounds Chemical compounds that vaporize (turn into gases) at relatively low temperatures.

Volcanic aerosol forcing A perturbation in the radiation budget caused by the presence of volcanic aerosols.

Vorticity The tendency of a fluid to undergo rotary motion. A tendency for counterclockwise motion is referred to as positive vorticity; a tendency for clockwise motion is negative vorticity.

Wavelength The distance between two adjacent wave crests.

Weakly reducing atmosphere An atmosphere composed mainly of N_2 and CO_2, plus small amounts of reduced gases and virtually no free O_2.

Weathering The physical or chemical break-up of rocks exposed at the Earth Surface.

White dwarf A small, compact star that represents the final stage in the evolution of a star such as our Sun.

Wien's law A relationship stating that the flux of radiation emitted by a blackbody reaches its peak value at a wavelength that depends inversely on the body's absolute temperature; derived from the Planck function.

Wilson cycle A plate tectonic cycle of supercontinent assembly and destruction; each cycle lasts about 500 million years.

Wind power The production of electricity from windmills, utilizing Earth's solar-energy-driven winds.

Wolf Minimum An interval of low sunspot activity between 1282 and 1342.

Younger Dryas A climatic reversal to glacial conditions that occurred 10,500 years ago after the end of the last glaciation. Evidence of climate change at this time is seen in various parts of the world, but the main effects of the Younger Dryas event appear to be centered on the North Atlantic region.

Zircon A mineral composed of zirconium silicate that has been used to date some of the oldest rocks on Earth.

Zooplankton Free-floating, marine consumers, including small invertebrates and microorganisms that cannot photosynthesize and therefore feed on phytoplankton.

Index

Absolute vorticity, 86
Absolute zero, 39
Absorption coefficient, 279
Accretion, planetary, 156, 157
Acid rain, 7–8
Acids, 143
Action spectrum, 279
Active continental margins, 113
Adaptations, 197
Aerobe, 134
Aerosols, 226
Aesthetic value of species diversity, 312
Age of Earth, 89, 153–55
Agriculture
    biodiversity and, 312–13
    slash-and-burn, 308
Air, movement of, 58–59, 61–63
Albedo, 24–25, 41
    cloud-albedo feedbacks, 225–27
    snow and ice albedo feedback, 52, 168, 220–21, 250
Alpha Draconis, 217
Altitude
    pressure and, 44
    temperature and, 44–46
Altman, Sydney, 176
Alvarez, Luis, 14–15, 201–4
Alvarez, Walter, 14–15
Amazon forest, development of, 309
Amino acids, 176, 177
Ammonia, faint young Sun problem and, 160–61
Amplitude of seismic waves, 100
Anaerobe, 134
Anhydrite, 205
Animal migration, global warming and, 273
Anions, 142
Anorthite, 143
Anoxic basins, 182, 190
Anoxygenic photosynthesis, 180
Antarctic Bottom Water (AABW), 92, 93
Antarctic Circumpolar Current, 93
Antarctic Ice Sheet, 220, 271
Antarctic ozone hole, 8–9, 10, 288–90
Anthropogenic changes, 2
Anticyclonic flow, 66
Aphelion, 67, 216
Arctic Ocean, sea-ice drift in, 249
Arecibo, Puerto Rico, radio telescope at, 326
Asteroid belt, 207
Asteroid impact, dinosaur extinction and, 15–16
Asteroids, 206–7
Asthenosphere, 107–8, 116
    lithosphere-asthenosphere interactions, 116–17
Atmosphere, 3
    carbon exchange between oceans and, 141
    formation of, 158–59
    weakly reducing, 175
Atmosphere (atm), 44
Atmosphere-centimeter, 281

Atmospheric carbon dioxide. *See under* Carbon dioxide
Atmospheric circulation system, 55–77
    Coriolis effect, 63–65
    equatorial atmospheric circulation, 244–45
    general gas law and, 59
    global circulatory subsystems, 56, 57–58
    global energy distribution, 60
    movement of air, 58–59, 61–63
    surface winds, distribution of, 64, 66–68
    temperature and rainfall, distribution of, 57–58, 68–77
        global patterns in, 70–77
        land-ocean contrasts, 68–70
Atmospheric composition, 42–44
    early, 158–59
    one-layer atmosphere, 43
Atmospheric evolution, 173–91
    effect of life on, 179–82
    origin of life and, 176–79
    oxygen, 174, 182–85, 187–88
        banded iron formations (BIFs) and, 182–84
        detrital uraninite and pyrite, 184–85
        modern controls on, 188–91
        paleosols, 185
        redbeds, 185
    ozone, 185–87
    prebiotic, 174–75
Atmospheric structure, 44–46
    pressure and altitude, 44
    temperature and altitude, 44–46
Atomic number, 11
Australia, southwestern, 311
Autotrophs, 301, 302
Autumnal equinox, 67
Avogadro, Amedeo, 132
Avogadro's number, 132
Axis
    major, 216
    semimajor, 216

Bacteria
    methanogenic, 180–81
    methanotrophic, 181
    thermophilic, 179
Banded iron formations (BIFs), 182–84
Barometric law, 44
Basalt, 103
Bases, 143
Bending mode of vibration, 47
Bicarbonate ion, 142
Bifurcation point, 235
Biodiversity, 2, 194–209
    defined, 195
    diversity/stability in, 196–97
    fossil record of, 197–200
    hot spots, 309–11
    human threats to, 296–315
        consequences of losses, 9–10, 312–15
        ecosystems and, 297–303
        modern extinction, 303–11

importance of, 311–15
mass extinctions and, 199, 200–209
    extraterrestrial influences on, 202, 206–9
    K-T, 201–6
    measures of, 195–96
Biological evolution, factors in, 325–28
Biological pump, 136, 206, 221
Biomass, 133, 303
Biomass-based fuels, 275
Biomes, 297, 298
    terrestrial, 300–301
Biosphere, 129, 297, 298
Biota, 3
Bioturbation, 258
Blackbody radiation, 39–41
Blackbody radiation curve, 39–40
Black smokers, 107, 179
Body waves, 99
Borneo, northwestern, 311
Bottom water, 91–92, 95
Boundary currents, 84–85, 87
Box models of ocean circulation, 259–60
Boyle, Robert, 59
Boyle's law, 59
Brazil, Atlantic coast of, 310
Broecker, Wallace, 92
Bromine cycle, 288
Bromine radicals, 285
Buffer, 258
Buoyancy, 58–59
Butterfly diagram of sunspots, 240–41

CAFE (Combined Automobile Fleet Emissions) standard, 275
Calcite, 141
California floristic province, 310
Calvin cycle, 272
Canary Current, 85
Captured rotation, 324
Carbon
    burial in sedimentary rocks, 138–40
    carbon isotopic evidence of high Mesozoic CO_2 levels, 168, 169
    inorganic, 129–30
    organic, 130
    oxidized, 133
    reduced, 133
    sedimentary C:P ratios, 190–91
Carbon-13, 168
Carbonate compensation depth (CCD), 145
Carbonate ion, 142
Carbonate metamorphism, 147–48
Carbonate rock, 14
Carbonate-silicate geochemical cycle, 17, 147–48
Carbonate weathering, 144, 146
Carbon cycle, 128–49
    carbonate-silicate geochemical cycle in, 147–48
    flux in, 255
    inorganic, 130, 141–47

carbonate mineral deposition in, 144–45
    carbon exchange, 141
    chemical weathering in, 142–44, 149
    net removal of carbon dioxide in, 145–47
    in water, chemistry of, 142
    organic, 133–41, 167
        long-term, 138–41
        short-term, 133–38
    organic and inorganic links in, 148–49
    systems approach to, 128–33
        carbon reservoir dynamics, 130–33
        mole, concept of, 131, 132
        oxidized and reduced carbon, 133
        terrestrial organic carbon cycle, 129–30
Carbon dioxide
    atmospheric
        biological influences on, 166–67
        future concentration of, 259–62
        glacial climate feedbacks and, 221–25
        measurements of, 4–5
        radiative effect of doubling, 51
        seasonal fluctuations in, 131
        steady state of, 131–32
        variations in, 13–14, 161, 166–67, 168
    chemistry of uptake of, 258
    data from ice cores, 5, 6
    in early atmosphere, 161
    long-term projections of, 261
    removal processes, speed of, 256–59
Carbon dioxide band, 47
Carbon dioxide compensation point, 319–20
Carbon dioxide fertilization, 132, 257
Carbon dioxide tax, 275
Carbon fixation, 272
Carbonic acid, 142
Carbon isotopes, constraints imposed by, 187–88
Carbon reservoirs, 254–56
    dynamics of, 130–33
    sedimentary organic, 140
Carrying capacity, 197
Carter, Jimmy and Rosalynn, 21
Catalysis, 176
Catalyst, 285
Catalytic cycles, 284–85
Cations, 142
Cech, Thomas, 176
Celsius scale, 38–39
Cenozoic Era, climate of, 169–70
Centrifugal force, 65
Change, time scales of, 229–31
Chaos theory, Younger Dryas and, 233–35
Chapman mechanism, 283–84
Characteristic response time, 133

Charles, Jacques, 59
Charles' law, 59
Chemical equilibrium, 142
Chemical weathering, 142–44, 146
  biological enhancement of, 149
  climatic factors regulating, 148
Chemosynthesis, 107
Chert, 183
Chicxulub, Mexico, impact crater in, 204–5
Chile, central, 310
Chlorine, reactive, 289
Chlorine catalytic cycle, 285
Chlorine cycle, 286–88
Chlorine monoxide (ClO), 9, 285
Chlorofluorocarbons (CFCs or freons), 5, 47, 286–88
  increasing concentration of, 262–64
  substitutes for, 293
Chlorophyll, 137
Chondrites, 104, 153
Cirrus clouds, 48
Class, taxonomic, 198–200
Climate, 2–3
  deep ocean circulation and, 95
  defined, 231
  Gaia hypothesis, 16–17, 31, 167
  global, 2
  volcanoes and, 236–40
Climate change
  K-T mass extinction and, 201
  projections of, 262–67
    general circulation models (GCMs) and, 264–68
    long-term warming, 268–69
    radiative forcing and, 262–64
Climate feedbacks
  glacial, 219–27
    atmospheric carbon dioxide, feedbacks affecting, 221–25
    cloud-albedo feedbacks, 225–27
    ice-albedo feedbacks, 220–21
    terrestrial biomass, changes in, 225, 226
  in greenhouse effect, 51–53
  IR flux/temperature feedback, 52
  snow and ice albedo feedback, 52, 168, 220–21, 250
  uncertain feedback caused by clouds, 52–53
  water vapor feedback, 51
Climate modeling, 50–51
Climate regulation, long-term, 152–70
  atmosphere and oceans, formation of, 158–59
  causes of, 164–67
    atmospheric carbon dioxide variations, 13–14, 161, 166–67, 168
    continental position changes, 164–66
    climate of past 200 million years, 167–70
      Cenozoic Era, 169–70
      Mesozoic Era, 167–69
    faint young sun paradox, 153, 159–61, 319
    long-term climate record, 161–64
      glaciation record, 162–64

paleoclimate, 151, 161, 162
Climate stability, 317–29
  climate evolution
    carbon dioxide compensation point, 319–20
    in distant future, 318–20
    on Mars, 321–22
    on Venus, 320–21
    Drake equation, 324–28
    habitable planets, of other stars, 322–24
    long-term survival and, 328–29
Climate system, hypothetical (Daisyworld), 23–31
  climate history of, 30–31
  couplings in, 24–26
  equilibrium states in, 26–28
  forcing, response to, 28–31
  increasing solar luminosity in, 28–31
  lessons of, 31
  perturbation, response to, 28
Climate variability, short-term, 229–50
  components of climate system, 243–44
  defined, 231
  Holocene, 230–43
    causes of climate change in, 240–43
      European Medieval Warm Period, 236–37, 242
      Holocene Climatic Optimum, 235
      Little Ice Age, 236–39
      Younger Dryas, 233–35
    present-day, 242–50
      El Niño and, 240, 243–48
      sea ice and, 247–50
    proxy climate data, 232
    trends in change, 231
Cloud-albedo feedbacks, 225–27
Cloud condensation nuclei (CCNs), 75
Clouds
  Earth's global energy budget and, 49, 50
  greenhouse effect and, 48–49
  interstellar, 155
  opposing climatic effects of, 48
  polar stratospheric (PSCs), 288
  types of, 48
  uncertain climate feedback caused by, 52–53
CO₂. *See* Carbon dioxide
Coal, 139, 167
Coastal zone color scanner (CZCS), 138
Coccolithophorids, 135, 136
Colombian Choco, 310
Column depth, ozone, 281–82
Comets, 206
Comet Shoemaker-Levy, 329
Commercial value of species diversity, 312
Community, 195, 297, 298
Companion-star hypothesis, 208, 209
Comparative planetology, "Goldilocks problem" of, 35, 42
Components, system, 20
Condensation, 74
Condensation point, 51

Conduction, 46
Constants, fundamental physical, 344
Consumer biomass, 133
Consumers, 133, 301
Continental/continental convergent margins, 112
Continental crust, 103
Continental drift, 98, 104, 119–21
Continentality, 69–70, 71
Continental margins, 113
Continental position changes, 164–66
Continental shelves, 222
Continuously habitable zone (CHZ), 323
Convection, 45, 46
  in mantle, 116
  radiative-convective model, 50–51
Convergence
  atmospheric, 61–62
  wind-driven surface-ocean currents and, 82–83
Convergent margins, 110–12
Copenhagen Accord (1992), 292–93
Coral reef hypothesis, 223–24
Core of Earth, 3–4, 101, 104
Coriolis effect (Coriolis force)
  atmospheric circulation and, 63–65
  ocean currents and, 80, 84
Corona of Sun, 159
Couplings in systems approach, 20
  in hypothetical climate system, 24–26, 29
C₃ photosynthesis, 272
C₃ plants, 272, 320
C₄ plants, 272, 320
Cratons, 117
Cretaceous-Tertiary (K-T) mass extinction, 201–6
  K-T Strangelove Ocean, 206
  possible causes of, 201–4
Crick, Frances, 177
Crust of Earth, 4, 99, 103–4
Cumulonimbus clouds, 48
Cumulus clouds, 48
Curie point, 106
Currents, ocean
  deep-ocean, 88–95
  surface-ocean, 80–88
Current shear, 86, 87
Cyanobacteria, 179
Cyclonic flow, 66

Daisyworld. *See* Climate system, hypothetical (Daisyworld)
Darwin, Charles, 197
Decomposers, 301
Decomposition, 131
Deep ocean, oxygenation of, 189–90
Deep ocean circulation, 88–95
  climate and, 95
  salinity and, 88–89
  thermohaline circulation, 89–92
  thermohaline conveyor belt, 92–95
Deep-sea cores, 162
Deforestation, 2, 9–10
  carbon reservoirs and, 256, 261
  tropical, consequences of, 305–9

Dendrochronology, 232
Denitrification, 182
Density
  buoyancy and differences in, 58
  in oceans, vertical profiles of, 90
Deoxyribonucleic acid (DNA), 176, 177, 178
Dependent variable, 26
Deserts, 75–77
Detrital mineral, 184–85
Detritus feeders, 301
Deuterium, 11
Diatomic molecules, 47–48
Diatoms, 135
Dimethyl sulfide (DMS), 226
Dinosaurs, asteroid impact and extinction of, 15–16
Direct greenhouse forcing, 264
Dissociation, 181
Dissolution of rock, 117
Divergence, 62, 83
Divergent margins, 110
Diversity of Life, The (Wilson), 310
DNA, 176, 177, 178
Dobson, George, 282
Dobson spectrophotometer, 282
Dobson unit (DU), 185, 280, 281–82
Doldrums, 66
Dolomite, 141
Dose rate of radiation, 279
Double helix, 177
Downwelling, 83
Drake, Frank, 325
Drake equation, 324–28
Dropstones, 162
Dry ice, 321
Dust Bowl period, 243

Early type stars, 319
Earth
  age of, 89, 153–55
  core of, 3–4, 101, 104
  crust of, 4, 99, 103–4
  generalized structure of, 99–103
  history of 119–25
    evolution of driving force, 119
    paleogeographic reconstructions and, 119–21
    Wilson cycles and, 121–25
  mantle of, 4, 99–101, 104, 116
  physical properties of, 344
  physiology of, 114–17
  seismic probing of, 98–103
  surface temperature of, 6–8, 35
Earthquakes, 98–99, 100, 108, 109
  at subduction zones, 109, 110
Earth system, 3
  redundancy in, 196
East African Rift zone, 125
East Antarctic ice sheet, 271
Eastern boundary current, 84–85, 87
Eccentricity, 216, 217–18
Ecosphere, 322–23
Ecosystems
  defining, 297–301
  global warming and, 272–73
  human threats to biodiversity and, 297–303
  species interactions in, 301–3
Ecotone, 300
Ecuador, western, 310

Eddies in atmosphere, 56
Effective radiating temperature, 42
Ehrlich, Paul and Anne, 314
Ekman, Walfrid, 82
Ekman spiral, 82–83
Ekman transport, 83, 84
El Chichón, eruption of, 248, 291
Electromagnetic radiation, 35–38
electromagnetic spectrum, 36–37
Electron capture detector, 287
Elliptical orbits, 216, 217
El Niño, 239, 244–48
El Niño–Southern Oscillation (ENSO), 240, 244–47, 303
climatic impacts of, 247
events, 244–46, 248
Endemic species, 304
Energy balance, planetary, 41–42, 49
Energy budget, global, 49, 50
Energy distribution, global, 60
Energy sources, alternative, 274–75
Enzymes, 134, 176
Eons, 11, 153
Epicenter of earthquake, 99
Epochs, 11
Equatorial atmospheric circulation, 244–45
Equilibrium states, 22, 26–30
Equilibrium surface warming, 265–66
Equinoxes, 67
Eras, 11
Erosion, 118
Errors, random and systematic, 7
Erythemal action spectrum, 279–80
Eukaryotes, 181, 182, 186
European Medieval Warm Period, 236, 242
Evaporation, 73
Evaporite deposits, 88
Evapotranspiration, 306
Evolution, 197. *See also* Atmospheric evolution
biological, factors in, 325–28
Exosphere, 45
Extinction, 197
hot spots in, 309–11
mass, 10, 14–16, 199, 200–209
modern, 303–11
present rate of, 297
tropical deforestation, consequences of, 305–9
Extraterrestrial influences on mass extinctions, 202, 206–9
Extratropical cyclones, 66
Extrusive igneous rocks, 103

Fahrenheit scale, 38, 39
Faint young Sun paradox, 153, 159–61, 319
Family, taxonomic, 198–200
Faults, 98–99
transform, 112–13, 114
Feedback loops, 21, 31
Feldspars, 103
Felsic igneous rocks, 103
Fermentation, 134
Ferric iron, 183
Ferrous iron, 179, 183

Fertilization
carbon dioxide, 132, 257
nitrogen, 257
Fission, nuclear, 274
Fixed nitrogen, 181, 285
Flux, 37–38
in carbon cycle, 255
Focus or earthquake, 98
Food chain, 301, 302
Food supply, loss of biodiversity and, 312–13
Food webs, 301
Foraminifera, 135, 136, 204
Forcings in systems approach, 22–23
in hypothetical climate system, 28–31
Forest clearance. *See* Deforestation
Forest fires, 188–89, 191
Forests, temperate vs. tropical, 307–8
Forest-soil system, nutrient flow through, 307
Formaldehyde, 178
Formic fuels, 254–55
formation of, 139–40
greenhouse gases from burning, 2
rates of burning, 255–56
Fossil record of biodiversity, 197–200
Fossils, shelly, 153
Freons. *See* Chlorofluorocarbons (CFCs or freons)
Frequency of wave, 36
Fusion
latent heat of, 72–73
nuclear, 16, 274–75

Gabbro, 103
Gaia hypothesis, 16–17, 31, 167
Galactic plane hypothesis, 208
Galaxy, number of stars in, 325
Galileo, 241
Gamma rays, 37
Gas chromatograph, 287
Gases
greenhouse, 2, 44, 46–48
reduced, 174
trace, 5, 47, 262–64
General circulation model (GCM), 50, 52–53, 240
climate change projections and, 264–68
parameterized clouds in, 265
General gas law, 59
Genes, 313
Genetic code, 177
Genetic diversity, 12
centers of, 313, 314
Genus, 198–200
Geologic time, 153–55
Geologic time scale, 11, 12
Geostrophic current, 84
Geothermal heat, 160
Geothermal heat flux, 114
Geothermal power, 275
Giant impact hypothesis, 158
Glacial climate feedbacks, 219–27
Glacial deposits, 212–13
Glacial-interglacial cycles, 11–14
oxygen isotope record of, 213–15, 218–19

Glacial periods, 3, 13–14
Glacials, 215
Glacial state, 218
Glacial striations, 162, 212–13
Glacial surge, 271
Glaciation. *See also* Pleistocene glaciations
evidence of past, 162, 212–15
low-latitude, 164, 165
Permo-Carboniferous, 167
runaway, 323
Glaciation record, 162–64
Glaciers
mountain, 270, 271
valley, 233
Global change, 1–17
on long time scales, 10–17
glacial-interglacial cycles, 11–14
mass extinction, 10, 14–16, 199, 200–209
solar luminosity, changes in, 16–17
on short time scales, 4–10
biodiversity, loss of, 9–10, 312–15
deforestation, 2, 9–10, 256, 261, 305–9
global warming. *See* Global warming
ozone depletion. *See* Ozone depletion
themes of, 2–4
Global circulatory subsystems, 56, 57–58
Global climate, 2
Global climate model. *See* General circulation model (GCM)
Global energy budget, 49, 50
Global energy distribution, 60
Global warming, 2–8, 10, 253–76, 328. *See also* Greenhouse effect
atmospheric carbon dioxide and future concentrations of, 259–62
measurements of, 4–5, 6
carbon dioxide removal processes and, 256–59, 261
carbon reservoirs and fluxes, 254–56, 261
climate change and, 6–8, 262–67
general circulation models (GCMs) and, 264–68
long-term warming, 268–69
radiative forcing and, 262–64
ecosystems and, 272–73
policies to slow, 273–76
sea level change and, 269–72
Glossary, 331–40
"Goldilocks problem" of comparative planetology, 35, 42
Granite, 103
Granodiorite, 103
Graphs in systems approach, uses of, 26
Greenhouse effect, 2, 4, 34–53, 268. *See also* Global warming
atmospheric composition and, 42–44
atmospheric structure and, 44–46

blackbody radiation and, 39–41
climate feedbacks in, 51–53
climate modeling and, 50–51
clouds, effect on, 48–49
direct greenhouse forcing, 264
electromagnetic radiation and, 35–38
faint young Sun problem and, 160
magnitude of, 42
of one-layer atmosphere, 43
orbital changes and variations in, 240, 41
physical causes of, 46–48
planetary energy balance and, 41–42, 49
runaway, on Venus, 320–21
temperature scales and, 38–39
Greenhouse gases, 2, 5, 44. *See also* Carbon dioxide
chlorofluorocarbons, 5, 47, 262–64, 286–88
methane, 180–81, 262–64
molecular motion and, 46–48
nitrous oxide, 262–64
Greenland ice sheet, 270–71
Groundwater, 73
Grypania, 186–87
Gubbio, Italy, iridium and K-T boundary at, 14–15, 204
Gulf Stream, 85
Gyre, 81–82, 87

Habitable planets, of other stars, 322–24
Habitable zone (HZ), 318, 322–24
around other stars, 323–24
around Sun, 322–23
Habitat, 301
Hadley, George, 62
Hadley cells, 62–63, 77, 169
Hadley circulation, 62–63, 169
Haley, Edmund, 89
Half-life, 114
Halocline, 89
Halons, 288
Hansen, James, 264
HCFCs, 293
Heat capacity, 69
Heat transfer, mechanisms of, 46
Heat transport, ocean, 95
Heavy bombardment period, 158, 159
Hematite, 183
Hertzsprung-Russell diagram, 318–19
Heterogeneous reactions, 288
Heterotrophs, 301, 302
High-latitude circulation, 63
Himalayan Mountains, 169
lower slopes of, 310
Holland, Heinrich, 185
Holocene, 230–43
causes of climate change in, 240–43
European Medieval Warm Period, 236–37, 243
Holocene Climatic Optimum, 235
Little Ice Age, 236–40
Younger Dryas, 233–35
Holocene Climatic Optimum, 235
Holocene Epoch, 211
Homogeneous reactions, 288

Horizontal air movement, 59
Hormones, 176
Hornblende, 143
H$_2$O rotation band, 46, 47
Human populations, spread of, 303–4
Humidity, relative, 51, 75
Hydrated sea floor, 188
Hydrogen cyanide, 178
Hydrologic cycle
    global, 71–73, 74
    tropical deforestation and, 306
Hydrosphere, 3
Hydrothermal vents, 107, 179
Hydroxyl (OH) radicals, 285

Ice
    dry, 321
    pack, 249
    sea, 247–50
Ice-albedo feedbacks, 52, 168, 220–21, 250
Ice-climate interactions, 250
Ice floes, 249–50
Ice-rafting, 162
Ice shelves, 271
Igneous rocks, 103
    dissolution of, 117
    weathering of, 117–18
Impact degassing, 158
Impact events
    frequency of large, 327, 329
    meteorite-impact theory of K-T mass extinction, 201–6
Independent variable, 26
India
    formation of, 169–70
    Western Ghats of, 310
Infrared (IR) flux/temperature feedback, 52
Infrared (IR) radiation, 2, 36
Inorganic carbon, 129–30
Inorganic carbon cycle, 130, 141–47
    carbonate mineral deposition in, 144–45
    carbon exchange between ocean and atmosphere, 141
    chemical weathering in, 142–44, 149
    net removal of carbon dioxide in, 145–47
    in water, chemistry of, 142
Insect migration, global warming and, 273
Instrumental value, 311–12
Intelligent life
    probability of capacity to communicate over long distances, 327–28
    probability of evolution of, 327
Interactions, diversity of, 196
Interferometer, 326
Interglacial period, 3, 13–14
Interglacials, 215
Interglacial state, 218
Intergovernmental Panel on Climate Change (IPCC), 265
Intermediate disturbance hypothesis, 196–97
International Storage Center for Rice Genes, 313

Interplanetary dust particles (IDPs), 178–79
Interstellar clouds, 155
Intertropical convergence zone (ITCZ), 62, 63, 68
Intrinsic value, 311–12
Inverse square law, 38
Invertebrates, marine, 198
Iridium, 14–15
    anomalously high concentrations of, 201–4, 207
Iron
    banded iron formations (BIFs), 182–84
    in core, 104
    ferric, 183
    ferrous, 179, 183
    oxidation states of, 183
Iron fertilization hypothesis, 223
Islands arcs, 111
Isochron diagram, 153–55
Isotopes, 11
Ivory Coast, southwestern, 310

Joly, John, 89
Jupiter
    formation of, 156–57
    impact flux hitting Earth and, 327

Keeling, Charles David, 5
Keeling curve, 4–5, 6, 255
Kelvin (absolute) temperature scale, 38, 39
Kelvin wave, 245
Kepler's laws, 216
Kerogens, 187
Keystone species, 303
Kimberlites, 104
Krakatau (Indonesia), eruption of (1815), 239
K-T boundary, 14–15
K-T mass extinction, 201–6

Land-ocean contrasts in temperature and rainfall, 68–70
Land use, changes in, 308–9
La Niña, 245, 246, 247
Latent heat, 45
    of fusion, 72–73
    of vaporization, 72
Late Proterozoic Period glaciation, 164, 165
Late-type stars, 319
Lava, 103
Lead-lead dating, 153
Leads (polynyii), 249–50
Lean, J., 306
Lichens, 149
Life
    effect on atmosphere, 179–82
    origin of, 176–79
Limestone, 103, 141
Lithification, 103
Lithosphere, 107–8, 109, 116
    lithosphere-asthenosphere interactions, 116–17
    recycling of, 117–19
Little Ice Age, 236–39
Littoral deserts, 77
Loess, 212–13
Logging industry, forest clearance by, 308

Logistic growth, 197
Loma Prieta earthquake (1989), 100
London Accord (1990), 292–93
Lorenz, Edward, 234
Lovelock, James, 16, 17, 23, 31, 287, 299
Low-latitude glaciations, 164, 165
Low-velocity zone (LVZ), 103
Luminosity, solar, 16–17, 28–31

Madagascar, 310
Magma, 103, 118
Magma ocean, 158
Magnetic dynamo, 104
Magnetic field, 104, 105, 119–20, 121
Magnetic stripes, 106–7, 119
Magnetite, 183
Main-sequence star, 318, 319
Major axis, 216
Malaysia, peninsular, 311
Mantle drag, 116
Mantle of Earth, 4, 99–101, 104
    convection in, 116
Mantle oxidation state, changes in, 188, 189
Marble, 103
Margulis, Lynn, 16, 17, 23
Marine ecosystems, global warming and, 273
Marine organic carbon cycle, 130
Mars
    climate evolution on, 321–22
    search for life on, 326
    surface temperature of, 34–35
Mass extinctions, 10, 14–16, 199, 200–209
    Cretaceous-Tertiary (K-T), 201–6
    extraterrestrial influences and, 202, 206–9
    Permian, 200
Mass number, 11
Mass spectrometer, 115
Maunder, E.W., 242
Maunder Minimum, 242
Medicine, species diversity for, 312
Medieval Warm Period, European, 235–36, 242
Meridional circulation 63, 64
Mesosphere, 45
Mesozoic Era, 165
    climate of, 167–69
Metabolism, aerobic vs. anaerobic, 134
Metamorphic rock, 103
Metamorphism, 118
    carbonate, 147–48
Meteorite-impact theory of K-T mass extinction, 201–4
    environmental consequences of impact, 204–6
Meteorites, 153
Methane
    increasing concentration of, 262–64
    production of, 180–81
Methane sulfonic acid (MSA), 225–27
Methanogens, 180–81
Methanotrophic bacteria, 181
Microfossils, 179, 180

Micrometers, 37
Microwaves, 37
Mid-Atlantic Ridge, 113, 119
Midlatitude circulation, 63
Midlatitude ozone depletion, 290–94
    long-term trends in, 290–92
    mechanisms for halting, 292–94
    variability in, 290
Mid-ocean ridges, 105–7
Mid-ocean ridge vents, deep-sea life at, 107
Milankovitch, Milutin, 215
Milankovitch cycles, 215–19, 328
    Kepler's laws, 216
    orbital theory and, 216–18
Miller, Stanley, 175–76
Miller-Urey experiment, 175–76
Minerals, 3, 103
    detrital, 184–85
    rock-forming, 103–4
    silicate, 103
Mixed layer, 89
Moho, 101–3
Mohorovicic, Andrija, 101
Mole, concept of, 131, 132
Molina, Mario, 287
Monsoons, 70, 72
Montreal Protocol, 292–93
Moon, formation of, 157–58
Moon rocks, 155
Moraines, 162, 212–13
Morphology, 197
Mountain belts, formation of, 112
Mountain glaciers, melting of, 270, 271
Mudstones, 103
Myers, Norman, 309–11

Nanometer, 36
Nansen, Fridtjof, 82
National Center for Atmospheric Research, 299
Natural selection, 197
Negative coupling, 20
Negative feedback loop, 21, 31
Negative relative vorticity, 87
Net radiation, 61
Neutrinos, 159
New Caledonia, 311
Niches, 198–99
    biotic, 303
Nimbus-7 satellite, 282, 288
Nitric acid, dissociation of, 181
Nitrogen
    in early atmosphere, 174
    fixed, 181, 285
    odd, 285
Nitrogen catalytic cycle, 284–85
Nitrogen cycle, odd, 285–86
Nitrogen fertilization, 257
Nitrogen fixation, 181–82
Nitrous oxide, increasing concentration of, 262–64
North American plate, 113
North Atlantic Deep Water (NADW), 92, 93
North Atlantic Deep Water (NADW) formation, 233, 250
North Atlantic Drift, 85, 88
North Equatorial Current, 83, 85
Northern hemisphere reforestation, 256–57

North Star (Polaris), 217
Nuclear fission, 274
Nuclear fusion, 16, 274–75
Nucleotides, 177
Number density, 281
Nutrient limitation, 136–38
Nutrients, 129
  nutrient flow through forest-soil system, 307
  in oceans, 138, 139
  shelf nutrient hypothesis, 221–23

Obliquity of Earth, 67, 165, 216
  Moon and, 157
  variations in, 217, 218
Ocean(s). *See also* Ocean circulation
  anoxic basins of, 182, 190
  carbon exchange between atmosphere and, 141
  dissolution of carbon dioxide in, 257, 261
  formation of, 158–59
  nutrient contents of, 138, 139
  nutrient throughput of, 222
  temperature distributions in, 90, 94
  vertical structure of, 89–90
Ocean circulation, 79–95, 244
  box models of, 259–60
  deep ocean circulation, 88–95
    climate and, 95
    salinity and, 88–89
    thermohaline circulation, 89–92, 95
    thermohaline conveyor belt, 92–95
  winds and surface currents, 80–88
    boundary currents, 84–85, 87
    convergence in, 82–83
    divergence in, 83
    geostrophic flow, 84
    surface ocean circulation, 85–88
    upwelling and downwelling in, 83–84
Oceanic/continental convergent margins, 110–11
Oceanic crust, 103
Oceanic lithosphere, 116
Oceanic/oceanic convergent margins, 111–12
Odd nitrogen, 285
Odd nitrogen cycle, 285–86
Odd oxygen, 284
Olivine, 103–4
Oort cloud, 206–7
Orbital changes, 216–17
  greenhouse variations and, 240
Orbital theory, 216–18
Order, taxonomic, 198–200
Organic carbon, 130
Organic carbon cycle, 133–41, 167
  long-term, 138–41
  marine, 130
  short-term, 133–38
Organic carbon reservoir, sedimentary, 140
Organic compounds, prebiotic synthesis of, 176–78

Orographic precipitation, 75
Oxidation state, 183
Oxidative weathering, 175
Oxidized carbon, 133
Oxygen
  atmospheric, 174, 182–85, 187–88
    banded iron formations (BIFs) and, 182–84
    detrital uraninite and pyrite, 184–85
    modern controls on, 188–91
    paleosols, 185
    redbeds, 185
    rise of, 327
  dissolved, and sedimentary C:P ratios, 190–91
  odd, 284
  prebiotic concentrations of, 174–75
Oxygenic photosynthesis, 179–80
Oxygen isotope record of glacial-interglacial oscillations, 213–15, 218–19
Oxygen minimum zone, 137, 190
Oxygen replenishment, 139
Ozone, 2, 47
  in atmosphere, 185–87
  destruction of, 284
  spatial distribution of, 282
  stratospheric, 8
  tropospheric, 8, 282–83
  vertical distribution of, 280–83
Ozone column depth, 8, 185
  measurements of, 281–82
Ozone-depleting substances (ODS), international controls on, 292–93
Ozone depletion, 8–9, 278–94, 328–29
  Antarctic ozone hole and, 8–9, 10, 288–90
  catalytic cycles and, 284–85
  Chapman mechanism and, 283–84
  midlatitude, 290–94
  ozone depleting compounds, sources and sinks of, 285–88
    bromine cycle, 288
    chlorine cycle, 286–88
    freons, 286–88
    odd nitrogen cycle, 285–86
  ultraviolet radiation effects, 279–80
  vertical distribution of ozone, 280–83
Ozone hole, 2
Ozone layer, 2, 174
Ozonesondes, 281

Pack ice, 249
Paleoclimate, 151, 161
  geological indicators of, 162
Paleogeographic reconstructions, 119–21
Paleolatitudes, 119–20
Paleontologists, 194
Paleosols, 185
Palynology, 232
Pangea, 98, 119, 120–21, 125
Partial pressures, 73
Parts per million (ppm), 5
Pascal (Pa), 44

Passive continental margins, 113
Perihelion, 67, 216
Period, 11, 215, 216
Periodicity, 207
Periodic table, 343
Permian Period, mass extinctions at end of, 200
Permo-Carboniferous glaciations, 167
Perturbations in systems approach, 22, 23
  response of hypothetical climate system to, 28
Petroleum, 139–40
pH, 142, 143
Phanerozoic Era, 162, 164, 167, 198, 199
  history of global biodiversity of, 207–9
Philippines, 311
  Mt. Pinatubo, 237–38, 290–91
Phosphorus, sedimentary C:P ratios and, 190–91
Photic zone, 136
Photochemical models, 185
Photochemical reaction, 174
Photolysis (photodissociation), 283–84
Photolyzed molecules, 174
Photons, 36
Photorespiration, 320
Photosphere, 40
Photosynthesis, 16, 131, 133
  anoxygenic, 180
  C3, 272
  in oceans, rate of, 255
  oxygenic, 179–80
Photovoltaic cells, 275
Phylum, 198–200
Phytoplankton, 135–36, 137
  carbonate producers, 144–45
Pinatubo, Mt., 237–38, 290–91
Pioneer effect, 5, 256
Plages, 241
Planck function, 39–40
Planck's constant, 36
Planetary energy balance, 41–42, 49
Planetary system, number of Earthlike planets per, 325
Planetary vorticity, 86
Planetesimals, 115, 156
Planets
  Earthlike, formation of, 322
  formation of, 155–57
  fraction of habitable, on which life evolves, 325–27
  fraction of stars having, 325
  habitable, of other stars, 322–24
Planet X hypothesis, 208, 209
Plankton, 135
Plants, C3 and C4, 272, 320
Plates and plate boundaries, 108–14
  convergent margins, 110–12
  divergent margins, 110
  forces acting on plates, 116–17
  interactions of, 113–14
  transform margins, 112–13
Plate tectonics, 97–125
  anatomy of Earth and, 98–104
  core, 3–4, 101, 104
  crust, 4, 99, 103–4

  mantle, 4, 99–101, 104, 116
    seismic probing, 98–103
  climate over long-term and, 169
  through Earth history, 119–25
    evolution of driving force, 119
    paleogeographic reconstructions and, 119–21
    Wilson cycles and, 121–25
  physiology of Earth, 114–17
    convection in mantle, 116
    heat in Earth's interior, 114–15
    plates and plate boundaries, 108–14
  rock cycle, 117–19
  theory of, 98, 104–8
    asthenosphere, 107–8
    lithosphere, 107–8, 109
    sea-floor spreading, 104–7
Pleistocene Epoch, 211, 212
Pleistocene glaciations, 211–27
  geologic evidence of, 212–15
    glacial deposits, 212–13
    oxygen isotope record, 213–15, 218–19
  glacial climate feedbacks and, 219–27
    atmospheric carbon dioxide, 221–25
    cloud-albedo, 225–27
    ice-albedo, 220–21
    terrestrial biomass, changes in, 225, 226
  Milankovitch cycles, 215–19, 328
    Kepler's laws, 216
    orbital theory and, 216–18
Polar front zone, 63
Polaris (North Star), 217
Polarity, magnetic, 106
Polarity chron C29R, 201, 202, 204
Polar stratospheric clouds (PSCs), 288
Polar vortex, 289–90
Polynyii (leads), 249–50
Population, 297, 298
Positive coupling, 20
Positive feedback loop, 21
Positive relative vorticity, 87
Prebiotic atmosphere, 174–75
Prebiotic synthesis, 175
  of organic compounds, 176–78
Precession, 217
Precipitation, 58
  global patterns of, 70–77
  orographic, 75
  saturation vapor pressure and, 73–77
Pressure, altitude and, 44
Pressure-gradient force, 84
Primary producers, 133
Primary productivity, 133
Producers, 301
Productivity, 133, 314
Prokaryotes, 181, 182, 186
Proteins, 176
Proto-Sun, 155
Proxy climate data, 232
P waves, 99, 101, 103
Pycnocline, 89
Pyrite, 179, 184, 185
  detrital, 184, 185
Pyroxene, 103–4

Quantum mechanics, theory of, 46
Quartz, 103
  shocked, 203, 204

Radiation, 46. *See also* Infrared
  (IR) radiation; Solar radia-
  tion; Ultraviolet (UV) radi-
  ation
  electromagnetic, 35–38
  net, 61
Radiative-convective model
  (RCM), 50–51
Radiative forcing, climate change
  projections and, 262–64
Radical, 174
Radioactive decay, 11, 114–15
Radiocarbon dating, 115
Radio communication, detecting
  life around other stars by,
  326
Radioisotope, 114
Radiolarians, 135, 136
Radiometric dating, 115
Radio waves, 37
Rainfall distributions, 58
  global patterns in, 70–77
  monsoons, 70
Random errors, 7
Rare earth elements, 183
Reactive chlorine, 289
Recreational value of species di-
  versity, 312
Redbeds, 185
Redfield, Alfred C., 136
Redfield ratios, 136–37
Red giant, 318
Reduced carbon, 133
Reduced gases, 174
Redundancy in Earth system, 196
Reforestation, northern hemi-
  sphere, 256–57
Relative humidity, 51, 75
Relative vorticity, 86, 87
Reservoirs, 131–32
  carbon, 130–33, 140, 254–56
  residence time of, 133
Residence time, 89, 132–33
  volcanic eruption, 237
Respiration, 131
  chemical reaction, 133–34
Rhyolite, 103
Ribonucleic acid (RNA), 177, 178
  RNA world, 176
Richter scale, 100
Ridge axis, 105
Rift, 105
Rift valleys, 110–11
RNA world, 176
Rock cycle, 117–19
Rock-forming minerals, 103–4
Rocks, 3
  igneous, 103, 117–18
  mantle, 116
  metamorphic, 103
  moon, 155
  sedimentary, 103, 138–40
  weathering of. *See* Weathering
Rosy periwinkle, 312
Rowland, Sherwood, 287
Runaway glaciation, 323
Runaway greenhouse, 320–21

Sagan, Carl, 160–61, 325, 328
Salinity in oceans, 94

deep ocean circulation and,
  88–89
  vertical distribution of, 90, 91
San Andreas Fault, 112
Sandstone, 103
Sargasso Sea, 85
Saturation point, 74
Saturation vapor pressure, 73–77
Schwabe, Heinrich, 241
Scientific value of species diversi-
  ty, 312
Sea breeze, 69, 70
Sea floor, hydrated, 188
Sea-floor carbonate dissolution,
  257–59
Sea-floor spreading, 104–7
Sea ice, 247–50
  seasonal distribution of, 248–50
Sea level change, 269–72
  future rise in, 270–72
  K-T mass extinction and, 201
  in twentieth century, 269–70
  west Antarctic ice sheet and,
  271
Seasonal temperature contrast,
  216
Seasonal variability
  of atmospheric carbon dioxide,
  131
  continentality and, 69–70
  in distribution of sea ice, 248–50
  monsoons and, 70, 72
  in surface winds, 66–68
Sea spray, 88
Sediment, 3, 103
  at convergent margins, 111–12
Sediment accumulation, 118
Sedimentary C:P ratios, 190–91
Sedimentary cover, 104
Sedimentary rock, 103
  carbon burial in, 138–40
  weathering of, 140
Seismic tomography, 99
Seismic waves, 99, 102
  amplitude of, 100
Seismograph, 99, 100
Seismology, 98–103
Self-regulation, 31
Semimajor axis, 216
Shales, 103
Shelf nutrient hypothesis, 221–23
Shelly fossils, 153
Shocked quartz, 203, 204
Shoemaker, Eugene, 202
Shoemaker-Levy comet, 202, 207
Siderophile elements, 188
Silicate minerals, 103
Silicate weathering, 144, 146
Simpson's diversity index, 195–96
Slab, 110–11
Slash-and-burn agriculture, 308
Slate, 103
Smog, 8, 283, 285
Snow and ice albedo feedback, 52,
  168, 220–21, 250
"Snowball Earth" model, 165
Solar luminosity, 16–17, 28–31
Solar maximum, 290
Solar minimum, 290
Solar nebula, formation of, 155,
  156
Solar-power satellites, 275
Solar radiation
  distribution of incoming, 60–61

meteorite impact and, 205
Solar system, formation of, 155–58
  moon, 157–58
  planets, 155–57
  solar nebula, 155, 156
Solar thermal power, 275
Solar variability, Holocene climate change and, 240–42
Solar wind, 159
Solar zenith angle, 281, 282
Solid Earth, 3
  circulations of. *See* Plate tectonics
South Africa, cape floristic
  province of, 310
South Equatorial Current, 83
Southern Oscillation Index (SOI),
  246
Southern Oscillation (SO),
  245–47
  El Niño-Southern Oscillation,
  240, 244–47, 248, 303
Speciation, 297
  within forests, global warming
  and changes in, 272–73
Species, 297, 298
  endemic, 304
  instrumental and intrinsic val-
  ues of, 311–12
  interactions in ecosystems,
  301–3
  keystone, 303
  tropical deforestation and loss
  of, 9–10, 12, 305–8
  values of, 311–12
Spectroscopy, 326
Spin axis, precession of, 217
Spörer, Gustav, 242
Spörer Minimum, 242
Sri Lanka, 310
Stability, loss of biodiversity and,
  312–13. *See also* Climate
  stability
Stable equilibrium state, 22
Stable isotopes, 11
State of system, 20
Steady state, 41, 131
  of atmospheric carbon dioxide,
  131–32
Stefan–Boltzmann law, 40–41, 241
Stomata, plant, 257
Strange attractors, 235
Strangelove Ocean, K-T, 206
Stratosphere, 8, 45
  ozone in, 8, 280
  polar stratospheric clouds
  (PSCs), 288
Stratus clouds, 48
Stromatolites, 179, 180
Subduction, 110–11
Subduction zone, 109, 110, 111
Subsidence, 59
Sulfate aerosol particles, 7
Sun, 319
  faint young sun paradox, 153,
  159–61, 319
  habitable zone around, 322–23
  proto-Sun, 155
Sunspot cycle, 290
Sunspots, 240–42
Supersaturation, 75
Supersonic transport (SST) air-
  planes, 286
Surface-ocean currents, wind-dri-
  ven, 80–88

boundary currents, 84–85, 87
  convergence and, 82–83
  divergence and, 83
  geostrophic current, 84
  surface ocean circulation, 81,
  85–88
  upwelling and downwelling,
  83–84
Surface runoff, tropical deforestation and, 307
Surface temperature, 34–35
  changes in, 6–8
  on hypothetical Daisyworld,
  24–26
Surface waves, 99
Surface winds, distribution of, 64,
  66–68
Surface zone of ocean, 89
S waves, 99, 101, 103
Symbiosis, 303
System, 3
Systematic errors, 7
Systems approach, 19–23
  to carbon cycle, 128–33
  couplings in, 20
  equilibrium states in, 22
  essentials of systems, 20
  feedback loops in, 21, 31
  forcings in, 22–23
  graphs, uses of, 26
  hypothetical climate system,
  23–31
    climate in history, 30–31
    couplings in, 24–26
    equilibrium states in, 26–28
    forcing, response to, 28–30
    increasing solar luminosity in,
    28–29, 30, 31
    lessons of, 31
  perturbations in, 22, 23
Systems diagram, 20

Tambora (Indonesia), eruption of
  (1883), 238–39
Tanzania, eastern arc forests of,
  310
Taxon, 199
Taxonomy, 198–200
Technological civilization, fraction
  of planet's lifetime supporting, 328
Tectonic activity, 108–9. *See also*
  Earthquakes
Tectonics, 98. *See also* Plate tectonics
Temperate forest, tropical forests
  vs., 307–8
Temperature
  altitude and, 44–46
  effective radiating, 42
  surface, 6–8, 34–35
    on hypothetical Daisyworld,
    24–26
  tropical deforestation and,
  306–7
  vertical temperature profile, 44,
  45–46
Temperature conversions, 342
Temperature distributions, 57–58,
  68–70
  global patterns in, 70–77
  land-ocean contrasts, 68–70
  in oceans, 90, 94
Temperature scales, 38–39

Terrestrial biomass, changes in, 225, 226
Terrestrial biomes, 300–301
Terrestrial organic carbon cycle, 129–30
Terrestrial planets, 156
Terrestrial radiation, outgoing, 61
Thermal conductivity, 69
Thermal drag, 267–68
Thermal expansion of surface ocean water, 270
Thermocline, 89, 90
Thermohaline circulation, 89–92, 95
Thermohaline conveyor belt, 92–95
Thermophilic bacteria, 179
Thermosphere, 45
Tibetan Plateau, 169
Tidal locking radius, 324
Tidal power, 275
Tidal waves, meteorite impact and, 205
Till, 212–13
Tillites, 162
Tilt. *See* Obliquity of Earth
Time scales, 3
    of change, 229–31
    global change on long, 10–17
    global change on short, 4–10
Time stability hypothesis, 196–97
TOMS (Total Ozone Mapping Spectrometer), 10, 282
Topography, 105
Trace gases, 5, 47
    increasing concentration of, 262–64
Tracer, chemical, 259
Trade winds, 66
Transform faults, 112–13, 114
Transform margins, 112–13
Transient response, 268

Triatomic molecule, 47
Tritium, 11
Trophic levels, 301
Tropical deforestation, 305–9
    impact of, 9–10, 12, 305–8
    reasons for, 308–9
Tropic of Cancer, 67
Tropic of Capricorn, 67
Troposphere, 44–45
Tropospheric circulation, 61–63
Tropospheric ozone, 8, 282–83
Tsunamis, meteorite impact and, 205
T-Tauri wind, 159

Ultraviolet (UV) radiation, 37, 186–87
    ozone depletion and, 279–80
    sunspot cycle and, 290
    UVA and UVC radiation, 279
    UVB radiation, 279–80
        UVB flux and stratospheric ozone, 280
Uniformitarianism, 232
Units and unit conversions, 341
Unstable equilibrium state, 22
Unstable isotopes, 11
Uplift, 75, 118
Upwelling, 83–84
Uraninite, detrital, 184–85
Urey, Harold, 175–76

Valley glaciers, 233
Vaporization, latent heat of, 72
Variables, 26
Vascular plants, 166
Vavilov, N.I., 313
Vavilovian centers, 313
Vents, hydrothermal, 107, 179
Venus
    climate evolution on, 320–21
    surface temperature of, 34, 35

Vernal equinox, 67
Vertical air movement, 58–59
Vertical temperature profile, 44, 45–46
Vibration, bending mode of, 47
Viking spacecraft, 326
Visible radiation, 36
Visible spectrum, 36
Volatile compounds, 156
Volcanic aerosol forcing, 237
Volcanic arc, 111
Volcanic eruptions
    climate and, 236–39
    K-T mass extinction and, 201, 204
Vorticity, 86–87
Vostok ice core, evidence from, 220–21, 225
Vostok ice-core temperature record, 13–14
Vrba, Elizabeth, 327

Walker, Brian, 314
Warrilow, D., 306
Water
    global hydrologic cycle, 71–73, 74
    H_2O rotation band, 46, 47
Water vapor feedback, 51
Watson, James, 177
Watt (W), 37
Wavelength, 35, 36
Wave-particle duality, 36
Weakly reducing atmosphere, 175
Weathering, 88
    chemical, 142–44, 146, 148, 149
    of continental rocks, carbon dioxide removal and, 259, 261
    of igneous rocks, 117–18
    of organic carbon in sedimentary rocks, 140

oxidative, 175
Weddell Sea, 92
Wegener, Alfred, 98
West Antarctic ice sheet, 271
Western Amazonia, uplands of, 310
Western boundary current, 84, 87
West-Wind Drift, 82
White dwarf, 318
Wien's law, 40
Wildfires, K-T event and, 203–4
Williams, George, 165
Wilson, Edward O., 305, 310
Wilson, J. Tuzo, 121
Wilson cycles, 121–25
Wind(s)
    ocean surface currents and, 80–88
    solar, 159
    surface, distribution of, 64, 66–68
Wind-drift currents, 80
Window region, 47
Wind power, 275
Wind stress, 80
Wolf, Rudolf, 241
Wolf Minimum, 242
Wollastonite, 143

X-rays, 37

Younger Dryas, 233–35
    chaos theory and, 235

Zero, absolute, 39
Zircons, 153
Zonal air flow, 70
Zooplankton, 136
Zooxanthellae, 311